The International Library of B

The *International Library of Bioethics* – formerly known as the International Library of Ethics, Law and the New Medicine comprises volumes with an international and interdisciplinary focus on foundational and applied issues in bioethics. With this renewal of a successful series we aim to meet the challenge of our time: how to direct biotechnology to human and other living things' ends, how to deal with changed values in the areas of religion, society, and culture, and how to formulate a new way of thinking, a new bioethics.

The *International Library of Bioethics* focuses on the role of bioethics against the background of increasing globalization and interdependency of the world's cultures and governments, with mutual influencing occurring throughout the world in all fields. The series will continue to focus on perennial issues of aging, mental health, preventive medicine, medical research issues, end of life, biolaw, and other areas of bioethics, whilst expanding into other current and future topics.

We welcome book proposals representing the broad interest of this series' interdisciplinary and international focus. We especially encourage proposals addressing aspects of changes in biological and medical research and clinical health care, health policy, medical and biotechnology, and other applied ethical areas involving living things, with an emphasis on those interventions and alterations that force us to re-examine foundational issues.

More information about this series at https://link.springer.com/bookseries/16538

Nico Nortjé · Johan C. Bester
Editors

Pediatric Ethics: Theory and Practice

Springer

Editors
Nico Nortjé
The University of Texas MD Anderson
Cancer Center
Houston, TX, USA

Johan C. Bester
UNLV School of Medicine
University of Nevada, Las Vegas
Las Vegas, NV, USA

ISSN 2662-9186   ISSN 2662-9194 (electronic)
The International Library of Bioethics
ISBN 978-3-030-86184-1   ISBN 978-3-030-86182-7 (eBook)
https://doi.org/10.1007/978-3-030-86182-7

This Springer imprint is published by the registered company Springer Nature Switzerland AG
The registered company address is: Gewerbestrasse 11, 6330 Cham, Switzerland

Preface

Dear reader,

It is often said in pediatrics that children are not just little adults. When this is said, the meaning is that children are physiologically and developmentally different than adults, get different kinds of diseases, and respond differently to medications and treatments. One cannot just apply concepts applicable in adult medicine directly to the treatment of children; children require special treatments and special considerations.

The same can be said for pediatric ethics. Children are not just little adults, and we cannot approach ethical issues that affect the care of children by merely extrapolating from adult medicine. Children are unique. They are vulnerable, dependent on others for their wellbeing. Children are in a state of development and require a variety of things to be in place to secure adequate development and flourishing. Children lack the capacity to make their own decisions and to advocate for their own interests. The parent-child relationship is important to children and to society. These various factors mean that children are uniquely situated and that the ethical issues that arise in pediatric medicine are therefore unique.

As an independent academic and clinical field of inquiry, bioethics is still relatively young. The pioneers of bioethics laid the foundations of the field in the 1970s and onwards and were initially very concerned about the implications of the concept of autonomy for health care. For the first few decades of its existence, bioethics was laser focused on topics related to informed consent, respect for autonomy, the tension between autonomy and beneficence, and the like. These are all topics related to adult medicine and assume a freely choosing adult patient as the focus of ethical reflection.

The work on questions surrounding autonomy continues within bioethics, but over the past 25 years there has been a gradual recognition that ethical issues in pediatrics require focused attention in its own right. Scholars and pediatricians recognized that the 4-principle model does not fit directly into pediatric practice like it does in adult medicine, and we cannot simply extrapolate from ethical issues in adult medicine to find solutions for ethical questions in pediatric medicine. Gradually, as more and more work were published that focused on ethical issues in pediatrics, the field of pediatric ethics as a specialized area of focus developed. And as this field continues to develop, it is clear that the ethical questions, frameworks, and solutions relevant to pediatric medical practice are unique as children are unique.

The challenge for scholars of pediatric ethics is similar to the challenge faced by bioethicists working in other focus areas. The ethical issues raised in pediatrics are complex; working through them requires sophisticated knowledge and skill. Thus, the need for experts in pediatric ethics. At the same time, ethical issues arise in everyday pediatric practice, in the frontlines where parents and clinicians actually have to make decisions and provide care for children. Thus, the need for pediatric ethics to be accessible and straight-forward enough that it can be applied in everyday practice. This is perhaps one of the most serious challenges confronting the field of bioethics: we need to have specialized knowledge and skills while being able to yield insights and frameworks relevant to the general public and practitioner. It is no good to live in an academic ivory tower where all these wonderful ideas are bandied about if it does not affect the practice of medicine. At the same time, it is no good to dilute ethical reasoning and analysis to the point where important ethical content is overlooked to do what is easy or expedient. Good ethics work is of high academic quality and rigor while being accessible and practical to a general audience.

It is with this in mind that we embarked on the project of compiling this textbook. Our goal is to present ethical reasoning and analysis that have real-world implications for the medical care of children. This book wrestles with serious and complex ethical questions, applying rigor and skilled analysis to various problem areas. But it does so with the practice of medicine in mind, seeking to translate ethical debates into practical guidance for use in practice. The hope is that this book will provide theory that can guide practice, substance that will be of use for pediatricians, nurses, ethicists, and policymakers.

The first section of the book is devoted to theory. Chapters in this section examine various ethical theories and approaches prominent in pediatric ethics. This includes an overview of the best interest standard and the recent debate (particularly in the academia in the United States) about the best interest standard. It includes chapters on the right to an open future, on the UN Convention on the Rights of the Child, and on the ethical implications of different stages of childhood development. The second section of the book is devoted to the implications of ethical theory and reasoning to specific questions in pediatric practice. The first focus area is about how we make decisions for children, the second about critically ill children in ICU or in Emergency Rooms, the third about new and controversial issues in pediatric ethics, and the fourth about society's obligations to children. In each of these focus areas chapters address questions that arise frequently in pediatric practice and in policymaking all around the world.

This book is an international textbook. The temptation is always, when writing an ethics book meant to provide practical guidance, to focus overtly on the laws and legal precedents of a specific jurisdiction of practice. While this is a worthwhile endeavor, the challenge is to illuminate ethical reasoning instead of merely providing guidance linked to the laws of a specific state or country. This is important because some issues arise everywhere, are universal, and require ethical reasoning for its resolution. Our very first chapter illustrates this point. This chapter captures contributions from all over the world, where ethicists describe the most pressing ethical issues in pediatrics from different parts of the world. There are differences between countries and continents, sure. But what emerges clearly is the similarity of the issues and challenges faced in various countries; how similar the questions are in different parts of the world, and how important it is to have a global dialogue on the implications of these issues for practice. This book has therefore been written by authors from all over the world and is meant for an audience all over the world.

It is our hope and wish that this book will aid you in your practice of pediatrics, in your making of policy, or in your scholarly and academic work. Wherever your work touches the lives of children, we hope this book will help you. Ultimately, it is our goal that the interests of children are advanced and protected, and that this will lead to more peaceful, just, and well-off societies all over the world.

Sincerely and with best wishes.

Las Vegas, USA   Johan C. Bester
Houston, USA   Nico Nortjé

Contents

Part I Theory

**1 The Main Challenges in Pediatric Ethics from Around
the Globe** 3
N. Nortjé, M. Kruger, J. B. Nie, S. Takahashi, Y. Nakagama,
R. Hain, D. Garros, A. M. R. Villalva, J. D. Lantos, J. P. Winters,
and T.-L. McCleary

**2 A Developmental Perspective on Pediatric Decision-Making
Capacity** 23
N. Hardy and N. Nortjé

**3 The Child's Right to an Open Future: Philosophical
Foundations and Bioethical Applications** 39
J. R. Garrett

**4 The Best Interest Standard and Its Rivals: The Debate About
Ethical Decision-Making Standards in Pediatrics** 57
J. C. Bester

**5 Two Ethical Foundations for Pediatrics: The United Nations'
Convention on the Rights of the Child and Bioethical Principles** ... 79
J. P. Spike

**6 A Contextual Architecture of Praxis in Pediatric Case
Consultation** 93
C. M. Nelson and R. Posen

Part II Practice

**7 Parental Permission, Childhood Assent, and Shared
Decision-Making** 111
S. L. Teti and T. M. Silber

**8   Telling the Child: Ethics of the Involvement of Minors
in Health Care Decision-Making and in Considering Parental
Requests to Withhold Information from Their Child** 127
J. M. Marron and K. O. Kennedy

**9   Parental Refusal of Beneficial Treatments for Children:
Ethical Considerations and the Clinician's Response** 143
J. C. Bester

10 Caring for Adolescents: Unique Ethical Considerations 155
S. Barone and Y. Unguru

**11 Demands for Harmful Treatments in Pediatrics
and the Challenge of Reasonable Pluralism: A Quasi-Clinical
Ethics Consultation** 171
G. Birchley and D. M. Hester

**12 Family or Community Belief, Culture, and Religion:
Implications for Health Care** 187
T. Rossouw, P. Foster, and M. Kruger

13 Children Requiring Emergency Health Care 203
I. Mitchell and J. Guichon

**14 Ethical Issues and Considerations for Children with Critical
Care Needs** 225
B. M. Morrow and W. Morrison

**15 End of Life: Resuscitation, Fluids and Feeding, and 'Palliative
Sedation'** 239
R. Hain and F. Craig

**16 Medical Futility in Pediatrics: Goal-Dissonance
and Proportionality** 253
I. D. Wolfe and A. A. Kon

17 Newborns with Severe Disability or Impairment 275
M. Devereaux and K. L. Marc-Aurele

18 Neonatal Euthanasia and the Groningen Protocol 291
Jacob J. Kon, A. A. Eduard Verhagen, and Alexander A. Kon

19 Genetic Testing and Screening of Children 313
M. B. Menzel and V. N. Madrigal

20 Enhancement Technologies and Children 329
J. T. Eberl

**21 Predicting Childhood Neurologic Impairments: Preparing
for or Prejudicing the Future?** 343
P. C. Mann

22   Ethics of Pediatric Gender Management 357
     K. Moryan-Blanchard, L. Karaviti, and L. Hyle

23   The Child with Cancer: Blurring the Lines Between Research
     and Treatment 379
     M. Kruger and N. Nortjé

24   Reproductive Controversies: Fertility Preservation 387
     J. Taylor, L. Shepherd, and M. F. Marshall

25   The Ethical Principles that Guide Artificial Intelligence
     Utilization in Clinical Health Care 403
     W. A. Hoffmann and N. Nortjé

26   When Should Society Override Parental Decisions?
     A Proposed Test to Mediate Refusals of Beneficial Treatments
     and of Life-Saving Treatments for Children 421
     Allan J. Jacobs

27   Vaccine Ethics: Ethical Considerations in Childhood
     Vaccination 437
     J. C. Bester

28   Society's Obligations to Children 453
     F. X. Placencia

29   Pediatric Resource Allocation, Triage, and Rationing
     Decisions in Public Health Emergencies and Disasters: How
     Do We Fairly Meet Health Needs? 465
     D. J. Hurst and L. A. Padilla

Index 479

Part I
Theory

Chapter 1
The Main Challenges in Pediatric Ethics from Around the Globe

N. Nortjé, M. Kruger, J. B. Nie, S. Takahashi, Y. Nakagama, R. Hain, D. Garros, A. M. R. Villalva, J. D. Lantos, J. P. Winters, and T.-L. McCleary

Abstract This chapter highlighted some salient trends in pediatric ethics from different parts of the globe. It is interesting to note that although diverse, there are many similarities between ethical challenges in pediatrics in different parts of the world.

Keywords International perspectives · Pediatric ethics · Vulnerability

N. Nortjé (✉)
Department of Critical Care Medicine, University of Texas MD Anderson Cancer Center, Houston, USA
e-mail: NNortje@mdanderson.org

M. Kruger
Department of Paediatrics and Child Health, Faculty of Medicine and Health Sciences, Stellenbosch University, Stellenbosch, South Africa
e-mail: marianakruger@sun.ac.za

J. B. Nie
Otago Medical School, Bioethics Centre, University of Otago, Dunedin, New Zealand
e-mail: jing-bao.nie@otago.ac.nz

S. Takahashi
Department of Biomedical Ethics, Graduate School of Medicine, The University of Tokyo, Tokyo, Japan
e-mail: shizukoima@m.u-tokyo.ac.jp

Department of Obstetrics and Gynecology, The Japanese Red Cross Hospital, Tokyo, Japan

Y. Nakagama
Department of Parasitology, Graduate School of Medicine, Osaka City University, Osaka, Japan
e-mail: nakagama.yu@med.osaka-cu.ac.jp

Department of Pediatrics, The University of Tokyo Hospital, Tokyo, Japan

R. Hain
All-Wales Paediatric Palliative Care Network, Swansea, Wales, UK
e-mail: richard.hain@southwales.ac.uk

College of Human and Health Sciences, University of Swansea, Swansea, Wales, UK

1.1 Introduction

The field of modern bioethics is roughly around 50 years old. Most of the first 3 decades was spent on adult issues, and it is only in the past 2 decades that special attention has been paid to ethical issues pertaining to children (Hester 2008). Cognizant to the major differences of approach where adult bioethics focusses predominately on respect, and pediatric bioethics on protection, it is also paramount to highlight that the field of pediatric ethics presents a different dynamic to that of adult bioethics, namely the interaction between patient-family/parents-state-physician (Hester 2008). This dynamic interaction brings about more complex variables to medical care, where one simply cannot apply adult-centered ethical theories to cases involving the pediatric population. As such the formation of the sub-specialty group identified as pediatric bioethics was required.

Although children around the world have similar basic medical needs, the discrepancy in access to health care brings its own unique challenges when comparing different parts of the world. What is possible in one part of the world is not necessarily viable in another. Consequently, authors from around the globe were approached to describe the most salient pediatric ethical issues pertaining to their regions. These are presented in the next section of this chapter. At the end a summary of the most striking themes will be offered, and a brief discussion on next steps.

D. Garros
Department of Pediatrics, University of Alberta, Edmonton, Canada
e-mail: dgarros@ualberta.ca

Jon Dossetor Health Ethics Center, Stollery Children's Hospital, Edmonton, AB, Canada

A. M. R. Villalva
Hospital Infantil de México, Mexico City, Mexico

J. D. Lantos
Glasnapp Family Foundation Chair in Bioethics, Children's Mercy Hospital, Kansas City, MO, USA
e-mail: jlantos@cmh.edu

J. P. Winters
Bioethics Centre, University of Otago | Te Whare Wānanga o Otāgo, Dunedin, New Zealand
e-mail: janine.winters@otago.ac.nz

T.-L. McCleary
Medical Sciences Campus, University of Puerto Rico, San Juan, PR, USA
e-mail: Tamralee.mccleary@upr.edu

1.2 Africa

Mariana Kruger

It is estimated that Africa will host the world's largest child population by 2055, while sub-Saharan Africa will report the highest number of births, which is expected to persist throughout this century (UNICEF 2019). Although childhood mortality rates have decreased for all age groups, 85% of all deaths still occur in children under five years of age due to common causes such as pneumonia, malaria and diarrhea (36% of all childhood deaths under 5-year of age). Children in Africa are particularly vulnerable as they are living in contexts challenged by poor socio-economic circumstances, limited resources, and in certain countries, major political instability. Although the world's countries agreed to Sustainable Developmental Goals (SDGs) in 2015, few countries in Africa have documented much progress towards achieving these goals five years later (Clark et al. 2020).

An important SDG is to improve the survival of children, and to accomplish these dedicated African pediatric clinical trials are necessary to investigate improved childhood prevention strategies and clinical trials with medicines with appropriate age-specific formulations. Focus would need to be on African pediatric burden of diseases such as lower respiratory infections, malaria and tuberculosis (El Bcheraoui et al. 2020). Furthermore, given the huge inequality of access to health care for children, as exacerbated by poor socio-economic circumstances and politically instability, children are also particularly vulnerable as a research population in Africa.

As elsewhere in the world voluntary informed consent is a major requirement for participation in a clinical trial. Usually, parents are the decision-making persons on behalf of their minor children, while children assent should be obtained. One of the major ethical issues affecting research in Africa is that it is often only through participating in the testing of a new medicine for a specific disease that children can access treatment for their illness, as the particular country may not provide universal access to government-supported health care and families often cannot afford the existing health care options (Rutherford et al. 2010). The lack of access to health care may therefore negatively impact on the voluntariness of research participation and should be considered in the recruitment process for clinical trials in Africa.

In conclusion, the primary ethical responsibility of the pediatric clinical trial research team should be to protect children from harm and exploitation. This is of importance in Africa, where issues of lack of access to health care, illiteracy or poor education, and poverty may impact on the family's voluntariness to participate in research. These factors create significant risk for therapeutic misconception, where parents enroll children for a research study because they confuse the study for clinical treatment of the child's illness. There should therefore be strict adherence to international research guidelines aimed to protect children in research, while at the same time taking into consideration the cultural and local resource context when planning such research. Prior engagement with community preparedness is essential and may benefit the research team with the support the community can provide with regards to explaining in the local vernacular the research process and assisting with informed consent. Ancillary care needs should be arranged prior to initiation of the clinical trial and will increase the trust of communities in future research projects.

1.3   Asia

1.3.1   China

Jing-Bao Nie

Pediatric ethics in China, the most populous country with its own internal plurality and profound diversity regarding economic and social development, ethnicity, cultural traditions, religious and moral beliefs, has many different challenges. These challenges are underlined when pediatric health care is addressed across all the different regions of China. With this said, a few macro social-cultural features unique to China will be highlighted, which have been found to shape the practices, moral quandaries and ethical responses in pediatric ethics.

Suffice to say that many ethical issues faced by other nations pertaining to pediatric health care are also prevalent in China. To this effect Chinese authorities signed the international Convention on the Rights of the Child in 1990 and ratified it in 1992, following the United National General Assembly resolution in 1989. The convention now endorsed by all 193 UN member states aims to safeguard welfare including "the best available medical care" for children worldwide.

Although many similarities exist, there are also many differences. Given the scope of this chapter, the author will elaborate on only two.

First, the *wellbeing of "little emperors"* due to the "one-child policy" (Cameron et al. 2013). Since the 1960s and 1970s, population control has been a global movement, as a response to the perceived "population explosion". But no country has carried out a national strategical program like China's "one-child per couple policy" which constitutes the most ambitious and intrusive state-propelled project of demographic engineering in human history. Although the policy was terminated in late 2015, it was subsequently replaced with a "two children" (2016) and "three children" (2021) policy. However, as a direct result of the one-child policy the immediate and long-term consequences (intended and unintended) is the phenomenon of single children (nicknamed "little emperors") whose wellbeing, particularly mental and social wellbeing, has long been a concern. Numerous studies in China and overseas have shown that, compared to the children with siblings, many single children are much less cooperative, less caring, less trusting, less trustworthy, less conscientious, less risk-taking, and more pessimistic (Cameron et al. 2013; Yang et al. 1995). These characteristics hugely influence medical decision-making of parents who would be prone to agree to more aggressive therapies to make sure their "little emperors" have access to anything to continue to live. More so than what would normally be acceptable elsewhere in the world.

Second, "left-behind children" in vast rural areas (Chang et al. 2011; Jia and Tian 2010). Every society faces the constant challenges of inequality, inequity and injustice, especially for vulnerable populations like children. Rural–urban disparity has been a serious problem worldwide. In China, rural–urban disparity has been far more structural than that in most other parts of the world. As a means of political and social control, the caste-like system—the *fukou* (household registration)—was introduced following the establishment of the People's Republic of China in 1949 and has existed as a basic institution in China ever since. People are divided into two civil status groups ascribed at birth places: rural and urban. In Mao's regime, rural people were not allowed to work and live permanently in cities. In the past four decades, millions of young rural people have migrated to work in cities and contributed enormously to China's remarkable economic development. But they have never been treated as equal citizens as compared to urban residents. Consequently, for the great majority of children of these migrant workers who have been left behind in vast rural areas to be cared for by their grandparents, about 70 million, it is difficult to access publicly-funded educational and health care services. In general, the standard of social welfare and health care including pediatric care in rural areas is markedly and structurally lower than that in cities.

1.3.2   Japan

Shizuko Takahashi andYu Nakagama

In clear contrast to societies where life-death related decision-making is regarded as the right and duty of an individual, what is unique to the Japanese is the "devaluation of autonomy". This is evident prominent issues in Japanese pediatrics health care and include areas such as: disclosure of illness; end-of-life decisions; resuscitating very premature babies; preimplantation genetic testing; and policies regarding COVID-19 for children. Of interest is the fact that although medical practice in Japan is predominately Western, there are many cultural dilemmas with respect for autonomy, often not being considered a virtue in Japan. Families make shared decisions in their child's best interest, but often under the consideration of a third party, the *Seken* (the eyes of the public). *Seken* refers to all the members of a social network which surround an individual. These individuals include family, friends, neighbors, larger community, and fellow countrymen. Since it refers to a collective consciousness, these cultural norms and values ultimately function to order social behavior and regulate relations between different members of this network (Kurihara 2007).

Collective identity had long been valued over the individual, thus *Seken* had strongly influenced an individual's decision-making. People had been expected to behave uniformly, and even discouraged from considering autonomous decision-making (Kimura 1995). Until the middle of the nineteenth century, Japan had been isolated culturally and geographically from the outside world for almost 200 years. As a result, the people became homogenous and learned to live together as a whole. Of note, the word "individualism" did not exist until the country opened up to the West.

Although Japan has been "open" to the rest of the world it is interesting to note that there is still a national pride to live in accordance with the *Seken's* expectations. This often times give precedence over the developing autonomy of the child. Many parents aim to raise a child so that they become mindful of their surrounding *Seken*. Without exception, such societal tendency is readily reflected in the health care practice: disclosing serious illnesses, making end-of-life decisions and even the more recent discussions over perinatal genetic testing issues. For example, death is never individualized but rather seen as a family affair, if not the affair of the community as a whole (Akabayashi et al. 1999). A study where physicians were interviewed on their view of informing children of their cancer status, more than half felt that they should not disclose the diagnosis and rather have an obligation informed by *Seken* not to disclose this information (Yoshida et al. 2018).

Such devaluation of personal autonomy and the strong influence of *Seken* in the Japanese society has had some benefits as illustrated during the Corona (COVID-19) crisis. When the COVID-19 infection rate was peaking, no lockdowns nor penalties were imposed to the Japanese. They were only asked to *jishuku* (self-restrain). Japanese society complied to this request on a voluntary basis—a perfect example where the belief in *Seken* played a role where behaviors and attitudes of the individual is seen as important to that of the larger society. On the other hand, implicit expectation to live obedient to the *Seken*, often risks the normal process of informed decision-making. Making daily decisions under the strong influences of the *Seken*, or public expectations, may threaten the sense of control of an individual, rendering vulnerable minors to feel like they are losing control over their own actions (Tapal et al. 2017).

While the Japan Pediatric Society stresses the need of societal respect for children's autonomy, *Seken* shapes and may even distort the child's best interest. With modernization, the Japanese are still struggling to balance the growing autonomy and their traditional values.

1.4 Europe

Richard Hain

Identifying the main challenge facing pediatric ethics in Europe is complicated by the sheer range of cultures that marks this ancient continent. Europe is not a single polis; it is made up of 28 different countries. While for reasons of pragmatic commercialism they may unite under the banner of a single market, those separate countries nevertheless retain laws which reflect mores that are distinct and diverse. In The Netherlands, there are some circumstances under which doctors may be permitted actively to end the life of a child if her parents ask (Verhagen and Sauer 2005) (refer to Chap. 18). The Charlie Gard case (Wilkinson 2020) illustrated that it may be acceptable in Italy to prolong a child's life at parents' request while, in the UK, doctors are expected to act in the interest of the child but in the face of opposition from a child's family sometimes feel they need the protection of the Courts.

Of course, parental preferences are important, and where more than one course of action is equally reasonable, parental choice should be determinative. But the importance given to parental preferences in these cases suggests a view among some that in order for a medical decision about a child to be morally right, it is only necessary that it be made by the child's parents. It is perhaps a reaction to fifty years of paternalism, in which what was right emerged *ex cathedra* from the arbitrium of the pediatrician's own reasoning. In the past, parents were often expected simply to underwrite what the pediatrician had advised without asking too many questions. Most societies in Europe have, quite rightly, abandoned that approach and most now invite parents to participate in medical decision-making over their child.

But they have not always stopped there, and a new authoritarianism may be on the way. Increasingly, doctors are asking parents to make medical decisions for them, rather than with them. A dialogue between doctor and family is essential in establishing what course of action is likely to be best for the child, but there is a danger that it becomes instead an opportunity for parents to instruct the doctors as to what they want for their child. That would mean one or both of two things about the child. Either that it is acceptable to subordinate a child's interests to the interests of her parents, which would characterize the child as a being of inferior moral value in her own right, or that the child's interests cannot be separated from those of her parents, which would characterize her as a possessed object. In retreating from the influence of the twentieth-century paternalist, we would have blundered back into the company of the Victorian *pater familias*.

One important solution is to look more often to the child to make decisions for herself. But it will always be necessary for some medical decisions about a child to be made by someone other than the child herself. The doctor cannot do that without the family. But nor can the family do it without the doctor. Alone, neither has enough information to know what is likely to do the child most good and least harm. The biggest ethical challenge in Europe, as I see it, is to ensure that neither doctors nor parents expect, or are expected, to have the final word on medical decisions about a child. That means establishing systems in which parents and doctors are supported and protected in their decision-making *to the extent that they are doing it together.*

1.5 North America

1.5.1 *Canada*

Daniel Garros

This section is written from the appreciation that all Canadians have a right of free access to the publicly funded health care system under the Canada Health Act. This right is an attempt to address health disparities, and the act encompasses 5 principles:

Public Administration, Comprehensiveness, Universality, Portability and Accessibility (Canada.Gov 2015). The act holds that no one should pay for essential urgent or non-urgent medical care, and that very few services are allowed to operate for profit. It is against this background that the author would like to describe some important ethical issues in pediatric medicine in Canada, which is forming the way medicine is being practiced in Canada. These issues are illustrated by means of prominent case law as well.

1.5.1.1 Consent/Autonomy

In 2009, a case was heard in the Supreme Court of a minor who was forced to receive medical treatment against her religious belief. The court ruled that specific age limits apply when people wish to refuse medical treatment because a child may not have the necessary intellectual appreciation of the consequences of refusing medical treatment (A.C. v Manitoba, 2009) (Bailey and Freedman 2012).

Notwithstanding, there is no universally accepted, legally defined age of consent in Canada. A minor is someone below age of 18 or 19 years of age, depending which province one lives (Quebec considers majority at 14 or higher), but being a minor does not necessarily preclude provision of consent for treatment. Therefore, the capacity of the pediatric patient to consent to a proposed treatment varies with age and circumstances and must be determined on a case-by-case basis.

In some jurisdictions, mature minor status is conferred as part of a formal legal process. In others, the designation is used informally for adolescents who have met the criteria for capacity according to their clinicians. Members of Ontario's Consent and Capacity Board (a unique body in the Country that has the authority to counsel and intervene before the courts) have developed basic guidelines for assessing capacity in young people. Pediatricians need to assess properly the context of the patient's emerging self-awareness, developing values and beliefs, maturing cognitive skills and, where present, provincial/territorial laws determining the age of majority or consent (Coughlin 2018).

1.5.1.2 Medical Assistance in Dying and Kids

The Supreme Court decision in Carter v. Canada (2015) has led to changes to the Canadian Criminal Code, such that medical assistance in dying (MAID) is now a legal option for consenting adult patients who have a 'grievous and irremediable medical condition' that causes 'enduring' and 'intolerable' suffering. This was enacted in June 2016 by the Canadian parliament. Even though the term adult was not defined, mature minors or anyone below 19 years of age are currently excluded under federal legislation. Hence, a lot of debate has ensued about the topic (Singh et al. 2020). At present, The Netherlands and Belgium are the only two jurisdictions where MAID is extended to infants and/or children (Davies 2018) (refer to Chap. 18).

Canadian pediatricians surveyed (574 responses, 29% response rate) revealed that 46% were in favor of extending the MAID option to mature minors experiencing progressive or terminal illness or intractable pain; 29% believed access should be extended to children or youth with an intolerable disability and 8% only for kids with intolerable mental illness as the sole indication; 33% said no to MAID under any circumstance (Davies 2018). Of interest is the fact that Toronto's Hospital for Sick Children has already published a draft policy outlining a process for MAID in those aged 18 years or older, as legally allowed. The authors argue that the mature minor's legislation already allows minors with adequate capacity to make decisions regarding life support, so the same should apply to MAID.

1.5.1.3 Gender Reassignment

The topic has been well researched and discussed in the medical and legal literature (Clark et al. 2020). In British Columbia, a father of a 19-year-old (still minor in the province) who "was assigned female at birth but self-identified as male since 11 years old" opposed his child's hormonal treatment in court. The appeal court ruled that since a medical team had assessed the boy as sufficiently mature to make the treatment decision on his own, the father's appeal failed (A.B. v C.D. (2020), 441 Dominion Law Reports (4th) 505 (British Columbia Court of Appeal).

In 2017, the Government of Canada (Minister of Diversity and Inclusion and Youth), added protection for trans and non-binary people in the *Criminal Code* and the *Canadian Human Rights Act*. In 2020, legislation was introduced to protect LGBTQ2 Canadians from conversion therapy, including coercive efforts to change a person's identity. And in 2021, the Population Census will include not only identification of sex at birth but also gender identity (Youth 2020).

1.5.1.4 When There Is No Agreement

In 2011, at 3 months of age, baby Joseph was diagnosed with Leigh syndrome. At the London (Ontario) hospital, his doctors diagnosed him to be in a persistent vegetative state and refused to perform a tracheostomy which the family was requesting to allow him to go home. The medical team did not want to cause what they saw as "needless suffering" with another surgical intervention (Friedrich 2020). Ontario's Consent and Capacity Board judged that the child's best interest was transferring him home as he was and extubate him there. The transfer would preclude performing a tracheostomy, which is not seen as a palliative procedure. A discussion about the best interested standard ensued. The family took the child to a center in the US and he received a tracheostomy and after 6 months he was transferred home and died beside his parents. The case generated a lot of media attention and pressure on the clinicians, who had suffered significant physical threats and social media harassment (Drake and Cox 2012).

1.5.1.5 Modes of Death and Organ Donation

The great majority of patients in Canadian NICUs and PICUs die after a decision to withdraw or limit life-sustaining treatment (Fontana et al. 2013). This pattern has been established now for over 20 years (Garros et al. 2003). A consensus guideline for organ donation after brain death, or NDD (Neurological Determination of Death) has been standardized in the country (Ball et al. 2020). As of late, Donation after Cardiac Death (DCD) has been discussed and practiced in many centers in order to increase the pool of organs, with Canadian guidelines already published (Weiss et al. 2017). However, this is not without controversy, since some prominent Canadian pediatric intensivists have refused to participate and have written a manifesto against it, along with other well-recognized international colleagues. They argue that DCD is not ethically allowable because it abandons the dead donor rule, has unavoidable conflicts of interests, and implements pre-mortem interventions which can hasten death. These important points have not been but need to be fully disclosed to the public and incorporated into fully informed consent, according to the authors (Joffe et al. 2011). Without proper public participation and evaluation this issue will continue to raise ethical tension among pediatricians in Canada.

1.5.2 Mexico

Ada María Ruiz Villalva

Latin America has the largest wealth gap in the world. México is part of the 25% of countries with the highest levels of inequality (Solt 2020). In this country, the 1% of the wealthiest individuals have 21% of total income (Campos et al. 2014). Its economic system only benefits a few. Consequently, an overwhelming majority of disadvantaged people exist side by side with a small elite group.

Cardinal bioethical issues are injustice and unfair resource allocation. 49.6% of the underage population in Mexico live in poverty conditions (19.5 million), 61% lack access to social security and 9.3% live in extreme poverty, which totals 3.7 million Mexicans (UNICEF Mexico, CONEVAL 2019). It is a complex problem, causing children to experience severe deprivations: in nutrition, public services, education, health, social welfare and housing. Scarcity of opportunities leads to an enormous challenge to improving the standard of living (Keeley 2018). According to the Organization for Economic Cooperation and Development (OECD 2019a), when a child is born in poverty in Mexico it requires 11 subsequent generations to reach the average income. Extrapolated from this, one can argue that there is a correlation between the level of poverty and an individual's health and wellbeing. Sadly, Mexico is regarded as having one of the least effective legal systems in the world for protecting children (UNICEF 2009). Inequality and injustice permeate medical care. In Mexico there is public health coverage; nevertheless, approximately 25% of children (9.9 million) are reported as either not having received vaccination or not having seen a doctor to treat a respiratory disease, or both. Almost 40% of the children have not received additional vaccines after the age of two (ENNViH 2002).

Injustice against minorities is another serious problem. On the one hand, we have the marginalization of indigenous communities, with all the poverty-linked issues already mentioned, the misunderstanding of the idiosyncrasies of their culture, and the clash of Western vs. traditional medicine. On the other hand: migrant children. Mexico is located between Central America and the United States, providing a land route for families displaced by violence and precariousness in their home countries. Many of them end up in a detention center. In the absence of a support network for migrants, children are especially vulnerable to all kinds of harmful situations impacting on their health and wellbeing.

In conclusion, the failures in social justice in Mexico reflect multiple factors that originate in an unbalanced distribution of resources as a consequence of a system that is essentially corrupt at many levels.

1.5.3 United States of America

John D. Lantos

There are issues in three domains: clinical ethics, research ethics, and advocacy.

The central issue in clinical ethics has not changed in fifty years. It is the question of whether certain treatments for specific patients should be classified as clearly beneficial, clearly futile, or somewhere in between. When treatments are clearly beneficial and therefore deemed to be in the child's best interest, then they should be provided whether the parents consent to the treatments or not. Children have a right to beneficial medical treatment independent of their parents' beliefs or preferences. The most dramatic illustration of a situation in which parents wishes may be legally and ethically overridden is the situation of blood transfusion for the children of parents who are members of the Jehovah Witness religion. The parents' religious beliefs dictate that blood transfusions are absolutely forbidden. As competent adults, they are free to refuse blood transfusions, even if it means that they will die. For their children, however, doctors are obligated to seek (and will usually be given) a court order for treatment, based on the principle of beneficence. When treatments are clearly futile, the situation is murkier. There are powerful ethical arguments to buttress the belief that doctors should not provide futile therapy. However, it is often difficult to define futility in an uncontroversial way. Thus, in most cases, parents who legally challenge futility determinations in the United States usually win and the courts order that treatments be continued. That is not true in some other countries, for example the UK. The more interesting and controversial situations are those that arise when treatments are neither clearly beneficial nor clearly futile. This domain of ambiguity has been called "the gray zone" or the "zone of parental discretion." Most frameworks for decision-making stipulate that, within that zone, parental preferences should determine treatment choices (also refer to Chap. 4). It is not clear, however, just how to help parents make decisions in the gray zone. Doctors can influence parental preferences. So can digital media, other family members, information from television or movies, attitudes of clergy, or financial considerations. Much research in clinical ethics in the USA today focuses on the processes of shared decision-making and the appropriate ways to shape the choice architecture within which decisions are made.

The domain of research ethics is changing rapidly. Fifty years ago, when the Belmont Report shaped the regulatory oversight of biomedical research, the ethical issues seemed clear. The Belmont report clearly distinguished "clinical practice" from "research." Clinical practice was for the benefit of the patient. Research, on the other hand, was seen as an exploitative activity that was done to research subjects for the benefit of future patients and society. Thus, the regulatory oversight of research was designed to ensure that potential research subjects were fully informed of the risks of research participation, that their consent was voluntary, and that there be attentive monitoring of the project to prevent any adverse events. In the intervening years, we have learned that the lines between clinical practice and research are not always as distinct as they once seemed. We have learned that among clinicians idiosyncratic practice variation is ubiquitous so that, often, there is no consensus about which treatments are best. Furthermore, patients are seldom informed of these intra-professional disagreements and so their consent for treatment is not fully informed consent. On the flip side, we have learned that participants in well-designed clinical research studies have outcomes that are similar to outcomes for patients who are not in studies. Thus, the perceived risks of participating in research may be overstated and the perceived risks of receiving treatment outside of research studies understated. This led to a recent national controversy about a research study in premature babies designed to determine the optimum level of supplemental oxygen that should be provided. The controversy focused on the level of risk associated with being in the study compared to not being in the study and the way that the consent form should have described those risks. Such controversies continue.

The most important issue in child health advocacy in the United States today is the issue of health disparities among different racial groups (https://datacenter.kid scount.org/about). Black infants have an infant mortality rate that is 2.3 times the rate for white children. Among older children, mortality rates in Blacks are 60% higher than in Whites. Mortality rates closely track rates of poverty. Black children are three times as likely as White children to grow up in poverty. These health disparities have been highlighted by and amplified by the COVID-19 pandemic. White people are much less likely to be hospitalized or die from COVID than are people of color (https://www.cdc.gov/coronavirus/2019-ncov/covid-data/investigations-dis covery/hospitalization-death-by-race-ethnicity.html). Black children are much more likely to have had adverse childhood experiences that are associated with health problems later in life. There are things to do. We can advocate for better access to health care, childcare, and better food and nutrition. We can participate in programs that provide books to children. We can work to diversify the health care workforce and to educate ourselves about the toxic effects of systemic racism.

1.6   Oceania

1.6.1   New Zealand

Janine Penfield Winters

New Zealand (NZ) is a small and geographically isolated country of 5 million people that has universal health coverage. It was colonized by the UK in the eighteenth and nineteenth century and is heavily influenced by both historical and current British systems and values. NZ has a large public health system that provides the vast majority of care, but there is also a small private health system running alongside. With a broad lens, NZ is an OECD country with health care spending per person just below the OECD mean (OECD 2019a, b). NZ is very much aware that it is not as wealthy as the countries it compares itself to, which are principally the UK and Australia, and to a lesser extent Canada and the USA. Like the UK, pediatricians here are specialty providers and do not provide primary care.

New Zealand's legal and ethical systems for pediatric decision-making are most influenced by the UK. For example, child competency is based on the UK's 1985 House of Lords' Gillick decision (Grimwood 2009). On the other hand, NZ's medical cultural is also unique. NZ's small population means that the specialty pediatric workforce is well connected and culturally unified. There is one dedicated pediatric hospital, Starship, located in the largest city. Starship provides direct services to children, guidance for other pediatric specialty providers and is the hub for child and youth specialty networks (Starship Hospital 2019). While there are specialty pediatric services and inpatient units in several smaller cities, the pediatricians working in NZ are well connected to each other.

The treatment of patients and families in pediatrics reflects a unique combination of a British-centric medical system superimposed on the indigenous Māori culture. It is not possible to summarize this combination fairly in a few sentences, but the Māori cultural conception of *whānau* (family) has been a central tenant to care, well before the "family-centered care" model became popularly articulated in other countries.

New Zealand's resource constraints, linked to a focus on population health, are consistently articulated and acknowledged. This everyday open acknowledgement is a major contrast to wealthier comparison countries, especially the USA. NZ practice guidelines are very clear about resource allocation with both the Ministry of Health (Medical Council of New Zealand 2018) and the New Zealand Medical Association providing frank guidance that: "While doctors have a primary responsibility to the individual patient, they have a concurrent responsibility to all other patients and the community. Doctors therefore have an ethical responsibility to manage available resources equitably and efficiently (New Zealand Medical Association 2020)." As a result, there are few children with extreme medi-technology dependence here in NZ.

A final difference is that clinical ethics committees or consultation models are not funded or available to clinicians in most areas. When dilemmas and disagreements arise, there is a patchwork of methods for approaching them. Mostly, clinicians try to assist each other, but sometimes academic departments are consulted. This is an area of development for NZ.

In summary, New Zealand's unique features regarding ethical challenges for pediatric care and decision-making include four major features:

1. A culture of family centered care due to the long-standing, though restrained, influence of Māori values regarding *whānau* in shared decision-making;
2. Unfettered family directed care is limited due to an acknowledgement in law and guidance documents that resources are limited, and doctors make the ultimate decisions about which measures can be offered to individuals in this publicly funded system;
3. A relatively small population allows relationships between pediatricians and other doctors so that decision-making consensus can often be reached;
4. A paucity of funded ethics committees or ethics advisory groups requires clinicians to rely on each other and their teams for support. Ethical challenges are seldom reviewed by a formal ethics process.

1.7   South America

Tamra-Lee McCleary

South America has a total of twelve sovereign countries and three dependencies (WorldAtlas 2021). This continent is as diverse as the many languages that are spoken there. With five official languages, Spanish, Portuguese, English, Dutch and French, along with dozens other languages -including indigenous languages—the cultural scenery ought to be complex. Like every other region of the world, the customs, beliefs, and socio-political constructs of the many societies within South America give rise to the nuances of the ethical challenges encountered during medical interactions with its pediatric population. The bioethics themes constantly at play in this region are not much different from those all over the world. At an individual level, ethical challenges can range from autonomy and consent for treatment, to disclosure of information, big data, and privacy. At the collective level, ethical challenges range from public health ethics, and health policy, to research ethics. Aside from these commonalities with the rest of the world, there are some unique ethical challenges that are recurrent among South American countries. Two of the most salient bioethical issues are:

(1) The access to health care resources—this refers not only to physical and economic access but also to the integration of intercultural competences to clinical interactions. There are diverse indigenous and religious communities in this region that highly influence the response of parents to health care recommendations. Recognizing and working with this cultural aspect eliminates a barrier for an efficient therapeutic relationship. For example, Colombia used an intercultural health model to effectively implement COVID-19 pandemic

response efforts. By aiming resources and innovative interventions, the government was able to reach native communities and improve their adherence to public health recommendations (UHC 2020).

(2) Progressive autonomy of the child or adolescent; the family nucleus is of much importance in many South American cultures. In some instances, elderly members of the family have more authority over their younger members. Child participation, comprehension and the level of ascent required from them to ensure a shared decision-making process can be overshadowed by customs, violence or poverty (Modocar and Ubeda 2017).

Pediatric ethical issues do not exist in the abstract. There is a high prevalence of economic, education, and health care disparities throughout many countries in South America. For the current 2021 fiscal year, most of the countries are classified by the World Bank as Upper-middle-income economies or high-income economies. Despite this, the wealth gap is impressive, and many could argue that there is a significant number of communities whose lives are more in line with those living in low-income economies or lower-middle income economies. As a result, major health disparities can be observed throughout different sectors of the population. Children, adolescents, and their families face common ethical challenges of health care systems plagued by scarcity of resources, while living within highly industrialized countries.

The importance of the dynamic between the social constructs of health and ethical issues within health care cannot be ignored. Duncan Pedersen (2007) stated in his essay titled: "At the crossroads between global health and local cultures: a critical perspective" that the social determinants of health are often underestimated by researchers and health practitioners, and that generally there has been an inadequate articulation of the macro-social dimensions with the micro-social experiences, and the biological, in attempting to explain the construction of illness and its opposite, the production of health. This statement remains true almost 15 years later. Despite significant advances there is still much to study and implement.

Efforts are being made to strengthen research ethics systems, and systematically integrating ethics into ongoing public health work and decision-making process (PAHO 2018). Research ethics and clinical ethics committees have increased their presence in South America. They have highlighted that it is imperative to build capacity for ethics review, analysis, and oversight locally to effectively tend to ethical challenges at the individual clinical interaction and at the population level. Many States have made remarkable progress in bioethics, improving the frameworks for ethical standards, guidelines for analyses and collaborations for continuous improvement of care in pediatrics aiming for an inclusive decision-making process and ethical deliberations.

1.8   Conclusion

Although diverse in health care access and funding, it is interesting to note that there are indeed many similarities between the regions and countries addressed above. What is of importance is that access to health care for children continues to be a challenge all over the world, which greatly impacts ethical choices when limited resources are available. Many children are still marginalized, and their vulnerability exacerbated. This is especially true of regions where poverty and political instability plays a role. For many of these children health care is not even a basic offering and consequently they may be exploited even more.

Some countries also report on lowering the age of consent for pediatric patients while others underline the fact that children are a part of a bigger whole and consequently do not have the ability to share in choices which influence their health and lives. Children ought not to be seen as property of parents and therefore have no say in their own health care. Respect should be afforded to each child as a unique human being whom should be treated with dignity. Although the United Nations has called for equitable access and protection of children under its Convention on the Rights of the Child (refer to Chap. 5), it would be fair to say that as a collective society we are still faced with challenges when it comes to pediatric health care and addressing the ethical issues surrounding participation of children in their own care.

As a community, ethicists and clinicians need to become advocates for better access and equitable distribution of limited resources. It is also necessary for this community to be at the table when policy discussions are held to make sure children are not marginalized even more. The chapters following this one are good indications of where we can get involved and it is the hope of the authors, that each person who reads this book will have a renewed energy to stand up and continue to develop pediatric bioethics as a discipline, on issues that resonates with them.

References

Akabayashi, A., M.D. Fetters, and T.S. Elwyn. 1999. Family consent, communication, and advance directives for cancer disclosure: A Japanese case and discussion. *Journal of Medical Ethics* 25 (4): 296–301. https://doi.org/10.1136/jme.25.4.296. PMID: 10461591.

Bailey, P.G., and B. Freedman. 2012. Medical ethics. https://www.thecanadianencyclopedia.ca/en/article/medical-ethics. Accessed 15 Nov 2020.

Ball, I.M., L. Hornby, B. Rochwerg, et al. 2020. Management of the neurologically deceased organ donor: A Canadian clinical practice guideline. *Canadian Medical Association Journal* 192 (14): E361–E369.

Cameron, L., N. Erkal, L. Gangadharan, and X. Meng. 2013. Little emperors: Behavioral impacts of China's one-child policy. *Science* 339 (6122): 953–957.

Campos, R., E. Chávez, and G. Esquivel. 2014. *Los ingresos altos, la tributación óptima y la recaudación posible [High income, optimal taxation and possible collection]*. Mexico: Centro de Estudios de las Finanzas Públicas, Cámara de Diputados.

Canada. Gov. 2015. Canada Health Act Report. https://www.canada.ca/content/dam/hc-sc/mig
ration/hc-sc/hcs-sss/alt_formats/pdf/pubs/cha-ics/2015-cha-lcs-ar-ra-eng.pdf. Accessed 15 Nov
2020.

Chang, H., X.Y. Dong, and F. MacPhail. 2011. Labor migration and time use patterns of the left-
behind children and elderly in rural China. *World Development* 39 (12): 2199–2210.

Clark, B.A., A. Virani, S.K. Marshall, and E.M. Saewyc. 2020. Conditions for shared
decision-making in the care of transgender youth in Canada. *Health Promotion International*
daaa043. https://doi.org/10.1093/heapro/daaa043.

Coughlin, K.W. 2018. Medical decision-making in Paediatrics: Infancy to adolescence. *Paediatrics & Child Health* 23 (2): 138–146.

Davies, D. 2018. Medical assistance in dying: A paediatric perspective. *Paediatrics & Child Health*
23 (2): 125–130.

Drake, M., and P. Cox. 2012. Ethics: End-of-life decision-making in a pediatric patient with SMA
type 2. *The Influence of the Media* 78 (23): e143–e145.

El Bcheraoui, C., P.H. Mimche, Y. Miangotar, V.S. Krish, F. Ziegeweid, K.J. Krohn, et al. 2020.
Burden of disease in francophone Africa, 1990–2017: A systematic analysis for the Global Burden
of Disease Study 2017. *The Lancet Global* 8 (3): E341–E351. https://doi.org/10.1016/S2214-109
X(20)30024-3.

ENNViH. 2002. Encuesta Nacional sobre Niveles de Vida de los Hogares [National Survey on
Household Living Standards]. http://www.ennvih-mxfls.org/. Accessed 22 Oct 2020.

Fontana, M.S., C. Farrell, F. Gauvin, J. Lacroix, and A. Janvier. 2013. Modes of death in pediatrics:
Differences in the ethical approach in neonatal and pediatric patients. *The Journal of Pediatrics*
162 (6): 1107–1111.

Friedrich, A.B. 2020. The suffering child: Claims of suffering in seminal cases and what to do about
them. *Pediatrics* 146 (Supplement 1): S66–S69.

Garros, D., R. Rosychuk, and P. Cox. 2003. Circumstances surrounding end of life in a pediatric
intensive care unit. *Pediatrics* 112 (5): e371–e371.

Grimwood, T. 2009. Gillick and the consent of minors: Contraceptive advice and treatment in New
Zealand. *Victoria University of Wellington Law Review* 40: 743–769. http://www.nzlii.org/nz/jou
rnals/VUWLawRw/2009/38.pdf.

Hester, D. M. 2008. Pediatric ethics—Why it deserves special attention. *Practical Bioethics* 3 (4):1,
5–7.

Jia, Z., and W. Tian. 2010. Loneliness of left-behind children: A cross-sectional survey in a sample
of rural China. *The Child* 36 (6): 812–817.

Joffe, A.R., J. Carcillo, N. Anton, et al. 2011. Donation after cardiocirculatory death: A call for a
moratorium pending full public disclosure and fully informed consent. *Philosophy, Ethics, and
Humanities in Medicine: PEHM* 6: 17.

Keeley, B. 2018. *Desigualdad de ingresos. La brecha entre ricos y pobres. Esenciales OCDE
[Income inequality. The gap between rich and poor. OECD Essentials].* París: OECD Publishing.

Kimura R. 1995. *Contemporary Japan. Encyclopedia of bioethics*, revised ed., vol. 3. 1995 by
Warren T. Reich, 1496–1505.

Kurihara, T. 2007. Seken. In *Blackwell encyclopedia of sociology*, ed. G. Ritzer, 4162–4165. Malden,
MA: Blackwell Publisher.

Medical Council of New Zealand. 2018. *Safe practice in an environment of resource limitation.* https://www.mcnz.org.nz/assets/standards/ca25302789/Safe-practice-in-an-environment-
of-resource-limitation.pdf. Accessed 19 Nov 2020.

Modocar, C., and M.E. Ubeda. 2017. *La violencia en la primera infancia. Marco Regional de
UNICEF para America Latina y el Caribe,* 3–7. https://www.unicef.org/lac/media/686/file/PDF%
20La%20violencia%20en%20la%20primera%20infancia.pdf . Accessed 1 Feb 2021.

New Zealand Medical Association. 2020. *Code of ethics.* https://www.nzma.org.nz/publications/
code-of-ethics Accessed 19 Nov 2020.

Chapter 2
A Developmental Perspective on Pediatric Decision-Making Capacity

N. Hardy and N. Nortjé

Abstract Decision-making capacity (DMC) for pediatric patients can be difficult to determine and is influenced by a myriad of developmental considerations. This chapter begins with a discussion concerning the nature of decision-making and what constitutes competency. The "rule of sevens" framework is then used to explicate pertinent developmental milestones for children, dividing pediatric development into 0–7, 7–14, and 14+ years of age. In particular, the authors highlight the most important cognitive, social, and emotional considerations in each of these periods and how they pertain to a child's ability to make important medical decisions.

Keywords Development · Decision-making · Cognitive · Social · Emotional

2.1 Introduction

Montgomery (2009) holds that childhood should not be seen as a stage of physical change only but should also be seen as a socially constructed idea which can change depending on the historical and cultural setting (also refer to Chap. 12). As authors we agree with this statement and recognize that different authorities throughout the world have demarcated different rights and responsibilities ascribed to those under the age of 21. However, since the focus of this chapter is to introduce the ability of a child to make decisions in her own medical care which develops as maturation allows from a state of limited understanding and complexity to more advanced understanding and computational complexity, no reference will be made to specific legal requirements from any jurisdictions or social contexts. Rather the rule of sevens framework will be used. This framework is based on English Common law, where those under the age of

N. Hardy
Albert Gnaegi Center for Health Care Ethics, Saint Louis University, St. Louis, MO, USA
e-mail: natalie.hardy@slu.edu

N. Nortjé (✉)
Department of Critical Care Medicine, University of Texas MD Anderson Cancer Center, Houston, USA
e-mail: NNortje@mdanderson.org

© Springer Nature Switzerland AG 2022   23
N. Nortjé and J. C. Bester (eds.), *Pediatric Ethics: Theory and Practice*,
The International Library of Bioethics 89,
https://doi.org/10.1007/978-3-030-86182-7_2

7 were argued to have no capacity to consent or partake in any decision-making, those 7–14 were presumed not to have the capacity to consent or make good decisions, and those older than 14 were acknowledged to have capacity to consent but may still not fully understand and appreciate the consequences of their decisions (Schlam and Wood 2000).

As per the American Academy of Pediatrics' Committee on Bioethics' (2016) policy statement, patients should participate in decision-making in their own care to the extent allowed by their level of development (also refer to Chap. 7). It is the authors' opinion that parents and health care providers need to be cognizant of all the developmental changes, including cognitive, emotional, and social, that influence a patient's autonomy, and how these considerations could potentially impact decision-making.

The rest of this chapter will focus on the nature of decision-making and how this develops during specific periods (based on the rule of sevens).

2.2   The Nature of Decision-Making

When discussing the developmental implications of pediatric decision-making, it is important to first understand the nature of decision-making and how such considerations shed light on a patient's capacity. In the context of this discussion, decision-making capacity (DMC) can best be understood as, "the ability of a patient to understand the benefits and risks of, and the alternatives to, a proposed treatment or intervention (including no treatment)" (Barstow et al. 2018, p. 40). Consider a physician who proposes radiation treatment for a cancer patient. The radiation could instantiate several lifelong side effects including loss of bladder control, permanent alteration of taste, development of cataracts, and difficulty swallowing, which are known to be common after effects of radiation therapy (*Radiation Therapy Side Effects* 2018). However, suppose that this treatment is very successful in approximately 80% of patients. The question for the patient then becomes whether surviving outweighs the burdens of living with potentially low quality of life. A competent decision-maker would understand the severity of the side effects, weigh the benefits of survival against such effects, and evaluate other relevant treatment options to reach an appropriate decision—a complex decision that must also be considered in relation to the values and goals of the individual.

Though there are many frameworks outlining the nature of DMC, there seems to be a general paradigm. According to Ross (1981), there are five core skills that must be mastered by an effective decision maker, which include (1) the ability to identify a set of alternative courses of action, (2) identifying appropriate criteria for considering alternatives, (3) assessing alternatives by criteria, (4) summarizing information about alternatives, and (5) self-evaluation. Appelbaum (2007) outlines similar criteria, stating that DMC involves (1) understanding relevant information, (2) appreciating the situation and its consequences, (3) reasoning about treatment options, and

(4) clearly communicating a choice. Mann et al. (1989) include Ross and Appelbaum's criteria but add that DMC involves creative problem-solving, compromise, consistency, and commitment.

Given that the nature of decision-making is complex and necessitates "knowledge, affect, and skill" (Ross 1981, p. 279), DMC is not an "all or nothing" ability (Ruhe et al. 2015; Mccabe 1996). Rather, DMC is best viewed in terms of a "sliding scale" model, meaning that a child could have the capacity to make some decisions and not others (Drane 1985). Precisely *where* a patient falls on this scale depends on a multitude of factors including age, social/contextual factors, cultural influences, and cognitive development. As Bester et al. (2017) note, "Familiarity with the condition and the intervention, the frequency and duration of the treatment, and the benefits and burdens involved may also contribute to the complexity of the decision" (100). An adolescent who needs a standard laparoscopic cholecystectomy to treat recurrent gallstones arguably faces fewer complex questions and considerations than someone who needs a life-saving treatment known to cause serious and chronic side effects; the former patient will likely have DMC whereas the latter patient's capacity might be questioned.

Clearly, pediatric patients vary in their DMC. An individual can range from (1) being incompetent due to severe cognitive impairment, psychiatric disorders, or young age, (2) having the ability to assent or dissent, or (3) having full rational decision-making abilities in complex cases (Drane 1985). However, even if a pediatric patient is not the primary decision-maker, there are varying degrees to which she can be involved. Alderson (2003) outlines four levels of participation in pediatric decision-making: (1) being informed, (2) expressing a view, (3) influencing a decision, and (4) being the main decision-maker. Physicians must acknowledge that children can, and oftentimes do, participate in decision-making to some degree, which necessitates that providers actively involve their patients as much as possible. However, in order for children to be "optimally included," their caretakers, physicians, and family must "recognize how to involve [them] in a developmentally appropriate fashion" (Ruhe et al. 2015, p. 779). A physician must be cognizant of the cognitive, social, and emotional factors marking each stage of development. As children grow, learn, and develop, it is presumed that their DMC continuously evolves as well (Ruhe et al. 2015), and it is the goal of this chapter to shed light on how this may be so from a developmental perspective.

To be clear, it is not the goal of the authors to argue for or against the DMC of anyone, nor to imply that the factors mentioned here are the only markers of adequate decision-making. It is quite possible that a child or adolescent does not meet all presumed DMC "criteria" yet is able to make reasonable and sound medical decisions. Regardless of one's DMC, every patient ought to be recognized as "a moral being with all of the appropriate dignity and rights" and a person who is "a more vulnerable decision maker than adults" (Katz and Webb 2016, p. e8). Pediatric patients, no matter what the extent of their DMC, are inherently vulnerable and must be supported as they attempt to comprehend their medical circumstances. A developmental perspective sheds light on such vulnerabilities in an insightful manner, as each stage of life introduces important milestones that must be recognized as such.

2.3 0–7 Years Old

2.3.1 Cognitive Development

According to the 7-7-7 framework, children under seven do not have DMC (Katz and Webb 2016). However, a cognitive and social/emotional developmental analysis show that children can participate in medical decision-making to a certain extent, when they reach the latter stage of this period. "Developmental maturation of the child allows for increasing longitudinal inclusion of the child's voice in the decision-making process" (Katz and Webb 2016, p. e8)—the degree of which depends on the nature of each individual situation.

A child's understanding of her medical condition, treatment options, and the relevant risks and benefits depends largely on her cognitive abilities (Mccabe 1996). From ages 0–7, infants and young children experience cognitive developmental milestones in which synaptic connections and gray matter density increase (Arnett and Jensen 2019). The brain reaches about 70% of its adult weight by age three and 90% by age six. It does so by undergoing *transient exuberance,* a neuronal process characterized by the creation of innumerable synaptic connections. Myelination increases, which is known as the "formation of the myelin sheath surrounding neuronal axons" (Chevalier et al. 2015, p. 2), as well as synaptic pruning during toddlerhood, the process of selectively removing neuronal connections which results in more effective interneuronal communication (Arnett and Jensen 2019). These occurrences are associated with increased processing speed (Chevalier et al. 2015), attention span (Mahone and Schneider 2012; Arnett and Jensen 2019) and improved long-term memory (Arnett and Jensen 2019). Though a child's memory is still fleeting during this phase of life (Arnett and Jensen 2019), working memory begins to develop and improve substantially beginning at the age of six (Grooten-Wiegers et al. 2017)—skills that lay the foundation for complex reasoning (Edin et al. 2007) and decision-making. Language proficiency and verbal communication abilities are also important for the purposes of expressing preferences and are initiated around the age of five (Grootens-Wiegers et al. 2017, p. 3; Shaffer and Kipp 2007).

Though early childhood is characterized by such remarkable cognitive milestones, children under the age of seven still understand illness in a very nonspecific and universal sense (Kuther 2003). The prefrontal cortex, responsible for executive functions and balancing risks and benefits, is one of the last brain regions to fully develop and does not mature fully until young adulthood (Katz and Webb 2016). In addition, young children are less attentive than older individuals and are more likely to be distracted in an environment with extraneous stimulation (Higgins and Turnure 1984)—an important consideration regarding medical decision-making, which often occurs in hospitals or in contexts with distracting stimuli. Thus, compared to young children, older individuals are "more able to compare the relative virtues of two multidimensional decisions…[and] more likely to discover the best option in a set of complex options" (Byrnes 2002, p. 212). Although every child is different, children under seven likely do not possess complex reasoning abilities, comprehensive understanding of illness, or executive functioning necessary to have full DMC.

2.3.2   Social and Emotional Development

Social factors—including a patient's sense of self, her relationship with authority, and the level of stress in the environment—also play an influential role in pediatric DMC (McCabe 1996) by "influenc[ing] the stability of…choices" (Mccabe 1996, p. 510; Damon 1983). Katz and Webb (2016) add that "A coherent sense of identity and stable, deep-seated values are key to making reflective, autonomous decisions required for true informed consent" (p. e8). Such values include religious ideologies, cultural norms, or beliefs about health, which typically develop and are instilled well into adulthood. Children under the age of seven have hardly had time to develop their core values, nor do they have past experiences to guide decision-making (Byrnes 2002). Thus, such young children inherently rely on older adults who have made medical decisions to understand, anticipate, and advocate in ways that they are not yet capable of doing.

Emotional development also impacts a child's ability to make informed decisions. Children under the age of three have little to no emotional self-regulation (Ziv et al. 2017). Emotional understanding and regulation begin to develop between the ages of three and four (Cole et al. 2009), though development is not yet complete. By the age of five, a child's *theory of mind*, (the ability to describe the emotions of a situation as well as the emotions of others) significantly develops (Arnett and Jensen 2019). However, although emotions are helpful in guiding patients towards an appropriate decision, self-regulation ensures that emotions do not cloud one's rational judgement. These skills rapidly develop between the ages of 0 and 7, but children are still far from having full emotional regulatory abilities.

In light of these considerations, existing data suggests that pediatric patients between the ages of 0 and 7 experience significant cognitive, social, and emotional development. Such maturation allows most children to express assent or dissent, even if they cannot understand the true breadth of their medical circumstances. Parents and providers must therefore be "active in soliciting children's preferences" (Mccabe 1996, p. 514) and should always aspire to maintain open communication with the child. As Leiken (1983, p. 174) argues, mere relaying of information does not necessarily suffice in fostering an ethical relationship with minors—rather, an active "collaborative effort" is necessary to ensure that the child's values—no matter what the level of their DMC—remain at the locus of important medical decisions to be made.

2.3.3 Implications

Children of seven years of age or younger should ordinarily be presumed to lack the capacity to make autonomous and independent decisions. Clinicians should not seek informed consent or expect autonomous decisions at this age but should by default seek to protect the interests of the child. However, children develop the capacity to be involved in information-sharing and to provide various levels of assent depending on whether they perceive the locus of control to be internal or external—according to Leiken (1983), minors are less likely to publicly express dissent because they perceive their parents as being the decision-makers regarding treatment and goals of care. Thus, in order to authentically understand a minor's wishes, a social worker or psychologist ought to be consulted so that the child has a space to express preferences, aside from individuals who are directly involved with the patient's care (Leiken 1983).

2.4 7–14 Years Old

According to the rule of sevens, "a lack of capacity is presumed but may be rebutted with appropriate evidence between the ages 7 and 14 years" (Katz and Webb, 2016, p. e7). A developmental analysis sheds light on how some children can be presumed to have capacity in terms of exceptional cognitive or social and emotional development. Such assessments should be made on an individual basis and depend on each patient's intellectual capacity.

2.4.1 Cognitive Development

Pediatric patients between the ages of 7 and 14 experience significant maturation of cognitive abilities. Whereas children under seven have difficulty focusing their attention, patients in this developmental period develop selective attention—the ability to focus on relevant information and disregard distracting stimuli (Arnett and Jensen 2019, p. 287). In addition, processing speed exponentially decreases throughout childhood, meaning that it takes less time to "initiate responses in neuropsychological or problem-solving tasks" (Luna et al. 2004, p. 1367). This is a valuable skill to have when medical circumstances necessitate timely decision-making.

Developments in attention span are interrelated with significant improvements in working memory (Arnett and Jensen 2019; Olson and Luciana 2008), which underlies "the emergence of complex mental abilities" (Luna et al. 2004, p. 1357). 9-year old children can comprehend basic facts and express preferences regarding treatment options in a way that is "similar to those of adults" (Weithorn and Campbell 1982, p. 1596). Children also mature in inhibitory control and deploy greater problem-solving strategies during this phase of life (Arnett and Jensen 2019; Eccles

1999), completing the "ability to inhibit attention to irrelevant stimuli" by age 10 and mastering it by age 12 (Romine and Reynolds 2005; Passler et al. 1985). And, frontal lobe operations, which include self-regulatory behavior and cognitive functioning, mature to a "medium and large" degree between the ages of 8 and 11 (Romine and Reynolds 2005, p. 193). Clearly, children in this developmental phase are more focused, cognitively mature, and intellectually capable when it comes to DMC.

However, children between the ages of 7 and 14 do not yet make decisions in the same way as adults, which can be attributed to the continuing maturation of the prefrontal cortex into adulthood (Olson and Luciana 2008; Hooper et al. 2004). It has been found that adults are much more likely to consider the risks and benefits of a decision than 6th and 8th graders (~11 and 13 years of age), and they are also much more inclined to seek second opinions regarding medical decisions (Halpern-Felsher and Cauffman 2001). Nonetheless, it is important to reiterate that medical decision-making is situation-specific, and the complexity, nature, and details of a case greatly influence a patient's DMC.

2.4.2   Social and Emotional Development

Middle childhood (ages 9–11) is often characterized as a "golden age" emotionally, due to it being a time of "high wellbeing and low volatility" (Arnett and Jensen 2019, p. 306). This might be due to increases in autonomy, as children exhibit increased self-directed behavior and require less parental monitoring (Arnett and Jensen 2019). In addition, development of one's self-concept also becomes more complex—for instance, 7 to 9-year old children often describe themselves concretely (in terms of gender and possessions), while adolescents include "beliefs, characteristics, and motivations" in a self-concept that is more abstract in nature (Markus and Nurius 1984, p. 154). Such considerations are important, as one's self-concept and evolving values directly play into the sorts of medical decisions she makes.

However, as puberty approaches (typically around the age of 12, thought onset depends on the individual), early adolescence evolves into a period of "low [emotional] stability"; youth experience many emotional fluctuations as opposed to individuals in late adolescence (Larson et al. 2002). This is important to acknowledge for DMC because emotions such as sadness, anger, and fear can affect one's depth of information processing, risk perception, and interpersonal relationships in the clinical setting (Ferrer et al., in press). Nonetheless, children and adolescents *want* to be involved in making difficult decisions, even when they are "emotionally taxing" and when the decision involves "serious negative consequences" (Ruggeri et al. 2014, p. 5). However, it is also important to acknowledge how children differentially perceive illness at this age—compared to healthy minors, children with a higher level of anxiety are more prone to viewing illness as a punishment for misbehavior or as a "disruptive force" (Brodie 1974). Thus, to avoid provoking and intensifying illness-related anxiety, engaging conversations must be had with the child to understand how they view their illness and to attenuate any self-blaming tendencies (Leiken

1983). In addition, as Leiken argues, involving children in an "ethical, humanistic, and therapeutic" fashion means that a *collaborative* effort is facilitated between the child and involved parties to ensure that the child's concerns and values are integrated as part of the decision-making process (p. 174). A lack of refusal does not mean that minors *assent* to treatment, as there are many reasons that a patient might not wish to be involved in decision-making (such as denial or reluctantly submitting to authority figures) (Leiken 1983). Thus, minors should be intentionally included in discussions whenever possible, even if they express dissent, and should always be made to feel that their opinions, desires, and values are being prioritized.

In sum, existing data suggests that pediatric patients between the ages of 7 and 14 vary widely in their cognitive, social, and emotional development, depending on their particular age. Younger children have yet to experience full cognitive maturation, yet they are less likely to experience emotional volatility, while emerging adolescents are more prone to emotional highs and lows yet are more cognitively mature. As always, DMC must be assessed on a case-by-case basis as patients develop cognitively, emotionally, and socially.

2.4.3   Implications

Children of the age of 7–14 should be considered to lack decision-making capacity to engage in fully autonomous decision-making, but this presumption may be set aside by compelling circumstances. These circumstances would depend on an assessment of the individual development and experiences of the child, and of level of capacity required for the specific decision at hand. If there is adequate evidence that the minor "has sufficient knowledge and intelligent understanding of the illness and its treatment," then dissent of the child should be honored, just as assent is (Leiken 1983, p. 174). Children of these ages should be involved to an even greater extent in information sharing and decision-making than younger children, but always in an age-appropriate manner.

2.5   14+ Years Old

This period is often referred to as late adolescence and based on the Latin word *adolescere* —to grow up (Sawyer et al. 2018), however the age allocation is not without dispute. Per Montgomery's (2009) earlier argument this phase is greatly influenced by cultural definitions. Some cultures ascribe adult roles to adolescents earlier than others (Jones 2010; Sawyer et al. 2018). *Vis-vis*, this period is regarded as period of immense biological development and major social role transition.

2.5.1 Cognitive Development

Feldman (2011) indicates that one specific area of the brain that undergoes considerable development throughout adolescence is the prefrontal cortex. This area is important for executive functions such as logical thinking, evaluations, and making complex judgments. Synaptic exuberance, characterized as remarkable neuronal generation, also continues from its initiation in childhood; however, many more neurons are born during this developmental period (Lu and Sowell 2009) and continue to be myelinated (Spear 2013). Given that neurons are naturally overproduced during a person's lifetime, synaptic pruning occurs at this stage—the "hallmark" of adolescent brain transformation involving highly specific loss of neuronal connections (Spear 2013, p. S8). As a result, the prefrontal cortex becomes increasingly efficient in communicating with other parts of the brain, enabling it to process and integrate information more effectively (Feldman 2011; Newman and Newman 2012). Another important function of the prefrontal cortex is impulse control, which is important during decision-making, as decisions during this phase are not made reactively, but rather based on information. However, since pre-frontal maturation occurs around the age of 25, the adolescent may better integrate and process information, compared to younger individuals, however they could still act impulsively, which could influence DMC. Pertinent to this discussion some authors argue that adolescents, even though they may develop cognitive maturity, may still be faced with a *two-brain system* (Diekema 2011) or have a *dual systems model* for DCM (Katz and Webb 2016). On the one hand they develop cognitive control (as discussed earlier), but on the other hand there is development of the socioemotional system. The socioemotional system, which is made up of limbic and paralimbic structures (i.e. the amygdala) and increase the dopaminergic activity in the brain, focus on more reward-seeking and risky behavior (Christakou 2014; Diekema 2011; Katz and Webb 2016). This may therefore influence an adolescent's capacity to make fully informed decisions.

According to the Piagetian approach to cognitive development, adolescents enter the formal operational stage, which is characterized by the ability to think abstractly. Given that many medical decisions need to be made on the abstract thought of an internal disease, Piaget (1973) is of the opinion that adolescents are able to transcend the concrete thinking phase and are able to systematically evaluate problems and situations and make decisions accordingly.

Summative, some authors (Alderson et al. 2006; Raymundo and Goldim 2008; Weithorn and Campbell 1982) argue that since DMC is defined, at least in part, in terms of cognitive processes, it was found that adolescents who have had experience with chronic illness, made relatively similar choices compared to adults.

2.5.2 Social and Emotional Development

Adolescents progressively seek more autonomy from their parents, independence in their decisions, and a sense of control over their lives. This often frustrates parents as children who have previously accepted their judgments, declarations and guidelines begin to question (and sometimes rebel against) their parents' view of the world (Feldman 2011). As Newman and Newman (2012) argue these differences are normal and is both an area of growth for the parents and the adolescent. Most conflict takes place during the early stages of adolescence as differing definitions of, and rationales for, appropriate and inappropriate conduct exists between the parent and the child. As the adolescent's arguments become more compelling and less reactive, many parents realize that the child's behavior is not unreasonable and oftentimes more independence is encouraged by the parents. During this time, peer relations also become important to the adolescent and when they need to decide, it is often times influenced by the opinions of their friends.

Of note it is important to mention that although parents' consent and decision-making is often sought during this period, there are 3 broad criteria in the USA (which could be a guideline internationally) in which a minor may be allowed to make a medical decision for herself. Diekema (2011) and Katz and Webb (2016) identifies:

Exceptions based on specific diagnostic/care categories—as it relates to sexual activity which could include seeking treatment for sexually transmitted diseases (STDs), contraceptives; prenatal care, and abortion.

Mature minor exception—which recognizes that a subset of adolescents has the maturity and capacity to understand the risks, benefits, potential outcomes, and alternatives of medical interventions. In this judicial determination the age, overall maturity, cognitive ability, and social context of the minor is considered.

Legally emancipated—those adolescents who live separately from their parent and are self-supporting; who are married; or those on active military duty, are legally emancipated and competent to make their own medical decisions.

For a more elaborate discussion of the aforementioned please refer to Chap. 10.

In the United Kingdom, the legal standard to test a minor's capacity to consent is known as the Gillick test (Diekema 2011) (also refer to Chap. 13). According to this test a minor needs to demonstrate sufficient understanding and intelligence in order for her to fully understand what the decision is about.

The aforementioned USA and UK exceptions are all based on the general assumption of capacity to make a decision, which includes the ability to understand the information and facts relevant to the issue; appreciate the situation and consequences of the decision; ability to manipulate the information in a rational way; and the ability to communicate a choice (Diekema 2011). Based on this definition it would suffice to say that adolescents then also would have the ability to refused treatment on an informed basis. For a detailed discussion on this please refer to Chap. 9.

2.5.3   Implications

Adolescents over 14 years of age usually have cognitive capacities similar to those of adults and can be included in decision-making in a more far-reaching way than younger children. Adolescents may be able to make some decisions for themselves, depending on the particular circumstances. As Leiken (1983) notes, minors over the age of 14 are "more likely to believe that the locus of control is internal and that they are in control of their health and of illness outcomes" (p. 173). However, adolescence is still a time of cognitive development; adolescents typically employ their decision-making differently than do adults. They weigh immediate considerations higher, are more prone to peer pressure and to impulsive decision-making and are still forming long-term values and life plans. Consequently, for decisions that require very high levels of cognition and that have significant long-term implications, adolescents may not yet be ready to engage in fully autonomous and independent decision-making. For example, it should be presumed that adolescents cannot refuse life-saving treatments unless compelling considerations exist that rebut this presumption.

2.6   Conclusion

It is important to reiterate that although DMC and developmental changes were discussed in broad terms in this chapter, any interaction with a patient is (and ought to be) situation-specific, and the complexity, nature, and details of a case greatly influence a patient's own individuality and ability in having informed DMC. Parents and health care providers should take cognizance of the research (Alderson et al. 2006; Raymundo and Goldim 2008; Weithorn and Campbell 1982) which indicates that those children who have lived with illness and chronic conditions for most of their lives, often have a very good understanding of their situation and is often times very well informed—to such a level that they would be emotionally and cognitively able to make their own health care choices.

Fundamental to this discussion one would have to always keep in mind that health care providers have a fiduciary duty towards their patients to keep their trust and this is built through respecting their opinions. To facilitate a trusting and therapeutic physician–patient relationship, providers have an obligation to openly convey information to their pediatric patients (as appropriate), while also actively and collaboratively engaging them in the decision-making process (Leiken 1983).

Guiding Principles for the Ethical Care of Children 0–7 Years Old

Lack of Capacity and Inability to Consent
- Acknowledge that each child's cognitive abilities are unique and have different developmental trajectories
- Acknowledge that children under seven are limited in their ability to comprehend the nature of their diagnosis, treatment options, the medical establishment, and their role in such, meaning that providers, families, and caregivers must communicate clearly, directly, and supportively in a stress-free environment with minimal distractions. Facilitate active engagement with minors to ensure their values and preferences are at the center of such discussions (Leiken 1983)
- Ensure that minors are not self-blaming for their illness and make special efforts to monitor and control illness-induced anxiety (Leiken 1983; Brodie 1974)
- Maintain honest and transparent dialogue with the minor, especially regarding potential side effects and how a given treatment might affect their values (Leiken 1983)
- Protect the minor's best interests by collaborating with them through discussions and careful integration of assent and dissent. Respect the minor's vulnerability and be cautious of potentially abusive situations (Leiken 1983)
- Seek consultation from a social worker or psychologist to allow for a minor to candidly express preferences to a professional who is not directly related to the interests of the patient, such as a physician or parent (Leiken 1983)

Create a safe, private, and confidential space for minors to express concerns

Guiding Principles for the Ethical Care of Minors7–14 Years old

Developing Capacity and Consent
- Acknowledge that decision-making capacity varies on a case-by-case basis at this developmental stage; thus, assess capacity individually, taking into consideration the impacts of puberty, emotions, and cognitive development on decision-making capacity

- Promote the development of self-concept through facilitated conversations and value-focused dialogue. Encourage minors to openly discuss questions regarding their treatment and preferences and create safe and confidential spaces for them to do so

- Involve minors in decision-making to the greatest extent possible while being mindful that minors in this stage vary greatly in their decision-making capacity
- Create a safe, private, and confidential space for children 7–14 to express concerns

Guiding Principles for the Ethical Care of Adolescents

Emerging Capacity and Consent
- Appreciate the cognitive, social, and emotional developmental maturation of adolescents' whilst also acknowledging the role of peers in decision-making, and the potential for impulsive decision-making, and emotional volatility that characterize this developmental stage
- Integrate adolescents into care conversations and medical decision-making as much as possible by actively asking the adolescent about their values, preferences, and goals of care. Maintain regular conversations of this sort throughout treatment, acknowledging the dynamicity of adolescents' values and preferences as the nature of their medical circumstances evolves
- In the case of adolescent—provider and/or adolescent – parental conflict regarding treatment decisions or goals of care, seek to understand the nature of the conflict and the important implications surrounding growing autonomy of this developmental stage. Integrate ways that the adolescent can be autonomous and feel heard in the decision-making process while also making a medically appropriate decision in line with personal values and beliefs
- Appropriately acknowledge and apply the three broad criteria in the U.S. (and potentially internationally) which allow adolescents to make their own decisions: (1) Exceptions based on specific diagnostic/care categories, (2) Mature Minor exception, and (3) legal emancipation

---

Respecting privacy and confidentiality
- Create a safe, private, and supportive space for adolescents to express concerns related to their care independently from parental figures or caregivers
- Pay special attention to confidentially support adolescents' mental, reproductive, and sexual health. Follow-up with the adolescent regularly and encourage them to seek preventative care as well

---

References

Alderson, P. 2003. Die Autonomie des Kindes: über die Selbstbestimmungsfähigkeit von Kindern in der Medizin. In *Das Kind als Patient-Ethische Konflikte zwischen Kindeswohl und Kindeswille*, ed. C. Wiesemann, A. Dörries, G. Wolfslast, and A. Simon, 28–47. Frankfurt/Main: Campus Verlag.

Alderson, P., K. Sutcliffe, and K. Curtis. 2006. Children's competence to consent to medical treatment. *Hastings Center Report* 36 (6): 25–34.

American Academy of Pediatrics, Committee on Bioethics. Informed consent in decision-making in pediatric practice [policy statement]. Pediatrics. 2016.

Appelbaum, P. 2007. Assessment of patients' competence to consent to treatment. *New England Journal of Medicine* 357: 1834–1840.

Arnett, J.J., and L.A. Jensen. 2019. *Human development: A cultural approach*. Pearson Education.

Barstow, C., B. Shahan, and M. Roberts. 2018. Evaluating medical decision-making capacity in practice. *American Family Physician* 98 (1): 40–46.

Bester, J.C., M. Smith, and C. Griggins. 2017. A Jehovah's witness adolescent in the labor and delivery unit: Should patient and parental refusals of blood transfusions for adolescents be honored? *Narrative Inquiry in Bioethics* 7 (1): 97–106.

Brodie, B. 1974. Views of healthy children toward illness. *American Journal of Public Health* 64 (12): 1156–1159.

Byrnes, J.P. 2002. The development of decision-making. *Journal of Adolescent Health* 31: 208–215.

Chevalier, N., S. Kurth, M.R. Doucette, M. Wiseheart, S.C. Deoni, and D.C. Dean. 2015. Myelination is associated with processing speed in early childhood: Preliminary insights. *PLoS ONE* 10 (10): 1–14.

Christakou, A. 2014. Present simple and continuous: Emergence of self-regulation and context sophistication in adolescent decision-making. *Neuropshygologica* 65: 302–312.

Cole, P.M., T.A. Dennis, K.E. Smith-Simon, and L.H. Cohen. 2009. Preschoolers' emotion regulation strategy understanding: Relations with emotion socialization and child self-regulation. *Social Development* 18 (2): 324–352.

Damon, W. 1983. *Social and personality development*. New York: W. W. Norton.

Diekema, D.S. 2011. Adolescent refusal of lifesaving treatment: Are we asking the right questions? *Adolescent Medicine* 22: 213–228.

Drane, J.F. 1985. The many faces of competency. *The Hastings Center Report* 15 (2): 17–21.

Eccles, J.S. 1999. The development of children ages 6 to 14. *The Future of Children* 9 (2): 30–44.

Edin, F., J. Macoveanu, P. Olesen, J. Tegnér, and T. Klingberg. 2007. Stronger synaptic connectivity as a mechanism behind development of working memory–related brain activity during childhood. *Journal of Cognitive Neuroscience* 19 (5): 750–760.

Feldman, R.S. 2011. *Development across the lifespan*, 6th ed. Pearson: Boston.

Grooten-Wiegers, P., I.M. Hein, J.M. Broek, and M.C. de Vries. 2017. Medical decision-making in children and adolescents: Developmental and neuroscientific aspects. *BMC Pediatrics* 17: 1–10.

Halpern-Felsher, B.L., and E. Cauffman. 2001. Costs and benefits of a decision: Decision-making competence in adolescents and adults. *Applied Developmental Psychology* 22: 257–273.

Higgins, A.T., and J.E. Turnure. 1984. Distractibility and concentration of attention in children's development. *Child Development* 55 (5): 1799–1810.

Hooper, C.J., M. Luciana, H.M. Conklin, and R.S. Yarger. 2004. Adolescents' performance on the Iowa gambling task: Implications for the development of decision-making and ventromedial prefrontal cortex. *Developmental Psychology* 40 (6): 1148–1158.

Jones, G.W. 2010. Changing marriage patterns in Asia. 2010. Asia Research Institute—Working Paper 131. https://papers.ssrn.com/sol3/papers.cfm?abstract_id=1716533. Accessed 17 Feb 2021.

Katz, A.L., and S.A. Webb. 2016. Informed consent in decision-making in pediatric practice. *American Academy of Pediatrics* 138 (2): e1–e18.

Kuther, T.L. 2003. Medical decision-making and minors: Issues of consent and assent. *Adolescence* 38 (150): 343–358.

Larson, R.W., G. Moneta, M.H. Richards, and S. Wilson. 2002. Continuity, stability, and change in daily emotional experience across adolescence. *Child Development* 74 (4): 1151–1165.

Leiken, S.L. 1983. Minors' assent or dissent to medical treatment. *The Journal of Pediatrics* 102 (2): 169–176.

Lu, L.H., and E.R. Sowell. 2009. Morphological development of the brain: What has imaging told us? In *Neuroimaging in developmental clinical neuroscience*, ed. J.M. Rumsey and M. Ernst, 5–12. Cambridge University Press.

Luna, B., K.E. Garver, T.A. Urban, N.A. Lazar, and J.A. Sweeney. 2004. Maturation of cognitive processes from late childhood to adulthood. *Child Development* 75 (5): 1357–1372.

Mahone, E.M., and H.E. Schneider. 2012. Assessment of attention in preschoolers. *Neuropsychology* 22 (4): 361–383.

Mann, L., R. Harmoni, and C. Power. 1989. Adolescent decision-making: The development of competence. *Journal of Adolescence* 12: 265–278. https://pubmed.ncbi.nlm.nih.gov/2687339/

Markus, H.J., and P.S. Nurius. 1984. Self-understanding and self-regulation In middle childhood. In *Development During Middle Childhood: The Years from Six to Twelve*, ed. by W.A. Collins, 147–183. Washington, D.C.: National Academies Press

Mccabe, M.A. 1996. Involving children and adolescents in medical decision-making: Developmental and clinical considerations. *Journal of Pediatric Psychology* 21 (4): 505–516.

Montgomery, H. 2009. *An introduction to childhood—Anthropological perspectives on children's lives*. Chichester: Wiley-Blackwell.

Newman, B.M., and P.R. Newman. 2012. *Lifespan development: A psycosocial approach*, 11th ed. Sydney: Cencage.

Olson, E.A., and M. Luciana. 2008. The development of prefrontal cortex functions in adolescence: Theoretical models and a possible dissociation of dorsal versus ventral subregions. In *Handbook of developmental cognitive neuroscience*, ed. C.A. Nelson and M. Luciana, 575–590. The MIT Press.

Passler, M., W. Isaac, and G.W. Hynd. 1985. Neuropsychological behavior attributed to frontal lobe functioning in children. *Developmental Neuropsychology* 1: 349–370.

Piaget, J. 1973. *Main trends in psychology*. London: George Allen & Unwin.

Radiation Therapy Side Effects. 2018, May 1. National Cancer Institute. https://www.cancer.gov/about-cancer/treatment/types/radiation-therapy/side-effects.

Raymundo, M.M., and J.R. Goldim. 2008. Moral psychological development related to the capacity of adolescents and elderly patients to consent. *Journal of Medical Ethics.* 34: 602–605.

Romine, C.B., and C.R. Reynolds. 2005. A model of the development of frontal lobe functioning: Findings from a meta-analysis. *Applied Neuropsychology* 12 (4): 190–201.

Ross, J.A. 1981. Improving adolescent decision-making. *Curriculum Inquiry* 11 (3): 279–295.

Ruggeri, A., M. Gummerum, and Y. Hanoch. 2014. Braving difficult choices alone: Children's and adolescents' medical decision-making. *PLoS ONE* 9 (8): 1–7.

Ruhe, K.M., T. Wangmo, D.O. Badarau, B.S. Elger, and F. Niggli. 2015. Decision-making capacity of children and adolescents—suggestions for advancing the concept's implementation in pediatric health care. *European Journal of Pediatrics* 174: 775–782.

Sawyer, S.M., P.S. Azzopardi, D. Wickremarathne, and G.C. Patton. 2018. The age of adolescence. *Lancet Child Adolesc Health.* https://doi.org/10.1016/S2352-4642(18)30022-1.

Schlam, L., and J.P. Wood. 2000. Informed consent to the medical treatment of minors: Law and practice. *Health Matrix* 141 (10): 141–174.

Shaffer, D., and K. Kipp. 2007. *Developmental psychology*. Belmont: Thomson Wadsworth.

Spear, L.P. 2013. Adolescent neurodevelopment. *Journal of Adolescent Health* 52 (2): S7–S13.

Weithorn, L.A., and S.B. Campbell. 1982. The competency of children and adolescents to make informed treatment decisions. *Child Development* 53: 1589–1598.

Ziv, Y., M. Benita, and I. Sofri. 2017. Self-regulation in childhood: A developmental perspective. In *Handbook of social behavior and skills in children*, ed. J.K. Matson, 149–173.

Further Reading

Diekema, D.S. 2011. Adolescent refusal of lifesaving treatment: Are we asking the right questions? *Adolescent Medicine* 22: 213–228.

Raymundo, M.M., and J.R. Goldim. 2008. Moral psychological development related to the capacity of adolescents and elderly patients to consent. *Journal of Medical Ethics.* 34: 602–605.

Weithorn, L.A., and S.B. Campbell. 1982. The competency of children and adolescents to make informed treatment decisions. *Child Development* 53: 1589–1598.

Chapter 3
The Child's Right to an Open Future: Philosophical Foundations and Bioethical Applications

J. R. Garrett

Abstract The child's right to an open future (ROF) is a normative ideal that is frequently invoked in pediatric bioethics. At its core, ROF holds that when childhood decisions threaten the autonomy of the future adult, fiduciaries should defer and, hence, preserve the choice until children reach maturity and can choose for themselves. In this chapter, I explore several conceptual and normative issues raised by ROF. I begin by briefly summarizing how ROF came to dominate certain debates within pediatric ethics before mapping out the conceptual space that ROF occupies in relation to other rights attributed to children. I then reconstruct the prima facie case for ROF that has made this ideal attractive within pediatric ethics before analyzing some of its primary conceptual and normative challenges. Finally, I defend an alternative framework in which children's interest in preserving a relatively open future is regarded not as a strict right but instead as one (important) interest among many to weigh and balance in pediatric decision-making. This alternative interest-based framework allows for the open future to be treated within a broader pediatric ethical framework like the best interest standard rather than as an independently robust ethical constraint. I conclude by comparatively evaluating ROF with this interest-based framework when applied to three different pediatric bioethics controversies: (1) the sterilization of minors, (2) elective pediatric surgeries, and (3) predictive genetic testing.

Keywords Autonomy · Children's interests · Elective pediatric surgeries · Predictive genetic testing · Right to an open future · Sterilization of minors

J. R. Garrett (✉)
Children's Mercy Bioethics Center, Children's Mercy Kansas City, Kansas City, MO, USA
e-mail: jgarrett@cmh.edu

© Springer Nature Switzerland AG 2022
N. Nortjé and J. C. Bester (eds.), *Pediatric Ethics: Theory and Practice*,
The International Library of Bioethics 89,
https://doi.org/10.1007/978-3-030-86182-7_3

3.1 Introduction

Pediatric bioethics requires us to evaluate decisions made for children from both present-oriented and long-term perspectives. Consider three different scenarios:

(1) *Sterilization*: Katie is a 15-year-old girl with bipolar disorder who is undergoing a cesarean section. Custody of the baby will be assumed by the state following delivery. Katie's mother, citing her belief that Katie will never be "stable enough" to take care of a child and her desire to spare Katie the distress of separation from any future children, requests a tubal ligation during the surgery.

(2) *Elective Cosmetic Surgery*: Tony is a 3-year-old boy whose cleft lip and palate repair in infancy has left him with a large scar and strikingly atypical nose. Tony's parents are concerned that these facial abnormalities will negatively impact his emotional and social development and make him a target for bullying by others. They are requesting cosmetic surgery for Tony now as a preventive measure to lessen the likelihood of these concerns.

(3) *Predictive Genetic Testing*: Byron is a precocious 11-year-old boy whose paternal grandfather and father have recently been diagnosed with Huntington's disease. Byron is convinced that he also has the disease and, following discussions with his parents, pediatrician, and a genetic counselor, would like to receive testing to know for sure.

In each of these cases, parents and other fiduciaries must consider the immediate interests of children but also must not overlook the potential impacts of pediatric decisions for the future adults that children will (likely) become. According to many, the ethically obligatory decision in all three cases would be to defer the choices for the children to make for themselves as autonomous adults. They argue that children have a *right to an open future* (hereafter, ROF) that must be respected in a manner akin to respecting the autonomy of a competent adult. When childhood decisions threaten the autonomy of the future adult, then, fiduciaries must defer and, hence, preserve the choice until children reach maturity and can choose for themselves.

However, ROF remains a thinly scrutinized concept, despite the frequency with which it is cited as an important and even decisive normative consideration. Competing interpretations of ROF proliferate in the philosophical and bioethical literature, and each interpretation can yield significantly different normative implications in any given case. There is a critical need for further conceptual analysis regarding whether and how the concept of a child's open future might apply to any particular bioethical issue or case.

In this chapter, I explore many of these conceptual and normative issues. First, I briefly summarize how ROF came to dominate certain debates within pediatric ethics. Second, I map out the conceptual space that ROF occupies in relation to other rights attributed to children. Third, I reconstruct the prima facie case for ROF that has made it attractive within pediatric ethics. Fourth, I analyze some important conceptual and normative challenges confronting ROF. Fifth, I defend an alternative framework in which children's interest in preserving a relatively open future is regarded not as a strict right but instead as one (important) interest among many to weigh and balance in pediatric decision-making. This alternative interest-based framework allows for the open future to be treated within a broader pediatric ethical framework like the best interest standard rather than as an independently robust ethical constraint. Finally, I conclude by comparatively evaluating ROF with this interest-based framework when applied to the three different pediatric case scenarios described above: (1) the sterilization of minors, (2) elective pediatric surgeries, and (3) predictive genetic testing.

3.2   Historical Background

The "child's right to an open future" first emerged in philosophical and bioethical parlance in 1980. The philosopher Joel Feinberg coined the expression in his retrospective analysis of *Wisconsin v. Yoder*, 1972, the U.S. Supreme Court case which considered whether freedom of religion requires states to exempt religious communities (in this case, the Amish) from compulsory school attendance (Feinberg 1980). While the Court affirmed this exemption in its decision, Feinberg disputed the ethical permissibility of the Amish practice of foreshortening the formal education of their youth. As he saw it, by drastically limiting their children's education, the Amish were drastically restricting the life and career options that children would have available to them as adults, effectively violating their autonomy before it could be exercised.

In the decade following Feinberg's introduction of the concept, ROF remained largely dormant as an ethical ideal. However, in the 1990s, it gained traction as a concept pertinent to various debates in genetic ethics, most notably those concerning predictive genetic testing of children and reproductive decision-making. In these contexts, many understood ROF to prohibit any decision in utero or childhood that would fundamentally alter (certain features of) a human's biology or identity, thereby closing off important future options that should be left open for individuals to decide for themselves as adults. Dena Davis, for example, applied ROF as a constraint both against parents deliberately conceiving deaf children (Davis 1997) and against children being tested for adult-onset conditions or carrier status (Davis 2001). According to Davis (2001, p. 192), while the former pertains to "what parents can do" and the latter to "what they can *know*," both practices involve parents making decisions about genetics that radically limit a child's ability to define their own identity and choose among many possibilities. Meanwhile, numerous position papers among professional societies appealed to ROF in order to strongly discourage predictive genetic testing of children from the mid-1990s up to the present (Garrett et al. 2019, p. 2192). For example, the American Academy of Pediatrics claimed that predictive genetic testing "inappropriately eliminates the possibility of future autonomous choice by the person" to know (or not know) certain genetic information or have it known (or not known) by others; thus, "pediatricians should decline requests from parents or guardians ...until the child has the capacity to make the choice."

In the wake of ROF's resurgence within genetic ethics, many sought to apply the concept in numerous other bioethical contexts, including the sterilization of minors (Davis 1997), growth attenuation interventions (Wilfond et al. 2010), disorders of sexual development (Kon 2015), fertility preservation (Cutas and Hens 2015), non-therapeutic circumcision (Darby 2013), elective pediatric surgeries (Taylor 2018), and pediatric neuroenhancement (Graf et al. 2013). It is reasonable to expect that the concept may also find a toehold in many other debates within pediatric bioethics (e.g., gender-affirming care for transgender children).

3.3   Conceptual and Normative Analysis

An adequate conceptual and normative analysis of ROF must address two fundamental questions: (1) what does it mean to have an "open future," and (2) what is entailed by claiming a "right" to this condition? I will address each question in turn.

(1)   *What does it mean to have an "open future"?*

Addressing the first question requires exploring what precisely ROF is intended to protect against premature closure. According to Millum (2014, pp. 524–525), there are at least four categories of "actions and objects" that ROF "could range over," each of which "could be influenced by parents or other actors during a child's development and could make a substantial difference to the nature and quality of the child's life." The first category is *capacities*, which includes general abilities, such as means-end rationality and autonomy, that are important for almost all types of human activity and decision-making. Since they are global rather than domain-specific, capacities are fundamental components of ROF. Without these capacities, one arguably could not engage with anything in the other three categories.

The second category is *skills*, which are specific competencies and proficiencies, such as literacy or the ability to speak a second language, important for particular domains of human activity. Unlike capacities, skills are domain-specific, and one could live a recognizably human life without ever developing them (and indeed many have done so). Nonetheless, their possession can vastly expand one's set of options and preferences and enable a greater degree of wellbeing and variety of human experience than their absence.

A third category is *options*, which includes specific choices and opportunities, such as the option to go to college or be an Amish farmer. A given individual's options are constrained by the higher-level categories of capacities and skills—for example, the option of going to college is unlikely if one lacks means-end rationality or literacy—but also by "external circumstances" such as socio-economic status (Millum 2014, p. 525).

Finally, the fourth category is *preferences*. This category includes the underlying desires and values that shape one's decisions regarding the available options, such as preferences for being alone or among people.

Different accounts of ROF might include, exclude, and/or emphasize one or more of these categories. For the sake of simplicity, I will regard *options* as the primary category of direct concern when analyzing ROF, while treating capacities, skills, and preferences indirectly as shaping the options one has available. However, regardless of which category (or combination of categories) one favors, one can find in Feinberg's original essay and the work of other ROF proponents *at least three competing interpretations of what is important to focus on regarding it/them, each of which entails different moral prescriptions.* I call these interpretations (1) maximal quantity; (2) minimal threshold; and (3) vital quality.

The *maximal quantity* view attributes to children the right to "reach maturity with *as many* open options, opportunities, and advantages *as possible*" (Feinberg 1980, p. 130). The emphasis here is on the *quantity of options* left available to the child—an ideal maximum. Dena Davis, for example, claims that "deliberately [producing] deaf children...violates [the] child's own autonomy and *narrows the scope of her choices* when she grows up; in other words, it violates her right to an 'open future'" (Davis 1997, p. 9, emphasis added). According to the maximal quantity view, the wrong-making property of the ROF violation is the "narrowing" of a future adult's scope of options [presumably from a baseline of the full set of "possible" options].

At the other end of the interpretive spectrum is the *minimal threshold* view of ROF. According to this interpretation, children have the right to reach maturity without having their options, opportunities, and advantages *radically narrowed*. This view, then, identifies some minimally sufficient number of possible options that must be left open until the child is a fully formed autonomous adult. This second interpretation of ROF also emphasizes the *quantity of options* left available to the child, though here in the form of a basic minimum rather than an ideal maximum.

Many ROF proponents seem to adopt the minimal threshold view of an open future, including (problematically) those who elsewhere in their work seem to endorse a maximal quantity view. For example, Feinberg emphasizes that "an education that renders a child *fit for only one way of life* forecloses irrevocably his other options" (Feinberg 1980, p. 132). Meanwhile, Davis argues against parental choices that "take the form of a *radical narrowing of choices available to the child when she grows up*," (Davis 1997, p. 11).

Finally, a third interpretation of ROF is the *vital quality* view. According to this interpretation, children have the right to reach maturity with certain *vital* options, opportunities, and advantages left open. Rather than focusing on the quantity of choices, whether in terms of an ideal maximum or a basic minimum, the vital quality view instead focuses qualitatively on the *moral importance* of certain choices as they pertain to self-determination and self-fulfillment for one's adult life. Feinberg, for example, objects to others making "*certain crucial and irrevocable decisions*" (Feinberg 1980, p. 143) because such decisions "[guarantee] now that when the child is an autonomous adult, *certain key options* will already be closed to him" (Feinberg 1980, p. 126). Davis, meanwhile, describes ROF as including "virtually all the *important rights* we believe adults have, but which must be protected now to be exercised later" (Davis 1997, p. 9).

The adequacy of the vital quality view hinges, ultimately, on how the qualifier "vital" is defined and defended. Which options are so vital that they ought never to be closed off to a child prior to maturity? And what precisely is it that these options are vital *for*? In describing these options as "certain crucial and irrevocable decisions determining the course of his life," and as "certain key options," Feinberg and Davis identify vital options as *a subset of choices defined by their unique or fundamental moral importance and their immutable nature once made.*

But how do we identify options that have such moral importance and differentiate them from others less in need of absolute protection? Feinberg offers a promising explanation:

> ... *the two distinct ideals of sovereign autonomy (self-determination) and personal wellbeing (self-fulfillment) are both likely to enter, indeed to dominate, the discussion of the grounding of the child's right to an open future.* That right (or class of rights) must be held in trust either out of respect for the sovereign independence of the emerging adult (and derivatively in large part for his own good) or for the sake of the lifelong wellbeing of the person who is still a child (a wellbeing from which the need of self-government "by and large" can be derived), or from both. (Feinberg 1980, p. 145)

Two features of this passage are worth emphasizing. First, Feinberg views sovereign autonomy and personal wellbeing as providing the core basis for ROF. Hence, if the vital quality view is the best interpretation of that right, then what makes certain options "vital" in the relevant way is their importance for self-determination and self-fulfillment. Second, while these two ideals are distinct, they clearly are interrelated for Feinberg. Echoing John Stuart Mill in *On Liberty*, Feinberg asserts that leaving people free to decide among their options is not simply a matter of blind respect for autonomy but the surest and simplest way for individuals to realize their own good. And this insight would be even more salient for "vital" options, which are likely to affect a greater share of a person's good than non-vital options. Hence, if fiduciaries truly are concerned with promoting the child's greatest good, then, according to Feinberg, they would almost always do so by postponing choices among these vital options until the child is mature and can exercise this reliable mechanism of self-determination.

Perhaps the clearest example of a vital option being closed off prematurely occurs with the sterilization of minors (Davis 1997, p. 9). The choice of whether to reproduce and live life as a parent and caregiver is, unquestionably, among the most fundamental and defining decisions any individual will ever make, a vital option if ever there was one. Child rearing is a major project in its own right, and it also dramatically affects the time and energy that one will have to give to other projects and commitments. More fundamentally, becoming a parent also constitutes an elemental part of one's very identity. When parents, physicians, or society more generally decide to undertake sterilization in a child, the decision forever closes off the option for that child (and future adult) to become a parent through biological reproduction. Of course, it may be possible to reverse certain sterilization procedures, and sterilization does not itself preclude the possibilities of adopting or fostering children later in life. Nonetheless, sterilization is a clear paradigm example of a vital option being "closed" in childhood through the decision of others.

(2)   *What is entailed by claiming a "right" to an open future?*

Moral rights have a unique function within ethical theory: they "serve as one kind of constraint on the pursuit of social goals," thus "protecting their holders by imposing normative constraints on others,...[including] duties borne by these others" (Sumner 1987, p. 47). By treating an open future as a "right" that all moral agents have strict duties not to violate, Feinberg and other defenders of ROF are deploying the strongest ethical language of Western moral philosophy. This conceptual framing raises numerous concerns that will be explored later, but first I want to situate ROF within a larger framework of moral rights.

In his own account, Feinberg usefully distinguishes among three categories of rights (1980, pp. 125–126). First, there are rights held *only* by autonomous adults (e.g., the right to vote). Second, there are rights held by *both* adults and children (e.g., the right to bodily integrity). Finally, there are rights held *primarily* by children. This third category can be subdivided into two types. On the one hand, there are a host of dependency rights related to the *present wellbeing* of children (e.g., the right to be provided with basic necessities). On the other hand, there are so-called "rights-in-trust" related to the *future autonomy* that most children will eventually develop (Feinberg 1980, p. 125). These are rights "to be *saved* for the child until he is an adult, but which can be violated 'in advance,'...[guaranteeing] *now* that when the child is an autonomous adult, certain key options will already be closed to him" (Feinberg 1980, p. 126). The central idea of ROF is that the child has a right "while he is still a child is to have these future options kept open until he is a fully formed, self-determining adult capable of deciding among them" (Feinberg 1980, p. 126).

Distinguishing ROF from the class of dependency rights here is crucial. Not every decision that negatively impacts a child's future is ethically wrong *because* it closes off future options for an autonomous adult; rather, many such decisions are ethically wrong, first and foremost, because they harm individuals (during childhood and/or during adulthood). For example, children who are abused or neglected in childhood will almost certainly suffer certain negative impacts on their future compared with a counterfactual childhood in which such abuse and neglect did not occur. But the central wrong inflicted upon abused and neglected children pertains to the harm suffered (as children and then later as adults) rather than the advance violation of an (adult) autonomy right; to use Feinberg's categories, the abused or neglected child suffered a violation of dependency rights or the right to bodily integrity rather than a violation of ROF. With ROF, then, we are evaluating a *certain subset of decisions* that adults sometimes make for children: namely, *those decisions where the simple closing off of an option that properly should have been left open for children to make for themselves is the primary wrong-making property.* Importantly, even actions which *positively* impact a child's future wellbeing can violate ROF. For example, a parent might make a choice that better promotes the lifetime wellbeing of their child than the counterfactual choice that the child-as-future-adult eventually would have made; nonetheless, it may be ethically compelling to conclude that the choice still should have been left open for the future adult.

3.4   The Prima Facie Case for ROF

The ideal of an "open future" informs many cultural and ethical norms regarding children. Parents often teach their children that they can be anything they want to be when they grow up, while educators emphasize cognitive, communicative, and social skills that will equip students for any number of careers and lifestyles. Conversely, many are suspicious of parents who drive their children toward narrowly predetermined goals and/or shelter them from experiences that might threaten such goals—the stage parent who seems to be fulfilling his own dreams through his child or the much derided "helicopter parent" who infantilizes a child instead of encouraging their development of independence. In these and other examples, the ideal of leaving many choices "open" for children eventually to make for themselves is endorsed and even celebrated. The prima facie case for ROF is rooted in these cultural norms and buttressed by a host of liberal values, such as neutrality, tolerance, and respect for autonomy and diversity.

This vision of childhood development connects back to the twin ideals that, according to Feinberg, ground ROF: namely, self-determination and self-fulfillment. The degree to which one's overall life is self-determined will arguably depend, at least partially, on how one's childhood was shaped and governed by adults. While by no means incontrovertible, the proposition that self-determination and self-fulfillment will be maximized by deference to ROF has intuitive plausibility. By leaving many— especially important and irrevocable—decisions for individuals to make for themselves as autonomous adults, the total scope of self-determination is thereby widened. Similarly, if each individual is best positioned to understand and most motivated to pursue their own wellbeing, then the potential for self-fulfillment is thereby increased.

3.5   Challenges Facing ROF

Determining whether ROF can be defended *all-things-considered* requires (1) considering which of the three major views of ROF outlined earlier is most defensible, and then (2) evaluating rebuttals to the prima facie case as well as independent objections to ROF.

With respect to the first task, serious objections immediately confront the two quantitative interpretations of ROF. On the one hand, the maximal quantity view is simply too demanding and impractical. While the ideal of leaving maximal choices until adulthood has a certain surface appeal, this appeal readily dissolves when contemplating how one could ever satisfy its demands. Every choice a parent or pediatric clinician makes—including endeavoring to maximize open options—rules out others, closing certain doors while opening others. And each choice creates its own *path dependency* which makes certain future choices more likely and others less likely. Millions of choices will be made for children over the course of maturing into autonomous adults and we lack any practical way to determine at each point which options will result in maximal opportunities for the eventual adult. Moreover, parents cannot fulfill their obligations to a child while at the same time trying to keep all options open. Adequate parenting requires providing opportunities for children to develop skills and talents, but it is impossible to provide opportunity to develop *all* skills and talents, let alone in unison. For example, choosing to learn a musical instrument may preclude learning another language or developing competitive facility in a given sport. The reality is that endeavoring to keep all options open actually will close many doors for children.

Meanwhile, the minimal threshold view is too weak and limited in its application. Most would agree that parents and fiduciaries should not "radically narrow" a child's future; however, few—if any—*single* choices made for children actually have this effect (and those that do likely would be better characterized as violations of dependency rights or the basic rights that children share with adults). As such, the minimal threshold view of ROF may not actually apply to any standard pediatric practices, at least when considered in isolation.

These weaknesses with the two quantitative interpretations leave the vital quality view—with its grounding in the twin ideals of self-determination and self-fulfillment—as the most promising interpretation of ROF on offer. If so, then defenders of ROF must articulate the inclusion criteria for that *subset of options defined by their unique or fundamental moral importance and their immutable nature once decided*. While it is beyond the scope of this chapter to evaluate the prospects for such conceptual boundary setting here, several significant objections can be raised against any likely candidate since these objections target each of the three most basic components of ROF: (1) its singular focus on the future, (2) its conception of openness, and (3) its insistence upon a rights-based framing.

First, ROF's exclusive focus on the future is problematic. In multiple ways, ROF distorts, discounts, and marginalizes the nature and value of childhood. First, ROF encourages untenably rigid binary categories. It rigidly divides life into two categories—adulthood and everything prior to adulthood—and treats all relevant cases within the latter category identically. ROF functions effectively in the same manner regardless of whether a pediatric patient is a neonate or late adolescent, discounting the moral significance of development throughout childhood. Second, ROF obscures and discounts direct obligations to children-as-they-are-now by focusing entirely on the adults-they-will become. This shift in focus reinforces a "negative" conception of childhood, one where childhood is essentially a prospective state to be survived or solved—a time of deficits where children are merely "inferior adults"—and serves as a kind of "prep school" for "real life" as an adult (Brennan 2014, p. 37). However, most decisions that ostensibly threaten future autonomy make at least some connection with genuine values in the present, and pediatric bioethics will be richer and more compelling by openly considering these values rather than ruling them out categorically from the start.

A second major objection to ROF holds that its conception of openness and its aversion to closure are undeveloped, if not incoherent. First, ROF lacks an account of the necessary and/or sufficient conditions for some action to "close" an option. Do present actions have to make it *literally impossible* for a child to have a choice later? Or is ROF violated if we make having and exercising a choice considerably *more difficult*? Consider Feinberg's own claim that the Amish violate ROF by leaving their youth with no other open options except for becoming an Amish farmer. As Claudia Mills notes, this claim is highly suspect:

> In one sense, the worldly options…do remain open to the Amish child: He or she can certainly leave the Amish community at adulthood, and many Amish children do. Admittedly, it will be more difficult for such an adult to pursue certain careers than it would be for someone educated differently, but it is not impossible. And all of us, in whatever we pursue, face difficulties of one sort or another. (Mills 2003, p. 501)

Following up on this remark, Mills endorses a scalar view of an open future where options "are not properly viewed as open or closed, but as more or less encouraged or discouraged, fostered, or inhibited" (Mills 2003, p. 501).

This scalar view forces us also to assess the moral quality of decisions to foster or inhibit certain options to certain degrees. We might feel comfortable asserting a "rights violation" on the few occasions where the decisions of others thoroughly closed off an important option for a child, but how should we feel about a parent or other fiduciary making a decision that makes an important option 50% less likely to be available? Does such a decision violate ROF? If not, what about making the option 75% less likely? Or 95%? The scalar view of an open future, while more accurate in its description, is less obvious and straightforward in its normative implications.

Additionally, ROF dubiously assumes that more options entail greater self-fulfillment. As Mills notes, "no one, including Feinberg, would really endorse a completely open smorgasbord, exposing children in a nonjudgmental way even to options the parents consider seriously immoral or otherwise gravely flawed" (2003, p. 503). It is implausible, for example, to think that a child's prospects for self-fulfillment would be improved if the options of becoming a serial murderer or addicted to heroin are nonjudgmentally preserved rather than discouraged strongly. Moreover, the assumption favoring more options remains dubious even if we filter out seriously immoral or otherwise gravely flawed options. Arguably, self-fulfillment becomes more likely by focusing intently on certain opportunities (to the exclusion of others) and working to genuinely master the skills, knowledge, and general demeanor to seize those opportunities, rather than keeping lots of options open. A unique and important value of depth can be gained, for example, by giving oneself over to a craft and committing the time, patience, focus, grit, and so on to master it that cannot be gained by dabbling here and there in order to "not miss out on anything" or "have it all."

Third, and finally, ROF's invocation of "rights" injects an exceedingly strong, rigid, and individualistic ethical concept into an ill-suited context. Put simply, ROF lacks contextual tailoring. It inflexibly applies a constraint in the same way to *all* non-adults at *all* stages and in *all* contexts. However, pediatric ethics must account for the vast developmental transition between neonates and late adolescents, as well as the morally significant contextual differences among families and societies. Instead of the familiar balancing of numerous competing interests, both present and future, held by numerous stakeholders within a particular familial and cultural context, ROF functions to rigidly protect, in the strongest moral terms, a single future interest of a single stakeholder until adulthood.

Moreover, in all likely pediatric scenarios, a right overrides every other ethical consideration. Except in extreme circumstances, rights may not be violated. Since such extreme circumstances rarely arise in standard pediatric cases—we cannot save thousands of people by circumcising a young boy or testing a child for an adult-onset condition, for example—recognizing something as a right becomes dispositive in ethical reasoning.

And this overriding quality of rights leads to a final worry, which is that rights language has the practical effect of stifling conversation from the moment it is introduced. All other interests, values, principles, and other ethical considerations fall short of the moral currency of rights. This is not to say rights should never be invoked but rather that we should be exceedingly judicious of when and how we do so given the practical, conceptual, and normative consequences of rights language.

3.6 An Alternative Interest-Based Approach to the Open Future

Though an open future traditionally has been packaged as a "right," this tradition is not a conceptual necessity. Indeed, in recent work with colleagues, I have argued that an alternative interest-based approach better captures the moral salience of an open future in pediatric decision-making (Garrett et al. 2019). On this alternative account, children possess an *interest* in having important options and decisions preserved until they have developed sufficient maturity and autonomy to participate actively in making such choices themselves. A (sufficiently) open future is, thus, one element of their wellbeing, which is to say their lives generally will go better with many important choices preserved than without. The crucial point—and the point that distinguishes this interest-based approach from ROF—is this:

> The child's interest in an open future is one important, but not automatically the *most* important, interest to consider and balance in the process of shared decision-making. In other words, *an open future is best understood not as a separate principle of pediatric bioethics, but instead as one component of its traditional focus on interests and balancing benefits and harms to children and families.* (Garrett et al. 2019, p. 2193, original emphasis)

Unlike the rights-based framing of ROF, then, an interest-based approach does not strictly partition the open future and place it above children's other interests but instead encourages a holistic balanced evaluation of all relevant interests together.

Several competing lists of children's interests have been proposed, but Malek (2009) provides an especially compelling inventory (Table 3.1). Drawing on three independent statements of children's needs and interests, she identifies thirteen core interests to incorporate in pediatric decision-making, each of which is "a capacity, activity, or state of affairs that contributes to the wellbeing of children" (Garrett et al. 2019, p. 2193). No interest on the list is to be ranked categorically higher than any other, and "most medical decisions will involve some tradeoffs among these interests" (Garrett et al. 2019, p. 2193). The interest in an open future is not explicitly included in Malek's list but fits comfortably within the contours of the interest in autonomy (#13)—i.e., an interest in preserving *future* autonomy.

An interest-based framework has at least two significant advantages over a rights-based approach for pediatric clinical ethics. First, whereas a rights-based approach is wedded to a predetermined conclusion (i.e., that the right will—in all likely

Table 3.1 Proposed list of interests to evaluate in pediatric decision-making (adapted from Malek 2009)

#	Proposed interest
1	**Life**: To live and to anticipate a life of normal human length
2	**Health and health care**: To have good health and protection from pain, injury, and illness. To have access to medical care
3	**Basic needs**: To have an adequate standard of living, especially to be adequately nourished and sheltered
4	**Protection from neglect and abuse**: To be protected from physical or mental abuse, neglect, exploitation, and exposure to dangerous environments. To be secure that they will be safe and cared for
5	**Emotional development**: To experience emotion and have appropriate emotional development
6	**Play and pleasure**: To play, rest, and enjoy recreational activities. To have pleasurable experiences
7	**Education and cognitive development**: To have an education that includes information from diverse sources. To have the ability to learn, think, imagine, and reason
8	**Expression and communication**: To have the ability to express themselves and to communicate thoughts and feelings
9	**Interaction**: To interact with and care for others and the world around them. To have secure, empathetic, intimate, and consistent relationships with others
10	**Parental relationship**: To know and interact with their parents
11	**Identity**: To have an identity and connection to their culture. To be protected from discrimination
12	**Sense of self**: To have a sense of self, self-worth, and self-respect
13	**Autonomy**: To have the ability to influence the course of their lives. To act intentionally and with self discipline. To reflect on the direction and meaning of their lives. *To have "future autonomy" protected by having future options and opportunities kept open*

scenarios—override all other interests and considerations), the interest-based framework is flexible and can account for and integrate new evidence. Second, the interest-based framework provides both a comprehensive and balanced evaluation of children's many diverse interests. Ultimately, the interest-based framework promotes the full wellbeing of the existing child, rather than exclusively privileging the narrow interest in future autonomy of an ambiguously identified future adult.

This is not to say that a child's interest in preserving future autonomy is never their weightiest interest but rather that decision-making within an interest-based framework is sensitive and responsive to the circumstances of the particular child. In some cases, producing or preserving an open future may be in the child's best interest, all things considered. There are numerous decisions, like the choice of a spouse or a career, where the child's best interest is almost always served by keeping options open even when the decision comes at the cost of some opportunity (maintaining a strong childhood connection or providing financial security, for example). In other cases, though, foreclosing some options in childhood may better promote a child's overall interests. For example, while a parental decision to request surgical removal of an infant's supernumerary digit could be regarded as inappropriately foreclosing a choice that should be left for the future adult, this decision also can be viewed as a reasonable attempt to promote the overall best interest of the child (especially if the supernumerary digit is not functionally useful) and open other options that would be foreclosed by deferring the decision. Here pediatric bioethics must appreciate that "individual families working with their chosen care providers are best positioned to identify and balance competing interests in particular circumstances" (Garrett et al. 2019, p. 2194) (Table 3.2).

Table 3.2 Guiding principles for evaluating the future autonomy interests of children alongside other health-related and wellbeing related interests

Pluralism among values and interests
- Recognize that preserving a (sufficiently) open future for the adults-whom-children-will-become is one among many present and future interests related to the health and wellbeing of children (and other stakeholders) that must be weighed and balanced in pediatric decision-making
- Understand that the ethical weight carried by the interest in preserving a (sufficiently) open future is not fixed and universal across all choices but will vary in proportion to how vital a given choice is to self-determination and self-fulfillment and also will depend on a holistic assessment of the overall quantity and quality of choices that are being preserved for the future adult
- Appreciate that reasonable disagreements will frequently arise regarding how to weigh, balance, and prioritize interests in future autonomy vis a vis other present and future interests related to the health and wellbeing of children (and other stakeholders)

Respect for family context and local knowledge
- Demonstrate respect for the integrity, uniqueness, and expertise of each family by actively seeking their considered assessment of all relevant interests
- Assign strong prima facie justification to the family's particular weighting and balancing of interests appreciating that they will usually be (a) best positioned to understand the family context and the consequences that decisions will have within that context and (b) most directly and seriously impacted by decisions in both the short-term and long-term

3.7 Case Applications

With this sketch of an interest-based framework in hand, I will return to the three case vignettes introduced earlier in the chapter to contrast this framework with traditional ROF analysis.

(1) *Sterilization*

As discussed earlier, the sterilization of minors is *on its face* perhaps the strongest candidate to justify a robust "right" to an open future. In almost every case imaginable—including Katie's mother's request for tubal ligation following Katie's cesarean surgery—fiduciaries should be prohibited from undertaking permanent elective sterilization measures in a minor patient for whom the eventual development of adult autonomy is at all probable. If that presumption is correct, then one may understandably wonder whether there is any point in further analysis of such cases.

However, closer scrutiny reveals that *ROF is neither necessary nor particularly useful for understanding why the sterilization of minors is ethically impermissible.* Just as many different (and competing) ethical theories converge on specific prescriptions in particular cases (e.g., utilitarians and Kantians will usually agree that suicide bombings of innocent civilians are impermissible), so too will ROF and an interest-based approach converge in many cases involving open future considerations. What is ethically significant and interesting is the differing ethical reasoning leading to these common prescriptions, as well as analysis of cases where this differing ethical reasoning leads to divergent prescriptions.

When applied to cases involving the sterilization of minors, ROF directly and immediately lands on prohibition. For ROF, there are no relevant interests *even in principle* to consider on the other side of the ledger. ROF operates as a strict side-constraint on decision-making, requiring fiduciaries to defer reproductive decisions for minors to make for themselves in adulthood and to do so come what may.

The interest-based approach, meanwhile, will arrive at the same prohibition of sterilization in most such cases but in a fundamentally different manner than ROF. The interest-based approach regards other interests as relevant considerations *in principle* to weigh and balance against the interest in preserving reproductive choice. However, *in practice* this approach will almost always find these other relevant considerations to be lacking (even when jointly united) in moral significance to outweigh preserving this profoundly important choice.

In Katie's case, we can (and, I would argue, we should) regard many of her mother's concerns as *relevant* considerations *in principle*. Drawing from Malek's list (Table 3.1), we can remain open to claims that tubal ligation would potentially protect Katie from perceived danger and insecurity (#4) and promote her emotional development (#5), while still viewing these relevant interests as significantly outweighed by her autonomy interests (#13) given the vital contribution that preserving her reproductive choice will have for her overall life. And we can do all of this without committing to sweeping predetermined conclusions about *other* cases involving sterilization and much else.

One final point worth clarifying and expanding upon here is this: to the degree one regards the language of "rights" as warranted in this domain, a more parsimonious solution would be to conceptualize a (negative claims) right to retain reproductive capacity instead of a broader right to an open future that spills over into many other domains. Indeed, there is no obvious reason why the validity and strength of such a fundamental interest should be tied to other case domains where a "right" is less compelling.

(2) ***Elective Cosmetic Surgery***

In the second case vignette involving Tony, ROF and an interest-based approach diverge more significantly in terms of their recommended prescriptions. On a traditional reading of ROF, an elective cosmetic surgery (i.e., one lacking clear clinical therapeutic utility) should be deferred for children to make for themselves as autonomous adults. Here again the reasoning is simple and straightforward because ROF does not recognize the moral relevance of any other interests against which autonomy interests must be weighed and balanced. At the very least, ROF would prescribe that parents defer the decision until a child actually suffers some significant harm (Taylor 2018). A merely theoretical or anticipated psychosocial harm is insufficient warrant for compromising the child's (future) autonomy interest.

An interest-based approach likely would derive divergent prescriptions in many (though certainly not all) cases involving elective cosmetic surgeries. In Tony's particular case, an interest framework would recognize the parents' strong and prima facie reasonable case for preemptive surgical intervention to serve Tony's other interests. Again, drawing from Malek's list (Table 3.1), we can appreciate claims that cosmetic cranio-facial surgery likely will protect Tony from perceived danger and insecurity (#4), promote his emotional development (#5), and foster his likelihood for play, educational development, intimate interaction, and positive identity development (#6, #7, #9, and #11). However, in so doing, we can also understand why Tony's parents believe these combined wellbeing interests outweigh his autonomy interests (#13) in this case. Whereas the content and structure of ROF requires its proponents to evaluate Tony's case in the same direct (and prohibitive) manner as Katie's, an interest-based approach can recognize and respond differently (and, I would argue, more appropriately) to the specific contextual nuances of each case.

(3) ***Predictive Genetic Testing***

The third case vignette does not involve a bodily intervention but rather discovery and/or disclosure of predictive genetic information obtained in childhood. By definition, such information has no immediate clinical medical value for the child. Instead, whatever value this information has relates to promoting (1) future health interests the child will have as an adult, (2) wider wellbeing interests the child has now and in the future in terms of personal, familial, financial, or lifestyle planning, and/or (3) the health or wellbeing interests of important stakeholders such as parents or siblings.

Here, as in the prior two cases, a traditional application of ROF comes down simply and directly in favor of prohibition. Clinicians should neither test for nor disclose predictive genetic information in children, following the recommendations of numerous position papers among professional pediatrics and genetics societies over the past 25 years (Garrett et al. 2019, p. 2192). Again, the ethical reasoning here does not acknowledge the relevance of other interests that the child or other stakeholders may have. The "right" to have decisions about receiving this information and/or having it shared with others deferred until adulthood categorically constrains such interests from entering the ledger.

An interest-based approach, meanwhile, remains open to multiple possible prescriptions in cases involving predictive genetic testing. In some cases, such as where parental anxiety or curiosity is primarily driving the request for testing a young child, the child's interests in preserving autonomy likely will outweigh these relatively weak interests of a secondary stakeholder. In Byron's particular case, though, the reasons for acceding to the request for testing are much stronger. For one thing, as a precocious 11-year-old boy who has demonstrated considerable capacity for understanding the decision, Byron should be viewed as already having developed a certain degree of autonomy (even if not yet fully developed). Hence, our understanding of how to evaluate Byron's autonomy interests (#13) needs to be nuanced enough to recognize and appreciate this liminal space. An interest-based approach can accommodate such concerns much more readily than can the more rigid rights framework. Moreover, an interest-based framework can also incorporate and weigh other well-being interests that Byron and his family possess, including "Byron's emotional development (#5) and sense of self and identity (#11 and #12), his desire to plan for his education and career (#7 and #13), his relationship with his parents (#10), and his parents' desire to plan for long-term care (#2)" (Garrett et al. 2019, p. 2196) (Table 3.2).

3.8 Conclusion

While it has significant intuitive appeal and aligns with many widely shared cultural and political norms, ROF cannot adequately guide pediatric decision-making for particular children. It simply faces too many serious conceptual and practical flaws to be salvageable. Instead, the interest children have in preserving a reasonably open future can be protected and promoted without making this interest a strict and rigid right that overrides all other interests and considerations. Existing tools within pediatric bioethics, including interest-based standards as well as more practical and straightforward moral rights (e.g., dependency rights and rights against harm from others), are more than adequate for guiding decision-making in most cases.

References

American Academy of Pediatrics Committee on Bioethics. 2001. Ethical issues with genetic testing in pediatrics. *Pediatrics* 107: 1451–1455.

Brennan, S. 2014. The goods of childhood and children's rights. In *Family-making: Contemporary ethical challenges*, ed. F. Baylis and C. McLeod, 29–45. Oxford: Oxford University Press.

Cutas, D., and K. Hens. 2015. Preserving children's fertility: Two tales about children's right to an open future and the margins of parental obligations. *Medicine, Health Care, and Philosophy* 18 (2): 253–260.

Darby, R.J. 2013. The child's right to an open future: Is the principle applicable to non-therapeutic circumcision? *Journal of Medical Ethics* 39 (7): 463–468.

Davis, D.S. 1997. Genetic dilemmas and the child's right to an open future. *Hastings Center Report* 27: 7–15.

Davis, D.S. 2001. *Genetic dilemmas: Reproductive technologies, parental choices, and children's futures*. New York: Routledge.

Feinberg, J. 1980. The child's right to an open future. In *Whose child? Children's rights, parental authority, and state power*, ed. W. Aiken and H. LaFollette, 124–153. Totowa, NJ: Rowman and Littlefield.

Garrett, J.R., J.D. Lantos, L. Biesecker, et al. 2019. Rethinking the "open future" argument against predictive genetic testing of children. *Genetics in Medicine* 21: 2190–2198.

Graf, W.D., et al. 2013. Pediatric neuroenhancement: Ethical, legal, social, and neurodevelopmental implications. *Neurology* 80: 1251–1260.

Kon, A.A. 2015. Ethical issues in decision-making for infants with disorders of sex development. *Hormone and Metabolic Research* 47 (5): 340–343.

Malek, J. 2009. What really is in a child's best interest? Toward a more precise picture of the interests of children. *Journal of Clinical Ethics* 20 (2): 175–182.

Mills, C. 2003. The child's right to an open future? *Journal of Social Philosophy* 34 (4): 499–509.

Millum, J. 2014. The foundation of the child's right to an open future. *Journal of Social Philosophy* 45: 522–538.

Sumner, L.W. 1987. *The moral foundations of rights*. New York: Oxford University Press.

Taylor, M. 2018. Too close to the knives: Children's rights, parental authority, and best interest in the context of elective pediatric surgeries. *Kennedy Institute of Ethics Journal* 28: 281–308.

Wilfond, B.S., P.S. Miller, C. Korfiatis, et al. 2010. Navigating growth attenuation in children with profound disabilities: Children's interests, family decision-making, and community concerns. *Hastings Center Report* 40 (6): 27–40.

Further Reading

Davis, D.S. 1997. Genetic dilemmas and the child's right to an open future. *Hastings Center Report* 27: 7–15.

Feinberg, J. 1980. The child's right to an open future. In *Whose child? Children's rights, parental authority, and state power*, ed. W. Aiken and H. LaFollette, 124–153. Totowa, NJ: Rowman and Littlefield.

Garrett, J.R., J.D. Lantos, L. Biesecker, et al. 2019. Rethinking the "open future" argument against predictive genetic testing of children. *Genetics in Medicine* 21: 2190–2198.

Millum, J. 2014. The foundation of the child's right to an open future. *Journal of Social Philosophy* 45: 522–538.

Chapter 4
The Best Interest Standard and Its Rivals: The Debate About Ethical Decision-Making Standards in Pediatrics

J. C. Bester

Abstract The best interest standard (BIS) is the predominant ethical and legal principle in decision-making about children and in pediatrics. Over the past 25 years there has been a debate in the academic literature about the continued use of the BIS. Critiques of the BIS have been published, and alternative principles to replace or augment the BIS have been suggested. This chapter provides a review of the BIS, the functions it fulfills in pediatrics, the debate about the BIS, and alternative principles that have been suggested. The BIS is a robust ethical principle that fulfills many indispensable functions in pediatrics and is well able to overcome the objections of its critics. The BIS remains the best ethical standard to form the ethical basis of pediatrics.

Keywords Best interest standard · Harm principle · Constrained parental autonomy · Pediatric ethics · Decision-making for children

4.1 Introduction

Medicine is fundamentally a moral discipline with a moral goal: to provide benefits to patients and society, to heal, to relieve suffering. Medicine is practiced within the context of a therapeutic relationship, a professional relationship where the clinician's expertise is employed to serve the health and wellbeing of a patient. At its best such a relationship is characterized by mutual respect, good communication, and shared decision-making, where the patient and clinician work together to promote and protect the wellbeing of the patient.

When children are patients, the situation becomes a bit more complex. Children generally lack the capacity to make their own medical decisions or to protect and promote their own wellbeing. Children are vulnerable, dependent on others to meet their needs and to safeguard their wellbeing. For these reasons, those who stand in care relationships with children have special moral obligations by virtue of their

J. C. Bester (✉)
Kirk Kerkorian School of Medicine at UNL, University of Nevada, Las Vegas, USA
e-mail: johan.bester@unlv.edu

© Springer Nature Switzerland AG 2022
N. Nortjé and J. C. Bester (eds.), *Pediatric Ethics: Theory and Practice*,
The International Library of Bioethics 89,
https://doi.org/10.1007/978-3-030-86182-7_4

57

relationship to the child, to promote and protect the wellbeing of the child. When it comes to medical decision-making for the child, this is a central ethical principle: children are not property, are not objects, are not a means to an end, but have moral claims of their own (Bester 2019b; Bester and Kodish 2017; Woodhouse 1998). Decisions made about children should take these moral claims, the wellbeing of the child, as starting point and reference. Parents, clinicians, society cannot just do things to children as they wish; every decision about children should have the wellbeing of the child as its primary focus.

For many decades this idea has been expressed by reference to the best interest standard (BIS). The best interest standard has become the prevailing standard in decision-making about children, both in legal terms and in ethical analyses about issues affecting children. There is broad international consensus that the best interest of the child should be the central ethical and legal standard in all decisions about children. Article 3.1 of the United Nations Convention on the Rights of the Child reads:

> In all actions concerning children, whether undertaken by public or private social welfare institutions, courts of law, administrative authorities, or legislative bodies, the best interest of the child shall be a primary consideration. (UN 1990)

The UN Convention directs States (also refer to Chap. 5) to place the best interest of the child as primary consideration in policy decisions and actions that are aimed at children, and goes on to say that States have the obligations to provide for the welfare, protection, and development of children. The BIS had a global impact: many countries have adopted such measures, and many countries have instituted laws and procedures that makes the best interest of the child a primary consideration in decisions about children (Kohm 2008; UNICEF 2020). To date, 196 countries have ratified the convention (UNICEF 2020). There is a notable exception: The United States of America. The United States has signed the convention, indicating support for it, but has not ratified it. Still, in the United States the best interest standard has also become the prevailing legal and ethical standard in decision-making about children (Kohm 2008).

Beginning in the 1990s and going on to present day, a debate about the role of the BIS in decision-making for children has been taking place in the academic bioethics literature (Bester 2019a, b). Some have criticized the BIS severely, and some have suggested alternate decision-making frameworks to augment the BIS or some functions of the BIS. Some authors have defended the BIS against critiques and clarified the important role and place of the BIS; some authors have shown how the newly suggested replacements have problems of their own, making them insufficient to replace the BIS. Many of the most strident critiques against the BIS have come from scholars working in the United States, where there is a tension between the best interest of the child and the liberty rights of parents to decide for their own children as they see fit. Much of the debate has sought to respond to or clarify this tension, seeking the appropriate balance between parental authority and the interests of the child. This is perhaps a false dilemma, as parental authority is protected precisely because it serves the wellbeing of the child; as can be seen in the UN Convention, the role of parents is central to protecting and promoting the interests of the child, and States are to give support and protection to parents to further the interests of the child.

To this day, the BIS remains the prevailing and foundational decision-making standard with broad international consensus supporting its use. The debate about the BIS has clarified the role of the BIS and how it is used and has provided some insights into specific considerations affecting the authority of parents. This chapter will provide a brief overview of the BIS framework and its use in the medical care of children, the debate over the BIS, and alternate frameworks that have been suggested to replace or augment the BIS. It will ultimately conclude that the BIS should remain the primary ethical and legal standard in decision-making for children, but that there are insights in other suggested frameworks worth considering and exploring further within a BIS framework.

4.2   The Best Interest Standard in the Medical Care of Children

4.2.1   The Basic Claim of the BIS

The BIS places the focus primarily on the wellbeing of the child. The basic claim is that in all decisions or policies about children, the decision or policy should be followed that is most likely to protect and promote the wellbeing of the child (Bester 2019a, b). Thus, when a decision about a child is made, it is the wellbeing of the child about whom we are deciding that is of primary moral importance.

Now obviously, there are other moral considerations involved when such decisions are made as well. Children do not live in a vacuum, but within a parent-child relationship, a family, a community, a society. There are often other moral considerations that impinge on decisions about children. We are not to ignore the other morally relevant features of any given situation. But the BIS does ask us to place the wellbeing of the child as the first moral consideration we consider, the starting point in our deliberations. The BIS creates what are called *prima facie* obligations – things we must do for a child unless there are compelling reasons not to (Kopelman 1997). Often this means that we may have to do the best we can given countervailing moral considerations and practical constraints.

The BIS represents a bulwark that protects the wellbeing of the child against immoral and unreasonable compromise. This is important because children are vulnerable, completely dependent on others, and if we neglect the moral claims children have on us it will set back the wellbeing of children significantly. This will have serious consequences for the children in question, but also for their families and the societies they are part of.

The BIS speaks of the interests of a child. Interests are components of wellbeing or things that are necessary to reach wellbeing (Bester 2019b; Feinberg 1986; Malek

2009). For a child to develop and to flourish, to become an adult that can exercise her own moral agency and autonomy, many things need to be in place. Examples of such interests include: health with related components such as sufficient nutrition and sleep; opportunities for cognitive growth such as is afforded through play and education; the child's relationships and place in a culture and society, the idea of belonging; and safety from injuries, exploitation, and psychological harms. Interests are things that children need to flourish; if we provide for a child's interests, we promote and protect the wellbeing of the child, if we withhold a child's interests, we set back the interests of a child.

The BIS therefore states that the interests of a child is of moral importance and should be the starting point and central consideration in all decisions about the child. It is not the interests of parents, of governments, or society at large that is the starting point in decisions about a child; rather it is the interests of the child that is the morally appropriate starting point. This places moral claims on those who stand in significant relationships with the child, baring responsibility for the wellbeing of child: parents, clinicians, caregivers, society and its institutions. The wellbeing of the child should be protected through parental actions, and by society's policies and laws. The BIS also has a host of moral implications for clinicians who provide medical care for children.

4.2.2   The Role of the BIS in Pediatrics

The BIS fulfills a number of important roles in pediatrics (Bester 2019a, b; Buchanan and Brock 1990; Kopelman 1997, 2013). The clinician is at bottom tasked with attending to the welfare of the patient, so that all of a clinician's obligations can ultimately be traced back to the idea of promoting and protecting the interests of the child-patient. It is an important point that children are vulnerable, generally unable to make their own decisions or to advocate for their own interests, so that the function of the clinician to protect, promote, and advocate for the wellbeing of the child-patient is of great importance.

It is not so that the BIS only functions as a test to see what should be done in a specific case. Rather, the BIS fulfills a number of identifiable functions in pediatrics. I provide here a short description of some of these functions that demonstrate how indispensable the BIS is to pediatric practice and to pediatric ethics.

(1) The BIS grounds obligations of clinicians and decision-makers in pediatrics

In the principles-approach to bioethics, general principles form the ethical foundation of medical practice. More specific guidelines, moral rules, clinical obligations, laws, and policies are derived from the general principles through a process of specification (Beauchamp and Childress 2013). By way of illustration, consider an example from adult medicine. The principle of autonomy is an important ethical principle in adult medicine, and basically states that persons with adequate decision-making capacity should be free to rule themselves through making their own choices. It refers to self-rule, the freedom to choose for oneself in accordance with one's values and view of the good. This is a very general principle that in itself provides very little direct guidance in the clinical situation; it tells us that clinicians should respect patients' decision-making, but little beyond that. However, autonomy provides the foundation for a whole series of rules, policies, procedures, and clinician obligations that are central to medical practice. From autonomy is derived ethical norms and laws related to informed consent, patient confidentiality, privacy, shared decision-making, disclosure of relevant information to patients, and non-deception of patients. Each of these norms and the legal and ethical requirements built up around them are enormously complex and volumes have been written about each. They all are grounded in the principle of autonomy; what moral force they have relies largely on the moral force of the principle of autonomy itself. The principle of autonomy does not play a significant role in pediatrics, so this is purely by way of example of the role of general ethical principles in medicine.

The central ethical principle in pediatrics is the BIS (Bester 2019a, b; Lo 2013). The BIS is a general moral principle that grounds various specific norms, guidelines, laws, and policies related to the medical care of children. While the BIS itself is broad and general, from it we can derive more specific guidance and rules to govern clinical situations. This process is called specification, where a general principle of medical ethics is applied and interpreted with reference to a specific clinical context so that more specific guidance, rules, laws, and policies can be created. The moral force of rules, laws, and policies so created comes from the BIS. Examples of specific principles and rules that are derived by specification from the BIS include: obtaining parental permission for treatment, seeking childhood assent to treatment, sharing information with children in an age appropriate fashion, having parents make decisions on behalf of children rather than children making decisions for themselves, and allowing adolescents to seek medical care for sensitive concerns such as addictions and reproductive concerns (AAP 1995, 2016; Lo 2013). These more specific rules and guidelines are meant to serve the interests of children during the course of medical care, and they can all be traced back to the BIS.

Consider an example of how this may work in practice. During the general pediatric office visit, the clinician should work to forge a good clinician-parent-patient relationship with the parents and child. Good communication and good rapport engender trust, which helps parents bring forward concerns, ask questions, follow recommendations, and be active partners in the care of the child. This facilitates implementation of diagnostic and treatment plans that promote the child's wellbeing. Consider, for example, that a trust relationship with a pediatric care provider can help parents resist anti-vaccine misinformation and can increase vaccination uptake (Leask et al. 2006). Consider also that clear and open communication, continuity of care, and a trusting clinician-patient relationship appear to be important factors in parental acceptance of clinicians' recommendations regarding antibiotic treatment of childhood respiratory infections (Brookes-Howell et al. 2014). It is clear that these values central to good pediatric practice—shared decision-making, good communication, building trust—ultimately work to improve the health and wellbeing of the child-patient by improving adherence with beneficial treatment regimens. These things are good things because in general they lead to furthering the interests of the child. Thus, more specific ethical dimensions of pediatric care such as shared decision-making, good clinical relationships, clear communication, and promoting parental involvement are grounded by the more general guideline of serving the interests of the child.

The BIS defines roles and sets obligations for clinicians as well as for parents or other decision-makers in pediatrics. The clinician-patient relationship is often described as a fiduciary relationship. A clinician's chief duty is to further the interests of the patient as it relates to the health and wellbeing of the patient, and to divest herself from self-interest or other competing interests. Parents also have the responsibility to attend to the wellbeing of their child and have the authority to make medical decisions for their child and authorize medical treatments offered by clinicians. At its best, parents and clinicians form a partnership aimed at serving the wellbeing of the child in the clinical context, through shared decision-making and shared responsibilities working together to help the child flourish. This idea has been described as one of co-fiduciaries, where parents and clinicians each have a fiduciary obligation to work together to serve the interests of the child. The BIS therefore sets obligations and expectations of clinicians and of parents, a moral framework within which they fulfill their function in the care of the child.

The BIS sets obligations that are owed to children, speaking to both clinicians and parents. Obligations derived from the BIS are *prima facie*; they have to be weighed and balanced against competing moral considerations in order to arrive at one's actual duty (Bester 2018a, 2019a, b; Kopelman 1997, 2013). This is no different than any of the other principles of medical ethics, another reason why the BIS fits so seamlessly into the entire network of medical ethics principles (Beauchamp and Childress 2013).

(2) The BIS can function as guide to right choices in a difficult clinical case

The BIS can be used as a guide and deliberative tool in clinical cases to identify ethically supportable choices in the health care of children (Kopelman 1997). To function this way, the BIS would guide us to choose among available options the one that is most likely to successfully promote or protect the child's interests. Available choices are weighed in light of the impact they will have on the wellbeing of the child, and the ethically supportable choice is the one that is most likely to promote or protect the wellbeing of the child. Other countervailing ethical considerations and practical constraints should also be considered and weighed; there may be ideal options that we would like to pursue for the child but that are not practically possible or ethically feasible. Furthermore, it is important to consider all of a child's interests, or the net effect of a proposed intervention on the wellbeing of the child. It is too narrow to focus only on a specific set of interests. This means that a full BIS determination considers not only health related interests, but also interests related to the child's attachments and family, place in society, education, and future development.

The framework I described here is called the BIS *all things considered* (Bester 2018a, 2019a, b). It requires of decision-makers to consider all of a child's relevant interests as starting point, to weigh the various available choices in light of these interests, and then to take into account countervailing ethical considerations and practical constraints. It is a decision-making tool that is itself derived from the very general moral principle called the BIS, with the justification that it serves the wellbeing of children that such a decision-making guide is used in cases where moral deliberation is required.

There are occasions where it is mostly clear what the right option is for a child (Kopelman 2018). These may be situations where a treatment is known to be very effective, with little adverse effects, and foregoing the treatment may be detrimental to the child's wellbeing. In such cases, it is easy to see that providing the treatment is the best option, while foregoing it is hard to support ethically. However, there are in clinical medicine some cases that are complex, where it is not always clear what the single best option is. This may be because of the probabilistic nature of medicine, where it is not clear that there is a single best treatment option but there may be various promising ones. Or, where it is not clear which adverse effects from different treatment options are best to accept. This is no problem for the BIS; it does not like a tyrant demand from us that we absolutely must identify the one best option or face moral consequences. Rather, in such situations the BIS identifies a range of acceptable options, and decision-makers can choose from the options within this range after careful deliberation. Kopelman describes this as the BIS providing a standard of reasonableness (Kopelman 1997, 2018). The BIS sets limits to what can reasonably be chosen for a child in a specific case, and decision-makers can scrutinize these options to decide on a course of action that is mutually acceptable to all. Thus, there are times where the BIS can yield a range of ethically acceptable options rather than pointing to one, obviously best, option.

Using the BIS as decision-making guide in this way can be straightforward in a simple case and can be complex and demanding in a difficult and ethically nuanced case. This should not surprise us; the BIS as decision-making tool is itself a way to help us think through and manage the ethical content of a case involving a child; if the case is complex, of course using the BIS as tool would lead to complexity. One may be tempted to simplify complex cases by using a more simple or straightforward standard, one that does not look at quite so many values or ethical considerations in the decision-making process. It is true that this would be less taxing and more straight-forward. But it would neglect to account for important moral content within a case, meaning it cannot be ethically justified. The BIS *all things considered* is well suited to function as ethical guide and decision-making tool in clinical cases involving children.

(3) The BIS performs a limiting function

The BIS places limits on the actions and decisions of those who are tasked with protecting the wellbeing of children, including clinicians, parents, and the state (Kopelman 1997, 2018; Pope 2011). It does so in two ways. First, it grounds a variety of limiting principles, more specific ethical rules that indicate when a decision-maker's actions or decisions should be overruled or interfered with. These limiting principles derive their moral power from the best interest standard; it is to protect and promote the wellbeing of children that such limiting principles exist (Kopelman

1997). The scope and focus of limiting principles are defined by the wellbeing of the child, so that the interests of the child is the central focus of any limiting principle.

The BIS plays a further role in limiting decision-makers. Once it has been determined that a specific decision or action by a parent or clinician should be overruled, the BIS serves as guide to determine what should be done to protect the wellbeing of the child (Kopelman 1997). The BIS therefore grounds various limiting principles, and then functions as a guide to the course of action that should be taken once limiting has happened.

It is an important point to appreciate that limiting principles may apply to all who are tasked with making decisions or instituting policies for and about children. Parents, clinicians, and government do not have limitless authority. They have to make decisions that serve the wellbeing of children, and if they do not, the limiting function grounded by the BIS cuts down their authority. Much focus falls on limiting parental decisions, but it should be remembered that other decision-makers are subject to the moral force of the BIS as well in their decision-making about children. States cannot do what they wish to children; they must follow the guidance of the BIS in instituting policy and making decisions for children. If policies that impact children fall outside of the range identified by the BIS or violates a limiting principle grounded by the BIS, the state's authority also ends and must be overruled. The same goes for decisions by parents, and decisions by clinicians. No-one has absolute authority over children; all authority is checked by principles that protect the interests of children. Society, families, decision-makers cannot do things to children simply because they want to or for the benefit of others; the interests of children are always the first and most important consideration in any decision and policy about children. Limiting principles are meant to cut down decisions or policies that fail to appropriately prioritize the interests of children in decision-making.

(4) The BIS clarifies ethical reasoning

Using the BIS as foundation for more specific obligations or as a decision-making tool in the ways described can serve to illuminate the values and ethical considerations underlying eventual decisions and ethical judgments (Bester 2018a, 2019b). The BIS forms an anchor, a basic reference by which we can ground and clarify our reasoning and decision-making. When decision-makers deliberate, they can frame deliberations and arguments by reference to the various obligations resting on the BIS, or by reference to the BIS as decision-making tool. For example, someone may say something of this nature: "Treatment X will advance the wellbeing of the child by doing A and B. Treatment Y will advance wellbeing by doing A, but not B. Omitting treatment X will compromise wellbeing by leading to C. Therefore, treatment X is the ethically supportable choice in this situation." Clarifying one's reasoning in this way by reference to the BIS makes it possible for others to understand and scrutinize the underlying ethical reasoning, and offer counterarguments also tied to the wellbeing of the child. If the speaker in this example left out important ethical content, it is clear and can be challenged in the deliberation. For example, someone may respond, "you have considered that treatment X will do A and B. But you did not consider the effects on another set of interests, D." Again, when a case is simple and it is obvious what is best to do, this level of reasoning may not be helpful or necessary. But it is in complex cases, where many interests are at stake and many ethical considerations have to be balanced, the illuminating of ethical reasoning in these ways are helpful in moving towards an ethically supportable decision.

4.3   Critiques and Defense of the BIS

A variety of critiques of the BIS have appeared over the past few decades (Diekema 2004; Dresser 2003; Rhodes and Holzman 2014; Ross 1998, 2019; Salter 2012; Veatch 1995; Winters 2018). A number of ethicists (including myself) have continued to defend the BIS as the appropriate ethical and legal standard in decision-making regarding children (Bester 2018a, 2019a, b; Buchanan and Brock 1990; Kopelman 1997, 2018; Pope 2011, 2018).

Defenders of the BIS have responded by pointing out that critiques of the BIS tend to be mistaken or misplaced. Some critiques are aimed at a strawman version of the BIS that no-one actually defends, some blame the BIS for ambiguity in value judgments that is the result of the pluralistic society in which we live, and some miss the point as to the role and function of the BIS in pediatrics. I will here briefly mention some lines of critique and describe how the BIS is defended against them.

4.3.1   The BIS Cannot Fulfill Two Functions

Some have argued that the BIS cannot fulfill two functions: it cannot serve as both guidance principle and limiting principle. This line of thinking draws on Buchanan and Brock's work on surrogate decision-making and argues that we need principles to act as a guide and principles to act as interventional or limiting standard to challenge a surrogate decision-maker's authority, and that the same principle cannot do both things.

Even if one accepts the framework of needing guidance principles and intervention principles, it is not clear why one would think the same principle cannot fulfill both functions. In bioethics, a single principle often fulfills many different functions and can ground many different principles, guidelines, or rules. Think of the principle of autonomy in adult medicine. It is a general and broad principle that states that clinicians should respect the sovereignty of the individual over her own body and her own health care. From this broad principle we derive many specific guidelines and principles that fulfill a whole host of different functions. This includes bedrock ethical norms and values in medicine like informed consent, shared decision-making, respecting refusals of beneficial treatment, disclosure of information to patients, veracity or non-deception, and protecting confidentiality. The same principle that provides the moral force for informed consent and shared decision-making, is the same principle that sets limits to the actions of clinicians in cases of informed refusal of life-saving treatment, and is the same principle that limits the clinician's sharing of patient information with third parties. There is no problem here with the same principle fulfilling different functions or grounding different ethical standards that apply within the same case. The BIS works the same: it is a broad and general ethical principle with central importance in pediatrics. It can ground guidance and interventional principles and can stipulate different norms and values within pediatrics that can fulfill different functions, even within the same case.

There is therefore no need to undercut the BIS or limit it to a specific function, when it is clear that it can ground a full gamut of ethical values and norms within pediatrics. There is also no need to incorporate a separate and unrelated interventional principle while retaining the BIS as a guide only. Proceeding in this latter fashion would indeed be very confusing; what exactly would the relationship be between the BIS and this new interventional principle, and in cases of conflicts between the two, how would we mediate disputes between two unrelated principles that fulfill different functions?

4.3.2   The BIS Is Too Demanding

Some critics maintain that the BIS asks too much of clinicians and parents. According to this line of criticism, the BIS demands that we identify the one best option in every case, and that we inflexibly do whatever is necessary to follow that one best option. We must do what is literally the best for the child, no matter the consequences, costs, or ethical claims of others. This levels three charges against the BIS: (1) that it is inflexible, and cannot accommodate countervailing ethical obligations or values present in a case (such as value trade-offs within families or between different patients); (2) that it demands the identification of one literally best option that must be discharged, making all other courses of action unethical; and (3) that it asks too much of parents and clinicians, who must set all things aside in their own lives and the lives of others to continually maximize the wellbeing of an individual child whatever the cost.

The overview of the BIS in the earlier parts of this chapter described the BIS as a broad and general principle that provides grounding for more specific obligations, norms, and values within pediatrics. Duties generated by the BIS are *prima facie*, meaning that they are things that should be done unless there is a good reason not to. Obligations created by the BIS should be weighed against countervailing ethical values and obligations, and should be considered within the boundaries of existing practical constraints. Furthermore, it is in the nature of medicine that there sometimes may be a range of acceptable treatment options rather than one best option that must be followed. The BIS sometimes stipulates a range of acceptable options or sets parameters for a range of reasonable diagnostic and treatment actions. Such considerations are case specific, and draws on the nature of the BIS to be adaptable to the specifics of a clinical case.

The way in which the BIS actually works provides ample room for considering value trade-offs, the ethical claims of others, practical constraints in a situation, and ambiguity as to which option is better than another.

4.3.3 The BIS Is Too Vague

Some maintain that the BIS is too vague and ill-defined. It is not clear which values should be used to judge what best serves the interests of a child, and differences in value judgments about a life worth living means that there is no way to mediate disagreements about how to proceed in complex clinical cases. This critique essentially says that parents and clinicians have different ideas about what matters most when it comes to children, and because of these differing value judgments the BIS cannot be used as ethical standard. It should be pointed out that it is often the same critics who maintain that the BIS is too vague and ill-defined that also argue that the BIS is too demanding and asks us to do what is literally the best for the child. These two critiques seem mutually exclusive; if the former holds, it is not possible to do what is literally the best for a child. On the other hand, if we are able to recognize and do what is literally the best for the child, then the BIS is not too vague and ill-defined to provide ethical guidance.

If the claim is that we can never know which of a set of alternative best serves the wellbeing of a child in a given situation that seems clearly false. There are situations where it is very clear which alternative serves the wellbeing of a child best. It also seems too much of a stretch to say that doctors and parents can never agree about which option or options should be instituted to protect and promote the wellbeing of the child in their care. Rather, in the course of medical practice it seems that there is generally large agreement between clinicians and parents; every day there are millions of treatment decisions for children all over the world where parents and clinicians reach agreement on a course of treatment or diagnosis for a child. When parents and clinicians disagree, often these disagreements can be resolved through deliberation, compromise, clarification, and meeting in the middle to find a way forward that preserves the most ethical value.

The critique is probably aimed at complex cases, those difficult clinical situations of the kind where ethical and clinical uncertainty predominate. In cases where children are desperately ill, for example, different decision-makers may have different ideas about what constitutes a risk worth taking, suffering worth putting up with, harms that are acceptable and unacceptable, or interests of the child that should be prioritized over another. There may be disagreement about whether parents or clinicians should have the final say over a specific value judgment or decision. If the critique is aimed at these situations, one would have to agree that instances arise where there are bitter disagreements between parents and clinicians that involve both clinical uncertainty and differing value judgments. But this is surely not the fault of the BIS, nor a particular feature of the BIS. Rather, this is a feature of the kind of pluralistic society common to western democracies, where different views of the good and different value judgments co-exist in the same society. That is why part of the project of democratic societies is precisely to find ways to manage and mitigate differing value judgments in ways that are neutral between different conceptions of the good.

One feature of the BIS that sets it apart as an ideal ethical principle in such a kind of society, is that it places the focus on the interests of the child rather than the interests of parents, society, government, or clinicians. This is important because of the vulnerability of children, and the great risk that the wellbeing of children may be compromised in the muddle of sorting out differing value judgments. The BIS forms a bulwark of protection for children, protecting their wellbeing, and creates minimum standards from which we can proceed in situations where differing value judgments create disagreement.

It is true that the BIS is a general guideline or principle that lacks specific content. But that is how all the principles of bioethics work. The general principles ground more specific norms and values by applying them to specific situations and adding in specific content. Over time, a set of recognized specified principles and norms develop that can be applied directly to specific situations. For example, obtaining parental permission for treatment is derived from the best interest standard through a process of specification, with a set of rules and exceptions that guides deployment in practice. With regards to use of the BIS as limiting or interventional principle, Pope describes that a set of mediating maxims have been developed that guide the use of the BIS (Pope 2011).

It does not seem to be the case that the BIS is uniquely vague or ill-defined among our ethical principles, and the BIS seems ideally suited to act as the central ethical principle in pediatrics.

4.4 Suggested Alternatives to the BIS

4.4.1 The Harm Principle

Some ethicists have suggested that harm should form the basis of ethical decision-making for children. This has found its clearest articulation in the Harm Principle (HP) defended by Diekema (2004, 2011, 2019). The HP states that the only reason to overrule the decision-making of a parent is to avoid imminent and serious harm to the child. It draws on harm principle of John Stuart Mill, a liberty limiting principle which describes when a government has the moral authority to interfere with the liberty of its citizens. In short, a government can only limit the free actions of people if such actions place others at the risk of serious harm. Based on the HP, Diekema formulated a set of eight requirements that must be present before parental decision-making can be overruled (Diekema 2004). In brief, these criteria stipulate that:

1.   the parental decision must place the child at significant risk of serious harm;
2.   the harm is imminent, requiring immediate action to prevent it;

3. the intervention is necessary to prevent serious harm;
4. the course of action that parents refused or chose against is effective and backed by sound evidence;
5. the intervention proposed does not also place the child at risk of serious harm, and anticipated benefits outweigh anticipated burdens;
6. there is no other option to prevent serious harm to the child that is less intrusive on parental autonomy;
7. the governmental intervention can be generalized to similar situations;
8. most parents would agree that state intervention was reasonable.

Understood along these lines, parents would have virtually unlimited authority over their children, parental authority being treated similarly to any individual liberty right, with the only limits to parental authority being the risk of serious harm to the child. Since the HP was originally suggested, HP proponents have at times somewhat toned down this extreme view of parental authority, stating that the BIS should be retained as general ethical guideline in pediatrics and the HP only fulfilling a limited function to mediate a specific kind of case (Hester et al. 2018). Thus, if parental decision-making places a child at risk of serious and imminent harm, understood as a setback to the child's interests, the State should intervene and overrule the decision. Examples that fall under this sort of category is a parent that refuses a life-saving blood transfusion for their child or a parent that refuses treatment for meningitis for their toddler. In such cases, withholding the treatment will potentially lead to severe complications or death of the child. Since these are significant harms, State institutions (such as child protective services) should be engaged to overrule the parental decision.

In some other places HP proponents also state that the function of the HP should be broadened to apply to a greater scope of parental decision than parental refusals of treatment, which was the original focus of the HP (Shah et al. 2017). Understood along these lines, the HP would not only be relevant in limiting treatment refusals, but also when parents demand treatments that are non-beneficial. It appears in this instance as if HP proponents argue that parents should be allowed to demand non-beneficial treatment as long as it does not harm the child; if a parental request for treatment does carry risk of harm to the child, such requests should be overruled. A case appealed to is the terminally ill child at the end of life in intensive care, such as the recent prominent case of Charlie Gard in the United Kingdom (Shah et al. 2017). The difficulty here is that it is not always clear what counts as a harm in these kinds of cases (Bester 2018b). Some may see death as the ultimate harm that outweighs any other potential harms, so that withdrawal of life-support is harmful and any intervention that may have an outside chance of prolonging life is justifiable. Some may see continued treatment as prolonging suffering when death is inevitable and view this as the greatest harm, and therefore view as unacceptable continued treatment or any intervention that may prolong suffering or the dying process. It is not clear how we should use the harm principle to adjudicate between such value differences.

4.4.1.1   Assessment of the Harm Principle

There is merit to the idea central to the harm principle; that those who stand in morally significant relationships with children should protect children against harm seems almost to be a truism. The BIS gives rise to a whole host of other ethical concepts and rules, and one of these is that parents cannot refuse life-saving treatment for their children. This appears to be doing the same kind of work the HP proponents envision the HP doing. Furthermore, parental discretion is important in pediatrics and in broader society. Parents are tasked with rearing children, the parent-child bond is a morally significant bond, and the relationship with their parent is an important childhood interest. Parental authority exists to protect the wellbeing of the child, and others should intervene if parental authority acts in ways that is not consistent with the wellbeing of the child.

There are a number of issues with the HP which raise questions about the use of the HP as ethical principle in pediatrics.

(1)   The HP sets the bar of acceptable decisions made about children too low (Bester 2018a, 2019b). The BIS places the focus on the interests of the child as primary concern in decisions about a child, directing the decision-making of parents, clinicians, and governments to prioritize the wellbeing of the child. If instead we follow the HP, and serious and imminent harms to children is the only reason to interfere with or regulate parental authority, we would have to accept a whole range of shoddy decisions regarding children that do not rise to the level of serious and imminent harm. While it is important to protect children against serious harms, harm is not the only thing that matters morally in the life of a child. The moral claims of children on parents, family, community, and society go beyond the avoidance of harming the child. Children are vulnerable, their wellbeing depending on the actions of other to secure their interests, so that parents, clinicians, governments, and society at large must do more than merely refrain from harming the child. This is particularly so in pediatrics, where a clinician is duty-bound to protect and promote the wellbeing of the child. Focusing exclusively on harm and letting parents do whatever they wish beyond that threshold appears to remove a number of guardrails necessary to protect and promote the wellbeing of children. HP proponents may respond that the HP is only a limiting principle in a select number of cases, while the BIS remains the guidance principle. But even here, if parents decide they do not wish to do anything beyond harm avoidance for their child, there is no recourse available for those who want to promote the wellbeing of the child. A much better approach is to have a series of limiting standards that are grounded in the BIS, where the focus of limiting standards is to promote the interests of children overall.

(2)   Harms as limiting standard do not solve the problem of vagueness and indeterminacy (Birchley 2016; Pope 2011). Those who criticize the BIS may think that we find a more objective standard in harm, but the problems of vagueness persist. In a complex clinical case different decision-makers may come

to different conclusions about what counts as a serious harm. Consider the suffering child in the ICU, close to the end of life. Some may think that continued treatment only prolongs the child's suffering and dying, so that the serious harm to be avoided is continued treatment. Some may think that death itself is the ultimate harm, so that continued treatment can be justified to avoid the serious harm that is death. How do we adjudicate between these different value judgments, and who gets to decide? What we see is that the value conflict in the case is related to different value judgments, a product of the pluralistic society in which we live. There are differences in how we view the relative goodness or badness of death, suffering, treatment at the end of life, and the scope of parental authority. The HP does not solve these problems. The BIS is instead much better equipped to highlight the differences in value judgment, help us work through the dilemma, and focus decision-making on doing what we can to promote the wellbeing of the child. The BIS is also well suited for the development of more specific guidance in a case through a process of specification; indeed, in legal decision-making a series of mediating maxims have been created to guide the application of the BIS (Pope 2011). Nothing similar exists for the use of the HP, and if such maxims are to be developed it is not clear how these would then relate to the already existing framework of pediatric ethics that is grounded in the BIS.

(3)   The HP as based in the work of Mill treats parental authority as a liberty right, and smuggles in the assumption that parents have the same authority over their children as individuals have over their own bodies and care (Bester 2018a). This liberty right is then only constrained by serious harms to the child. Moral considerations are therefore framed in terms of what is owed to parents, removing the focus from the wellbeing of the child. This seems inappropriate given the central assumptions in pediatrics, where the child is the patient and the wellbeing of the child is the primary moral focus.

In conclusion, it is clear and obvious that we should protect the wellbeing of the child against serious harms. This flows directly from a consideration of how to best protect and promote the interests of the child; the existence of a series of limiting principles that protect the wellbeing of the child clearly is in the best interest of children. But the HP on its own is not a sufficient account of pediatric ethics, and can only play a limited role as one of a series of limiting principles in pediatrics. It is important to retain a robust focus on the variety of obligations that parents, clinicians, and society at large have to secure the wellbeing of the children in their care.

4.4.2   Constrained Parental Autonomy

Constrained Parental Autonomy (CPA) is a theory about the moral obligations owed to children developed by Lainie Ross (Ross 1998, 2019). According to this approach, parents have the freedom (autonomy) to make decisions for their children, and this autonomy should be respected. Parental autonomy is only constrained by the basic interests of the child, which are those interests that are fundamental to the wellbeing of the child. CPA is a theory that is overtly based on Kant's ethics. Children are to be seen as having inherent worth, not full Kantian persons but as beings who have developed some aspects of Kantian personhood. The interests of children therefore matter morally. We see two different moral principles at work in this theory, providing the moral grounding of the arguments: (1) the assumption that parental authority is an extension of individual autonomy, a liberty right, and should therefore be afforded the same respect we afford other liberty rights; and (2) the idea that children are partial Kantian persons who should be treated as worthy of moral consideration and respect. The basic interests of children then place constraints on the liberty rights of parents. Parents can do what they wish in regards to their own children, as long as the basic interests of children are provided for. Furthermore, other people and society cannot do things to children that set back the basic interests of the child; in all our dealings with children, we need to uphold their basic interests.

CPA has some clear potential applications in pediatric ethics. It is clear to see that CPA can provide grounding for the ethical concept of parental permission. Parents are the decision-makers for their children, and parental autonomy should be respected. This means that parents have large latitude in making decisions for their children as long as the basic needs of children are met, but also that clinicians should seek permission from parents to authorize treatment. Under CPA, parents have a larger role in determining the acceptable scope of treatment options, and the wishes of parents should generally be followed as long as a set of the child's basic interests remains intact.

CPA can also function as a limiting standard, providing ethical grounds for overruling parental decisions that compromise the basic interests of the child. At minimum, when a parental decision compromises the health and wellbeing of a child by withholding a beneficial treatment necessary to protect the child's basic interests or by demanding an intervention that threatens to set back a child's basic interests, clinicians and the state should intervene to overrule such parental decisions.

Perhaps the most appealing feature of CPA for many is the way in which it allows for making ethical trade-offs when navigating complex decision-making about a child, especially when different values collide with one another. Under CPA parents have the latitude to weigh competing obligations and family dynamics, obligations owed to other children and other family members. One example is in the area of intra-family organ donation (Ross 1998, pp. 112–115). If one of the children in the family requires an organ transplant, and another child in the family is a potential live organ donor, the parent can make trade-offs in the health-related interests of the donor child by taking into account additional considerations that are ethically relevant such as family dynamics and family intimacy in the decision-making, as long as the basic interests of the donor child are met. It is appropriate under CPA to accept a small risk of donation related harm to the donor child, outweighed by the benefit of keeping the family intimacy and relationships intact through preserving the life and health of the child requiring the donation. Ross argues that the BIS cannot accommodate such family dynamics. According to Ross, the BIS disregards some of the complex moral components impinging on a decision about a child, because the BIS demands that parents must maximize the wellbeing of every child. So, the parent cannot consent to one of their children being a donor, because the interests of the potential donor child may be set back by donation, even if the donation would further the interests of the other child in the family that is sick and in need of donation. I disagree with this assessment of how the BIS works; as I've argued, the BIS asks of us to weigh all of a child's interests in the equation, so that it can incorporate the impact of decisions on significant relationships, and the BIS asks as to weigh countervailing ethical obligations (such as obligations to other family members) in the moral calculus. Be that as it may, it is clear that one major focus of CPA is that it allows parents great latitude in incorporating interests of other family members in decision-making for a child and allows parents to make trade-offs when weighing ethical considerations outside of the parent-child relationship, as long as the basic interests of the child are intact. In this way, it provides a mechanism for working through ethically complex situations where various ethical values and obligations are in tension with one another.

4.4.2.1 Assessment of Constrained Parental Autonomy

CPA seems to be a theory about the good of a child more generally. It is undoubtedly so that parents should be given a large amount of latitude in decision-making about children. Parent-child relationships have inherent moral worth, and are important childhood interests. However, a number of issues and questions remain that makes it unclear how suitable CPA is as replacement for BIS.

As limiting standard CPA seems to adopt a similar sort of approach as the HP. It is superior to the HP framework in that it includes a broader array of moral considerations than mere harm avoidance. At bottom, though, parents still have a lot of leeway to do as they wish unless their choices for the child fall below some minimal standard, defined by the basic interests of the child. How low this bar of acceptable parenting is, depends entirely on how one defines the concept of basic interests. If we say that parents can do whatever they wish as long as they meet some very low, basic threshold, it seems to place us in similar territory as we saw with an expansive view of the harm principle as primary ethical standard in pediatrics. We seem the same wide-ranging, extreme view of parental authority as a liberty right, and a very low bar that sets limits to parental authority. Again, this places the focus on what is owed to parents, rather than what is owed to children, which is problematic in pediatrics.

It is also not quite clear how CPA would be applied to pediatrics beyond the limiting function (Bester 2019b; Salter 2019). Pediatric ethics has over many decades built up a robust set of rules, standards, and ethical values that all rest on the BIS. If CPA replaces the BIS, all of these would have to be reworked and re-grounded. Incorporating CPA would not just represent a minor tweak; it would be a rethinking of the fundamental assumptions of pediatric ethics with seismic consequences which we cannot even imagine. If we are to undertake such a task, it should be clear why it is necessary, and how the proposed new theory surpasses the old one. And with CPA, it is not clear.

A further set of issues should be considered. CPA is based on Kant's theory of ethics, meaning that those who hold to other theories of ethics besides Kant's are unpersuaded by it. CPA coins the idea of "basic interests", but it is not clear what these are, how they differ from a child's interests overall, or how we would know once a child has enough of them. It leaves much open to the judgment of the individual to decide which interests are properly basic, and when a child's interests are sufficiently met to further allow parents to do as they wish.

Overall, CPA makes a valuable contribution in reminding us that parents should be given latitude in decision-making, and that meeting of a child's interests are important independent of whether a parent wishes it to be so. Its greatest strength is the scope it provides for navigating ethical trade-offs in complex situations, where parents have to navigate a host of competing ethical considerations. As a comprehensive theory of pediatric ethics, it is not clear that it is superior to the BIS, or that it can replace the work done by the BIS.

4.5   Conclusion

The BIS remains the predominant ethical standard in pediatrics. It grounds the obligations of the clinician, the decision-making authority of parents, and a variety of rules, laws, and values in pediatrics. Although some have criticized the BIS as being inadequate or flawed, the BIS is able to withstand these critiques well. Suggested alternatives to the BIS do not solve the identified problems, and do not seem to be sufficiently equipped to replace the many functions of the BIS in pediatrics.

Guiding principles: The Best Interest Standard in pediatrics

The basic ethical implications of the Best Interest Standard
- In all decisions or policies about children, the option should be followed that is most likely to protect and promote the wellbeing of the child
- A child's interests are components of the child's wellbeing. The child's interests should be the starting point and central consideration in all decisions about the child
- Those who stand in morally significant relationships with the child, such as parents and clinicians, have the responsibility to protect and promote the child's wellbeing (or interests)
- The effect of a specific decision on ALL of a child's interests must be considered when weighing decision options
- The Best Interest Standard creates *prima facie* obligations that must be weighed against countervailing ethical considerations and practical constraints

---

The role of the Best Interest Standard in pediatrics
- The Best Interest Standard is a general ethical principle that grounds more specific obligations of clinicians and decision-makers in pediatrics
- The Best Interest Standard can function as practical guide to decision-making in complex clinical cases
- The Best Interest Standard performs a limiting function, describing the limits of appropriate and reasonable treatments
- The Best Interest Standard clarifies and illuminates ethical reasoning

---

References

American Academy of Pediatrics, Committee on Bioethics (AAP). 1995. Informed consent in decision-making in pediatric practice. *Pediatrics* 95 (2): 314–317.

American Academy of Pediatrics, Committee on Bioethics (AAP). 2016. Informed consent in decision-making in pediatric practice. *Pediatrics* 138 (2): e20161484.

Beauchamp, T.L., and J.F. Childress. 2013. *Principles of biomedical ethics*, 7th ed., 15–21; 101–140. New York: Oxford University Press.

Bester, J.C. 2018a. The harm principle cannot replace the best interest standard: Problems with using the harm principle for medical decision-making for children. *The American Journal of Bioethics* 18 (8): 9–19.

Bester, J.C. 2018b. Charlie Gard and the limits of the harm principle. *JAMA Pediatrics* 172 (3): 300–301.

Bester, J.C. 2019a. The best interest standard and children: Clarifying a concept and responding to its critics. *Journal of Medical Ethics* 45 (2): 117–124.

Bester, J.C. 2019b. The best interest standard is the best we have: Why the harm principle and constrained parental autonomy cannot replace the best interest standard in pediatric ethics. *The Journal of Clinical Ethics* 30 (3): 223–231.

Bester, J.C., and E. Kodish. 2017. Children are not the property of their parents: The need for a clear statement of ethical obligations and boundaries. *American Journal of Bioethics* 17 (11): 17–19.

Birchley, G. 2016. Harm is all you need? Best interest and disputes about parental decision-making. *Journal of Medical Ethics* 42: 111–115.

Brookes-Howell, L., F. Wood, T. Verheij, H. Prout, et al. 2014. Trust, openness and continuity of care influence acceptance of antibiotics for children with respiratory tract infections: A four country qualitative study. *Family Practice* 31 (1): 102–110.

Buchanan, A.E., and D.W. Brock. 1990. *Deciding for others: The ethics of surrogate decision-making*, 215–266. Cambridge: Cambridge University Press.

Diekema, D.S. 2004. Parental refusals of medical treatment: The harm principle as threshold for state intervention. *Theoretical Medicine and Bioethics* 25: 243–264.

Diekema, D.S. 2011. Revisiting the best interest standard: Uses and misuses. *The Journal of Clinical Ethics* 22 (2): 128–133.

Diekema, D.S. 2019. Decision-making on behalf of children: Understanding the role of the harm principle. *The Journal of Clinical Ethics* 30 (3): 207–212.

Dresser, R. 2003. Standards for family decisions: Replacing best interest with harm prevention. *American Journal of Bioethics* 3 (2): 54–55.

Feinberg, J. 1986. *The moral limits of the criminal law: Volume 1, harm to others*, 33–45. New York: Oxford University Press.

Hester, D.M., K.R. Lang, N.A. Garrison, and D.S. Diekema. 2018. Agreed: The harm principle cannot replace the best interest standard… but the best interest standard cannot replace the harm principle either. *American Journal of Bioethics* 18 (8): 38–40.

Kohm, L.M. 2008. Tracing the foundations of the best interest of the child standard in American jurisprudence. *Journal of Law and Family Studies* 10: 337–376.

Kopelman, L.M. 1997. The best interests standard as threshold, ideal, and standard of reasonableness. *The Journal of Medicine and Philosophy* 22: 271–289.

Kopelman, L.M. 2013. Using the best interest standard to generate actual duties. *AJOB Primary Research* 4 (2): 11–14.

Kopelman, L.M. 2018. Why the best interest standard is not self-defeating, too individualistic, unknowable, vague, or subjective. *The American Journal of Bioethics* 18 (8): 34–36.

Leask, J., S. Chapman, P. Hawe, and M. Burgess. 2006. What maintains parental support for vaccination when challenged by anti-vaccination messages? *A Qualitative Study. Vaccine* 24 (49–50): 7238–7245.

Lo, B. 2013. *Resolving ethical dilemmas*, 5th ed., 263–270. Philadelphia, PA: Lippincott, Williams, and Wilkins.

Malek, J. 2009. What is really in a child's best interest? Toward a more precise picture of the interests of children. *The Journal of Clinical Ethics* 20 (2): 175–182.

Pope, T.M. 2011. The best interest standard: Both guide and limit to medical decision-making on behalf of incapacitated patients. *The Journal of Clinical Ethics* 22 (2): 134–138.

Pope, T.M. 2018. The best interest standard for health care decision-making: Definition and defense. *American Journal of Bioethics* 18 (8): 36–38.

Rhodes, R., and I.R. Holzman. 2014. Is the best interest standard good for pediatrics? *Pediatrics* 134 (S2): S121–S129.

Ross, L.F. 1998. *Children, families, and health care decision-making*. New York: Oxford University Press.

Ross, L.F. 2019. Better than best (interest standard) in pediatrics decision-making. *The Journal of Clinical Ethics* 30 (3): 183–195.

Salter, E.K. 2012. Deciding for a child: A comprehensive analysis of the best interest standard. *Theoretical Medicine and Bioethics* 33: 179–198.

Salter, E.K. 2019. When better isn't good enough: Commentary on Ross's "Better than best (interest standard) in pediatric decision-making." *The Journal of Clinical Ethics* 30 (3): 213–217.

Shah, S.K., A.R. Rosenberg, and D.S. Diekema. 2017. Charlie Gard and the limits of best interest. *JAMA Pediatrics* 171 (10): 937–938.

United Nations (UN). 1990. Convention on the rights of the child. https://www.unicef.org/child-rights-convention/convention-text. Accessed 14 Oct 2020.

United Nations Children's Fund (UNICEF). 2020. History of child rights. https://www.unicef.org/child-rights-convention/history-child-rights. Accessed 14 Oct 2020.

Veatch, R.M. 1995. Abandoning informed consent. *Hastings Center Report* 25 (2): 5–12.

Winters, J.P. 2018. When parents refuse: Resolving entrenched disagreements between parents and clinicians in situations of uncertainty and complexity. *The American Journal of Bioethics* 18 (8): 20–31.

Woodhouse, B.B. 1998. From property to personhood: A child-centered perspective on parents' rights. *Georgetown Journal on Fighting Poverty* 5 (2): 313–320.

Further Reading

Bester, J.C. 2019a. The best interest standard and children: Clarifying a concept and responding to its critics. *Journal of Medical Ethics* 45 (2): 117–124.

Bester, J.C. 2019b. The best interest standard is the best we have: Why the harm principle and constrained parental autonomy cannot replace the best interest standard in pediatric ethics. *The Journal of Clinical Ethics* 30 (3): 223–231.

Diekema, D.S. 2004. Parental refusals of medical treatment: The harm principle as threshold for state intervention. *Theoretical Medicine and Bioethics* 25: 243–264.

Kopelman, L.M. 1997. The best interests standard as threshold, ideal, and standard of reasonableness. *The Journal of Medicine and Philosophy* 22: 271–289.

Chapter 5
Two Ethical Foundations for Pediatrics: The United Nations' Convention on the Rights of the Child and Bioethical Principles

J. P. Spike

Abstract This chapter presents a summary of two valid sources of ethical guidance for pediatrics. One is the United Nations' Convention on the Rights of the Child which has been ratified by almost every country in the world. It covers a wide range of important topics, and while it is distinctly unphilosophical in origin, it provides sufficient material to construct a complete pediatric ethics. The second one is the philosophical approach of using mid-level principles that is common in many fields of applied ethics. The essay proposes that pediatrics cannot use the exact same set of mid-level principles as adult medicine, and proposes a set of six ethical principles for Pediatrics. These are: Patient Autonomy, Respect for Children, Limited Parental Discretion, Beneficence, Non-Maleficence, and Justice. Both approaches are valuable, and there is no reason to select one and reject the other. The chapter includes brief discussions of the shortcomings of virtue ethics and the usefulness of casuistry, and of the principles found in research ethics and The Belmont Report.

Keywords Ethical principles · Mid-level principles · Clinical ethics · United Nations (UN) · Children's rights · Mature minors · Teenagers · Consent · Assent · Decision-making capacity · Adolescents · Respect for children · Parental discretion · Best interest · Casuistry · Research ethics · Belmont report

5.1 Introduction

Pediatric ethics is a relatively new field. It is a branch of bioethics, but it has as many differences from the usual models of biomedical ethics as adult medicine has from pediatric medicine.

There are two distinct sources that can be used as a foundation of pediatric ethics. This chapter will give a brief summary of both. While one can benefit from asking if one of the sources is superior to the other, there is no need to see it as a competition,

J. P. Spike (✉)
Children's National Hospital, Washington, D.C., USA
e-mail: spike@email.gwu.edu

© Springer Nature Switzerland AG 2022
N. Nortjé and J. C. Bester (eds.), *Pediatric Ethics: Theory and Practice*,
The International Library of Bioethics 89,
https://doi.org/10.1007/978-3-030-86182-7_5

choosing one and excluding the other. However it does help to be familiar with one or both of the groundworks or foundations that supports the more specific arguments for deciding what is the most ethical or best justified stance to take on common clinical controversies.

Bioethics taught in the US and Canada often begins with the conclusions of the influential book *The Principles of Biomedical Ethics* by Beauchamp and Childress (1979). This chapter will give a synopsis of that approach, and show how it must be modified in order to be relevant to pediatric ethics. Then the chapter will give a synopsis of an alternative approach to grounding pediatric ethics that may appeal to those with more of a global or international perspective, based on the UN Charter on the Rights of the Child (UNCRC 1990).

These are the two most powerful approaches to ground pediatric ethics, meaning have the strongest claim to universality, as well as the strongest justification. There are other philosophical theories of ethics that appeal to some people which will not be covered here as they are less influential, and (more importantly) have less claim to universality and greater weaknesses in their philosophical or logical justifications.[1]

5.2   The Six Principles for Pediatric Ethics

The four principles of Beauchamp and Childress began the dominance of the use of principles for biomedical ethics. Their choice of using principles followed the approach of the Belmont Report, which identified three principles for research ethics. One will also occasionally find a text in medical or nursing ethics that use the four

---

[1] This chapter does not include what is sometimes referred to as the "virtue ethics" approach to bioethics. It also leaves out the closely related approach known as casuistry. This is deliberate. The virtue ethics and casuist approaches both start with an assumption that well-established traditional practices, fundamental beliefs, and paradigm cases have normative validity without providing further empirical or logical justification for them. That makes them inadequate as a foundation for ethics.

Casuistry and virtue ethics are nevertheless often still popular with individuals and institutions affiliated with a religious tradition. All three Western religions were deeply influenced by Aristotle's virtue ethics via their own theological progenitors in the Middle Ages (e.g. Maimonides, Aquinas, and Averroes). All were brilliant minds, of course. But all three were pre-Enlightenment, and were unable to find a foundation for ethics without presuming authoritative value to their religious tradition (in the place of what Aristotle held to be "necessary" and "natural" in his teleological biology). These approaches may seem adequate to the facts when applied to a homogeneous culture that disregards opinions that deviate from their fundamental beliefs. However, in the modern world ethics must be capable of universality and applying equally fairly to believers and non-believers, men and women, gay and straight, able-bodied and disabled. Applied to a clinical setting, virtue ethics can lead to an inability to see the failures of the past, a defense of paternalism, or unjustified deference to physician authority. Casuist reasoning as a process or method is the basis for legal reasoning in case law based on precedent. But this chapter concerns itself only with the foundations of *ethical* reasoning, and not *legal* reasoning. Ethics must incorporate critical thinking, rejecting any uncritical presumptions. That said, one might wish to argue that casuistry is a useful method for weighing and balancing how each of the principles apply to a case analysis. That heterodox approach is a way to recognize a role for casuistry in pediatric ethics (and biomedical ethics generally) within the structure of ethical principles.

principles of Beauchamp and Childress, but adds a fifth or even a sixth principle. But in essence they are still adopting the principles approach. At a minimum, that approach entails identifying a list of principles which is widely agreed upon by a profession.

Ethical principles are called mid-level principles because they are more specific to the practice of the field and less general than a philosophical theory would be. Theories of ethics generally attempt to have a single criterion to decide right and wrong, such as the theory of Utilitarianism or Kantian theory. Valid principles should be more concrete than any theory. More importantly, each of the principles should be justifiable by any strong and sound theory. If it wasn't, there must be either a problem with the principle or with the theory.

Thus, the four principles in the received view of bioethics can each be justified by a Utilitarian philosopher, and each can also be justified by a Kantian philosopher. But, of course, the justifications would be different. Utilitarians will argue that following each principle will lead to better outcomes and helping more people, while Kantians will argue that pure reason (or our 'rational soul') dictates that all persons have a duty to follow each principle as if it were a law of logic.

To borrow and repurpose a common concept in modern medicine, one might say Utilitarian ethics is EBM, or evidence-based morality, while Kant would insist it is not just universal but categorical and discovered exclusively by reflection and rationality. Utilitarian theory likens ethics to science; Kantian theory likens it to math. (In philosophical terminology: Utilitarians see ethics as empirical and a posteriori, while Kantians see it as true a priori.)

In conclusion: the principles approach is powerful because it rests on this convergence in our lived-life of the two epistemologically incompatible yet universal theories of ethics. In other words, the interminable debate about which theory is right can be safely left to others (i.e. professors in the philosophy department), while bioethics forges ahead with its mid-level principles.

Thus, the principles have been most productive to the enormous progress of bioethics in the past 40 years. However, the usual set of principles must be revised to be useful in Pediatrics. A few other points about the principles are essential to understanding them. First, the principles are not ranked. Despite the protestations of some people who clearly do not understand them, no one principle is the most important. And to be certain a case has been completely analyzed, one must be sure to check to see how each of the principles applies. No case is just "an Autonomy case," without consideration of the relevance of the other principles.

Second, the principles are independent, meaning they can conflict with each other. Indeed, one of the most valuable aspects of having multiple principles (rather than pledging support for a single theory) is they do a very good job of explaining ethical challenges. A dilemma is when two different and conflicting choices of action are supported by two different principles. Principles clarify why there are so many dilemmas in clinical care, even if they do not provide an easy 'one best answer' to them. That, it seems to me, is not a failing of the principles approach, but simply a recognition of complex reality.

These two points are nicely summarized by saying all the principles are *prima facie*, or presumed initially to be relevant. In other words, each of the principles is true, but its relevance must be tested when one first looks at a case (at first glance, as they say). Which principles are most important for any particular case will depend on the particulars of the case, including the stakeholders who are involved and the empirical and contextual facts such as the diagnosis, what therapeutics are reasonably available, and family dynamics, economics, and religious beliefs.

The definitions of the six principles in this chapter are mine. They should not be considered either exact representations, quotations, or paraphrases of the original four principles by Beauchamp and Childress.

Despite my defense of the principles, however, it is clear that the principle that is usually introduced first (lexically) in any presentation is problematic in Pediatrics. It can be called either patient autonomy, or respect for autonomy. The problem is that it is based on the legal presumption in adult medicine that all patients have decision-making capacity (unless assessed otherwise). The principle was developed to recognize the important of recognizing patient's rights, including the right to be informed of all of the reasonable treatment choices and the right to choose the one most in agreement with the patient's values.

Autonomy was developed largely out of cases where a patient did not want what physicians would normally recommend, such as a refusal of a blood transfusion by a Jehovah's Witness. But by making it a principle, it was recognized that every one of us has the right to refuse what physicians recommend and that right does not have to be based on any religious belief. It might just be a person who prefers not to take a chance to live longer if it requires many months of debilitating treatment with many unpleasant side-effects. One might say it represents the value of pluralism. It also protects against paternalism, i.e. it prevents allowing the physician's values to overrule the patient's values.

A few common misconceptions about the Principle of Autonomy need to be corrected. It is not more important than the other principles. It also does not give a patient the right to demand unreasonable treatments, only to refuse any treatment and to choose from the reasonable options. The single most common ethical error made in Pediatrics is to transfer the rights of adult patients to the parents of a pediatric patient. The fallacy is tempting: one assumes Autonomy applies to the adult decision maker, and that is usually the parent. But Autonomy in adult medicine does not apply to any adult decision-maker, it only applies to the adult patient.

There are two different scenarios that must be defined in Pediatric ethics to cover the territory of Autonomy. One is to define when a child or minor is capable of making adult or mature decisions; the other is to define the rights of the parent as decision-maker when the patient is not capable of decision-making.

For a textbook with an international audience, this is not a simple question. Different countries have different ages for legal adulthood. In the US it is always 18, although by setting it at 18 all 50 states end up adding various legal carve-outs, usually including a right to birth control, treatment for STDs, and for mental health treatment. (Abortion is a topic that bitterly divides the 50 states in the US, with some including it with birth control, and others requiring either parental consent or at least parental disclosure.)

Thus, Patient Autonomy for Pediatrics is best divided into three principles as follows:

Patient Autonomy (similar to the original principle): Pediatric patients who are of legal age of adulthood have all the rights of an adult. That principle, while legally uncontroversial, can be easily violated by (usually well-meaning, loving) parents of teen patients. College students will often continue to see their pediatrician, and it is tempting for both pediatricians and parents to treat 18 and 19-year-olds as if they were still children, i.e. minors.

In pediatrics there is another very important clarification of Autonomy. Ethically this right is premised on the patient's capacity to make decisions, and thus may also apply to patients who have not yet reached the legal age of adulthood. In the US where adulthood is age 18, it is not uncommon for patients as young as 14 to be fully capable, and thus autonomous (also refer to Chap. 2). The only thing that changes at the eighteenth birthday is that the *presumption* of lacking capacity changes to the *presumption* of having capacity. However, anyone who has decision-making capacity should have the right to make their own decisions; when a person is capable is a clinical judgment, not a birthday.

The topic of decision-making capacity can lead to complex debates because it is at the core of deciding when (or to what degree) a patient has Autonomy (or the right to make her own decisions). It is thus also at the core of the concept of a mature minor, i.e. someone who has Autonomy but has not yet reached the age of 18. The American Academy of Pediatrics has suggested "most minors probably have capacity at age 14" (Weithorn 1982; Leikin 1983; AAP 1994; Appelbaum 2007; Spike 2017; Weithorn 2020).

This clinical judgment is evidence-based, using psychological data about the ability to reason, think abstractly, conceptualize hypotheticals, and weigh future outcomes. Capacity is especially likely for minors who have faced serious illness and have already experienced the treatments for their illness which are being discussed. Adults quite often hold an unfairly jaundiced opinion of teenagers, despite having once been teenagers themselves. Ethics often needs to help redress that bias.

Respect for Children: Patients who have not yet achieved the legal age of adulthood still deserve to be treated with respect. That includes the right to be fully informed, in an age-appropriate way, of their medical condition and the treatments being planned; and to be allowed to participate in the decision-making process to the degree they are capable and wish to be involved. This means in some cases that a legal minor can be assessed to be fully capable and given the full rights of an adult with capacity. (In those cases, giving the minor the final say could be justified by both the principles of Patient Autonomy and Respect for Children.)

Respect for Children also supports the doctrine of assent, often taken to include minors starting at age six or seven years old. Assent grew out of research ethics and the requirement for NIH funding in the US that research involving children include their assent starting at age 6 or 7. But clinical practice ought to follow the same rule.

Each visit should include some private time with the minor where you explain your role as their doctor, looking out for them, and giving them a chance to talk to you privately. Children's fear of going to the doctor can be mitigated considerably if they feel they are being treated like an intelligent partner by a respected adult, rather than being forced (sometimes using physical restraint) to submit to painful experiences while being treated without respect.

Limited Parental Discretion: When patients are not capable of complex medical decision-making, such as infants and young children, parents or guardians must make decisions based on the best interest of their child. Their role is analogous to that of any surrogate for an incapacitated adult patient with no advance directive to guide them. In other words, there will be restrictions on their right to refuse treatment. They cannot, for example, refuse life-saving blood transfusions for their child, even if they (the parent or guardian) are a Jehovah's Witness (also refer to Chap. 9). This "gap" in the available choices to the parents is preserved by the greater duty of clinicians (and even the state) to protect the child's best interest. Thus, one of the most important lessons of the changes to the principles required for Pediatrics; *there is no principle of parental autonomy* (or parental rights).

However, there will be difficult decisions in a "grey zone" where different parents will make different choices, and clinicians will have to leave room for some degree of *reasonable* (defensible, understandable) variation. How aggressively to treat a 'micro-preemie' is the most common example. Parents will be assuming a lifetime of responsibility for the care of such an infant, and so deserve a substantial say in the decision about whether to institute comfort care. Nevertheless, their choices are only open to reasonable options. In particular, the right to refuse treatment (which is unlimited for autonomous adult patients) is clearly much more restricted for parents and guardians (Gillam 2016).

The remaining three principles do not require major revisions:

Beneficence: Clinicians ought to do what is in the patient's best interest, determined by balancing all of the likely benefits and burdens of each treatment alternative. This is a very high standard, and *altruistic* in nature because it doesn't allow self-interest or third-party interests (such as an insurance company's) to interfere with the decision process. It is also known as a *fiduciary* requirement, meaning the clinician is exclusively devoted to the patient's interests. (That concept is common in banking, where a trust must be run by a fiduciary who is very restricted in the types of risks she can accept with her client's money.) Occasionally one might see this referred to as the principle of best interest, conflating the principle of beneficence with the best interest standard which should guide the decision-making of both clinicians and surrogates (including parents).

Non-Maleficence: Clinicians must consider quality of life as well as length of survival and be especially cautious to not allow their interventions to make things worse. It is a translation of the Hippocratic maxim that one must always, at the very least, *Do No Harm*. The traditional maxim can be thought of as a warning against *hubris*, and its modern meaning as the medical equivalent of the precautionary principle in science and public health: avoid high-risk, low probability interventions or 'heroics' that may cause suffering that is disproportionate to its benefits. In modern medicine it also means aggressive treatment of pain is always required, and one ought to be careful to not overlook the value of adjuvant palliative care or a transition to hospice. Indeed, if one principle is inviolable, or first among equals, it is non-maleficence. Non-maleficence is an ethical principle analogous to John Stuart Mill's Harm Principle, which Diekema argues for as the correct threshold for when to refuse a parent's surrogate medical decisions and when to initiate a state investigation of child abuse or neglect (Diekema 2004).

Justice: Justice requires at least two elements be considered in clinical decisions: it is both a positive duty requiring that we give the same care to the poor and powerless, and members of all vulnerable, marginalized, uninsured, and undocumented groups as to the members of dominant social groups; and a negative duty that we be responsible stewards of resources so that they are used to maximum benefit without harm to society. This principle is further justified by the idea that every profession has a social contract, central to communitarian ethical theory.

To fully fill out this principle requires considerable social policy work, for example to decide whether it is even ethical for a society to have people who are uninsured. In the US it can even include individual decisions about where to locate a private practice, or what insurance to accept.

Thus, pediatric ethics is best seen as resting on six principles rather than four, if one is going to use an approach based on principles. Together these six principles provide a foundation for ethical practice in Pediatrics as sound as the four principles for adult medicine.[2]

---

[2] In research ethics a slightly different model is used, which specifies only three principles. It is based on the Belmont Report. The Belmont Report (1978) was a direct result of the public outcry in the U.S. after the public learned about unethical research that had been done by American physicians, particularly the Tuskegee Study of Untreated Syphilis in the Negro Male that had ended in 1972. The Belmont Report and accompanying Commission reports led to making what had been internal NIH rules more clearly defined, turned them into strict federal regulations, and required every study protocol be approved by an institutional review board (IRB) before it can even be submitted for funding to the NIH or other federal agency. The Belmont Report has three basic principles for research ethics: Respect for Persons, Beneficence, and Justice. Non-maleficence is not mentioned, and its requirements can be thought of as subsumed by either Respect for Persons or Beneficence. Respect for Persons focuses on not just lack of coercion in recruitment of research subjects, but full informed consent. Beneficence focuses on good study design which could yield valuable data with minimal risk to research subjects. And Justice is concerned with selecting subjects in a fair way, including people from populations most likely to benefit from the research and not deliberately choosing subjects from vulnerable or marginalized groups simply because they are easy to enroll. The Belmont Report also includes additional protections for minors, pregnant woman, prisoners, and the disabled for any study with more than minimal risk. While Tom Beauchamp and Jim Childress proposed the four principles in their textbook, Beauchamp was also one of the staff authors of the Belmont Report. However, Beauchamp joined the staff when a few drafts had already been proposed and discussed. The Report was the summary of a years' long government commission, and it had numerous other important contributors, including both Stephen Toulmin and Albert Jonsen, who later became known as the primary defenders of casuistry for bioethics. Thus, the hypothesis in

5.3   Children's Rights as the Foundation of Pediatric Ethics

The United Nations Convention on the Rights of the Child is another source that can serve as a broad and objective consensus statement for Pediatric Ethics (UN CRC). Rather than a philosophical justification for a small set of basic biomedical principles, it is a document written by an international organization. Part I contains 40 "articles" which are claims to substantive rights, and then ends with article 41 that says nothing in the Convention should prevent a state from having higher standards than are represented in the UN CRC articles.

The CRC was first adopted by the UN in 1989 and was ratified by the required minimum number of nations on September 1990. It has since been ratified by 196 nations, including every member of the UN except the United States. (The other two holdouts were Somalia and South Sudan, who both signed in 2015.)

The history of the UN CRC began with the work of Eglantyne Jebb, a woman who founded the organization "Save the Children" in 1919 following World War I. It started in England with a concern primarily for German and Austro-Hungarian orphans and refugees. Her draft of a Children's Charter was the basis of the League of Nations' Declaration of the Rights of the Child (or Declaration of Geneva) in 1924.

Both the principles of biomedical ethics and the CRC can claim to have started with a consensus that had been already accepted by many professionals for a substantial period of time and in many countries, and that then went through a process of expert verification, the former a philosophical justification and the latter the ratification by most of the nations on earth. The CRC differs from the biomedical principles by being concerned with a wider range of issues than clinical care alone, including civil, political, social, and cultural rights. This seems like a natural result of it being produced by political entities rather than by medical professionals. But that can also make it more readily applicable to a wider range of concerns.

One way to explain the difference in scope is to see it as a question of what "bioethics" is meant to cover: is it only clinical medicine (biomedical ethics) or is it everything that impacts life and health? In the latter case, while the principles can be used, they need to be interpreted generously and broadly. Many people will find it easier to refer to the CRC, which transparently are meant to apply to such additional issues as public health (including nutrition, exercise, preventive medicine) and environmental quality (at a minimum including safe air to breath, safe water to drink, and safe public areas and parks for recreation).

In addition, while the principles of biomedical ethics have been taught in medical schools throughout the English-speaking world and beyond, it is fair to say that

---

footnote 1 that one might use casuistry as a method of weighing conflicting principles could have been implicitly endorsed by the two people who revived interest in casuistry in bioethics. (This is all documented in the book *Belmont Revisited*, edited by JF Childress, EM Meslin, and HT Shapiro (Georgetown University Press, 2005). In Europe, the Declaration of Helsinki covered similar ground to The Belmont Report, with its first edition in 1964 (and five revisions since). And it is worth noting that The Nuremberg Code served as a very important (and very succinct) historical precedent for both Belmont and Helsinki.

it is clearly recognized as guidance primarily by physicians, surgeons, nurses, and hospitals. In contrast, the UN CRC has been recognized by governments, and so carries ethical weight for all aspects of society, including the judiciary, the military, and education.

The UN Convention assumes a basic understanding and acceptance of the idea of universal human rights, largely a creation of the Enlightenment in Europe, and extends them to minors (persons under 18). That assumption of human rights, endorsed by virtually all nations, serves more or less the same role as (or, is isomorphic to) the justification of the biomedical principles by both Utilitarian and Kantian philosophical ethical theories insofar as it includes ratification by a range of countries from egalitarian secular and socialist nations to religious theocracies.

It is important to understand that including minors as having human rights, while it might seem 'obvious' and uncontroversial to some people today, represents hard-fought progress just as much as (say) realizing that women deserve human rights (or 'women's rights are human rights'), or that slavery is incompatible with any definition of universal human rights (or more recently, 'Black Lives Matter'), or that Shudras and Dalits (untouchables) should have the same rights as Brahmins, Kshatriyas and Vaishyas. All of those assertions, though now obvious to many people, required decades (or even centuries) of political organizing to achieve long after the great philosophical works of the eighteenth-century Enlightenment. And there is (always) still ethics work to be done (Wilkerson 2020).

In other words, the general precept can sound simple, but it can still require a lot of time and education and political work to get a population to fully accept the logical ramifications of a simple precept. Even today many adults still question the idea that children have or deserve universal human rights—whether or not their parents, their nation's laws, or their religious tradition recognize them.

Many presentations of the UN CRC recently like to start with what have come to be called four core principles, but these should not be confused with the four principles of biomedical ethics from Beauchamp and Childress. (I am not sure if this approach to the presentation of the CRC was influenced by the success of Beauchamp and Childress, but if so only supports the claim that imitation is the sincerest form of flattery.) The UN CRC four core principles are:

- **non-discrimination** whether based on the child's or the child's parents' race, sex, political beliefs, language, ethnicity, or disability: Article 2. Of the Beauchamp and Childress principles, this may be closest to Justice. Discrimination based on skin color, gender, sexual orientation, caste, or religion are still widespread and so this is an important and controversial principle to defend in many countries.
- devotion to the **best interest** of the child by parents as well as by guardians, courts, and welfare institutions: Article 3 (and 18). The article explicitly includes the safety and health of the child, and suitability of staff and competent supervision. This is most like the principle of Beneficence, but it also justifies the principle of Limited Parental Discretion. Article 18 adds the recognition of the right to childcare services for working parents. For more on the best interest standard, please refer to Chap. 4.

- the **right to life, survival, and development**: Article 6. For Americans it is worth noting that this "right to life" is not an anti-abortion screed but means the right to not be killed by parents (e.g. infanticide of newborn girls) or by warring tribes or clans (e.g. indoctrinated to be child soldiers). The idea of a right to development can include nutritional health and social health development as well as education. An updated definition might be the right to grow up and have a realistic opportunity to realize your potential. If one were to map this onto the six principles, it most closely fits with Non-Maleficence.
- respect for the views of the child or **right to be heard** in accordance with the age and maturity of the child: Article 12. While the article focuses on judicial and administrative proceedings, it is easily applied to medical decisions as well. This corresponds best to the suggested principle of Respect for Children as well as the principle of Patient Autonomy.

In addition, as a pediatric ethicist, some of the other CRC articles can be useful to highlight:

Article 7. Right to a name and nationality, and to know and be raised by your parents (which may suggest forbidding anonymous gamete donors).

Article 9. Forbids separation of children from their parents unless it for the best interest of the child. Even when it is decided by a court that it is in the child's best interest to be separated from their parents, it is the right of the child to maintain personal relations and direct contact with them. If the state has imprisoned, deported, and executed a parent, the child has the right to know the facts. (Many of the ICE policies under the Trump administration likely violated this Article.)

Article 18.3. Which may require child-care be provided for working parents. (Something that has been yet to be achieved in the US as of 2021.)

Article 24.2c. Which may forbid exposing children to malnutrition, but also to many types of environmental pollution including unsafe drinking water or air pollution. All of these may increase the risk of asthma, developmental delays, or premature birth, and all have higher prevalence in poor and marginalized communities.

Article 28.2. Which may forbid many types of school discipline or corporal punishment (some of which are still legally practiced in parts of the US). This article includes the state ought to encourage secondary education that is free (publicly funded) and encourage attendance and the reduction of drop-out rates.

Article 29.1 d–e. Which requires childhood education include teaching equality of the sexes, respect for indigenous peoples, and tolerance to prepare them for life in a free society (d) and respect for the natural environment (e).

Article 32. Requires setting minimum ages and work hours and other workplace protections for children so that work does not endanger them or their education and development (which may forbid practices still common in much of Asia and Africa).

Article 34. States and parents must protect children from sexual abuse, trafficking, and use for child pornography and child prostitution.

Article 37. There should be no death sentence *or* life in prison without parole for crimes committed by children, and a right to a lawyer, a fair trial, and to see a judge (*habeas corpus*).

Article 38. Nations must protect children from armed conflicts, and forbid child soldiers younger than 15 (and preferably 18).

5.4   Preliminary Comparisons and Conclusions

Physicians in the US will be more familiar with the principles of biomedical ethics since the US is the one country that has resisted signing the UN CRC. The principles approach also predominates in most of the English-speaking world. On the other hand, physicians throughout the rest of the world may find the UN CRC more familiar. But it is not a competition, and both are valuable ways to accomplish an important goal: to have an external higher-order (or objective) values-check on the biases we all tacitly inherit from our own particular religious, cultural, and family background.

Using the UN CRC helps fill in the many ways that pediatrics can be seen as an ethical enterprise, and the many ways that a society may still treat minors as second class citizens who can be easily abused and denied their rights. It should not be a difficult ethical change in professional identity to progress from thinking "I'm 'just' a pediatrician" to "As a member of my professional community I have a responsibility to speak out for justice, including better treatment for all children, rich or poor, boy or girl, black or white."

While Justice is one of the four principles of Beauchamp and Childress, their narrow biomedical focus on clinical medicine leads to seeing Justice as primarily concerned with the right to health care, while using the UN CRC as a foundation for Pediatric ethics gives a far wider sense of the many ways that children need and deserve our protection and active advocacy besides care in the clinic.

One might also fairly conclude that both systems will yield similar recommendations for how to act or what policies to support, and so there is no dichotomy here. Though arrived at by different means, ethics should always lead to one set of answers, and so the significant overlap between these two approaches lends reassurance that both are on the correct path. One approach, the principles, may be more useful for individual decisions about individual patient care decisions, while the other, the rights of children, may be more useful for social policy level decisions. But ethics should be concerned with both domains.

Children as a class are all vulnerable or marginalized, and because of developmental biology and psychology all harms to them risk lifelong damage. Childhood is a perilous time, and if there are many adults who are ignorant or unethical it is most likely because they were themselves subject to misinformation, brutality or abuse as children. Hence to improve any society, including our own, we must start with improving how we treat children and until we have the perfect society there will always be resistance, sometimes fierce, from people who have drawn the wrong lessons from the failures of their own parents and social institutions.

Guiding principles for the ethical care of school age children (pre-teens and teens)

Emerging capacity and consent
- Respect the wishes of children as they develop and mature
- Respect includes both honest information and gentle non-coercive interactions to the degree possible starting at age 6 or 7 (assent)
- Evaluate decision-making capacity for all teens on a case-by-case basis, taking into consideration the patient's age, developmental level, and ability to *understand and appreciate* (1) the risks and benefits of the recommended treatment (2), any available alternatives, and the (3) consequences of their decision
- According to the AAP many or most minors have substantial capacity to make their own medical decisions by age 14
- Even before that, minors have the right to participate in the decision-making in a developmentally appropriate way, and not be excluded or deceived

---

Privacy and confidentiality
- Ensure a private environment for some portion of each patient encounter starting at age 6 or 7
- Meet with teens (children 13 and over) alone for a confidential interview (and physical exam, if appropriate); start to do this as part of routine care during early childhood (age 6 or 7) to normalize it and build trust in the confidentiality of the physician–patient relationship
- Explain to teens their right to confidential health services for issues involving mental health, reproductive and sexual health, and substance abuse
- Remember the child is your patient, not the parent/s or guardian/s
- Explain to parents the importance of confidentiality in order for their child to talk honestly with you, and inform them that you are allowed to share information with them only if the teen agrees or if you judge the minor is at risk of imminent and irreversible harm

---

References

American Academy of Pediatrics (AAP). 1994. AAP, guidelines on forgoing life-sustaining treatment. *Pediatrics* 93 (3): 532–536.

Appelbaum, P.S. 2007. Assessment of patients' competence to consent to treatment. *NEJM* 357 (18): 1834–1840.

Beauchamp, T.L., and J.F. Childress. 2019. *Principles of biomedical ethics*, 8th ed. Oxford: Oxford University Press. (The first edition was published in 1979.)

Diekema, D. 2004. Parental refusals of medical treatment: The Harm Principle as a threshold for state intervention. *Theoretical Medicine* 25: 243–264.

Gillam, L. 2016. The zone of parental discretion. *Clinical Ethics* 11 (1): 1–8.

Leikin, Sanford L. 1983. Minors assent or dissent to medical treatment. *Pediatrics* 102: 2.

Spike, J.P. 2017. Informed consent is the essence of capacity assessment. *Journal of Law, Medicine, and Ethics* 45: 95–105.

UNCRC. 1990. Convention on the rights of the child. https://www.ohchr.org/Documents/Professio nalInterest/crc.pdf. Accessed 9 September 2020.

Weithorn, L.A. 2020. When does a minor's legal competence to make health care decisions matter? *Pediatrics* 146 (s1): 25–32.

Weithorn, L.A., and S.B. Campbell. 1982. The competency of children and adolescents to make informed treatment decisions. *Child Development* 53 (6): 1589–1598.

Wilkerson, I. 2020. *Caste: The origins of our discontents*. London: Random House.

Further Reading

Fleischman, A.R. 2016. *Pediatric ethics*. Oxford University Press.

Lo, B. 2013. *Resolving ethical dilemmas: A guide for clinicians*, 5th ed. Philadelphia: Wolters Kluwer Health and Lippincott Williams and Wilkins.

Chapter 6
A Contextual Architecture of Praxis in Pediatric Case Consultation

C. M. Nelson and R. Posen

Abstract The relationship between ethical theory and practical judgment has been a focus of concern for many centuries. This chapter will outline a description and explanation of an approach to pediatric bioethics that will give a big picture involving both theoretical wisdom (*sophia*) and practical wisdom (*phronesis*) concerning moral beliefs, attitudes, and behavior. It will draw on Jonsen's simile concerning balloons and bicycles, viewing the moral landscape from above as if in a hot air balloon. The balloonist can communicate to the earthbound specifics about topography and the general landmarks one might encounter. The cyclist has detailed insight into navigating the bike trail taking into account the rough terrain, the negotiation of sharp curves and hills, hairpin turns, and unexpected forked intersections. Pediatric cases will be examined employing critical elements or tools of ethics that include process vision, neo-casuistry, human dignity and organizational integration. We intend our analysis of pediatric bioethics to sharpen the view of the horizon far and wide like the balloonist surveying impassable rivers, roads that take one nowhere and like the bicyclist on the rugged trail who has to constantly maneuver impending cliffs and sheer drops.

Keywords *Sophia* · *Phronesis* · Process vision · Neo-casuistry · Human dignity · Somebodyness · Organizational integration · Community of concern

6.1 Introduction

This chapter will emphasize the need to develop both theoretical and practical wisdom to investigate ethical issues in pediatrics. A description and explanation of this practical wisdom will be provided that helps clinicians and ethicists embrace an approach

---

Ideas in this chapter are developed from the dissertation completed by Nelson (1997).

C. M. Nelson (✉) · R. Posen
St. Bernardine Medical Center, San Bernardino, CA, USA

R. Posen
e-mail: rposen@att.net

© Springer Nature Switzerland AG 2022
N. Nortjé and J. C. Bester (eds.), *Pediatric Ethics: Theory and Practice*,
The International Library of Bioethics 89,
https://doi.org/10.1007/978-3-030-86182-7_6

to pediatric bioethics that will give a big picture concerning moral beliefs, attitudes, and behavior much like viewing the moral landscape from both a hot air balloon and a bicycle, described by Jonsen in his simile concerning balloons and bicycles.[1] The balloonist can communicate to the earthbound specifics about topography and the general landmarks one might encounter. The cyclist has detailed insight into navigating the bike trail taking into account the rough terrain, the negotiation of sharp curves and hills, hairpin turns, and unexpected forked intersections. This essay will describe both the balloon of moral imagination, *sophia*, and the practical feel of the road in cycling the path of moral action, *phronesis*. Albert Jonsen reminds us that Plato worried about the problem when he described it in his sixth book of the Republic (Jonsen 1991). Those who have perfect vision or *theoria* have an understanding of the outer world "that orders the laws about beauty, goodness, and justice in the world" (Plato 1945). Aristotle also distinguished between theoretical wisdom (*sophia*) and practical wisdom (*phronesis*). Aristotle noted that philosophers Anaxagoras and Thales had *Sophia* "of extraordinary, wonderful, difficult and super-human things, but such knowledge is useless because the good they are seeking is not human. *Phronesis* on the other hand is concerned with human affairs and with matters about which deliberation is possible" (Aristotle 1962).

Pediatric cases will be examined employing critical elements of consultation that include process vision, neo-casuistry, human dignity and a systematic organizational integration. These critical elements are likened to both the balloonist's view forming a moral imagination or the *Sophia* of our ethical approach and the more detailed rough terrain or the *phronesis* of our ethical approach. We intend our analysis of the pediatric cases examined to sharpen the vision of the horizon far and wide like the balloonist surveying impassable rivers, roads that take one nowhere or impending cliffs with sheer drops while taking seriously the detailed contextual particulars of each case. The context of each particular case will determine the way in which **process vision** was helpful, the way in which **neo-casuistry** was practical, the way **human dignity** as **somebodyness** was privileged, and the way **organizational integration** served case review. These tools of ethics will be contextual and be likened to the bicyclist feeling the grinding bumps and turns of the bike path. They form the tools of practical wisdom so needed to maneuver over rough terrain and form the wisdom or phronesis of our ethical endeavor.

6.2   Conceptual Analysis and Clarification

A process vision emphasizes connectedness and a web of relationship (Suchocki 1992; Whitehead 1978). Parsing an ethics case involves moving the focus beyond a single moment of the patient's life but connects moments that have succeeded

---

[1] The philosopher is hoisted high above the landscape and is above the fray and able to see far into the distance. The physician petals furiously on a bicycle negotiating the ethical curves and potholes that suddenly appear in clinical practice.

from one another in an ongoing manner. By connecting the moments of a patient's history we come to understand how to approach the ethical question emerging from a patient's unresolved situation. Often, the moments of a patient's history present like the many facets of a diamond. We must broaden our perspective to take in all the facets. It is an assembly of facets that present a clearer vision of the diamond. It takes all important stakeholders—a community of concern (Glaser 1994)— to work together in a connected web of relationship to have a clear vision of the patient's many components that are present within the ethics case.

A process model further helps us understand that ethics cases often become dissected and need to have a methodology to interconnect what tends to become separate and independent. Sometimes in medicine we have dealt well with treating symptoms in the attempt to eliminate disease at the expense of being person-centered. A process model reminds us that the parts of a patient's hospital course and medical history are no longer the basic elements of which the patient's world is composed but comprise the processes and relations that form a synthesis of the patient's complete narrative (Whitehead 1978).

Modern casuistry, also known as neo-casuistry is a method of moral analysis using reasoning, paradigm cases, and analogies to help formulate moral obligations like the ones present in pediatric bioethics. Neo-casuistry's contextual flexibility, its dependence on a synthesized understanding of experience, a reliance on narrative, and an emphasis on negotiation for moral insight facilitates the ethical inquiry to remain focused on concrete particularities of the case. Neo-casuistry is person-centered in supporting strategies of life in concrete situations and celebrates interconnectedness as it stresses an ethics of discretion. This ethics of discretion forms a careful balance between a narrow dogmatism and a shallow relativism (Kirk 1927). The ethical challenge of every case needs to be discussed in the concrete, in relation to a particular paradigm example with assurance that ethical principles involved are adapted to the specific circumstances of the case.

Neo-casuistry also stresses the importance of analogy. We do not abandon the guidance of principles of ethics (i.e. autonomy, justice, beneficence, fairness, truth-telling, non-maleficence) but strive to decide what is to be done in this case or that case guided by the discussion of cases in the light that is cast upon the principle from the ultimate particular of the case. Discussion of the case should help to ground speculative questions, include succinct presentation of principles, acceptance of probable opinions in certain circumstances, and resolution by solid argument.

The heart of the moral experience is not found in the mastery of general rules and principles, however sound and well-reasoned they may appear. It is rather to be found in the wisdom that comes from seeing how the ideas behind the rules and theoretical principles work in the course of patients' lives and in particular seeing what is involved in insisting on waiving this or that rule in one or another recombination of circumstances. Moral knowledge is essentially particular and sound resolutions of moral problems and dilemmas are rooted in a concrete understanding of specific cases and circumstances.

Human dignity and its demands and the way it is privileged as a tool of ethics deliberation forms the next step in developing the moral imagination of the balloonist for our approach to pediatric cases. Human dignity is certainly a multivalent word as is both something we become and something we are. We attempt to give human life dignity through the personal formation of character and the constructive work of character building. Human dignity is also an innate quality, the 'somebodyness' (Baker-Fletcher 1993) that is a quality of being human. Each person possesses a dignity as a quality of humanity that could grow and be diminished but could never be extinguished regardless of the state of development that has been cultivated in individual character-building. In pediatric patients we are focusing on fostering the dignity of becoming with the therapeutic purpose emphasizing a certain treatment course that would be therapeutically successful and help this child potentially pursue whatever gifts or talents that are self-fulfilling and meaningful. We approach the place of human dignity in our discussion by being more succinct about the demands of human dignity. This dignity or somebodyness does not come from a list of givens or universals for all times and ages but the understanding that dignity is a mystery that unfolds in significant ways in all times and places. Jack Glaser reminds us that the ethical wisdom flowing out of respect for human dignity is the kind of knowing we have of friendships and sunsets. It is prized and protected. All its dimensions come twisted and interwoven being both personal and impersonal, individual and social, emotional and mental, spiritual and physical (Glaser 1985). The human flourishing at the heart of human dignity is multifaceted as it forces open all forms of oppression that degrade, defile, and negate. The many facets of human dignity challenge us to live and relate to other people while realizing that the challenge involves genuine and fundamental understanding between particularities. One must not abandon their particular individual way of belief or expression but strive to crossover and connect with the mindful importance of Martin Buber's "I-Thou" (Buber 1996).[2]

The community of concern (Glaser 1994) is the primary vehicle of discernment that helps discover the many facets of human dignity and employs and develops the wisdom that other people bring as gift to the pediatric case consult. The community of concern, a primary agent of ethical discernment, includes those stakeholders who have insight into the many-sided reality of the case's ethical question. The rich complexity involved in answering ethical questions of any pediatrics case exists beyond any one person's grasp as there are many facets of human dignity. A moral tragedy occurs when you have acted in a certain way and after you have acted, you come to the conclusion that upon reflection, had you thought about it before you acted, you would have acted differently (Dyck 1977). The many facets of human dignity and its demands call us to live and relate to each other while realizing that the call involves genuine and fundamental understanding between particular points of view and so we are challenged to not abandon the individual particularity of the individual but in humility connect with the other viewpoints with the integrity of Buber's I-Thou. The community of concern reminds us that no one person is

---

[2] For Martin Buber, *Thou* addresses another person not as an object but as a presence and never as a means to one's own end. The I-Thou relationship is characterized by mutuality, directness, transparency, presentness, intensity and a deep connection distinguished by joy and meaning beyond words.

capable of seeing the facets of the many-sided reality of human dignity. There is a rich complexity at stake beyond one person's grasp. Ethical wisdom flows from a reflective community while ethical expertise is more the domain of the ethicist. One vital aspect of this domain consists of enabling, promoting, and facilitating the reflection and discernment of that community of concern whose recommendation is determined by such reflection (Schuller 1980). If there is an ethical challenge, we gather the appropriate community of concern—all treatment team and family stakeholders who have important insight to offer for the case consult.

Organizational integration or system draws heavily from a sapiential dialogue that highlights the encounter between process vision, neo-casuistry, human dignity. It is not a simple endorsement of these tools of ethics but a recognition that all of these components working together in the crucible of ethical deliberation help the community of concern form a moral sense that draws upon wisdom of the heart also understood as *sapientia cordis* (Steinbock 2014). The heart of the moral experience is found in the wisdom that comes from seeing how the ideas behind the rules and theoretical principles work in the context of the pediatric patient's life in the now and foreseeable future. The heart of the moral experience is not found in the mastery of general rules and principles however sound and well-reasoned they appear (Nelson 2012). Moral knowledge is always particular. We were reminded earlier that sound resolutions of moral problems are rooted in a concrete understanding of specific cases and circumstances. The use of analogical reasoning, paradigm cases, and practical judgments go beyond the situations covered by any single rule. Insights from cases and case discussion must be connected to appropriate sets of concepts, principles, and theories. Landmark bioethics cases are instructive and form a shared resource that becomes integral to the way the community of concern thinks and draws conclusions of how to draw on concepts, principle, and theories. Cases also profoundly influence our standards of fairness, negligence, paternalism, and the good of the patient.

There are vital tools for discerning human dignity and its demands in a disciplined and systematically consistent manner. These include fact vetting, gathering the appropriate participants for discerning the rich complexity and many sided reality of human dignity, calling special attention to words, definitions, and language within the setting of our moral space, employing distinctions that assist us in describing realities that are different but appear the same and the use of principles that clearly help us lay out key value assumptions that are critical for our understanding of human dignity. Finally, there is a level of organization that ethics brings to the efforts of the community of concern and to the service of human dignity. This ethical protocol provides a way of touching and integrating all the critical elements or tools of ethics and combines expertise in theory deployment and moral insight. There is not an over-reliance on code-like theories and law-like principles. Margaret Urban Walker stresses the importance of weaving both narrative and negotiation of which our model is open. Story and individual history of every pediatric patient help highlight a holistic and inclusive approach for moral decision-making (Walker 1993).

6.3 Practical Application

The next section of our discussion will examine two cases brought before local ethics committees. As we begin our case discussions, we need to establish an understanding of which kind of case presents the need of an expanded community of concern beyond the attending physician and the parents of the pediatric patient.

When there is a patient care conflict in a clinical setting and the matter is brought to the attention of an ethics committee member, it must be decided if the conflict is an ethical problem or an ethical challenge. An ethical problem would be understood as a situation that did not pose any dilemma or conflict but was more a matter of physician or patient education. Examples of ethical problems would be when can pediatric patients speak for themselves? Who is the decision-maker for a pediatric patient? How do we document surrogate selection? Is it against the law to remove a respirator or a feeding tube from a pediatric patient?

On other occasions principles sometimes conflict and cause an ethical challenge. These are situations when values cannot all be respected to the extent that we would like them to be. Hard choices are required to resolve dignity conflicts when not all dignity can be honored. Principles themselves do not attempt to dissolve the ambiguity and tension in tough cases. They often help dissect the complexity of challenging cases and help form a recommendation that promotes value priorities. The following two cases are examples of two ethical challenges where a collegial systematic approach was guided by the critical elements or tools of ethics discussed in this essay. The methodology employed in examining the two pediatric cases emphasized process vision, neo-casuistry, human dignity understood as 'somebodyness' and a systematic organizational integration.

Case 1

The details of the following case have been changed to mask the identity of the individuals involved. The patient was a 16-year-old single female who was 31–32 weeks pregnant. A recent sonogram indicated the unborn baby presented with a diagnosis of Arnold Chiari Malformation[3] possibly involving massive hydrocephalus and spina bifida.

The expectant patient and her mother were informed of the baby's diagnosis and the attending neonatologist reviewed several treatment options with both the mother and grandmother. Although informed that the child may live, the expectant patient wanted to wait and see if a more accurate prognosis for her baby could be better ascertained after delivery. The patient's mother urged her to request no surgical intervention once the baby was born. The patient's mother, the expectant newborn's grandmother, was originally demanding an abortion but this treatment option was not entertained as the pregnancy was beyond the statutory limit for abortion. The father of the unborn baby was still in relationship with the young mother and was a soldier stationed about 150 miles away. The father was over 18 and was not informed about

---

[3] Arnold Chiari Malformation is a condition in which brain tissue extends into the spinal canal present at birth. The condition occurs when part of the skull is abnormally small or misshapen.

the condition of his pregnant girlfriend. The patient wanted the biological father to be included in the decision-making process. Once informed, the biological father fully supported the wishes of the patient to wait and see if a more accurate prognosis could be ascertained after delivery.

The prognosis in these cases vary. Without surgical treatment less than 20% survive for as long as two years. For newborns with severe hydrocephalus the natural mortality is even higher as 50% of patients die within the first month of life and the remainder within six months. The management of Arnold Chari Malformation is a complicated matter and there are no easy answers. Treatment plans depend on the form, severity, and associated challenges of Arnold Chiari malformation (Sastry et al. 2020; Talamonti et al. 2020).

The attending obstetrician recommended inducing labor. In her view, this would prevent the patient from undergoing a C-section due to concern about the head size of the infant. There was also concern that the sack surrounding the protruding spinal cord may rupture causing severe complications. The neonatologist discussed life support measures with the patient and grandmother as a ventilator probably would be required due to prematurity. The possibility of respiratory distress syndrome was also a concern. The prognosis for this baby included the discussion about the percentage probability of mental disability and the belief that a more accurate prognosis for this baby could be better ascertained after delivery.

6.4   Conceptual Analysis and Clarification

The context of each particular case will determine the way in which *process vision* was helpful, the way in which *neo-casuistry* was practical, the way *human dignity* as *somebodyness* was privileged, and the way *organizational integration* served case review. These tools of ethics will be contextual and be likened to the bicyclist feeling the grinding bumps and turns of the bike path. They form the tools of practical wisdom so needed to maneuver over rough terrain and form the wisdom or phronesis of our ethical endeavor.

Process Vision

Process thought plays a vital role in encouraging a rich and meaningful quality of relating. This understanding advocates for an inclusive approach to ethics that includes making moral choices in a particular situation partly dependent in how the situation is illustrated. The situational perspective helps in determining the substance of the ethics recommendation as the grounds of moral judgment have as much to do with how a recommended choice is made as with what the choice is. Gray reminds us that the ethical chooses for the present while maintaining that the future is open (refer to Chap. 3) and that ethical decisions cannot be mere repetitions of the past but need to embody interpretations of the past that are influenced by the demands of the present and the hopes of the future (Gray 1983).

For this case, a process vision would include the value positions of all stakeholders in the community of concern involved in this family. What value concerns would the presumed father offer in the deliberation? For the unborn and the mother an important focus would be the most reachable, hopeful, and possible outcome for both of them. Can a better prognosis for the expected newborn can be ascertained after delivery? Care should be taken to present an honest and thorough description about the clinical precedents involved in treatment outcomes for cases such as these.

Neo-casuistry

Neo-casuistry assists a process vision by offering a way to express an interrelated and contextual bioethical deliberation. It employs rhetoric and argument, analogical imagination as a tool for paradigm comparison, plays an important role for placing paradigms on a similar plane, and stresses options and recommendations. A most helpful way to honor the specific context of case one would be to maintain an open posture for deciding on treatment or no treatment by remaining open to what is learned from observing the infant's clinical presentation post-delivery.

An important recognized California Supreme Court paradigm case allowed parental refusal of risky treatment (In Re. Philip B., California, 1979; [Civ. No. 44291. First Dist., Div. Four. May 8, 1979.]). In this case it was decided that the state may insist on medical care only after these are considered:

1. The seriousness of the harm the child is suffering or the substantial likelihood that he or she will suffer serious harm.
2. The evaluation and recommendation of the treatment team.
3. Risks involved in the treatment of surgery.
4. Preferences of the child through substituted judgment of the parents.

In addition, important maxims and circumstances involved in reviewing case one would need to be used to address the issue of the moral worth of the baby's life weighed against the cruel prolongation of dying as well as intrinsic and extrinsic probabilities of benefit and burden on the baby's life as a result of the course of treatment discussed. Also, the patient wanted the input of the biological father. The opinion of the patient's mother may carry some moral weight, but must not override the wishes of the patient.

The extrinsic probability of a moral judgment in this case involves referencing the community standard of care based upon the valued experience of the medical community who have treated similar cases and documented their results and opinions. This involves what treatments might be deemed beneficial and non-beneficial from a benefit/burden perspective.

In case one it would be important to broaden the community of concern to include the medical expertise of neonatologists in the ethical deliberation. The ethics committee as a moral community can help the patient articulate her wishes with the guidance of clinical experts who have case experience in similar cases. Forming guiding paradigms that assist in clarifying patient values can also be helpful. The conclusion of the recommendation involves a resolution disclosing the moral appropriateness of acting in one particular way or another taking into account all of this information.

In summary, the neo-casuistical method of examining this case involves an ordering of cases by paradigm and analogy, appeals to maxims and analyses of circumstances, the qualification of opinions, the accumulation of multiple arguments, and the statement of practical moral problems in light of this list of considerations.

6.5   Human Dignity as *Somebodyness*

Human dignity as somebodyness holds an esteemed position within the holistic nature of ethical deliberation. In pediatric cases somebodyness reminds the ethical community that parsing a case always needs to remain life centered. In addition to being life centered, reflecting on somebodyness adds perspective and dimension to clinical ethics and shapes the ongoing quest of examining important questions about human nature, human purpose, and human destiny. Being life centered in an existential and contextual way will add clarity to what degree of care required for the newly born patient with complex medical challenges.

Important to case one is the obligation to honor the dignity of the most immediate community of concern which includes the infant, mother, and father. How would privileging human dignity help form the appropriate rhetoric for a casuistical argument? We must entertain likely treatment scenarios that provide a successful outcome for the infant including both optimum outcomes for mentation and ambulation. Are there candidate therapies or specific invasive treatments that could help this child lead a good quality of life to pursue self-fulfillment? Would the possible treatment options burden the parents with ongoing calamitous medical expense? Do the parents have any particular spiritual tradition guiding their value judgments? How do the parents understand the enduring questions about purpose and destiny for their child? We must strive to honor their held values and honor the good of this patient (Pellegrino 2008)?

6.6   Organizational Integration

The level of integration or ethical process serves to coalesce the critical elements that assist the moral conversation. Have we gathered all the pertinent medical information? How do we ascertain the patient preferences by best interest standard and the helpful insight offered from the harm principle (Bester, 2019)? Does the family system of the patient (in this case the unborn child) need evaluation? What have we learned about this case from treatment team input? Are there any economic considerations to discuss? Is there an ideal picture that describes this case outcome? What are some of the "howevers"—things that make this case distinctive? Values, principles, and existential situations all help dissect the complexity of case one and hopefully help us move toward recommendations that promote thoughtful options.

The ethics committee recommendation for case one included waiting until delivery and evaluate appropriate treatment plans. Included in this recommendation would be to abide by the parents' wishes for their newborn. The benefit/burden of treatment options depend on the prognosis for the newborn after birth. Important principles such as beneficence, autonomy, and non-maleficence all come into discussion: autonomy of the parents to choose from the treatment options presented by the medical team and severity of the child's medical picture will define the beneficence or non-maleficence involved in this case.

After reaching consensus the following recommendation was presented to the treatment team by the ethics committee:

1. Encourage counseling for both parents.
2. Keep parents updated frequently regarding all medical facts, treatment options, and surgical interventions available; inform the parents that a more accurate prognosis will be ascertained after delivery.
3. Abide by the parents' wishes.
4. Maintain supportive care for the infant until treatment or non-treatment plans were chosen by the parents.
5. Contact the biological father as requested by the patient.

> After delivery both parents agreed to have their baby undergo corrective surgery due to the better than expected prognosis given by the treatment team. More accurate information was ascertained about the child's condition. At last contact with the family, it appeared the child would have normal intellect and would have minimal ambulatory challenge.

Guiding principles for the ethical care of adolescents

Emerging capacity and consent
- Recognize and respect the wishes of adolescents as they develop and mature evaluate decision-making capacity on a case-by-case basis, taking into consideration the patient's age, developmental level, ability to understand risks, benefits, the and consequences of a decision and its alternative

Privacy and confidentiality
- Ensure a private environment for the patient encounter
- Meet with adolescents alone for a confidential interview and physical exam (if appropriate); start to do this as part of routine care during early adolescence
- Deliver confidential health services for issues involving mental health, reproductive and sexual health and substance abuse to consenting adolescents

Case 2

The following case has been changed to mask the identity of the individual described in the ethics consult. An ethical challenge involved a pregnant 22-year-old who had a history of fetal demise at 12 weeks from a prior pregnancy. She had an amniocentesis at 32 weeks resulting in a discovery of Patau Syndrome, also known as trisomy 13.[4] In addition, a focused ultrasound revealed holoprosencephaly with cleft lip and palate. The expectant mother was admitted in labor at 35 weeks. More than 80% of babies with Trisomy 13 die in the first month of life, 99% in the first year of life. Surgical interventions are generally withheld for the first few months of life because of the high mortality rates of babies with trisomy 13, parents and medical personnel must carefully weigh decisions about extraordinary life-prolonging measures against the severity of the neurological and physical defects that are present after birth and the likelihood of postsurgical recovery or prolonged survival.

Medical management of children with Trisomy 13 is considered on a case-by-case basis and depends on the individual circumstances of the patient. Treatment of Patau Syndrome focuses on the particular physical problems with which each child is born. Many infants have difficulty surviving the first few days or weeks due to severe neurological problems or complex heart defects. In addition, surgery may be necessary to repair heart defects or cleft lip and cleft palate. Physical, occupational, and speech therapy often help individuals with Patau syndrome reach their full developmental potential.

The fetal diagnosis was thoroughly explained to both parents. They insisted on full resuscitative measures after birth including mechanical ventilation, fluids, cardiopulmonary support and surgery as indicated. Comfort measures with supportive care was offered by the treatment team as a recommendation in place of the treatment course the parents desired, but they declined.

Process Vision

Within a process approach to case two the combination of situational perspective, grounds for moral judgment and future hope for the infant together help from an integrated understanding of how to shape an ethical recommendation in this case. The situational perspective along with how the ethics committee as the community of concern comes to form a recommendation is important. As we attempt to give guidance for recommending the most ethically appropriate action for this newborn, we must be open to what future this infant would likely have. We must be honest about the experiences of the treatment team as they have had with prior cases similar to this one. We must also consider what expectations and hopes the parents of the newborn hold. Are the parents' hopes based upon the what they wish in the best of circumstances or on what the demands of the newborn's condition will present after birth?

Neo-casuistry

The insight from neo-casuistry highlights the importance of context in ethical deliberation by emphasizing an analogical approach for paradigm comparison clarifying

---

[4] Trisomy 13, also called Patau syndrome, is a chromosomal condition associated with severe intellectual disability and physical abnormalities in many parts of the body. Individuals with Trisomy 13 often have heart defects, brain or spinal cord abnormalities, very small or poorly developed eyed, extra fingers or toes, an opening in the lip with or without an opening in the roof of the mouth, and weak muscle tone.

similarity of different but comparable cases, the use of rhetoric and argument, and the description of options and recommendations. A recognized analogical paradigm for case two, the Baby Doe Regulations (Espinoza et al. 2010; Baylis and Hellman 2002), draws its insight from U.S. Law for the Prevention of Child Abuse and Treatment of Children (Paris 2003). Some highlights include:

1. Physicians must not recommend a treatment that is against their professional conscience.
2. Physicians are not obligated to provide a treatment that only prolongs death.
3. Ethics Committee diligence to prevent abuse.
4. Following current community standards that emphasize potential medical benefit.
5. That the suspension or denial of proposed medical treatment is not prohibited by law.

Maxims and circumstances involved in clarifying ethical decision-making for case two would focus on respecting quality of life versus the prolongation of the dying process. Circumstances and probability would help discern the likelihood of proposed treatment options bringing about the desired outcome for the newborn. In addition, the extrinsic probability involved in forming a recommendation for case two would involve seeking agreement with respected medical expertise. The wisdom of the medical community who have treated similar cases and professionally codified their results and opinions should be embraced. Valuable guidance can also be acquired from the work of the legal community's codified legal decisions.

6.7   Human Dignity as *Somebodyness*

Human dignity as somebodyness holds a prominent place within the holistic nature of ethical deliberation in case two. To embrace somebodyness as an integral part of parsing this case celebrates the importance of being life-centered in an existential and contextual way for this unborn baby diagnosed with trisomy 13. Sometimes pediatric cases seem challenging as no clear best answer seems apparent. Potential inappropriate treatment may include treatments that have a small chance of accomplishing the physiological effect sought by the parents of an infant or when clinicians believe that competing ethical considerations justify not providing them (Bosslet et al. 2015). Treatment requests can be characterized into four categories:

1. Non-beneficial: not able to achieve desired physiological goal.
2. Legally proscribed: prohibited by law (i.e. euthanasia).
3. Legally discretionary: physicians can refuse to administer treatments based upon laws or precedents.
4. Potentially inappropriate: may prolong suffering or the dying process.

Parents are morally entitled to request for their newborn's appropriate medical treatment, especially in emergent circumstances involving an unclear prognosis.

Physicians are also moral agents and should not be required to provide clinical treatment contrary to indicated medical treatment (i.e. community standard of care) as this upholds their vow to do no harm (Detora and Cummings 2015).

6.8   Organizational Integration

As in case one, the level of organization or ethical protocol serves to integrate critical elements of ethical deliberation to assist the moral conversation. Have we gathered all the pertinent medical information? How do we ascertain the patient preferences by substituted judgment, best interest standard, or the harm principle? What have we learned about this case from treatment team experience from other similar cases? Are there any ongoing economic considerations to discuss that may deeply impact the parents? Is there an ideal picture for this case outcome and are they realizable? What are some of the "howevers"—things that make this case distinctive? Values, principles, and existential situations all help dissect the complexity of case one and hopefully help us move toward recommendations that promote thoughtful options.

The infant's clinical course after birth presented numerous challenges. Upon birth the infant was limp, apneic, cyanotic, and bradycardic. She was suctioned, intubated and underwent chest compressions including one dose of epinephrine. She had a low APGAR score (1–1-4) with poor reactivity. The baby had a severe cleft lip and palate, no nasal columella, hypotelorism, slanted eyes and low set rotated ears. Her presentation also included overlapping fingers, semi-lobar holoprosencephaly and she experienced ongoing seizures. She also had a double outlet right ventricle with a mal-aligned ventral septal defect and hydronephrosis. In the NICU she continued to have gastritis and bloody stools. TPN (total parenteral nutrition) progressed to NGT (nasogastric tube) feeds for her and she required elemental formula eventually requiring the use of diuretics. She was on a ventilator for 3–4 weeks and eventually weaned to room air. Her renal insufficiency continued with worsening somnolence, apnea, desaturation even on combined nasal cannula and face mask. Vital signs deteriorated progressing to cardiopulmonary arrest.

In this case the treatment team believed that comfort measures were indicated and not heroic interventions. The parents persisted in wanting full resuscitative measures including supportive surgery. One issue in case two became the treatment team's concern that the parents' wishes may include treatment options that are non-beneficial or ineffective. The benefits/burdens of treatment options depended on the severity of trisomy 13 complications after birth. Important principles such as beneficence, autonomy, and non-maleficence all come into discussion: autonomy of the parents to choose from the treatment options presented by the medical team and severity of the child's medical picture is an important indicator in how to seek beneficence and avoid non-maleficence. As the infant's hospital course progressed, her parents requested no reintubation and no CPR (cardiopulmonary resuscitation). During the ethics consultation it became apparent that extended family was an influence in maintaining the choice for heroic measures for this newborn. The infant's parents continued to refuse the initiation of comfort measures and opposed the use of low-dose narcotics to reduce end of life anxiety and pain until near the very last day of their daughter's life when it was apparent her demise was inevitable.

In this case process vision helped the ethics committee focus on the present and future expectations for this infant as well as the professional expectations of the experienced treatment team. The analogical approach of neo-casuistry helped ground the ethical recommendation by stressing the importance of the outcomes of similar cases while helping to clarify insight from the probabilities and specific circumstances of this infant's clinical experience. Somebodyness stressed how being life-centered is often contextual in upholding the vow to first, do no harm. And although organizational integration helped the medical team promote a thoughtful option for this infant, there was an enduring tension between what the medical team felt was appropriate and what the family embraced as the direction of treatment and care for their child.

6.9   Conclusion

This chapter outlined a descriptive approach to pediatric bioethics that included a big picture viewed from the hot air balloon of moral imagination that included process vision and neo-casuistry or *sophia*. We also examined the granular topography and the specific landmarks highlighted by somebodyness and organizational integration or *phronesis*. Together this *sophia* and *phronesis* formed a contextual architecture of praxis in pediatric case consultation.

References

Aristotle. 1962. *Nicomachean Ethics*. (Martin Ostwald, Trans.). The Bobbs-Merrill Company Inc. 6.1141b.
Baker-Fletcher, G. 1993. *Somebodyness*, 45, 56. Fortress Press.
Baylis, F., and J. Hellman. 2002. Ethics in perinatal and neonatal medicine. In Fanaroff AA, Martin JR. *Neonatal–Perinatal Medicine*. Mosby, 37.
Bester, J. 2019. The best interest standard and children: Clarifying a concept and responding to critics. *Journal of Medical Ethics* 45 (2): 117–124.
Bosslet, G. et al. 2015. Policy statement: Responding to requests for potentially inappropriate treatment in intensive care units. *Am J Respir Crit Care Med*
Buber, M. 1996. *I and Thou*. Touchstone, 51–180.
Detora, A. and Cummings, C. 2015. Ethics and law: practical applications in the neonatal intensive care unit. *NeoReviews* 384–392.
Dyck, A. 1977. *On Human Care: An Introduction to Ethics*, 28. Abingdon.
Espinoza, A. et al. 2010. Ethical challenges and decision-making at Neonatal Care Units. *Bol Med Hosp Infant Mex* March 9, 2010, 252.
Glaser, J. 1985. *Caring for the Special Child*, 1ff. Leaven Press.
Glaser, J. 1994. Three realms of Ethics. *Sheed and Ward* 26: 1–38.
Gray, J. 1983. *Process Ethics*, 48–71. University Press of America.

Jonsen, A. 1991. Of balloons and bicycles: the relationship between ethical theory and practical judgment. *Hastings Center Report.* September/October, 14–16.

Kirk, K. 1927. *Conscience and its Problems: An Introduction to Casuistry.* Longmans, Green and Company. 150–212.

Nelson, C. 1997. *Becoming Architects of a Morally Sensitive Postmodern Bioethics* (Publication No. 9805066) (Doctoral Dissertation, The Claremont Graduate University). UMI Dissertation Services.

Nelson, C. 2012. The familiar foundation and the fuller sense: Ethics consultation and narrative. *The Permanente Journal.* Spring, 60–63.

Paris, J. et al. 2003. Ethical and legal issues. In Goldsmith, P. and Karotin, H. (Eds.) *Assisted Ventilation of the Neonate.* Saunders, 81–90.

Pellegrino, E. 2008. *The Philosophy of Medicine Reborn.* University of Notre Dame Press.

Plato. 1945. *The Republic of Plato.* Oxford University Press. 6.484.

Sastry, R., R. Sufianov, et al. 2020. Chiari I malformation and pregnancy: A comprehensive review of the literature to address common questions and to guide management. *Acta Neurochirurgica.* https://doi.org/10.1007/s00701-020-04308-7.

Schuller, B. 1980. *Die Begrundung Sittlicher Urteile.* Second Edition. Patmos Verlag.

Steinbock, A. 2014. *Moral Emotions: Reclaiming the Evidence of the Heart.* Northwestern University Press.

Suchocki, M.H. 1992. *God, Christ, Church: A Practical Guide to Process Theology.* Crossroad Publishing Company.

Talamonti, G., E. Marcati, G. Gribaudi, et al. 2020. Acute presentation of Chiari 1 malformation in children. *Childs Nervous System* 36: 899–909. https://doi.org/10.1007/s00381-020-04540-7.

Walker, M. 1993. Keeping moral space open: New images of ethics consulting. *Hastings Center Report.* March–April, 33–40.

Whitehead, Alfred. 1978. *Process and Reality.* Reprint Edition. The Free Press.

Further Reading

Baker-Fletcher, Garth Kasimu. 1993. *Somebodyness: Martin Luther King and the Theory of Dignity.* Fortress Press

Jonsen, Albert and Stephen Toulmin. 1988. *The Abuse of Casuistry: A History of Moral Reasoning.* University of California Press.

Pellegrino, E. 2008. *The Philosophy of Medicine Reborn.* University of Notre Dame Press.

Schuller, Bruno. 1986. *Wholly Human.* Georgetown University Press.

Part II
Practice

Introduction to Part II

The following section of the book is dedicated to the implications of ethical theory and reasoning to specific questions in pediatric practice. Four different focus areas address pertinent issues in pediatric bioethics.

Focus area *One* looks at *Making Decisions about and for Children*. Chapters 7–12 address issues such as parental permission; childhood assent; shared decision-making; disclosing information to the child; parental refusal; requesting harmful treatments by parents; and the influence of religion and culture on decision-making.

The *Second* focus area will focus on *The Critically Ill Child*. Chapters 13–18 are devoted to issues related to pediatric bioethics during emergency care; critical needs; end-of-life; severe disabilities or impairments; receiving non-beneficial treatment; and euthanasia.

Focus area *Three* examines *New and Enduring Ethical Controversies*. The authors of chapters 19–25 examine issues such as genetic testing and screening; enhancement technologies; gender management; reproductive issues; AI and health; challenges between research and treatment; and predicting neurological impairments.

The *Fourth* and final focus area of this section is dedicated to *The Child and Society*. Chapters 26–29 look at issues pertaining to societal obligation to override parental decisions; childhood vaccination; overall obligation of society towards children; and resource allocation.

In each of these focus areas, chapters address questions that arise frequently in pediatric practice and in policymaking all around the world.

Chapter 7
Parental Permission, Childhood Assent, and Shared Decision-Making

S. L. Teti and T. M. Silber

Abstract Most minors are not able to give informed consent to be treated. Thus, for most of the pediatric years, the question is not about respecting patient autonomy, but rather determining who has the authority to decide *for* the child. This is accomplished through a therapeutic partnership: a parent or duly appointed surrogate representing the child's interests makes decisions for the child and gives permission for the child to be treated. The health care provider likewise has a duty to the child. Together, parents and providers are co-fiduciaries of the child's health care and wellbeing. However, the ethical authority to decide for another is substantially less certain than is self-regarding decision-making. In practical terms, certainty in this sense translates to the degree to which the decision is reviewable by others; the more harm that may be imposed in relation to the expected net benefit, the more the decision is justifiably questioned by others. Shared decision-making (SDM) attempts to address this concern by mediating authority among the co-fiduciaries. It is the role of the parents to express their values and normative preferences for their child's care, and the role of the health care team to present the best treatment options available, and make recommendations in light of the values and goals expressed by the parents. To ameliorate the ethical problems inherent in deciding for another, both parents and providers should involve the minor patient in health care decisions to the extent possible in an age-appropriate manner. Assent should be sought from adolescents and teenagers with capacity, and any decision to override those wishes should meet high standards of justification.

Discussion of ethical work done by parental permission and childhood assent, role of parents as co-fiduciary, apply theoretical discussion to scope and limits of parental decision-making.

S. L. Teti (✉)
Center for Clinical and Organizational Ethics, Inova Fairfax Medical Campus, Falls Church, VA, USA
e-mail: Stowe.Teti@inova.org

T. M. Silber
Children's National Medical Center, Washington, DC, USA
e-mail: TSILBER@childrensnational.org

© Springer Nature Switzerland AG 2022
N. Nortjé and J. C. Bester (eds.), *Pediatric Ethics: Theory and Practice*,
The International Library of Bioethics 89,
https://doi.org/10.1007/978-3-030-86182-7_7

Keywords Assent · Authority · Co-fiduciary · Permission · Shared decision-making

7.1 Introduction

7.1.1 The Ethical Structure of Therapeutic Relationships

In order to understand the ethical issues at work regarding parental permission, childhood assent, and shared decision-making one must examine the ethical structure of therapeutic relations generally. In the simplest terms, *informed consent* and *trust* form the bedrock of the therapeutic relationship between health care providers and patients (O'Neill 2002). Drawing on English common law, (Birchley 2014) patients have the right to self-determination, to be 'autonomous' (Whitney et al. 2004) However, patients are generally not sufficiently knowledgeable to make well-informed health care decisions due to the complexities of modern medicine; patients must trust that health care providers will advise, and act, in their best interest (Teti et al. 2017). Indeed, two of several concepts at work in this bilateral relationship are one's *best interest* (also refer to Chap. 4), and one's *capacity* (refer to Chap. 10) to make health care decisions.

Both are also centrally relevant in pediatric ethics. But as most minor patients are not legally competent to make most health care decisions,[1] even if some are mature enough to do so, the therapeutic alliance in pediatrics involves a trilateral relation: the minor patient, the parent or other duly appointed surrogate (henceforth, "parents"), and the medical provider(s). As the parents and providers are working together on behalf of another person, the child, an additional concept ranges over this relation: *authority*.

In some senses, while the child is too immature to contribute to making his or her health care decisions, *authority* does to the ethical work in pediatrics that *autonomy* does in adult medicine; the question around both being who is the appropriate decision-maker of matters that affect this patient? In other words, who can justifiably give informed consent to treat the child patient? (Katz and Webb 2016).

However, there are some important shortcomings in attempting to graft the notion of informed consent onto pediatric decision-making. First, informed consent is grounded in the patient's autonomy rights, which are generally not applicable to pediatric patients. Second, informed consent is a self-regarding notion; it involves

---

[1] There are two caveats to this statement. (1) Emancipated minors can consent to their own treatment. Being emancipated is a legal process which has varying requirements from state to state, but generally the minor lives separately from his or her parents and support themselves financially. (2) Under the doctrine of *parens patriae*, the State has made some treatment decisions that affect public health, such as contraception/family planning and mental health care, which adolescents may be reluctant to discuss with their parents, available directly to the minor. The reason for this is the public interest is served by these minors accessing care they might not otherwise seek if parental permission and consent were required.

normative judgments about the risks and benefits of the available options. These judgments reflect the wishes, values and goals the patient has for and about their own future. It is thus more appropriate to speak of *parental* permission, rather than informed consent, to treat the minor child.

For older children who can meaningfully participate in their own health care decisions, part of the uneasiness posed by substituting authority for autonomy is answered by the doctrine of pediatric *assent* (AAP Committee on Bioethics 1995). In all matters, the practice of *shared decision-making* is considered the gold standard in health care today (Sandman and Munthe, 2009). Each of these issues and the related concepts will be discussed in turn.

7.1.2 The Ethical Work Done by Parental Permission

In adult medicine today, persons by default are considered the *authority on themselves*; They alone have first-person epistemic knowledge of their own wishes, values, and goals, and will have to live with the results of their decisions in a way no one else will. Their decisions are self-regarding, in that an adult patient is making choices for oneself. Thus, unless deemed medically incompetent, adult patients, in being the authorities on themselves, are said to be autonomous—something others should respect *in virtue* of that first-person authority and the self-regarding nature of the decision. In bioethics this concept is embodied in the principle of *respect for persons* (Lysaught 2004). As a result of this status, a core tenet of bioethics is that a person must give fully informed consent, free of coercion or influence, before their body is interfered with. Moreover, that consent is revocable at any time; competent patients are free to refuse treatment (Kottow 2004).

There are two important differences in pediatrics. The first, particularly regarding very young children, is that *no one* has first-person epistemic knowledge of the child's own wishes, values, and goals—and these children will not yet have formed any such normative preferences. The wishes, values, and goals of older children may be only partially developed. Adolescents may have developed some normative preferences, and be developing the capacities for autonomy, but still lack the experience adults can draw upon in making important decisions. Moreover, there are important limitations to adolescent decision-making abilities, such as the influence of an imbalance or overactivation of risk-reward brain systems (Johnson et al. 2009). Exceptions to this, which will be addressed in the discussion of assent, are those children who, in virtue of their long experience with chronic disease, have developed a maturity and knowledge of their condition that belies their age.

Second, while the degree the child has developed normative preferences and the capacities for autonomy varies, the child will live with the results of the health care decisions made on her or his behalf; the decisions are not self-regarding, but other-regarding. It is therefore not clear how to honor the principle of *respect for persons* in the treatment of minor patients.

Thus, for most of the pediatric years, the question is not about respecting autonomy, but rather determining who has the authority to decide *for* the child (Katz and Webb 2016). It is for this reason that the preferred term in pediatrics is parental *permission* rather than parental *autonomy,* as the latter term suggests a self-regarding relation that is, in this case, absent. Since no one has the first-person vantage point, nor will live with the outcome of any health care decisions in the way the child will, the ethical authority to decide for another is substantially less certain than is self-regarding decision-making (Buchanan and Brock 1990). In practical terms, certainty in this sense translates into the degree to which the decision is reviewable by others; the more harm that may be imposed in relation to the expected net benefit, the more the decision is justifiably questioned by others (Teti et al. 2017; Diekema 2018).

While shared decision-making is the standard practice today, the complication remains that both parents and health care providers have the same epistemic standing—second-person knowledge of the child's own wishes, values, and goals. While both are dedicated to the wellbeing of the child, neither will live with the results themselves; both the provider's recommendations and the family's wishes are other-regarding. This second-person, other-regarding status also sets up the possibility of one party coming to conclude the other is mistaken about what is best for the child, not acting in the child's best interest, or disagreeing about which interests are paramount. Neither parent nor provider has the first-person authority to decide for another in the way a person can decide for oneself (Katz and Webb 2016). As a result of the second-person, other-regarding status of parents and providers, Micah Hester notes "the surrogate's legitimacy is not derived by expressing the young child's *own* values but merely by having authority to decide *for* the child, thus creating a different' moral space.'" (Hester 2012).

Notably, these tensions are not relieved by any of the positions taken by ethicists about what 'best interest' means; the problem is inherent to making decisions for someone other than oneself. As noted, the default assumption in pediatrics is that since the child's parents and the child's physician are co-fiduciaries of the child, each is obligated to prioritize the child's interests in health care decision-making (Miller 2010). In general terms, the best interest's standard serves as an ideal from which to balance one's *prima facie* duties with one's actual duties. The determination that follows from that should meet a standard of reasonableness that others can evaluate (Kopelman 1997).

7.2   The Co-fiduciaries: The Roles of Parents and Providers

7.2.1   The Role of Parents

The family is the basic building block of society, and as such is accorded special respect. One aspect of this is the default assumption that parents are in the best position to make decisions for their children. It should be remembered that parents are more than custodians throughout the child rearing years. Children are part of a family's generational project, stretching backward and forward in time, encompassing a history, culture, and values unique to them (Diekema 2018; Derrington et al. 2018). Familial relationships are generally life-long, and while discerning the interests of parent and child is important, that should not necessarily imply separating those interests; they are usually considered inseparable, or not separable without causing great harm. For that reason, every effort should be made to accommodate and respect the existing relations.

The role of the parents is providing protection and rearing the child, so that at the age of consent the child is able to make decisions for her- or himself. This ideally involves the parent titrating the freedom of the child to act independently in developmentally appropriate ways while modeling good decision-making practices, so the child develops the capacities of self-governing necessary for making decisions in her own interests and in accord with her own values (Katz and Webb 2016). For more on the discussion of capacity please refer to Chap. 2.

When the child is young and incapable of expressing her or his wishes, the parent makes all decisions on the child's behalf. As the child matures into her or his early childhood years, parents continue to make all health care decisions but may include the child in some limited aspects of health care. As this maturation continues into adolescence, the child begins to express her or his will, and to develop wishes, values, and goals. It is during this phase that the parent should involve the child in age-appropriate decisions that can affect the course of care (Katz and Webb 2016). When and how this occurs can vary widely with culture and child rearing philosophy.

However, parental authority is not absolute; from the time the child is born, under the doctrine of *parens patriae* (refer to Chap. 26), the state has an interest in the child's wellbeing (Katz and Webb 2016). While parents are assumed to be the appropriate authorities on what is best for their children, there are limitations to the decisions parents can make on their child's behalf. Parental requests regarding treatment or refusal of treatment for their children must be reasonable, in a way they would not have to be about decisions for their own health care.[2] Parental authority is also limited by the standards of professional practice to which health care providers must adhere; health care providers have no obligation to provide care that is not medically indicated (Teti et al. 2017).

These limits on parental authority are ethically justified by the limitations of the second-person, other-regarding nature of the pediatric health care decision-making process. Parents can at their best approach the tenets of informed consent—knowledge of the child's own wishes, values, and goals. However, the child's parents will not live with the outcome of health care decisions in the first-personal way that generates the moral authority one has to make decisions for oneself. An exception to this principle that warrants consideration is parents of children with serious cognitive limitations who are, as a result, lifetime caregivers of their children. In such situations it becomes more difficult to determine where the child's interests differ from the

---

[2] The foundational caselaw cited in bioethics in. this regard isPrince v. Massachusetts.

parents. However, even in extreme cases the decisional status of the parents remains other-regarding.

Throughout the developmental process, the child is 'developing autonomy' in the sense of not just being able to express normative preferences, but also highlighting the ethically uncomfortable space inherent in making other-regarding decisions. There are emergent considerations for parents as this process unfolds, such as the need for privacy and confidentiality as a matter of maturation and development (Blustein 1996). The state also recognizes this maturation, and has made certain public health matters, such as access to birth control or family planning and mental health services, available to minors without a parent having to provide consent (Katz and Webb 2016).

7.2.2   The Role of Providers

Across the spectrum of the child's development, the provision of medical care must remain sensitive to three components of ethical health care: the parent's authority to make decisions for their child aligned with their/their family's values, respecting the child's own wishes appropriate to the present state of developing autonomy, and the physician's obligations to provide medically appropriate care in the interests of their patient and in accordance with professional rules of conduct (Teti et al. 2017; Katz and Webb 2016; Sandman and Munthe 2009). This often occurs in the setting of multiple, complex decisions made over time, which may vary with differing cultural practices and child rearing philosophies (Derrington et al. 2018).

The rubric for balancing and aligning this trilateral relation has changed over the years. While there has been a move away from the paternalistic practices of clinicians, there is a debate about the role and extent of parental authority. The argument made on this latter trend is that the authority of the parent and the interests of the child may sometimes be in conflict, or minimally, are not coextensive (Derrington et al. 2018). Opposing this is the view that (a) the parents know the child and the child's interests better than a health care team, and (b) it is the parents who will be caring for the child. As noted, the inherent difficulties of making decisions for another are not relieved in these debates; rather the contours of what is at stake are debated (refer to Chap. 4).

Through state licensure, physicians and other health care providers are obligated to assess and report child abuse or neglect. However, in those instances when questionable parental decisions do not rise to the level of constituting abuse, providers can nonetheless be placed on the horns of a dilemma. Respecting parental authority may involve continuing non-advised treatment (e.g. withholding of vaccinations); upholding the patient's best interest may indicate another course of care (e.g. refusing to provide a bridge therapy where no curative or palliative options exist). The physician's primary duty is to the minor patient, not the child's parents, and health care providers are often in the position of having to adjudicate the 'reasonableness' of parental requests (Teti et al. 2017; Kopelman 1997).

Health care providers are structurally positioned to evaluate the actions of parents as they pertain to the child's health and wellbeing. While some circumstances are clear-cut in this regard, often they are not, and may be all the less so as a result of prognostic uncertainty, where the relationship between values, decisions, and what counts as rational is difficult to assess and can turn on non-objective factors, a fact true of physicians and laypeople alike (Groopman and Hartzband 2012). Providers need to keep in mind that they have both legal and ethical duties to their child patients to render the care the child needs, not what another—even a parent—requests (Teti et al. 2017; Katz and Webb 2016).

7.3 Shared Decision-Making and Pediatric Assent

7.3.1 The Foundations of Pediatric Assent

There are two sorts of shared decision-making dynamics. The first sort of SDM is bilateral, and pertains to the periods during which the child patient is an infant or young child and cannot meaningfully contribute to decision-making. This sort also includes children who have cognitive deficiencies that preclude involvement in decision-making. The second sort of SDM is trilateral, and pertains to the periods during which the child patient can participate in decision-making. As noted, this will vary with maturity and the age-appropriateness of the decision under consideration; capacity is decision-specific, so the ability to make one kind of decision does not imply the same is true of other decisions. It is this second sort of SDM to which *pediatric assent* is relevant, but as the two sorts occupy a spectrum, the discussion of SDM must include some comments about assent. Much has been said about the dynamics of the first type, so before formally exploring SDM, a discussion of assent is in order.

It was with the rise of shared decision-making in the late 1980's and early 1990's that the American Academy of Pediatrics Committee on Bioethics (1995) acknowledged problems with "proxy" consent. No one else could give consent for the child patient because, as the committee noted, "consent (which literally means "to feel or sense with") expresses something for oneself." It is in the terminology used here, first-person and self-regarding. Consent is not fully transferrable because doing so deprives it of the two ethically relevant senses of its authority: (1) being guided by one's own wishes, values, and goals, and (2) living with the results of the decision oneself. For that reason, the concept of parents giving permission to treat is more accurate than "giving consent" (Katz and Webb 2016; AAP 1995).

7.3.2 Pediatric Assent

The concept of *pediatric assent* expresses in spirit what the concept of informed consent enshrines in law; an informed willingness to be so treated (Leikin 1983; AAP 1995). The principal difference is that consent is both a legal and an ethical requirement, while assent is an ethical and pragmatic one. Assent is a legal requirement only in the context of research, and even there it can be waived under certain circumstances. Pediatric assent becomes important as children develop their sense of self as an independent being, especially as their wants and opinions may either positively or deleteriously affect efforts to treat or participate in medical research.

The American Academy of Pediatrics Committee on Bioethics endorsed the notion of pediatric *assent* to help relieve the ethical discomfiture noted above (AAP 1995). As the young child matures into the early years of childhood, her or his preferences will emerge, and there exists the opportunity to include the child in the health care decisions that affect her. In childhood proper and through adolescence, the child's own preferences become a matter of personal interest. At early stages, the child's expressions may be no more than refusals; not wanting a needle stick, for example. To what extent those expressions should be respected varies with circumstance. In a therapeutic setting, a child needing a blood test but not wanting to be poked has little choice. On the other hand, a child disinclined to have the same needle stick for a non-beneficial research project should not be forced to comply.

Thus far we have noted the dynamics of the therapeutic relationship as primarily involving the parents and health care providers. As the child matures, what had been the potential for bilateral disagreement between parents and providers, can—and to the extent possible, should—become trilateral: not only may parents disagree with the provider, but the child may disagree with either, or both.

While a child agreeing with the parent and provider is very satisfactory, it is when disagreements take place that the child's ability to participate in medical decisions that affect her tends to be questioned by those in disagreement. The appropriateness of a medical provider taking sides with the patient "overruling the parents" or going "behind the parent's back" is problematic and a subject of concern (Diekema 2018). Cultural factors about the role of children in families may lessen or exacerbate such situations. The AAP recommends that all providers acknowledge their own cultural beliefs and values, and those of the field of medicine, in order to maintain awareness of the assumptions and potential biases they may bring to the clinical encounter (Derrington et al. 2018).

As an ethical matter, the child's role and wishes in making decisions should be increasingly respected by both providers and parents as the child matures. As a practical matter, as the child ages, her or his ability to interfere with a given intervention increases, so that procedures such as bone marrow or solid organ transplant, which require complicated and involve long-term follow-up, would be difficult to carry out and be successful without the child's collaboration. Moreover, as it is the child who will have to live with the outcome of the intervention or therapy, it is appropriate to accord the child some degree of authority in decision-making appropriate to the child's age and ability (Katz and Webb 2016; U.N. Convention on the Rights of the Child 1989). Foreman argued, "informed consent in children should be regarded as shared between children and their families, the balance being determined by implicit, developmentally-based negotiations between child and parent." (Foreman 1999).

With minor patients who have capacity to make a given decision, assent should be sought and respected to the extent reasonable. Here a balance must be struck between fostering developing autonomy to prepare the child to make medical decisions at the age of majority and the parallel responsibilities of providers and parents to look out for the child's interests. In the balance may be the risk of the child feeing excluded or alienated from matters that will affect them more than either parent or provider. At a time when a teen may be wrestling with his or her identity and independence, such a result can threaten the therapeutic enterprise.

Children and teens who live with chronic illness can become deeply informed about their disease and the risks and benefits of its treatment. A teenager who has been through one or more recurrences of cancer, for example, can exhibit a rationality and wisdom that belies the child's age (refer to Chap. 2). In addition to the impracticality of attempting to force a child through a long therapeutic regimen, one must consider the epistemic burden of overruling the patient's first-person, self-regarding assessment of the decision. In such situations, often parents take a "do everything" approach to the teenager having "had enough." Every effort should be made to bring the family together around a decision they all can accept.

Providers can also face difficulty when they witness parents insisting their adolescent children follow a course of action that may seem dubious. The foremost difficulty is determining at what point, if any, should the medical provider challenge the parent's authority to make the decision for their child. In some instances, there may be a clear-cut medical rationale, but when the disagreement reflects differing values, the determination may not be so clear. The role alignment of SDM should guide both providers and parents, and has been shown to reduce some decisional conflict (Wyatt et al. 2015). In cases of persistent disagreement or uncertainty, a clinical ethics consult may help find a course to consensus and resolution.

7.3.3 Shared Decision-Making

Building on their status as co-fiduciaries, shared decision-making (SDM) aims to establish a role alignment among the members of the therapeutic alliance for the purpose of making medical decisions for the minor patient. SDM involves the health care provider's use of the best available evidence, aligned with the parent/patient's values, preferences, and treatment goals (Sandman and Munthe 2009). It is the role of the parents to express their values and normative preferences for the care of their child. As noted, it is generally presumed parents are in the best position to know of, and act, in their child's best interest. It is the role of the physician and medical team to present the best treatment options available, based on science and standard of medical care, and to make recommendations taking into account the values and goals expressed by the parents, and the child—to the extent the child expresses age-appropriate wishes.

This latter element, the child's involvement in decision-making, develops over a period of years. This bilateral relationship between parents and providers becomes trilateral as the child patient matures and begins to express interests, preferences, and goals of her or his own. The relationship between the wishes of the parents and their child scales with two factors: the significance of the decision to be made, and the maturity of the minor patient (See Fig. 7.1).

At its best, shared decision-making enables the parent's wishes and goals and the provider's professional obligations to mesh into a decisional unit that can effectively address the complexities of modern health care. In the vast majority of cases, SDM works well, yielding a role clarity that improves agreement. However, this ideal-ized relation is in reality quite complicated. In theory, as co-fiduciaries, parents and providers each bring particular expertise in differing but important domains. Parental expertise relates to education, enculturation, family's values, and the child's long-term interests. Provider expertise relates to the best therapeutic approach to achieve the goals of care and meeting a minimum standard of care. Disagreement can arise when one party feels the other is not acting in the child's best interest or not respecting the expertise the other brings to the matter. Parents may feel the medical team is not respecting their moral or religious commitments. Medical teams may feel that the parents are not acknowledging the medical realities that render some of the family goals unachievable.

Complicating matters further, each has to rely on the other in order to do their own part. A parent who is not adequately informed of the medical options is not in

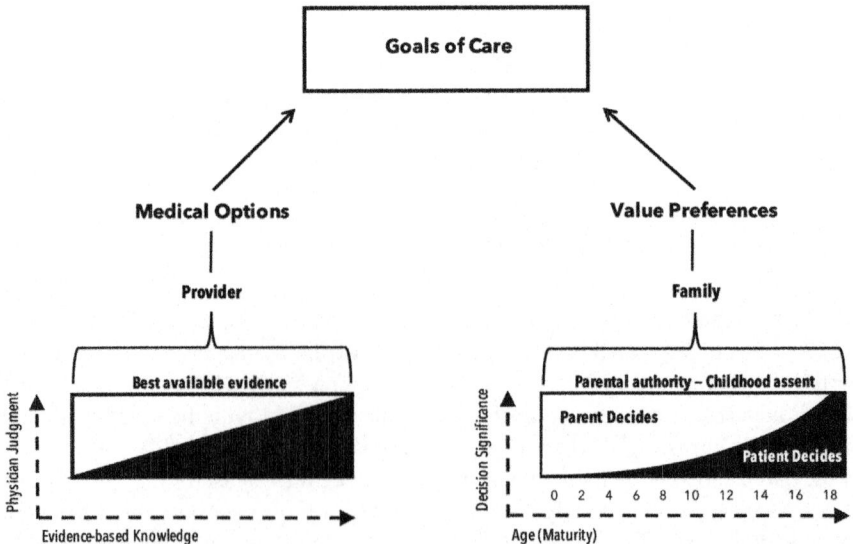

Fig. 7.1 Shared decision-making

a position to evaluate which among them will best realize value-aligned long-term goals. In turn, a provider who is unable to elicit value preferences from parents has a difficult time determining what therapeutic options to recommend.

This is all further made more difficult by two sets of unknowns. First, the uncertainty inherent in medical diagnosis, and more acutely, in medical prognosis. As a result, the disagreements that arise, for example, regarding a proposed plan of care, are not based in what *will*, in fact, happen with or without the therapy under consideration, but rather what is *expected* to happen. Beyond the uncertainty of the outcome, the very nature of the trilateral therapeutic relationship may introduce obstacles. As noted, parents are not generally experts in the condition their child suffers from. They must make decisions of ultimate importance about matters for which they seldom have expertise themselves; they must decide to trust, or not trust, the physician or health care team treating their child. The ability of the parents and medical team to work effectively together depends on that trust, and it can be won or lost (O'Neill 2002).

Problematically, even extremely competent physicians acting according to the best available evidence, and with the greatest wisdom, can get things wrong. The vagaries of modern science can thus rend the trust between parent and clinician. Notably, this can occur without anyone making an error in judgment (although that may occur too). Modern medical care for serious conditions involves scores of individual decisions made over time, and parents, by and large, are focused on matching what was prognosticated with what actually came to be. There is little else they can assess on their own. While diagnosis, prognosis and treatment are iterative processes, from the parent's point of view, new information can appear to be a rolling disclosure or constantly moving target that gives them little certainty, and may undermine trust in the provider.

The aforementioned also presumes the parents are able to discern for themselves which, among many competing values, is paramount. The reality is that parents may not be able to choose between two or more competing value interests; they may disagree with one another, or not be able to bring themselves to face the ramifications of a course of action that makes the most sense (Teti et al. 2017). The stress and emotional toll of having a child in intensive care, suffering from a chronic condition, having an ailment for which a diagnosis has yet to be found, or facing an end-of-life decision can deleteriously affect a parent's ability to make decisions (Meyers 2004). Providers take varying stances on how much they should attempt to influence parental decision-making (Lantos and Meadows 2006). Some feel their role is to clearly articulate the available options, pros and cons to the best of their ability, and leave the rest up to the parents. Such a provider may feel that doing any more is, if not coercive, nonetheless exerting an improper influence that strays beyond the medical facts into normative considerations—perhaps the provider's own preferences. Others feel that not offering parents advice on what to do is effectively stranding them with a difficult, complex decision of a type they may have never faced (and which the physician has at least witnessed perhaps many times) at the worst possible time (Meyers 2004).

In most cases, such disagreements can be resolved through open discussion between the parties. Family meetings are often useful to share information, explain emergent issues, and address concerns. Both parents and providers share the overall goal of caring for the patient; that common ground is generally sufficient repair breakdowns in provider-parent interactions. When issues remain, an ethics consult may help elicit differing perspectives, identify competing values, and help pave a way back to a good working relationship.

It should also be noted that while SDM has been found to be helpful for providers and parents, some research indicates SDM has not succeeded in bringing children into the decision-making process (Wyatt et al. 2015). Research on this matter continues. The ill fit of informed consent in the pediatric setting as a result of the shortcomings of second-person other-regarding decision-making, should encourage health care providers and parents alike to make efforts to include children in the decisions that affect them.

7.4   Conclusion

As minors are not able to give informed consent to be treated, the therapeutic decision-making alliance in pediatrics is comprised of the child's parents and health care providers. They are co-fiduciaries of the child, dedicated to prioritizing the child's interests above other concerns. Thus, for most of the pediatric years, the question is not about respecting patient autonomy, but rather determining who has the authority to decide *for* the child. However, the ethical authority to decide for another has less *prima facie* legitimacy than self-regarding decisions. Shared decision-making (SDM) can help to address this concern by mediating authority among the co-fiduciaries, and may relieve potential tensions between parents and providers. However, while shared decision-making is useful for role delineation and consensus building, it must be remembered that deciding for another is a poor substitute for deciding for oneself. As the child matures, both parents and providers should involve the minor patient in the decisions that affect her or him to the extent possible. Through adolescence and the teenage years, pediatric assent should be both sought and respected. This serves the dual goals of ensuring the minor patient is involved in the decisions that will affect him or her, and in modeling the skills and attitudes the minor needs to develop to make informed decisions as an adult.

Guiding Principles in this Chapter

Authority and Permission
- The ethical foundations of informed consent and autonomy are ill-fitted to pediatrics, where decisions are made from a second-person epistemic standing on behalf of the child patient. Such decision-makers derive authority from their status as legal guardians of the child patient. It is therefore more appropriate to speak of having *authority* and giving *permission* to treat the child patient

---

(continued)

(continued)

Guiding Principles in this Chapter

Informed Consent and Autonomy

- *Informed consent* presumes first-person epistemic standing and self-regarding decisions. The freedom to choose and refuse medical care is justified by the capacitated person knowing their own wishes, values, and goals better than anyone else, and living with the results of the decision themselves. It is in virtue of this standing that *autonomy* should be respected

---

Parental Authority

- Parents have wide discretion, but it is not without limits. Because parental decisions are based on their second-person epistemic standing and are other-regarding, and the State has an interest in the protection of children, parental authority is subject to review and may be overridden. The more risk imposed in relation to the expected net benefit, the more a parent's authority is justifiably questioned by others. Parental authority may be overridden if the decision puts the child at immediate risk of serious, preventable harm (Harm Principle)

Pediatric Assent

- Paralleling informed consent but without legal standing, assent expresses an informed willingness to be treated. Beginning with the age at which the child can meaningfully participate in health care decisions, assent should be increasingly sought in an age-appropriate way. If assent is sought, refusals should generally be respected; if the refusal is not to be respected, the reasons for doing so should be explained and the patient should have the opportunity to respond. As the child approaches the age of majority, the child's input should be given increasing weight so that the young adult is experienced with speaking to providers and weighing medical options

---

Shared Decision-Making

- Specifies a role division between providers, parents, and patients in medical decision-making. Providers are responsible for providing the range of medically appropriate options and a recommendation based on the parent's expressed wishes, values, and goals. Parents are responsible for making and adjusting the goals of care based on the child's interests and in alignment with their wishes, values, and goals. Both providers and parents are responsible for encouraging age-appropriate participation of the child in decision-making to foster developing autonomy so that at the age of majority the young adult can make informed decisions in her best interests

---

References

American Academy of Pediatrics Committee on Bioethics. 1995. Informed consent, parental permission, and assent in pediatric practice. *Pediatrics* 95 (2): 314–317. (Reaffirmed (2007) 119(2): 405). https://doi.org/10.1542/peds.2006-3222.

Birchley, G. 2014. Deciding together? Best interest and shared decision-making in paediatric intensive care. *Health Care Annals* 22 (30): 203–222. https://doi.org/10.1007/s10728-013-0267-y.

Blustein, J. 1996. Confidentiality and the adolescent: An ethical analysis. In *Pediatric ethics—From principles to practice*, eds. R.C. Cassidy and A.R. Fleischman, 83–96. Harwood Academic Publishers.

Buchanan, A.E., and D.W. Brock. 1990. *Deciding for others: The ethics of surrogate decision-making*. Cambridge University Press, 215–266. https://doi.org/10.1017/CBO9781139171946.008

Derrington, S.F., E. Paquette, and K.A.Johnson, 2018. Cross-cultural interactions and shared decision-making. *Pediatrics* 142 (Supplement 3): S187–S192. https://doi.org/10.1542/peds.2018-0516J.

Diekema, D.S. 2018. When parents and providers disagree. *Pediatric Ethicscope* 31 (2): https://ped iatricethicscope.org/issue-and-article-archives/

Foreman, D.M. 1999. The family rule: A framework for obtaining ethical consent for medical interventions from children. *Journal of Medical Ethics* 25 (6): 491–496.

Groopman, J., and P. Hartzband. 2012. *Your medical mind: How to decide what is right for you.* Penguin Books

Hester, D.M. 2012. Ethical issues in pediatrics. In *Guidance for health care ethics committees*, eds. D.M. Hester and T. Schonfeld, 121. Cambridge University Press.

Johnson, S.B., R.W. Blum, and J.N. Giedd. 2009. Adolescent maturity and the brain: The promise and pitfalls of neuroscience research in adolescent health policy. *J Adolescent Health.* 45 (3): 216–221.

Katz, A.L., and S.A. Webb. 2016. Committee on bioethics. Informed consent in decision-making in pediatric practice. *Pediatrics* 138 (2): e20161485. https://doi.org/10.1542/peds.2016-1485.

Kopelman L.M. 1997. The best interests standard as threshold, ideal, and standard of reasonableness. *The Journal of Medicine and Philosophy.* 22 (3): 271–289. https://doi.org/10.1093/jmp/22.3.271.

Kottow, M. 2004. The battering of informed consent. *BMJ Journal of Medical Ethics* 30 (6): 565–569. https://jme.bmj.com/content/medethics/30/6/565.full.pdf.

Leikin, Sanford L. 1983. Minors' assent or dissent to medical treatment. *The Journal of Pediatrics* 102 (2): 169–176. https://doi.org/10.1016/S0022-3476(83)80514-9.

Lysaught, M.T. 2004, 2010. Respect: or, how respect for persons became respect for autonomy. *Journal of Medicine and Philosophy* 29 (6): 665–680.

Meyers, C. 2004. Cruel choices: Autonomy and critical care decision-making. *Bioethics* 18 (2): 104–119. https://doi.org/10.1111/j.1467-8519.2004.00384.x.

Miller, G. 2010. Pediatric bioethics. *Cambridge University Press.* 11–12: 17–20.

O'Neill, O. 2002. *Autonomy and trust in bioethics.* Cambridge University Press, 12–13.

Sandman, L., and C. Munthe. 2009. Shared decision-making and patient autonomy. *Theoretical Medicine and Bioethics* 30 (4): 289–310. https://www.ncbi.nlm.nih.gov/pubmed/19701695.

Teti, S.L., K. Ennis-Durstine, and T.J. Silber. 2017. Etiology and manifestation of iatrogenesis in pediatrics. *AMA Journal of Ethics* 19 (8): 783–792. https://doi.org/10.1001/journalofethics.2017. 19.8.stas2-1708.

United Nations. 1989. *Convention on the rights of the child.* http://www.ohchr.org/en/professional interest/pages/crc.aspx.

Whitney, S.N., A.L. McGuire, and L.B. McCullough. 2004. A typology of shared decision-making, informed consent, and simple consent. *Annals of Internal Medicine* 140 (1): 54–59. https://doi. org/10.7326/0003-4819-140-1-200401060-00012.

Wyatt, K.D., B. List, W.B. Brinkman, G.P. Lopez, N. Asi, P. Erwin, et al. 2015. Shared decision-making in pediatrics: A systemic review and metal-analysis. *Academic Pediatrics* 15 (6): 573–583.

Further Readings

Diekema, D.S. (2018). When parents and providers disagree. *Pediatric Ethicscope* 31(2). https:// pediatricethicscope.org/issue-and-article-archives/.

Katz, A.L., and S.A. Webb. 2016. Committee on bioethics. Informed consent in decision-making in pediatric practice. *Pediatrics.* 138 (2): e20161485. https://doi.org/10.1542/peds.2016-1485.

Lantos, J., and W.L. Meadow. 2006. *Neonatal bioethics: The moral challenges of medical innovation,* 111–119. The Johns Hopkins University Press.

Silber T.J. 2014. Chapter "pediatrics, adolescent". In *Encyclopedia of bioethics*, 4th Ed. eds Bruce Jennings. Farmington Hills, MI: Macmillan Reference USA.

Weithorn, L.A. 2020. When does a minor's legal competence to make health care decisions matter? *Pediatrics* 146 (s1): e20200818G.

Chapter 8
Telling the Child: Ethics of the Involvement of Minors in Health Care Decision-Making and in Considering Parental Requests to Withhold Information from Their Child

J. M. Marron and K. O. Kennedy

Abstract Truth-telling is a core value in medical ethics, and its importance has only grown in recent decades with the increased focus on patient-centered care. How and whether to inform children about aspects of their health care (and how/whether to include them in health care decision-making), however, is even more complex. In this chapter, using an example case, the authors describe some of the ethical underpinnings of truth-telling both in pediatrics and more generally, as well as important features of the disclosure of health care information to children. They then provide arguments both for and against withholding information from children when that is requested by parents, examining some common arguments made for each of these strategies in the literature and in clinical practice. After touching upon unique considerations regarding disclosure of different types of information to minors, the authors close by providing practical guidance on how clinicians can respond when parents explicitly request to not tell their child information about their health and/or health care.

Keywords Truth-telling · Consent · Assent · Decision-making · Disclosure · Pediatric ethics

J. M. Marron (✉)
Dana-Farber/Boston Children's Cancer and Blood Disorders Center, Harvard Medical School, Boston, MA, USA
e-mail: jonathan_marron@dfci.harvard.edu

J. M. Marron · K. O. Kennedy
Boston Children's Hospital, Boston, MA, USA
e-mail: Kerri.Kennedy@childrens.harvard.edu

© Springer Nature Switzerland AG 2022
N. Nortjé and J. C. Bester (eds.), *Pediatric Ethics: Theory and Practice*,
The International Library of Bioethics 89,
https://doi.org/10.1007/978-3-030-86182-7_8

8.1   Introduction

Truth-telling is among the core ethical values in health care, but it has a particularly unique (and complex) role in pediatrics. In this chapter, we explore practical and ethical considerations in truth-telling in pediatric health care. First, we introduce a case to provide a foundation for these challenging, nuanced concepts. Next, we describe some of the ethical underpinnings of truth-telling in medicine, both generally and specifically in the care of children, followed by a description of some of the important features of information disclosure to children and of involving minors in health care decision-making. Then, we provide several arguments in support of withholding information from children if such is requested by parents, followed by several arguments against doing so. Finally, we conclude with practical guidance for clinicians when faced with a parent requesting to withhold information from their child, referring back to our case and providing concrete steps for those caring for children and their families.

8.2   Case Description

Billy is a 9-year-old, previously healthy boy who has just been diagnosed with leukemia (a cancer of the blood cells), which was diagnosed based on a screening blood test by his pediatrician. He is asymptomatic and feels very well. His health care team plans to run additional tests in the next few days, but Billy likely will require admission to the hospital to start his chemotherapy treatment course. His mother and father, who are both understandably distressed by the news of Billy's diagnosis, wish not to tell him that he has cancer. "He just got treated with antibiotics for an ear infection," they state. "Can we just tell him that he needs medicine for another infection? He won't understand what cancer is anyways, and telling him will just make him scared. What's the point of that?".

8.3   Ethical Underpinnings of Truth-Telling

Truth-telling ("veracity") is considered a core principle in modern medical ethics (Beauchamp and Childress 2009). In health care in general, as it relates to adult medicine, truth-telling typically refers to complete, comprehensive, and accurate provision of information about a patient's diagnosis, treatment option(s), and prognosis. Chapter 4 discuss the ethical theory that underpins the obligations of clinicians and parents to children receiving health care services, but several points about truth-telling and disclosure of information warrant mentioning here. First, the obligation to tell the truth to patients ultimately is grounded in the respect owed to them as persons (Beauchamp and Childress 2009; Jonsen et al. 2015). Even those who cannot legally make their own decisions (e.g., those without capacity and/or minor children) deserve respect, including being told the truth. Further, individuals cannot make well-informed choices about their care without information about their options; this relates to the individual's ability to provide informed consent (or assent), but also more fundamentally to their autonomous right to control what is done to their body. Pediatric and adolescent assent are discussed in greater detail elsewhere in this text (See Chaps. 2 and 7), but for children, adolescents, and adults, informed decisions rely on the provision of adequate information. Truth-telling also is closely connected to other fundamental ethical principles, such as fidelity. The clinician-patient relationship implicitly assumes—and relies upon—truthful delivery of information by both the clinician and patient. Without this, one of the most fundamental aspects of health care can be endangered: trust. Finally, actively engaging patients in the decision-making process by providing them with information about their diagnosis, treatment options, and prognosis gives them the opportunity to take an active role in their treatment. This may potentially improve health care outcomes (a beneficence-based argument) (Lo 2009). That is not to say, however, that truth-telling is obligatory in all circumstances. Below, we will explore whether/when non-disclosure of information, limited disclosure, or even outright lying might be ethically justifiable.

It is also worth distinguishing between truth-telling as a legal entity versus an ethical one. Here, we will be focusing on ethical considerations around truth-telling. Lying is only rarely explicitly illegal in the health care setting (for example, when testifying under oath), but the threshold for criminality is a very different standard than that for ethical obligation/expectation (and certainly not to say that a physician who lied to a patient or parent might not be subject to civil action). Importantly, legal standards regarding decision-making for children, such as statutes addressing mature minors and emancipated minors, are complex and discussed elsewhere in this text. While comprehensive truthful disclosure has not always been considered ethically obligatory, most codes of medical ethics now advocate for the fundamental importance of telling the truth. The American Medical Association, for instance, had previously supported "therapeutic privilege" (American Medical Association 2017; Beauchamp and Childress 2009), which is the idea that a clinician can withhold information they feel would only harm a patient. However, they now state that, as long as a patient has capacity, "withholding information without the patient's knowledge or consent is ethically unacceptable" (American Medical Association 2017, p. 31).

8.4 Ethical Underpinnings of Truth-Telling in Pediatrics

In many ways, the ethical underpinnings of truth-telling in pediatrics align with those in the adult health care setting. The American Academy of Pediatrics states that, for both children and adults, the informed consent process is grounded in the same fundamental ethical principles of beneficence, justice, and respect for autonomy (Katz and Webb 2016).

While there is significant overlap in ethical decision-making between pediatric and adult health care, there are also notable differences. One example is that, while adults with decisional capacity provide first-person informed consent for treatment, children have not yet developed the autonomy to do so. As described in greater detail in Chap. 7, in pediatrics, the informed consent process requires seeking the "informed permission" of the child's parent(s). This is achieved through a process of shared decision-making between parents and the child's clinician, ideally with assent from the child in accordance with their age, maturity, and developmental capacity (Katz and Webb 2016). This creates a trilateral clinical relationship structure which differs from that of the clinician-patient dyad that is typical in the adult population, lending added complexity to the pediatric health care encounter.

In addition, as described elsewhere in this text, children, by virtue of their age, are at varying stages of development and maturity. This, in part, informs the degree to which children can be substantively involved in decision-making, both in terms of providing assent (a concept that has evolved over time from absence of explicit dissent to one of active engagement) and more generally (Leikin 1983). In turn, this variation necessitates highly nuanced approaches to caring for them, and it can be challenging to discern how and to what extent children should be engaged in health care decisions that are being made by others on their behalf. There is wide agreement in bioethics, however, that, while clinicians and parents may weigh numerous factors when considering health care decisions for a child, the child's interests must remain central (See also Chap. 4) (McCullough 2009). This being said, clinicians often must balance competing ethical obligations when caring for pediatric patients—those to the child (beneficence/nonmaleficence), to parents (autonomy/parental authority), and to clinicians themselves by not acting in ways which violate their own professional/personal integrity. When these ethical obligations come into conflict, it can be difficult to weigh them. In some circumstances, clinicians may determine that their obligations to the child outweigh those to parents (Diekema 2004).

8.5   Important Features of Information Disclosure in Pediatrics

There are several unique considerations around information disclosure in pediatrics. Perhaps most important are the child's age and developmental capacity, which greatly inform the degree of detail and manner in which information will be shared. Preschool-aged children, for example, do not possess the maturity required to participate in health care decisions to the degree that older children do, such as typically developing adolescents. Even so, as noted above, all children, regardless of age or developmental capacity, are deserving of respect and generally should be afforded the opportunity to participate in their medical care to the extent possible, with guidance from skilled clinicians and input from parents. As described further in our practical guidance below, the participation of multidisciplinary members of the health care team often are helpful in these discussions.

Historically, many parents have worried that disclosure of diagnostic or prognostic information would cause undue psychological or emotional distress to their child (Sisk, 2016). In our case scenario, Billy's parents cite this as their reason for not wanting him told about his leukemia diagnosis. It is important to keep in mind that the disclosure of health-related information to children does not necessarily need to be an "all or nothing" proposition. Approaches can be nuanced and tailored to best meet the needs of each individual child and family. In some situations, parents and clinicians might agree that particular aspects of diagnostic or prognostic information will be shared with the child and not others. Or, depending on the child's maturity, they might agree to let the child "take the lead" by providing information only in response to the child's queries.

Families' cultural, religious, or philosophical values also may influence their preferences for treatment decision-making, including information sharing (for further information on this topic, see also Chap. 2). Rosenberg and colleagues thoughtfully explored these issues, emphasizing the importance of eliciting and respecting cultural differences and working together with families to proceed in ways which respect their values while also respecting the child's developing autonomy (Rosenberg et al. 2017). The authors suggest language to help guide conversations with parents where a request for nondisclosure appears to be "culturally mediated." Of course, no culture or religion is monolithic in nature, and families from the same cultural or religious background may have different interpretations of what the culture or religion requires of them. It is also important to keep in mind that, while some societies such as the United States prize patient autonomy and parental authority, others are more communitarian in nature and may weigh more heavily the interests of the family as a whole, for example (See Chap. 12). Humility is important when working with families from varied sociocultural backgrounds and navigating differences between cultural norms. This includes norms around telling information to a child and, more generally, norms of an individual family. When these conflict with the values of the clinical team, treatment decision-making becomes even more complex (Rosenberg et al. 2017).

8.6 Arguments for Withholding Information from Children

Occasionally, parents will express interest in keeping information from their children for various reasons. Typically, this is not malicious in nature, but rather a well-intentioned attempt to protect children from harm. In our case, for example, Billy's parents state their concern that being told his cancer diagnosis will only scare Billy. Here, we explore several common arguments in support of withholding information from children when requested by parents. Importantly, we are only discussing here nondisclosure of information when such is explicitly requested by the minor's parents/guardians, not due to the clinician's own prerogative. The latter is quite infrequent and beyond the scope of this discussion. We also do not intend here to argue that any of these reasons carries greatest valence, but only to familiarize readers with them.

8.6.1 Harm Avoidance

A common reason that parents might wish to withhold information from their child is to avoid hurting them. As in Billy's case, a parent may feel that telling their child about their diagnosis (or prognosis, or upcoming treatment, or other aspect of care) will cause them physical or psychological harm (Bluebond-Langner 1980). This rationale is not unlike that of clinicians who invoke therapeutic privilege to withhold information. Further, it is neither easy nor straightforward to provide complex, distressing information to a child in an age- and developmentally-appropriate fashion. Parents may report concern that the process of delivering the information itself may be harmful, whether due to an impact on trust or other aspect of the relationship with the child. Finally, sometimes information might be withheld in an attempt to protect others. Consider, for example, if Billy's mother wished to not tell Billy or others in their family about his diagnosis for fear of the impact that information would have on his elderly grandmother. Importantly, many parents understandably worry about information such as this harming their child while considering less the counterfactual—the potential harms from withholding the information (harms that we will discuss further below).

8.6.2 "It's My Right"

While there is significant sociocultural variability (as introduced above and described further below), many Western countries and cultures are quite individualistic. This is demonstrated, in part, by a Western focus on patient autonomy, and in pediatrics, on parental authority. Many parents feel strongly about their right and responsibility to care for their children as they best see fit, both regarding health care and more generally. As a result, it is common for parents to reference rights-based arguments for desiring to withhold information from their child (See also Chap. 11). Importantly, this typically is a well-intentioned desire, aimed at doing what they feel to be best for their child. Billy's father, for example, might consider it his paternal right (or even his paternal duty) to protect his son in the best way he knows how, even if the clinical team disagrees with his reasoning. Parental authority, however, is not absolute, as described throughout this text, so this right also is not absolute.

8.6.3   Sociocultural Considerations

Truth-telling practices in Western medicine have evolved greatly in recent decades, from a prior era of nondisclosure as standard to now, when nondisclosure is, as described above, only rarely ethically supported. Research demonstrates that shifts in favor of disclosure may have not been as significant in other areas of the world (including Asia and the Middle East, for example), (Rosenberg et al. 2017) and that disclosure practices in many cultures/communities vary. Parents may sometimes, as a result, request not to disclose information to their child based on their sociocultural beliefs (particularly if those differ from those of the clinical team). Interestingly, however, globalization and the near-universal access to information via the internet may have decreased variability in such practices across geographic and sociocultural lines. In Billy's case, it is conceivable that his cultural or religious background considers it unnecessary, or even inappropriate, to deliver news of a cancer diagnosis to a child, hence his parents' reticence. These sociocultural considerations are incredibly important, and further research is needed to fully understand how best to balance respecting sociocultural beliefs with optimally caring for—and informing—children.

8.6.4   Right to an Open Future

Joel Feinberg first described a child's right to an open future, arguing that children will one day be autonomous adults, so their future opportunities ("open future") should be limited as little as possible (Feinberg 1980). Chapter 3 of this text provides a more detailed discussion of the concept of a child's right to an open future, which is often referenced as a benchmark for how to make decisions for children who cannot do so themselves. In this setting, it could be argued that providing information might prevent a child from being able to "be a kid," much like how arguments have been made for withholding information from adult patients according to therapeutic privilege (American Medical Association 2017; Beauchamp and Childress 2009). Here, one could argue that telling Billy he has cancer would impose an unnecessary burden on him and risk taking away his opportunity to continue to live a free and happy childhood. That said, however, as discussed further below, in not providing this information, clinicians also limit certain future opportunities to the child (e.g., having a voice in treatment decisions, being involved in legacy-building activities, etc.).

8.7 Arguments Against Withholding Information from Children

In addition to arguments supporting a parent's request to withhold information from their child, there are important arguments in favor of disclosure of diagnostic, prognostic, and other health care information. Here, we provide several common considerations.

8.7.1 Deontological Versus Consequentialist Perspectives

There has been a growing trend in pediatrics in favor of disclosure of health-related information to children (Committee on Hospital Care 2012; Sisk et al. 2016). Theoretically, from a deontological perspective, most would argue that telling patients the truth is simply "the right thing to do." This rule-based ethic holds that the rightness or wrongness of an action is grounded in the action itself, regardless of the action's outcome (e.g., "it is always wrong to lie.") (Beauchamp and Childress 2009). Therefore, a strict deontologist would not be likely to honor Billy's parents' request to withhold his cancer diagnosis from him, because this would violate the rule of truth-telling and therefore be considered ethically unacceptable.

In contrast, a consequentialist approach evaluates an action's rightness or wrongness based upon its resulting consequences. In this view, withholding information from a child could be justified if doing so would further the child's interests in important ways, or achieve some 'good' outcome, such as protecting the child from harm. A consequentialist clinician might either honor or not honor Billy's parents request to withhold his cancer diagnosis from him, depending upon their assessment of the likely consequences. Billy's parents' request to withhold his diagnosis from him on the basis that this information would harm him reflects a consequentialist approach.

8.7.2 "It's the Child's Right"

In addition to the benefits discussed previously, engaging children in health care decisions through information disclosure supports their burgeoning autonomy. Most children ultimately will become future autonomous decision-makers. Withholding medical information from them in childhood could deprive them of important opportunities to develop skills for medical decision-making over time—skills they will need once they transition into adulthood and begin making their own health care decisions (American Medical Association 2017). For Billy, having the opportunity to know of his diagnosis and participate in his health care decision-making early on (via provision of assent or otherwise) may help improve the quality of the therapeutic relationship by fostering his trust in both his parents and providers. In addition, allowing him to assent (or to dissent, if appropriate) to certain treatments conveys respect to him and also may reduce feelings of powerlessness about his condition by providing him as much control as is clinically reasonable. Demonstrating respect for the child and fostering trust early in the clinical relationship is important, and especially so when the child has a condition which will require frequent or long-term encounters with the health care system.

8.7.3 Right to an Open Future

When responding to parental requests for nondisclosure it is also important to consider the possible trade-offs associated with the decision. For example, if the request to withhold information is honored, are there opportunities that will be foreclosed to the child that might otherwise have been important to them? Though arguments (as above) can be made for *withholding* information in order to protect a child's open future (Feinberg 1980) (See also Chap. 3), arguments can also be made for *providing* that information to protect a child's right to an open future. For example, for children with incurable illness, withholding prognostic information could foreclose the opportunity to say goodbye to loved ones and friends, or to participate in legacy-making activities. Research examining the impact of legacy-making for children with serious illness has identified numerous benefits for both children and their family members, including giving ill children the chance to communicate about death and do and/or say something that will help them see that they will be remembered and providing a coping strategy for patients and families (Foster et al. 2012).

Billy is very early in his diagnosis. Hopefully he responds well to treatment, in which case legacy-making will not need to be a consideration in his care. However, for all options to remain open to him, he must be made aware (in an age- and developmentally-appropriate fashion) of his diagnosis and what choices he is empowered to make, especially given what might be at stake for him, and that he is the one who will have to bear both the benefits and burdens of treatment. As we have discussed, an open future argument could be made either in favor of or in favor against information disclosure. The best approach for an individual child is context-specific and will be dependent on a number of factors, such as the child's age and maturity level, the nature of the child's illness and their previous experience with it, patient and family dynamics and values, and sociocultural considerations.

8.7.4 Children Often Are Already Aware

Research also has shown that, particularly in the setting terminal illness, a child may already be aware of their poor prognosis but not "let on" out of a desire to protect their caregivers. Children and their caregivers may engage in a behavioral dynamic known as "mutual pretense," where all parties are aware of the child's poor prognosis, but none acknowledges this openly (Sisk et al. 2016). As a result, attempts to withhold potentially harmful information actually may be based on a false premise—that the child is unaware of the information. It is certainly conceivable that in our case Billy already is aware that he has cancer (or at least that he has something other than "another infection") such that withholding that information from him—no matter the rationale for doing so—cannot achieve the parents' goal of harm avoidance. In fact, it is possible that harm could be compounded by nondisclosure if Billy already knows about his diagnosis or has sensed that caregivers are not being truthful with him. For example, it is possible that without disclosure of his diagnosis he could imagine that he is far sicker than he actually is, causing him the fear and anxiety that his parents are hoping to avoid. The very nature of mutual pretense, when this is present, makes it difficult to discern precisely how much a child understands about their medical condition. However, this warrants consideration for discussion with parents so that they can be best equipped to make health care decisions that they believe to be in their child's best interest.

8.7.5 Trust

A final important consideration, as described further above, is that of trust. Trust is a fundamental component of health care; and withholding information—particularly if that information could later be discovered—runs the risk of irreparably harming the trilateral clinical relationship (Katz and Webb 2016). This is a particularly compelling argument, as once trust is broken, it can be very difficult to repair. In Billy's case, it is conceivable that he would later learn the truth about his diagnosis, which could greatly harm his relationship both with his parents and the health care team and possibly have a negative impact on his willingness to engage in his care going forward.

8.8 Differences Based on Type of Information

Health care decisions rest on numerous types of information: diagnosis, prognosis, treatment options, potential treatment toxicities, etc. Parents may prefer not to disclose any or all of these information types to their minor children throughout the illness trajectory. The details and potential implications of non-disclosure vary somewhat with the type of information being withheld. Some pediatricians and ethicists might approach this case differently if Billy's parents wished to withhold his new cancer diagnosis from him than they would if his parents wished not to tell him that, for example, his cancer had returned after multiple courses of treatment and he was now expected to die. It is important, when considering both how best to integrate a minor into their health care decision-making and whether/how to provide information to them, to attend to the type of information to be disclosed and the child's place in their illness trajectory. As described by anthropologist Myra Bluebond-Langner regarding her research with children with leukemia throughout the illness trajectory, it may well be that the greatest challenge "is not whether to tell, but how to tell, in a way that respects the children and all of their many, often conflicting needs" (Bluebond-Langner 1980, p. 235).

8.9 Practical Guidance for Navigating Parental Requests for Nondisclosure

Every patient is unique, as are the circumstances surrounding their care. As a result, all instances of parental requests for nondisclosure should be considered on a case-by-case basis to ensure that the unique considerations of the patient, their family, and their socio-clinical circumstances are adequately considered. Here, we provide some general practical guidance for considering these challenging situations, summarized in Table 8.1, alongside general guiding principles when considering truth-telling and information disclosure in this setting.

Table 8.1 Guiding principles for truth-telling and information disclosure with minors

Truth-telling is a core, but complex, ethical value in health care
- The obligation to tell the truth is grounded in the respect we have for the people (children or otherwise) to whom we are telling the truth. In pediatrics, truth-telling is nuanced due to the variability in age and developmental status of children. Parents occasionally wish to withhold the truth from their children, typically in an attempt to protect them or otherwise serve their children's best interests

Arguments for withholding information from children
- Common arguments for withholding health care information from children include the avoidance of harm, rights-based arguments, sociocultural considerations, and arguments based in the child's right to an open future

Arguments against withholding information from children
- Common arguments against withholding health care information from children include rights-based arguments, those based in the child's right to an open future, the practical point that children often are already aware of the truth, and considerations related to supporting and maintaining trust

Practical guidance
- Information disclosure should be situation-dependent, taking into account factors related to the child, the parents, the information being disclosed, etc. Disclosure is not an all-or-nothing concept, and it may be personalized to the situation as needed
- Clinicians should always clarify why parents are requesting that information be withheld from their child and inquire about parents' values/beliefs/hesitancies. They can then help to correct any misunderstandings and engage in discussion about potential benefits of disclosure
- When exploring how/whether to disclose information to a minor (particularly over the objections of a parent), it is important to maintain collaborative rapport with the parents while still maintaining one's sense of professional integrity
- Seek support and engagement of other members of the multidisciplinary health care team, particularly in the setting of conflict and/or uncertainty

Ultimately, all things being equal, clinicians have a *prima facie* obligation to tell children the truth about their diagnosis, prognosis, and treatment in an age- and developmentally-appropriate manner, unless there is a compelling, morally relevant reason to do otherwise. The questions of how and when to do so can be tremendously challenging to resolve; optimal solutions will likely depend on a number of factors, described further below.

8.9.1   Explore Rationale for the Request for Nondisclosure

First, parents may have reasonable and ethically justifiable reasons for requesting that certain information be withheld from their child. Clinicians should seek to understand why parents are requesting to have information withheld. For example, as discussed earlier in this chapter, is the request based on a concern that the child will be harmed by the information? Or are there other factors at play, such as parental misunderstanding or cultural/religious values? As appropriate to the child's situation, parents should be made aware of the possible *benefits* of disclosure. These could include the potential for a child's increased buy-in to the treatment plan, improved quality of the therapeutic relationship with providers, and the opportunity to empower a child in his or her own health care (Hudson et al. 2019; Mack and Joffe 2014). Understanding the rationale behind a request to withhold the truth may help inform how best to proceed and whether involvement of additional staff and/or resources might be beneficial. Such queries may even lead to the parents changing their mind about withholding information from their child without needing further intervention. In Billy's case, the health care team should ask Billy's parents why they wish to withhold information about his cancer diagnosis. They should further ask about the parents' beliefs, values, and hesitancies as a way to better understand their perspectives (and hesitancies), and to serve as a foundation for further conversations.

8.9.2   Maintain Collaborative Rapport

Superseding parental refusals to disclose information to their child over the explicit objection of parents carries the possibility of eroding parents' trust in the clinical team or otherwise harming the clinician-parent-child relationship. Harming this relationship rarely if ever serves the best interest of the child and should be avoided if possible. Ideally, parents and the clinical team will come to consensus about not only what is told to the minor, but when and how. It is critical that clinicians strive to maintain the collaborative rapport with parents that is so fundamental to providing high quality pediatric health care. Building upon the initial questions aimed at understanding Billy's parents' perspectives, in this case, the health care team should continue to work with them, in hopes of finding a path forward that enables the delivery of age-

and developmentally-appropriate information to Billy in a fashion agreeable to his parents while maintaining the therapeutic alliance.

8.9.3 Uphold Professional Integrity

At the same time, the moral integrity of clinicians is an important consideration in this calculus and should be promoted and supported. Clinicians should not be expected to abdicate their own professional morals/responsibilities. For example, even if a parent wishes to wait to tell their child about the results of a new test, if the child explicitly asks the clinician, they should not be expected to lie in order to support the wishes of the parents. In such circumstances, a clinician might determine that a shift in favor of disclosure to the child is ethically justified. This may be based on various reasons, including the desire to respect the child's developing autonomy and/or to preserve the clinician's professional integrity. A "compromise approach" could be to inform parents who have requested that information be withheld that clinicians will not lie in response to a child's direct question. But that they will, at least for the time being, strive to calibrate responses narrowly, in proportion to the child's query, while continuing collaborative communication between clinicians and parents about how best to discuss these issues with the child going forward. Clarifying expectations a priori may help to minimize damage to the clinician-parent relationship if such a situation were to arise. It also aims to strike a challenging balance among the interests and perspectives of the parents, clinicians, and pediatric patient. Clearly, these are exceedingly challenging situations to navigate for health care providers and parents alike, especially when agreement cannot be reached in advance about how best to proceed. In this case, while working with Billy's parents to determine a mutually agreeable plan, the health care team members should be clear with the parents about their own professional obligations, including what they will (and will not) say to Billy in order to ensure mutual understanding and expectations.

8.9.4 Employ a Multidisciplinary Approach

Finally, it is generally advisable to involve other members of the health care team to optimize the skillset and expertise at navigating these challenging dilemmas. This also carries the benefit of dispersing moral responsibility for a potentially distressing decision among members of the health care team. There are many clinical scenarios in which utilizing a multidisciplinary team approach is beneficial; this can be especially true when parents and clinicians are unable to agree about whether health-related information should be disclosed to a child. As such, depending on the institutional resources available, involvement of other skilled clinicians, such as social workers and psychosocial clinicians, palliative care specialists, clinical ethicists, chaplains, child life specialists, communication and developmental specialists, and others, can aid in decision-making. They also can provide additional support, as needed, to the child and parents, as well as to members of the primary clinical team. In addition, it can be helpful to seek consultation from community-based clinicians, such as primary care pediatricians, who may know the child and family longitudinally and be able to assist with communication and with insights regarding intrafamilial relational dynamics. Lastly, support and guidance from other institutional resources such as patient-family relations and legal services may also be beneficial. Such multidisciplinary collaboration can help support all involved stakeholders, provide alternative points of view, and supplement the expertise and skills of the clinicians tasked with navigating this challenging and distressing situation. In Billy's case, it is likely that the support of social work, psychosocial support, ethics, and/or Billy's primary care pediatrician could help the inpatient health care team as they proceed forward with Billy's parents.

8.10   Conclusion

Communication is a core component of modern health care, with the delivery of truthful, comprehensive, high-quality information at the core of compassionate communication. In pediatrics, the trilateral nature of the clinician-parent-child relationship makes this communication more complex, and sometimes more ethically fraught. Minors, though typically not legally empowered to make independent decisions about their health care, should generally be provided information in an age- and developmentally-appropriate fashion in a manner consistent with their parents' preferences. Occasionally, parents will request to withhold information from their child, typically wishing to do so in attempt to benefit (or protect from harm) their child. Such situations can be very challenging for clinicians, who should work to understand the parents' rationale for withholding the truth and then ideally work with a multidisciplinary team to find a disclosure plan that is amenable to both the parents and clinicians, while balancing competing ethical obligations. Though decisions about information disclosure in pediatrics are nuanced and should be considered on a case-by-case basis, these recommendations can help guide clinicians who are tasked with skillfully navigating these challenging circumstances.

References

American Medical Association. 2017. *Code of medical ethics of the American Medical Association.* Chicago: American Medical Association.
Beauchamp, T.L., and J.F. Childress. 2009. *Principles of biomedical ethics*, 6th ed. New York: Oxford University Press.
Bluebond-Langner, M. 1980. *The private worlds of dying children.* Princeton University Press.

Committee on Hospital Care and Institute for Patient- and Family-, Centered, and Care. 2012. Patient-and family-centered care and the pediatrician's role. *Pediatrics* 129 (2): 394–404. https://doi.org/10.1542/peds.2011-3084.

Diekema, D. 2004. Parental refusals of medical treatment: The harm principle as threshold for state intervention. *Theoretical Medicine and Bioethics* 25 (4): 243–264. https://doi.org/10.1007/s11 017-004-3146-6.

Feinberg, J. 1980. The child's right to an open future. In *Whose child? Children's rights, parental authority, and state power*, ed. William Aiken, and Hugh LaFollette, 124–153. Totowa, NJ: Littlefield Adams.

Foster, T.L., M.S. Dietrich, D.L. Friedman, J.E. Gordon, and M.J. Gilmer. 2012. National survey of children's hospitals on legacy-making activities. *Journal of Palliative Medicine* 15 (5): 573–578. https://doi.org/10.1089/jpm.2011.0447.

Hudson, N., M. Spriggs, and L. Gillam. 2019. Telling the truth to young children: Ethical reasons for information disclosure in paediatrics. *Journal of Paediatrics and Child Health* 55 (1): 13–17. https://doi.org/10.1111/jpc.14209.

Jonsen, A.R., M. Siegler, and W.J. Winslade. 2015. *Clinical ethics: A practical approach to ethical decisions in clinical medicine*, 8th ed. New York: McGraw-Hill Education.

Katz, A.L., and S.A. Webb. 2016. Informed consent in decision-making in pediatric practice. *Pediatrics* 138(2). https://doi.org/10.1542/peds.2016-1485.

Leikin, S.L. 1983. Minors' assent or dissent to medical treatment. *Journal of Pediatrics* 102 (2): 169–176. https://doi.org/10.1016/s0022-3476(83)80514-9.

Lo, B. 2009. *Resolving ethical dilemmas: A guide for clinicians*. Philadelphia: Wolters Kluwer Health/Lippincot Williams & Wilkins.

Mack, J.W., and S. Joffe. 2014. Communicating about prognosis: Ethical responsibilities of pediatricians and parents. *Pediatrics* 133 (Supplement 1): S24–S30. https://doi.org/10.1542/peds.2013-3608E.

McCullough, L.B. 2009. Contributions of ethical theory to pediatric ethics: Pediatricians and parents as co-fiduciaries of pediatric patients. In *Pediatric bioethics*, ed. G. Miller, 11–21. Cambridge: Cambridge University Press.

Rosenberg, A.R., H. Starks, Y. Unguru, C. Feudtner, and D. Diekema. 2017. Truth telling in the setting of cultural differences and incurable pediatric illness: A review. *JAMA Pediatrics* 171 (11): 1113–1119. https://doi.org/10.1001/jamapediatrics.2017.2568.

Sisk, B.A., M. Bluebond-Langner, L. Wiener, J. Mack, and J. Wolfe. 2016. Prognostic disclosures to children: A historical perspective. *Pediatrics* 138 (3): e20161278. https://doi.org/10.1542/peds.2016-1278.

Further Readings

Bluebond-Langner, M. 1980. *The private worlds of dying children*. Princeton University Press.

Hudson, N., M. Spriggs, and L. Gillam. 2019. Telling the truth to young children: Ethical reasons for information disclosure in paediatrics. *Journal of Paediatrics and Child Health* 55 (1): 13–17. https://doi.org/10.1111/jpc.14209.

Katz, A.L., and S.A. Webb. 2016. Informed consent in decision-making in pediatric practice. *Pediatrics* 138(2). https://doi.org/10.1542/peds.2016-1485.

Mack, J.W., and S. Joffe. 2014. Communicating about prognosis: Ethical responsibilities of pediatricians and parents. *Pediatrics* 133 (Supplement 1): S24–S30. https://doi.org/10.1542/peds.2013-3608E.

Rosenberg, A.R., H. Starks, Y. Unguru, C. Feudtner, and D. Diekema. 2017. Truth telling in the setting of cultural differences and incurable pediatric illness: A review. *JAMA Pediatrics* 171 (11): 1113–1119. https://doi.org/10.1001/jamapediatrics.2017.2568.

Chapter 9
Parental Refusal of Beneficial Treatments for Children: Ethical Considerations and the Clinician's Response

J. C. Bester

Abstract When parents refuse beneficial treatments for their children, it creates a unique set of ethical considerations for the clinician. This dilemma pits respect for parental authority and recognition of the parent–child relationship as an important childhood interest against the clinician's obligations to promote and protect the health-related interests and wellbeing of the child. This chapter examines the ethical considerations central to this dilemma and provides practical guidance for responding to parental refusals of treatment.

Keywords Parental refusal of treatment · Pediatric decision-making · Childhood wellbeing · Pediatric ethics

9.1 Introduction

9.1.1 How Pediatric Medicine is Different

Sometimes patients refuse beneficial treatments that, if accepted, would help them live better and live longer. In adult medicine it is fairly well established that patients have the right to refuse any treatment, no matter the implications to the patient, as long as the patient is making a truly autonomous decision and is not lacking in capacity in some way (Lo 2013, pp. 85–89). This reflects the respect we have for patients' autonomy: the freedom and right that patients have to make decisions about their own care and their own bodies require us to get a patient's authorization before providing treatment. In part, it also reflects the realization that what counts as a benefit in the first place is somewhat dependent on a person's life-plan and values; I decide for myself what counts as good for me, and I therefore have a large say in what counts as a benefit to me in the first place.

J. C. Bester (✉)
Kirk Kerkorian School of Medicine at UNLV, University of Nevada, Las Vegas, USA
e-mail: johan.bester@unlv.edu

© Springer Nature Switzerland AG 2022   143
N. Nortjé and J. C. Bester (eds.), *Pediatric Ethics: Theory and Practice*,
The International Library of Bioethics 89,
https://doi.org/10.1007/978-3-030-86182-7_9

The situation is very different in pediatric medicine. In pediatrics it is usually the case that the patient lacks the capacity for autonomous decision-making, and has not yet developed the maturity, the set of values, and the long-term life goals that would undergird the kind of reasoning that would allow the patient to decide for themselves what counts as a benefit to themselves (AAP 1995; AAP 2016). For this reason, the primary consideration in pediatric medicine is the best interest of the child and not autonomy; we first think of how to secure the child's wellbeing when making decisions about the child (refer to Chap. 4). This is usually articulated as the obligation resting on those providing care to the child and making decisions for the child to act in the best interest of the child. This is a general stipulation when making decisions or policy that directly affect children, but also specifically so in pediatric medicine. This means that doing right by the child is the primary focus of pediatric medicine, placing obligations on those who have authority to make decisions for children.

Straight-forwardly, then, parents and clinicians must do what would protect and promote the wellbeing of the child in the course of their decision-making and care. When parents (or other decision-makers)[1] refuse beneficial treatments for their child it sets up an ethical challenge, one that must be responded to by clinicians who care for the child and perhaps even by society itself. This dilemma is that on one hand, clinicians and society must respect parental authority and the parent's role in the child's life, while on the other hand they must ensure that children receive those treatments that would protect the child's wellbeing.[2]

9.1.2 How Children Are Morally Situated: The Nature of the Dilemma When Parents Refuse

Parents have authority to make decisions for their children with the idea that parents will protect and promote the wellbeing of the child.[3] There are good reasons for

---

[1] I will use the term "parents" to refer to those who stand in the role of primary caregiver and decision-maker on behalf of the child. We should recognize that though in a majority of cases parents are the decision-makers for their children, some children are raised by people filling this role that are not their biological parents—say, grandparents, foster parents, adoptive parents, or appointed guardians—who then also act as medical decision-makers for these children. I will use the catch-all phrase "parents" both for simplicity sake and to reflect that the decision-maker stands in the parental role.

[2] In this chapter I will focus on the ethical issues related to parental refusal for treatment. This assumes a situation where the child lacks the capacity for autonomous decision-making, and the parent is the designated decision-maker for the child. In cases where children are considered to be mature minors or to have the capacity to make their own decisions autonomously, the usual considerations familiar to clinicians apply. That is, patients may autonomously refuse beneficial treatments if they have the necessary decision-making capacity for an autonomous refusal.

[3] Wellbeing is an all-encompassing term, and does not just focus on health, but also on how well someone is doing overall. There are different theories of wellbeing in the philosophical literature; one group of theories equate wellbeing to fulfillment of wishes and desires, so that wellbeing is

this: parents love their children and would seek to protect the wellbeing of their children; parents know their children and the circumstances of children best, so that they know better than others what options are more feasible and workable in the lives of the child; parents have a stake in the outcome and will have to live with the effects of the decisions that are ultimately made (Buchanan and Brock 1990). Parental authority is not unlimited, but rests on the foundation of the child's wellbeing. Parental authority is meant to protect the wellbeing, or the interests, of the child. Parents cannot make arbitrary decisions about children but must be guided by the outcomes of their decisions on the wellbeing of the child. Parental authority that steps outside of the bounds of protecting and promoting the wellbeing of the child is illegitimate and should rightly be challenged and limited by society. There are a variety of ways to articulate these ideas. One way to say this is that the parents are custodians of the child's rights, that the parents hold the child's rights in trust, and must make decisions that protect the child's future rights (Millum 2014). Another way to say it is that parents must promote and protect the interests of their child, so that in all decisions made for the child the parent must consider the impact of the decision on the child's various interests, and choose the available option that has the best chance of protecting and protecting the child's wellbeing, all things considered (Bester 2018a, 2019; Kopelman 1997). One could appeal to the UN Charter on the Rights of the Child, which articulates a set of moral obligations owed to children by society, parents, and others who stand in morally significant relationships with the child (UN 1990) (also refer to Chap. 5). Or, one could say that parents have the freedom to make decisions for the child using whichever moral criteria they wish, but that this freedom is constrained by the basic interests of the child; parents must ensure that the basic interests of the child are met (Ross 1998).

Parental authority is one consideration, but there is another. The parent–child relationship is itself an important childhood interest (Malek 2009). Children are situated in families, grow up under the care of a parent, and much of the child's wellbeing depends on this relationship. It is a morally significant relationship, and to burden this relationship unduly is to harm the child. This is a very intimate relationship, where the interests of parents and children are intertwined in complex ways (Elliot 2001). To disrupt the parent–child relationship, to remove the parent from the child, may represent one of the worst attacks on the wellbeing of the child imaginable. For this reason, the UN Charter on the Rights of the Child is clear on the rights of children to be raised by their own parents, and on the protections that should be afforded the parent–child relationship by the state and by society in general (UN 1990, see articles 4, 9, 10). Yet, there may be times when disrupting such a relationship is

---

determined by the person's own view of the good. Other theories describe objective lists which can be used to measure someone's wellbeing using the items on the list (Bester 2020). Since children have not developed an independent view of the good and a life-plan, it makes sense to describe childhood wellbeing in terms of an objective list linked to childhood flourishing and development. See for example Powers and Faden (2006), who describe wellbeing as achieving a minimum level over six dimensions: health, personal security, attachments, reasoning, self-determination, and respect. From this concept of wellbeing one can develop a set of childhood interests, components of wellbeing that should be present to protect the child's flourishing and development.

morally required; when parents are abusive, or when parents themselves threaten the wellbeing of the child, it may be necessary to intervene to protect the wellbeing of the child.

Clinicians stand in morally significant relationships with their patients, and have the obligation to protect and promote the wellbeing of these children (Bester 2019). Society is charged with the responsibility of protecting the wellbeing of the vulnerable within society, which includes children: justice requires of society to ensure for children what is needed to flourish (Powers and Faden 2006). Indeed, this is one of the central assumptions within the UN Charter on the Rights of the Child. Clinicians and society must therefore underwrite for children those things that children need in order to flourish, to achieve wellbeing. Clinicians play an important role here as health caregivers—offering those treatments and interventions that protect the health of a child, a critical component in childhood flourishing.

It is therefore clear what is at stake. Clinicians must respect parental authority, and must recognize the parent–child relationship as an important childhood interest. Thus, clinicians must in general take steps to involve parents in the care of children, must ask for parental permission before treatment, and must promote the parent–child bond in all aspects of medical care. At the same time, clinicians must be advocates for the health and wellbeing of their child-patient. The child is vulnerable, dependent on others to protect the child's health and wellbeing. The clinician must therefore challenge decisions by the parent that step outside of the bounds of parental authority, or that are contrary to the wellbeing of the child. Beneficial treatments are those treatments that promote and protect the wellbeing of the child, so when parents refuse beneficial treatments clinicians recognize a duty to ensure that the child receives the treatment. We see here conflicting duties that have to be weighed and balanced, so that clinicians respond in ways that preserve their various obligations to the greatest extent and so that clinicians have sound reasons that justify their actions.

9.2 Ethical Considerations that Are Important in Weighing the Dilemma

When faced with a refusal of beneficial treatment, then, we confront an ethical challenge. To work our way through the dilemma, we must find an action that best resolves the dilemma and justify this action through sound moral reasons. In other words, we must find a way to act that is best given the moral complexity we face, and must give good reasons why this way of resolving the dilemma is better than others. In doing so, a number of ethical considerations impinge on our analysis of any given refusal of treatment, and I consider these briefly here.

9.2.1 How Beneficial the Treatment is: Weighing Benefits, Risks, and the Impact on the child's Interests

To say that something counts as a benefit to a patient is to say that this something improves the patient's wellbeing. In terms of health care, it is often the case that treatments remove disease, symptoms, or suffering that impinges on wellbeing, thereby advancing or protecting the patient's wellbeing (Bester 2020). When it comes to children, wellbeing is defined in terms of interests, identifiable components of wellbeing necessary for childhood development and flourishing. To say that we benefit the child is to promote the child's interests and ultimately the child's wellbeing. In various accounts of childhood wellbeing, the health of the child is an essential interest that occupies a central role in the child's overall wellbeing (Malek 2009; Powers and Faden 2006). Compromises in the child's health leads to compromise in other domains of wellbeing, and often locks in insufficient wellbeing life-long. For children to flourish they need to be healthy; and children need to flourish to ensure life-long wellbeing. Benefits that accrue from medical treatments are aimed at protecting the health of the child, and thereby protecting life-long wellbeing.

When weighing the benefits of a treatment, one has to account for the benefits and the risks. With any given treatment, there is potential for benefits and there is potential for adverse effects. But the risks of a treatment in a child may not always be health care related; treatments meant to advance the child's health may impact on a child's other interests. For example, while a lengthy hospitalization may be needed to treat an illness and restore a child to health, if a child's relationships and education are neglected during this time, it sets back the child's interests in important ways. Another way to say this is that any treatment may advance some of a child's interests, while setting back others.

To consider whether a treatment is beneficial or not, we need to weigh the expected outcomes of the treatment against a variety of a child's interests, and consider how the treatment course impacts the child's interests as a whole (Bester 2018a). Those treatments where the child's interests are overall promoted and protected would be considered beneficial. Here we may think of life-saving treatments or interventions where the benefit clearly outweighs any perceived risks to the child. In contrast with these, some treatments would have beneficial effects but would also carry the risk of setting back some of the child's interests, so that the risk–benefit ratio is less clear, but overall, we think children would benefit if the treatment were used.

These considerations influence how a clinician would work through the dilemma of parental refusal of treatment. Consider a treatment of the first kind, with significant benefits to the child, low risk of adverse effects, and not setting back other childhood interests in a significant way. Such a treatment is clearly beneficial, and the obligations of parents and clinicians to further the wellbeing of the child would stipulate that parents and clinicians must ensure that the child receives the treatment. Here we cannot talk of parental prerogative or parental discretion, but this is the arena of parental obligation.[4] If a parent refuses a treatment in such a situation, it amounts to doing wrong to the child, neglecting their parental duty to protect the child's wellbeing. Parents stray outside of the bounds of parental authority when they refuse beneficial treatments of this kind. If this happens, the obligation rests on the clinician to challenge the parental decision, and perhaps to use the avenues provided by society and its institutions to overrule the parental decision. Simply stated: if a parent refuses a treatment that is clearly beneficial, where the benefits clearly outweigh any risks to the child, this decision must be challenged and possibly overruled by clinicians and/or the state.

But not all treatments are of this kind. Some treatments are riskier than others in terms of health-related complications. There is a clear difference between a low-risk surgery that is highly effective (say, appendicectomy), and a risky cardiac surgery with a 60% chance of success but a 40% chance of serious adverse effects. Some treatments may impact other childhood interests, perhaps in an unintended fashion, so that it is not always clear how much the net benefit is over risk to the child's overall wellbeing with a given treatment. With treatments of this kind, where benefits are less certain and potential for risk is higher, reasonable people may come to different conclusions about how to weigh the risks and benefits. With such a treatment, it is less concerning if a parent refuses the treatment. Here we have good reasons to weight parental authority and discretion higher. In such a situation, parent and clinician together survey the options available, measure the expected impact on the child's various interests, and together decide about the best way forward, all things considered. Here is a place for true shared decision-making, where parent and clinician work together to promote and protect the interest of the child in the best way they can figure out together.

It is important when considering treatment refusal to consider the nature of the treatment and the expected benefits and risks, weighing these outcomes by the effect they will have on the child's wellbeing overall. The more beneficial and low-risk the treatment, the more the ethical calculus weighs towards ensuring that the child receives the treatment despite parental objection. The less certain the benefit, and the higher the risk to the child's various interests, the more the ethical calculus weighs towards shared decision-making and deference towards the parent's authority (Lo 2013, p. 267–268).

---

[4] To be clear, these are moral obligations. Parents have the moral responsibility to protect and promote the wellbeing of their children by virtue of the special role and relationship that exist between parents and children. These moral obligations can be grounded in a variety of ways, by reference to different moral approaches. For example, I've argued that these could be social obligations by virtue of the parental social role, with society's justice obligations to provide for childhood wellbeing resting in large part on parents (Bester 2018b). These are not to be understood as legal obligations, though it may be that society creates legal frameworks and policies to protect the wellbeing of children that may place similar legal obligations on parents.

9.2.2   The Parent–Child Relationship as an Important Childhood Interest

It is undoubtedly true that health is one of the most important aspects of securing childhood wellbeing, and that health therefore is a central childhood interest. But it is equally true that the parent–child relationship is an important childhood interest (Malek 2009; UN 1990). The parent is not just someone who makes decisions for the child; the parents is herself part of the child's overall wellbeing and represents an important childhood interest. This provides a good reason for clinicians to do their utmost to protect and promote the parent–child relationship. Including parents in the child's care is important for overall childhood wellbeing, not just because of parental decision-making authority.

This consideration guides and constrains the response of the clinician and of society to parental refusal of treatment. While a response to the refusal of beneficial treatment is morally required, careful consideration should be given to the type of response so that the child who is meant to be protected is not ultimately harmed. Consider vaccine refusal as an example. Vaccines are some of the most effective interventions of modern medicine. Benefits far outweigh risks, and there are good reasons to think that vaccines represent something that is morally owed to children (Bester 2017, 2018b; Powers and Faden 2006). Parents who refuse beneficial vaccines therefore wrong their child, withholding something that would work towards protection of the child's wellbeing and that the child should receive. I have argued that clinicians and the state should respond to parental vaccine refusal, since vaccines is something that we owe children, something children can claim from society and those who care for them as a moral right (Bester 2018b) (also refer to Chap. 27). But society's response must consider the role of the parent in the child's life. For example, if society decides to jail parents for non-vaccination, think of the consequences to children whose parents are jailed. Such a response would so set back the wellbeing of affected children that it cannot be justified as a response to non-vaccination. In recent times, there have been growing popularity among pediatricians for a policy of patient dismissal for non-vaccination. Reasons given for such a policy include that it would motivate parents to vaccinate children, functioning as a persuasive lever to ensure vaccination and thereby responding in a morally justified fashion to the parental refusal of vaccination (Diekema 2013, 2015). Further consideration illuminates that the consequences of dismissal from a pediatrician's practice will fall on the child: to the harm of being non-vaccinated is added the injury of losing access to beneficial medical care. Besides, it is not clear that such policies change the minds of parents in the first place.

If clinicians and governments must respond to parental refusal of beneficial treatment, as we have good reason to suppose they do, such a response must take care not to injure the interests of the very child that is meant to be protected by the clinician's or state's intervention. Every response to parental refusal of treatment must by itself be evaluated by use of the best interest standard. The various response options must be weighed by considering their impact on various childhood interests, and the option should be implemented that has the best chance of protecting and promoting the child's overall wellbeing. To this idea we must add that the parent–child relationship itself is also an important interest that should be weighed in this process. No doubt there are cases where removing a parent's authority or removing a child from a parent–child relationship is the only way to ensure the wellbeing of the child. But in most cases, the response to parental refusal should be such that it includes protections for the child-parent relationship.

9.3   Responding to Parental Refusals of Treatment: Practical Guidance[5]

Fortunately, when parents and clinicians focus on the best interest of the child they very often agree on the course of treatment for the child. Even if there are minor disagreements or differences in perspective, it is usually possible to reach a mutually acceptable agreement on a course of action that is focused on the child's wellbeing first and foremost. This underlines the importance of creating a good clinical relationship with parents, skilled communication on the part of the clinician, and the need to focus on a broad set of childhood interests.

If parents refuse a treatment, the first step is to engage in a discussion aimed at reaching agreement between parent and clinician. If the treatment is clearly beneficial and required to protect the health and wellbeing of the patient, the goal is persuasion of the parent so that the parent would ultimately accept the treatment. If the treatment is less clearly beneficial, or if there are significant potential risks, the goal may be more aligned with a shared decision-making process, where parents and clinicians work together to find a mutually acceptable treatment plan. The goal of this first step is to find a mutually acceptable plan, to engage parents with the treatment plan, and to foster agreement. It may be necessary to find a compromise that is acceptable to parents and clinicians, that at the same time preserves the wellbeing of the child to an optimal level. Such a plan protects a wide range of the child's interests. If the clinician is successful in bringing about agreement around a treatment plan, it ensures that the child receives treatment while also promoting the child's interests related to the parent–child relationship.

If disagreement persists and negotiations stall, a clinician should call in help. There are a variety of resources available to aid parents and clinicians working through difficult clinical decisions where disagreements persist. Clinical ethicists may be helpful in addressing ethical questions and uncertainty and may also help with conflict resolution where disagreements have become intractable. If religious

---

[5] This section is my own suggestion of how to respond to parental refusals of treatment, and it draws on personal experience, arguments presented in this article and my other scholarly works, but also on themes running throughout the following sources: AAP (1995), AAP (2016), Diekema (2005), Lantos (2015), Lo (2013) (Chap. 4, Promoting the patient's best interest and Chap. 37, ethical issues in pediatrics); Lee et al. 2020.

values are at stake, spiritual care personnel or some form of chaplain or pastoral leader may be valuable to help illuminate and negotiate value differences. It is important that all additional services and resources take as starting point the goal to serve the interests of the child in question. Ultimately, the goal is to find a way forward to do what is right by the child.

Ultimately it may emerge that disagreements become intractable and that parents become entrenched in their refusal. If a treatment is clearly beneficial, a treatment where the benefits to the child's wellbeing clearly outweigh any potential risks related to the treatment, this may be the time to call on society's resources designed to challenge or overrule specific parental decisions in order to uphold the wellbeing of the child. This is particularly so for life-saving treatments or treatments that remove illness or suffering that may have long-term consequences for the child's wellbeing. Steps taken may include petitioning a court to provide a court order to authorize treatment or involving social services or child protective services to intervene. The route to follow would depend on the setting and jurisdiction of the clinician's practice. These steps are justified by the obligation resting on clinicians and society to ensure that children have the social circumstances and goods required to protect childhood wellbeing. Clinicians play a particularly important role here, given the care relationship and related obligations clinicians have with their child-patients. It should be noted that clinicians should carefully weigh the potential impact of calling on state or societal intervention on the child's broader interests, and particularly in terms of the parent–child relationship. For example, calling in a state resource may erode trust between parent and clinician, which may negatively impact the child's health care in future and consequently burden the child's wellbeing. Nevertheless, if a compelling case exists that a beneficial treatment is morally necessary because of its importance to the child's wellbeing, a clinician must make use of every resource available to uphold the child's wellbeing.

To summarize this guidance in point form. If parents refuse a treatment:

(1)   Decide whether the treatment is clearly a beneficial treatment where benefits to the child's wellbeing outweigh any burdens on the child's overall interests. If yes, proceed. If not, consider a process of shared decision-making to decide together on a treatment plan.

(2)   If parents refuse a beneficial treatment, engage in discussion aimed to foster agreement between parent and clinician. This may include persuasion, and may include skilled communication to resolve differences. The goal is to reach a mutually acceptable treatment plan through clear communication and compromise, while at the same time optimally protecting the child's wellbeing.

(3)   If disagreement persists, consider involving other resources and services. This may include ethics consultation, conflict resolution, and spiritual or pastoral care.

(4)   If a treatment is clearly beneficial and morally owed a child and disagreement or refusal becomes intractable, make use of the resources made available by society to protect the wellbeing of the child. This may include petitioning a court or involving some form of child protection. This option should

be reserved for situations where a benefit of the intervention outweighs the potential negative impact on the child's interests.

9.4 Conclusion

These steps are aimed at resolving the dilemma of parental refusal in a way that preserves the most value, that protects the interests of the child in the greatest possible way. Each intervention to challenge parental refusal must itself be considered in light of the potential impact on the child's interests. It is much better to find agreement, preserve parent-clinician trust, engage parents and children, and communicate clearly than to simply overrule parents. The wellbeing of the child and the best interest of the child remains the first and most important consideration, and is the justifying moral grounding for each of these recommendations. Ultimately, it may be necessary to use state power to protect the wellbeing of the child. Even if this is done, the clinician should do the best they can to preserve trust with the child and parents, and to minimize the impact of this intervention on the wellbeing of the child.

9.5 Guiding Principles and Summary: Parental Refusal of Beneficial treatments for Children

General relevant ethical principles
- Generally, parents have decision-making authority for their children. Parental authority is meant to protect the wellbeing (or interests) of the child
- The parent–child relationship is an important childhood interest
- Clinicians have obligations to protect and promote the wellbeing of children who are their patients
- Clinicians must take a wide view of the wellbeing (or interests) of children

Ethical considerations in weighing a response to refusal
- Consider the effects of a treatment on the interests of a child, weighing benefits against potential risks to all of a child's interests, to determine to what degree a treatment is beneficial for the child
- If a parent refuses a treatment which has clear benefits and low risks to the child, this decision must usually be responded to, challenged, and/or possibly overruled by the clinician and/or state
- A treatment where benefits are less clear or risks are more significant may lead to different decisions by reasonable people. Parental discretion should weigh higher, and shared decision-making is essential

(continued)

(continued)

Practical guidance for clinicians—responding to parental treatment refusal of a beneficial treatment
- First step: Engage in discussion with parents. The goal is to find a mutually acceptable way forward, and to engage parents in the treatment plan. This may require either persuasion or mutually acceptable compromise, or both
- Second step: If disagreement persists, call in help to facilitate decision-making and negotiation. This may include consulting services such as clinical ethics, spiritual care, ombudsman, social work, or the legal office, depending on the situation
- Third step: If parental refusal is intractable and the treatment is clearly beneficial to the child, consider making use of society's resources to challenge the parental decision. This may include a court order to authorize treatment, or involving child protective services, depending on the situation
- At each step, the primary consideration is always the wellbeing (interests) of the child

References

American Academy of Pediatrics, Committee on Bioethics (AAP). 1995. Informed consent in decision-making in pediatric practice. *Pediatrics* 95 (2): 314–317.

American Academy of Pediatrics, Committee on Bioethics (AAP). 2016. Informed consent in decision-making in pediatric practice. *Pediatrics* 138 (2): e20161484.

Bester, J.C. 2017. Measles vaccination is best for children: The argument for relying on herd immunity fails. *Journal of Bioethical Inquiry* 14 (3): 375–384.

Bester, J.C. 2018a. The harm principle cannot replace the best interest standard: Problems with using the harm principle for medical decision-making for children. *The American Journal of Bioethics* 18 (8): 9–19.

Bester, J.C. 2018b. Not a matter of parental choice but of social justice obligation: Children are owed measles vaccination. *Bioethics* 32 (9): 611–619.

Bester, J.C. 2019. The best interest standard and children: Clarifying a concept and responding to its critics. *Journal of Medical Ethics* 45 (2): 117–124.

Bester, J.C. 2020. Beneficence, interests, and wellbeing in medicine: What it means to provide benefit to patients. *American Journal of Bioethics* 20 (3): 53–62.

Buchanan, A.E., and D.W. Brock. 1990. *Deciding for others: The ethics of surrogate decision-making*, 232–234. Cambridge: Cambridge University Press.

Diekema, D.S. 2005. Responding to parental refusals of immunization of children. *Pediatrics* 115 (5): 1428–1431.

Diekema, D.S. 2013. Provider dismissal of vaccine-hesitant families: Misguided policy that fails to benefit children. *Human Vaccines & Immunotherapeutics* 9 (12): 2661–2662.

Diekema, D.S. 2015. Physician dismissal of families who refuse vaccination: An ethical assessment. *The Journal of Law, Medicine, & Ethics* 43 (3): 654–660.

Elliot, C. 2001. Patients doubtfully capable or incapable of consent. In *A Companion to bioethics*, ed. H. Kuhse and P. Singer, 452–462. Oxford: Blackwell.

Kopelman, L.M. 1997. The best interests standard as threshold, ideal, and standard of reasonableness. *The Journal of Medicine and Philosophy* 22: 271–289.

Lantos, J. 2015. The patient-parent-pediatrician relationship: Everyday ethics in the office. *Pediatrics in Review* 36 (1): 22–30.

Lee, K.J., D.L. Hill, and C. Feudtner. 2020. Decision-making for children with medical complexity: The role of the primary care pediatrician. *Pediatric Annals* 49 (11): e473–e477.

Lo, B. 2013. *Resolving ethical challenges*, 5th ed. Philadelphia, PA: Lippincott, Williams, and Wilkins.

Malek, J. 2009. What is really in a child's best interest? Toward a more precise picture of the interests of children. *The Journal of Clinical Ethics* 20 (2): 175–182.

Millum, J. 2014. The foundation of the child's right to an open future. *Journal of Social Philosophy* 45 (4): 522–538.

Powers, M., and R. Faden. 2006. *Social justice*. New York: Oxford University Press.

Ross, L.F. 1998. *Children, families, and health care decision-making*. New York: Oxford University Press.

United Nations (UN). 1990. *Convention on the rights of the child*. https://www.unicef.org/child-rights-convention/convention-text.

Further Reading

Bester, J.C., M. Smith, and C. Griggins. 2017. A Jehovah's Witness adolescent in the labor and delivery unit: Should patient and parental refusals of blood transfusions for adolescents be honored? *Narrative Inquiry in Bioethics* 7 (1): 97–106.

Diekema, D.S. 2005. Responding to parental refusals of immunization of children. *Pediatrics* 115 (5): 1428–1431.

Lo, B. 2013. *Resolving ethical challenges*, 5th ed. Philadelphia, PA: Lippincott, Williams, and Wilkins. Chapter 4, Promoting the patient's best interest and Chapter 37, Ethical issues in pediatrics).

Chapter 10
Caring for Adolescents: Unique Ethical Considerations

S. Barone and Y. Unguru

Abstract Adolescence is a period of rapid physical, cognitive and psychosocial growth that represents a progressive transition from childhood to adulthood. With this transition come unique health care concerns, an emerging quest for autonomy, and desire and capacity to make one's own health care decisions. Health care providers (HCPs) caring for adolescents require a general framework for navigating issues of confidentiality and consent for this unique segment of the pediatric population. An adolescent's decision-making capacity depends not only on their age, but on their developmental level and maturity, emotional intelligence, personal experience making determinative decisions, and other factors including, but not limited to, influence by peers, family, and their dominant culture. Providers should use all the information at their disposal to evaluate decision-making capacity and, therefore, the adolescent's ability to meet criteria for the provision of informed assent/consent on a case-by-case basis. Given adolescents legal standing as minors, some level of parental involvement is typically expected, and HCPs must possess appropriate communication skills for managing situations when there is disagreement between an adolescent patient and their parents or where parental involvement represents a barrier to confidential care. In general, relying upon principles of shared decision-making (SDM), providers must strive to provide comprehensive, accessible care for adolescents that is respectful of their evolving autonomy and emerging capacity, while supporting and facilitating open and honest communication between adolescents and their parents.

Keywords Adolescence · Clinical ethics · Consent · Confidentiality · Decision-making capacity · Communication

S. Barone
The Lighthouse, Children and Families, McGill University, Montreal, QC, Canada
e-mail: sbarone@phare-lighthouse.com

Y. Unguru (✉)
Division of Pediatric Hematology/Oncology, The Herman and Walter Samuelson Children's Hospital at Sinai and Johns Hopkins University Berman Institute of Bioethics, Baltimore, MD, USA
e-mail: Yunguru@lifebridgehealth.org

© Springer Nature Switzerland AG 2022
N. Nortjé and J. C. Bester (eds.), *Pediatric Ethics: Theory and Practice*,
The International Library of Bioethics 89,
https://doi.org/10.1007/978-3-030-86182-7_10

10.1 Introduction

Adolescence is a period of rapid physical, cognitive and psychosocial growth that represents a progressive transition from childhood to adulthood. One of the main tasks of adolescence is identity development. During this time, older children begin to make decisions for themselves independent of their parents and learn to navigate increasingly complex social situations while taking on greater levels of responsibility for their own health and wellbeing. This increasing independence is counterbalanced by the fact that adolescents' decision-making processes and health are intimately tied to their families, peers and wider societal influences such as school, the dominant culture, socioeconomic class and political structures (Dahl et al. 2018).

Adolescents comprise 1.2 billion or one-sixth of the global population (World Health Organization 2020a). The health-related behaviors which often begin in adolescence (diet, exercise, sexual activity, experimenting with drugs and alcohol) can have a significant impact not only on adolescents' current health and wellbeing, but also on their future health. For example, cigarette smoking or obesity in early adolescence not only compromise adolescent health and development, but predict health-compromising tobacco use and obesity in later life, which has important personal and public health implications (Chassin et al. 1990; Simmonds et al. 2016). For this reason, it is especially important for adolescents to have access to equitable and effective health care dispensed in confidential and supportive environments.

Clinicians who provide care for adolescents must have a general framework for navigating issues of consent, assent, and decision-making capacity for this unique segment of the pediatric population that is not yet fully autonomous, but in a transitional period of emerging autonomy. One of the main challenges faced by health care providers (HCPs) caring for this population is the obligation to provide comprehensive patient-centered care in a manner that respects both the adolescents' need for confidentiality and autonomy with parents' desired level of involvement and their legal rights. This chapter will provide a foundation for navigating ethical issues related to the clinical care of adolescents. We will first discuss the basic notions of consent and assent and the special consideration these concepts afford adolescents. We then discuss the unique significance of respect for confidentiality in this population and examine emerging capacity in adolescents with regard to medical decision-making. Finally, we apply some of the basic principles discussed to specific clinical situations encountered with adolescents where ethical challenges may arise.

10.2 Definition of Adolescence and Adolescent Development

The World Health Organization defines adolescence as the phase of life between childhood and adulthood, spanning ages 10–19 years (World Health Organization

2020a). The American Academy of Pediatrics (AAP) identifies adolescence as 11–21 years of age, dividing the group into early (ages 11–14 years), middle (ages 15–17 years) and late (ages 18–21 years) adolescence (Hardin et al. 2017) (refer to Chap. 2). However, it is abundantly clear that age is an overly simplistic parameter used to delineate this period and that any chronological threshold between childhood and adulthood is, to a certain extent, arbitrary. Sawyer et al. (2018) proposed an expanded and more inclusive definition of adolescence, 10–24 years that accounts for evidence of earlier puberty in nearly all populations as well as continued growth, development and delayed timing of role transitions well into the 20 s. This proposed definition acknowledges that the social and emotional development that occur throughout adolescence depend on a wide variety of factors beyond age, including the immediate and larger socio-cultural environment. External influences, which both affect and are affected by the physical changes of adolescence, and differ among cultures and societies, include social values and norms and the changing roles, responsibilities, relationships and expectations of this period of life (World Health Organization, 2020b). Both the AAP and Canadian Pediatric Society (CPS) discourage arbitrary age limits on pediatric care by HCPs. The CPS, for example, proposes a definition of adolescence wherein adolescence begins with the onset of physiologically normal puberty and ends when an adult identity and behavior are accepted (Sacks and Canadian Pediatric Society Adolescent Health Committee 2003). Regardless of the definition employed, HCPs must be flexible accommodating specific situations such as the emancipated minor (discussed further below) or the young person with a chronic condition who may exhibit developmental delay and prolonged dependency or alternatively, heightened awareness and capacity for certain types of decisions.

10.3 Consent, Assent and Emerging Capacity

10.3.1 Consent

The pediatric model of patient- and family centered care recognizes that patients and their families are integral partners with the health care team and that the child and family's perspectives are essential components of high-quality clinical decision-making. With a few exceptions, parents are granted authority to make medical decisions on behalf of minor children, including adolescents. Parents are generally afforded a great deal of discretion in terms of the choices they make, so long as their choices do not place the child at risk of serious harm as compared with the alternatives.

By definition, children constitute a vulnerable population, relating directly to their limited decision-making capacity (i.e., the ability to make reasonable decisions). Adolescents, though potentially possessing significantly greater decision-making capacity than younger children, may have extra sources of vulnerability.

Some adolescents may experience emotional or psychological distress related to many of the health-related concerns that arise in this period of life, e.g., navigating romantic and sexual relationships and the potential for coercion or abuse; contraceptive care; experimentation with drugs and alcohol; and issues of peer pressure and bullying. As the appropriate surrogate decision-makers for minors, only parents (or legal guardians) can legally provide permission (consent) to treatment and procedures for their child. The goals of the informed consent process, which include protecting and promoting health-related interests and incorporating the patients and/or families values in health care decision-making apply to both the pediatric and adult population and are grounded in the same ethical principles of beneficence, justice, and respect for autonomy.

There are several exceptions that grant special status to minors under the age of 18, whereby they are able to provide consent for their medical care and services. Examples are the mature minor and the emancipated minor.

10.3.2   The Mature Minor

In recent years, legislators and clinicians have recognized the rights of a mature minor—i.e., an adolescent who possesses the necessary cognitive and emotional aptitude to provide his or her own informed consent. Empirical research dating to the 1980s demonstrates that by age 14 years, adolescents manifest emotional maturity, cognitive capabilities, and decisional competence comparable to adults. If an adolescent can exhibit the ability to understand factual issues, potential outcomes, the consequences of each alternative, and the meaning of the decision within the framework of their personal values that adolescent displays all the prerequisites to provide informed consent. The ability to provide informed consent is less a function of the adolescent's age and more a function of their individual level of emotional maturity and psychosocial development. Therefore, each specific patient's capacity to provide voluntary consent must be assessed within the context of the complexity of the decision and the youth's psychosocial level of development and current mental and emotional status. Furthermore, legal standards with regard to mature minors can vary widely not only between countries, but also between jurisdictions within the same country. In some jurisdictions, mature minor status is conferred as part of a formal legal process. In others, the designation is issued informally for adolescents who have met the criteria for capacity according to their HCP. For example, in the United States of America, all 50 states have enacted some form of legislation that grants mature minors the authority to consent to health care decisions related to contraceptives, pregnancy, sexually transmitted infections, mental health, or substance abuse (Belitz and Bailey 2009). Specific laws vary from state to state and many of the laws permit, but do not require, parental notification. In Canada, a patient need not reach the age of majority to give consent to treatment. In most Canadian provinces, the determinative factor in a child's ability to provide or refuse consent is whether the adolescent's physical, mental and emotional development allows them to meet the necessary criteria for informed consent—whether or not the patient has attained the age of majority (Canadian Medical Protective Association 2016). The exception is the province of Québec where the Civil Code generally establishes the age of consent at 14 years, below which the consent of the parent or guardian, or of the court is required. In Québec, adolescents over age 14 may seek confidential medical care, and HCPs are not required to notify parents. Parental notification is required, however, if medical treatment requires a hospital stay of more than 12 h. These examples highlight the importance of HCPs familiarizing themselves with local policies and legislation regarding the mature minor designation.

10.3.3 The Emancipated Minor

A number of laws in both Canada and the United States recognize the special status of emancipated minors. In general, minors who are married, active military duty, or living separately from their parents or legal guardians and independently managing their own financial affairs are considered emancipated (Davis and Fang 2020). Minors may or may not require a formal court declaration of emancipation to change their legal status, and states set a minimum age at which emancipation can be granted. Emancipated minors are free from parental or legal guardian control and parents and legal guardians are also free from responsibility for the emancipated minor. More specifically, emancipated minors can legally consent to or refuse medical care without parental permission and notification.

The age at which a young person can be considered emancipated and therefore responsible for his or her own financial affairs and medical decisions varies widely from country to country. HCPs must be aware of their local culture, traditions, and laws concerning the age of majority or the developmental level at which a young person is considered capable to make their own medical and financial decisions and face criminal responsibility for their actions. For example, some Saudi Arabian scholars have argued that a person has reached majority (*baligh*) based on known physical signs of puberty (*bulugh*), and therefore can be prosecuted and sentenced as an adult (Human Rights Watch Report 2008). In Scotland, the *Age of Legal Capacity Act* states that a person has full legal capacity, with some limitations, at the age of 16. One must be careful not to conflate the age of criminal responsibility with the voting age, the legal age for buying alcohol and tobacco, and the age of majority, which may be independent from each other and differ even within the same country. HCPs must familiarize themselves with the local laws and whether a specific age limit exists for consideration of maturity or emancipation.

10.3.4   Assent and Emerging Capacity in Adolescents

The principle of pediatric *assent* recognizes that children, and especially adolescents, are capable of participating in decision-making related to their care and provides a process by which to meet these goals for children and adolescents who cannot legally provide informed consent. In many jurisdictions in North America, 18 years of age is considered the age of maturity and the legal age at which adolescents are considered legally competent to make their own decisions. However, child health and legal experts acknowledge there is no specific age threshold at which a child or adolescent suddenly acquires full decision-making capacity. Rather, the assent requirement calls for the need to recognize and respect the wishes of adolescents as they develop cognitively and mature (Unguru 2011). When seeking assent, HCPs should help the patient achieve a developmentally appropriate awareness of his or her condition and perform a clinical assessment of the patient's understanding of the situation before they can solicit an expression of the patient's willingness to accept the proposed care. This process must also include an assessment of factors that may influence the patient's response, including inappropriate pressure to accept testing or therapy. Assent is not simply the absence of refusal, but rather represents active engagement of the pediatric patient in the decision-making process and acquiescence based on the patient's wishes, knowledge and understanding of the illness and its treatment (Leikin et al. 1983). A persistent refusal to assent (dissent) should be taken seriously and seen as an opportunity to further explore the patient's concerns regarding the proposed care.

As previously alluded to, empirical studies of cognitive development in children suggest that many minors reach the formal operational stage of cognitive development that allows abstract thinking and the ability to handle complex tasks by mid-adolescence. The most widely cited study on adolescents' capacity for rational decision-making was published nearly 40 years ago and concluded that by 14 years of age, adolescents are as able as adults to make rational and reasonable health care decisions (Weithorn and Campbell 1982). However, an adolescent's ability for adult-like decision-making does not guarantee such decisions. Advanced imaging techniques, such as functional MRI, and clinical neuropsychological evaluation have provided additional insight into decision-making capacity and processes. Diekema (2011) argues that although adolescents may be capable of adult-like decision-making, they do not perform at a level commensurate with their cognitive abilities. Adolescents are more affected by the influence of peers, are less future oriented, more impulsive, and differ in their assessment of risks and rewards as compared with adults. Although adolescents may have the right equipment to allow for decision-making capacity, they have yet to fully master its implementation.

Psychologists distinguish between cold and hot cognition. Cold cognitive abilities are those used in calm situations with little to no peer influence and time to deliberate and reason logically with facts. Studies of cold cognition have shown that the skills necessary to make informed decisions are firmly in place by 16 years of age (Steinberg et al. 2009). By that age, adolescents can gather and process information, weigh pros and cons, reason logically with facts, and take time before making a decision. This finding suggests that under ideal conditions, adolescents are often as able as their adult counterparts to make health care–related decisions; however, ideal conditions are rare. Hot cognitive abilities are those used to make decisions in situations of emotional arousal or conflict, when there is peer influence and/or time pressure. In these situations, the most critical skill is self-regulation, which enables an individual to control emotions, withstand pressures from others, and check impulses. Hot and cold cognition are subserved by different neuronal circuits and have different developmental courses (Steinberg 2005). From a neurobiological standpoint, in emotionally charged situations and decisions made under stressful conditions (e.g., whether to enter inpatient treatment for an eating disorder or whether to attempt a phase 1 clinical trial for cancer), adolescents may rely on their more mature limbic systems than the less mature prefrontal control system (Diekema 2011; Casey et al. 2011). The prefrontal cortex is the region of the brain responsible for high-level reasoning, executive function, weighing consequences, planning, organization, and emotional regulation, i.e., rational decision-making. Rational decision-making is the last faculty to mature, not finalized well into the 20 s (Steinberg 2013). These facts are important as they will inform the process of discussion with adolescents and families and the process of obtaining assent and informed consent, but they do not negate the duty to respect an adolescent's autonomy. Adolescents' participation in medical decision-making should be sought in proportion to their developmental capacity to understand the nature and consequences of their medical problem as well as the reasonably foreseeable risks and benefits of the treatment proposed. Every effort should be made to respect assent and dissent whenever possible, and to work with adolescents and their parents to reach medical decisions based on the patient's best interest that optimize respect for autonomy.

10.4 Confidentiality

Confidentiality protection is an essential part of any professional therapeutic relationship, but it takes on particular importance in health care for adolescents. Respect for confidentiality is a professional duty that derives from the moral tradition of physicians and is consistent with adolescents' development of maturity and autonomy. It is an essential component of the therapeutic relationship; if confidentiality cannot be ensured, some adolescents will forego care, particularly for more sensitive health care concerns including sexual health and mental health issues, potentially leading to both short- and long-term harm and worse health outcomes.

The care of adolescents sometimes involves "walking a fine line" between respect for the adolescents' confidentiality and deference to the parent's role and opinion about their child's health. Parents may express concern about "being left out" of the discussion about their child's health. Beginning in early adolescence, it is helpful to routinely spend a part of each visit alone with the patient. Establishing clear ground rules early on promotes shared decision-making (SDM) and minimizes potential conflicts. A general guiding principle is that no matter the situation (barring gross negligence or abuse on the part of the parent or guardian) clinicians must strive to encourage communication between adolescents and their parents or other trusted adults without betraying the adolescents' trust in their health care provider. Open communication between adolescents and their parents should remain an overarching goal of care.

Just as in the care of adults, confidentiality has its limits. It is acceptable and, in some cases mandated, to breech confidentiality in certain specific situations. Limits to confidentiality might include, for example, disclosure of suicidal or homicidal ideation or acts, serious chemical dependence, life-threatening eating disorders, and the legal obligation to report child abuse (Ford et al. 2004). If adolescents do not understand that respect for confidentiality is not absolute, they may understandably feel betrayed if confidentiality is breeched, which can lead to an ongoing lack of trust in the health care system and foregoing necessary care in the future. Accordingly, during the very first patient encounter HCPs should explain the meaning of confidentiality to adolescents and their families, the scope of confidentiality protection, and the limits to trustconfidentiality. This information is essential to building a therapeutic relationship and avoiding the erosion of that might otherwise occur if it is not explicitly communicated from the very beginning (Ford et al. 2004).

Some of the most difficult situations for HCPs to navigate arise when there is conflict between the needs of the adolescent and those of the parent. For example, parents and their adolescent children might disagree on the best course of treatment; parents might ask for testing to be done in a way that would subvert the adolescent patient's right to assent to the testing and/or limit the adolescent's right to an open future as in the example of drug testing and genetic screening for an adult onset condition; adolescents might require care for issues they are uncomfortable discussing with their parents. These situations are fraught with potential for conflict and confusion by the HCP as to the most ethical course of action. However, these situations also provide opportunities for honest and open discussions between HCPs, adolescents, and their parents that ultimately can lead to better communication and enhanced quality care.

10.5   Cases

The remainder of this chapter illustrates representative cases of paradigmatic ethical challenges that may arise in the care of adolescent patients and attempt to apply the ethical principles previously discussed to the commentary and/or resolution of the cases.

10.5.1 What Happens When Parents and Adolescents Disagree?

1. *Life-limiting illness and disagreement regarding experimental therapy or life-sustaining medical treatment (LSMT)*

Jill is a 17-year-old with an aggressive form of leukemia who has undergone 2 cycles of chemotherapy. Her cancer and its treatment have left her with distressing symptoms, including neuropathic pain, mouth sores, and extreme fatigue. Jill loves to dance, and watch TV cuddled up at home with her golden doodle, Moxie. Because of her illness, she no longer can dance. She lives with her parents and 14-year-old sister. Jill and her parents recently met with Dr. Milner, her trusted oncologist, who compassionately explained that Jill's cancer has relapsed. There are no further treatment options at Jill's home institution, but a clinical trial is taking place halfway across the country. The trial is extremely expensive and would require the family to move to a different state, away from home and their support system. Dr. Milner clearly states that the investigational intervention is unlikely curative and may, at best, extend Jill's life by a few months. There is also the real possibility that Jill will spend these last few months dealing with worsening nausea and fatigue. Jill's parents want to pursue this option; Jill does not. She understands that that the trial might extend her life by a few months, but she prefers to spend what time she has left at home with her family, friends, and support system. She states that she wants to "make the most of the time she has left." Her quality of life has already been significantly affected by her current symptoms, and she does not want to put herself through more chemotherapy with worsening nausea and fatigue. She understands that by not participating in the clinical trial, she will likely die sooner. She wants the doctors to focus on managing her fatigue and neuropathic pain so that she can play with her dog, see her friends, and spend time with her sister at home. Jill does not suffer from delirium or other medical condition that could negatively affect her ability to make decisions, nor is she suicidal. Jill's parents cannot believe what they are hearing—if there is any chance at all for a cure or prolonging life, shouldn't they take it? Jill's parents tearfully ask Dr. Milner to "talk some sense into her."

Does Jill have a right to refuse this treatment? Should Dr. Milner oblige her parents' request and attempt to convince her to pursue the treatment?

Several elements must be considered when determining whether Jill has a right to refuse the treatment and what role Dr. Milner should play in any attempt to sway her decision to be consistent with her parents' wishes.

First, there must be a factual and thoughtful consideration of the potential burdens and benefits of the proposed treatment. The clinical trial will not be curative and potentially may only add a few months to Jill's life. Furthermore, there are many potential burdens including financial barriers, time away from home and Jill's support system, and anticipated side effects/complications which could lead to significantly increased morbidity and worsening quality of life. Presented with these facts, it seems that both the decision to proceed with the clinical trial and the alternate decision to forego treatment and focus on comfort could be viewed as equally reasonable approaches. In this case, every effort should be made to respect Jill's decision so long as it is consistent with her values and wishes for end-of-life care. In this case, Jill is clear about what she considers important and her decision should be respected in keeping with respect for her autonomy and dignity at the end-of life.

A second important element is Jill's capacity to make her decision. Jill has shown that she understands the risks and benefits of the clinical trial and the risks and benefits of foregoing further treatment. She has articulated her thoughts, clearly stating she wants to prioritize her quality of life over prolongation of life. If Dr. Milner is able to determine that Jill has adequate decision-making capacity, she can be considered a mature minor whose decision should be respected.

What about a situation where the benefits of treatment clearly outweigh the risks? An example of this would be an adolescents' refusal of a standard course of chemotherapy for low-risk acute lymphoblastic leukemia with a high expectation of cure. Given the consequences of refusal (death) it is incumbent on the HCP to gain a better understanding of the patient's reasoning to assure it is a truly informed refusal, free from misunderstanding, undue pressure or other influences. This scenario illustrates that the magnitude of the consequence(s) of a medical decision must figure into the process for determining whether an adolescent's treatment refusal is to be respected. A decision by a generally healthy adolescent with a suboptimal diet not to take a multivitamin recommended by his family doctor, is more likely to be accepted as the consequences are (seemingly) trivial. However, in the case of the adolescent refusing life-saving chemotherapy, where the consequence of the decision represents the difference between life and death, the HCP must ascertain whether the adolescent truly understands the facts and unearth the reasons for refusal. This is a case where application of the best interest standard is quite clear: death in this previously healthy adolescent with a high chance of cure is not in the patient's (or anyone's) best interest (Bester 2019). More complex medical situations, especially those that involve decision-making with regard to experimental therapies or LSMT, require more rigour in determining whether the adolescent has decision-making capacity and the reasons for their decision. Involvement of a pediatric palliative care team or clinical ethics team can be extremely helpful for managing these extremely complex situations.

Although, Dr. Milner's duty is ultimately to Jill, and not her parents, he must strive to balance this duty while attending to her parents' wishes, accepting and validating their position. Ultimately, if Jill is capable and her refusal is reasonable, Dr. Milner must make every effort to support her decision. Communicating to the parents about Jill's values and how important it is for her to have some control over the last months of her life will be crucial to including parents in the discussion and helping them to find peace and mitigate feelings of guilt and regret. Simultaneously, Jill must appreciate her parents' position. Negotiation and compromise often ensue in such cases. These discussions are extremely challenging demanding patience, sensitivity, and nuance. Providers must be willing to seek the expertise of a specialized pediatric palliative care team and other members of the interdisciplinary team to provide support.

2. *Parental requests for drug testing*

Jordan is a 16-year-old boy cared for the past 7-years by Dr. Barnaby. He is an average student and loves playing basketball and baseball. Jordan has 2 younger siblings, a 12-year-old sister and 10-year-old brother. Jordan's parents, Claire and Michael, generally very involved, lately have been less present, and recently told their children that they are divorcing. Since learning of the divorce, Jordan's grades have dropped and he has been acting out: missing school, coming home late, yelling at his mom, and spending time with a new group of friends Claire thinks are "trouble." More withdrawn at home, Jordan spends most of his time in his room playing video games only coming out to grab a quick bite. When Claire recently collected laundry from his room, she noticed that they smelled like marijuana. When she confronted Jordan, he became defensive saying that some of the guys he was with were smoking, but that he had not. At Claire's request, Dr. Barnaby meets with Jordan and Claire. Claire is concerned about the changes in her son's behaviour and the decline in his academic performance. She is convinced that Jordan has been doing drugs and asks Dr. Barnaby to perform a drug test to prove her suspicion so she can get him the help he needs.

When Dr. Barnaby asks Jordan what he has just heard, he responds, "I don't do drugs. I won't take the test." Dr. Barnaby then meets with Jordan alone. Jordan is frustrated, but opens up and explains that he is not taking drugs. He is upset with his mother for not trusting him and says, "she just doesn't understand." Jordan states that he has been having a difficult time adjusting to his parent's divorce and is feeling more sad than usual. He used to love playing basketball, but has lost motivation to participate in this activity. He does not want to see his usual friends because they always ask questions related to the divorce ("are you going to live with your mom or dad?") so he has started hanging out with a new group of friends. He likes playing video games online because it is a way for him to escape what is going on. He is sleeping more and eating a lot more junk food. Jordan articulates his feelings clearly and has no suicidal ideation. With his permission, Dr. Barnaby performs a physical examination and everything is normal.

When Dr. Barnaby brings Claire back in the room, she says, "he told you he's taking drugs, didn't he?" Jordan denies drug use and becomes upset, he continues to refuse the drug test. His mother insists that it is not his decision to make and asks Dr. Barnaby to go ahead with the test despite her son's protest.

What is the pediatrician's duty? Should he perform the drug testing? What is the best way to handle the situation?

The request to test an adolescent for drugs without his/her permission raises ethical issues involving decision-making locus of control, consent, confidentiality, and trust. According to the AAP policy statement on drug testing in adolescents: "involuntary testing is not appropriate in adolescents with decisional capacity—even with parental consent—and should be performed only if there are strong medical or legal reasons to do so." As a pediatrician, Dr. Barnaby knows that changes in behavior, grades and school performance are suggestive of possible drug use. However, during their discussion, Jordan was able to clearly articulate his feelings and explain his actions.

His behavior and physical examination do not suggest he is under the influence of drugs. Therefore, if Jordan has decisional capacity, his consent is required for drug testing.

Respecting Jordan's right to confidential care recognizes his rights while bolstering the therapeutic relationship Dr. Barnaby has built with him over the past 7 years, making him more likely to continue to seek health care in the future when he needs it. Although Jordan's thoughts and behaviors raise concern for a possible adjustment disorder or depression, Jordan does not seem to pose an imminent threat of harm to himself or others and there is no compelling medical or legal reason to breech confidentiality. Alternatively, an adolescent with impaired mental status or one who has been involved in trauma, violence, or overdose should be tested for drug use.

Another factor to consider is the utility of drug testing in this scenario. A urine drug test could be falsely negative and not eliminate concerns regarding a substance abuse problem. Furthermore, a positive screening test would not provide information on the frequency, pattern, or extent of drug use. Jordan's mother has expressed concern for her son and stated that her goal is to get him the help he needs. If the goal of the drug test is to proceed with an intervention to help him, testing against his wishes seems counterproductive. If the test is positive, the adolescent may feel betrayed by his parents and HCP and it is not clear that the proposed intervention would be successful.

In summary, Jordan is presumably competent and does not appear to be acutely intoxicated, accordingly, there is little to warrant an involuntary drug test. Since there is no risk of imminent harm, there is no justification for violating his autonomy. *How should Dr. Barnaby respond?* In this case, Dr. Barnaby has a responsibility to help Jordan manage this difficult time. She should continue to meet alone with Jordan to obtain more information about possible risky behaviours. Dr. Barnaby can encourage Jordan to share this information with his parents, with her help as a facilitator. Since the brief confidential interview revealed some depressive symptoms and difficulty adjusting to the current situation, Dr. Barnaby should consider referral to a qualified mental health professional to determine the need for further evaluation or treatment. Such an approach, predicted on trust and respect for Jordan's emerging autonomy, while simultaneously respecting his mother's concerns exemplifies SDM.

10.6   Conclusion

The ethical challenges raised in the care of adolescents are not merely the domain of the adolescent medicine specialist. They resonate with parents and providers from a range of medical specialties and scope of practice. Adolescents' health concerns often are sensitive in nature and demand respect for confidentiality. Adolescents with chronic illness or life-limiting illness represent a particularly vulnerable subgroup, as they struggle to attain autonomy and carve out an identity within the reality of their illness. Navigating the adolescent patient-parent-provider triad can be challenging when there are competing needs and responsibilities. Sigman (2005), states:

> the most common paradigmatic ethical challenge that arises in the health care of adolescents is a struggle between the adolescent's need and request for autonomy, the moral imperative of a clinician to respect a person's autonomy, and the parents' need and requirement to know about and understand their children's thoughts, attitudes and behaviors.

In all clinical situations, reasonable efforts should be made to include the adolescent patient in health care decisions and encourage honest and open communication between adolescents and their parents. In situations where there is disagreement, it may be helpful to consult with colleagues specialized in adolescent medicine or a clinical ethics consultation team.

The following table presents a summary of guiding principles for the ethical care of adolescents

10.7   Guiding Principles for the Ethical Care of adolescents

Emerging capacity and consent
• Recognize and respect the wishes of adolescents as they develop and mature
evaluate decision-making capacity on a case-by-case basis, taking into consideration the patient's age, developmental level, ability to understand risks, benefits, the and consequences of a decision and its alternative

Privacyand confidentiality
• Ensure a private environment for the patient encounter
• Meet with adolescents alone for a confidential interview and physical exam (if appropriate); start to do this as part of routine care during early adolescence
• Deliver confidential health services for issues involving mental health, reproductive and sexual health and substance abuse to consenting adolescents

Communication
• Explicitly delineate the limits of confidentiality to the adolescent and parents at the very beginning of the patient encounter
• Encourage open and honest communication between adolescents and their parents
• Facilitate disclosure of information-sharing between adolescents and their parents
• If it cannot be facilitated, every effort should be made to ensure confidentiality within the limits of legal and ethical standards

Laws and requirements
• Be knowledgeable about local laws and reporting requirements
• Be aware of institutional policies regarding mature minors
• Know what resources are available for complex situations (adolescent medicine specialist, ethics consultation team, pediatric palliative care team)

References

Belitz, J., and R.A. Bailey. 2009. Clinical ethics for the treatment of children and adolescents: A guide for general psychiatrists. *Psychiatric Clinics of North America.* 32: 243–257. https://doi.org/10.1016/j.psc.2009.02.001.

Bester, J.C. 2019. The best interest standard is the best we have: Why the harm principle and constrained parental autonomy cannot replace the best interest standard in pediatric ethics. *Journal of Clinical Ethics* 30 (3): 223–231.

Canadian Medical Protective Association. 2016. Duties and responsibilities: Can a child provide consent? Retrieved from https://www.cmpa-acpm.ca/en/advice-publications/browse-articles/2014/can-a-child-provide-consent.

Casey, B., R.M. Jones, and L.H. Somerville. 2011. Braking and accelerating of the adolescent brain. *Journal of Adolescent Research* 21 (1): 21–33. https://doi.org/10.1111/j.1532-7795.2010.00712.x.

Chassin, L., C.C. Presson, S.J. Sherman, and D.A. Edwards. 1990. The natural history of cigarette smoking: Predicting young-adult smoking outcomes from adolescent smoking patterns. *Health Psychology* 9 (6): 701–716. https://doi.org/10.1037//0278-6133.9.6.701.

Dahl, R.E., N.B. Allen, L. Wilbrecht, and A. Ballonoff Suleiman. 2018. Importance of investing in adolescence from a developmental science perspective. *Nature* 554 (7693): 441–450. https://doi.org/10.1038/nature25770.

Davis, M., and A. Fang. 2020. Emancipated minor. In *StatPearls.* Retrieved from https://www.ncbi.nlm.nih.gov/books/NBK554594/.

Diekema, D.S. 2011. Adolescent refusal of lifesaving treatment: Are we asking the right questions? *Adolescent Medicine: State of the Art Reviews* 22 (2): 213–228.

Ford, C., A. English, and G. Sigman. 2004. Confidential health care for adolescents: Position paper of the Society for Adolescent Medicine. *Journal of Adolescent Health* 35 (2): 160–167.

Hardin, A.P., and J.M. Hackell. 2017. Committee, and on practice and ambulatory medicine. Age limit of pediatrics. *Pediatrics* 140 (3). https://doi.org/10.1542/peds.2017-2151.

Human Rights Watch Report. 2008. *Adults before their time: Children in Saudi Arabia's criminal justice system.* Retrieved from https://www.hrw.org/report/2008/03/24/adults-their-time/children-saudi-arabias-criminal-justice-system.

Leikin, S.L. 1983. Minors' assent or dissent to medical treatment. *The Journal of Pediatrics* 102 (2): 169–176. https://doi.org/10.1016/s0022-3476(83)80514-9.

Sacks, D., Canadian Pediatric Society Adolescent Health Committee. 2003. Age limits and adolescents. *Paediatrics & Child Health* 8 (9): 577. https://doi.org/10.1093/pch/8.9.577.

Sawyer, S.M., P.S. Azzopardi, D. Wickremarathne, and G.C. Patton. 2018. The age of adolescence. *The Lancet Child & Adolescent Health* 2 (3): 223–228. https://doi.org/10.1016/S2352-4642(18)30022-1.

Sigman, G. 2005. An adolescent with an eating disorder. *Virtual Mentor* 7 (3): 213–217. https://doi.org/10.1001/virtualmentor.2005.7.3.ccas2-0503.

Simmonds, M., A. Llewellyn, C.G. Owen, and N. Woolacott. 2016. Predicting adult obesity from childhood obesity: A systematic review and meta-analysis. *Obesity Reviews* 17 (2): 95–107. https://doi.org/10.1111/obr.1234.

Steinberg, L. 2005. Cognitive and affective development in adolescence. *Trends in Cognitive Sciences* 9 (2): 69–74. https://doi.org/10.1016/j.tics.2004.12.005.

Steinberg, L., E. Cauffman, J. Woolard, S. Graham, and M.T. Banich. 2009. Are adolescents less mature than adults? Minors' access to abortion, the juvenile death penalty, and the alleged APA "flip-flop." *American Psychologist* 64 (7): 583–594. https://doi.org/10.1037/a0014763.

Steinberg, L. 2013. Does recent research on adolescent brain development inform the mature minor doctrine? *The Journal of Medicine and Philosophy* 38 (3): 256–267.

Unguru, Y. 2011. Making sense of adolescent decision-making: Challenge and reality. *Adolescent Medicine: State of the Art Reviews* 22 (2): 195–206.

Weithorn, L.A., and S.B. Campbell. 1982. The competency of children and adolescents to make informed treatment decisions. *Child Development* 53 (6): 1589–1598.

World Health Organization. 2020a. *Adolescent health: overview and global situation.* https://www.who.int/health-topics/adolescent-health#tab=tab_2.

World Health Organization. 2020b. *Maternal, newborn, child and adolescent health: adolescent development.* https://www.who.int/maternal_child_adolescent/topics/adolescence/development/en/.

Further Reading

Ford, C., A. English, and G. Sigman. 2004. Confidential health care for adolescents: position paper of the Society for Adolescent Medicine. *Journal of Adolescent Health* 35(2): 160–7.

Silber, T., A. English (eds). 2011. AM: STARs: ethical and legal issues in adolescent medicine. *American Academy of Pediatrics* 22 (2). ISBN: 978-1-58110-647–3

Chapter 11
Demands for Harmful Treatments in Pediatrics and the Challenge of Reasonable Pluralism: A Quasi-Clinical Ethics Consultation

G. Birchley and D. M. Hester

Abstract This chapter explores demands for harmful treatment through the lens of a case study. Therein, an adolescent with terminal cancer, supported by her mother, demands treatment that her doctors believe will be harmful. Besides analyzing the ethics of demands for harmful treatment, we also consider the challenges placed to ethical deliberation by the fact of reasonable pluralism, where persons disagree about ethical issues despite taking reasonable and thoughtful approaches. Treating the chapter as a quasi-clinical ethics consultation, we jointly consider some important concepts within the case, before separately applying different frameworks that emphasize our individual concerns. For Birchley, the importance of focusing on the equal humanity of the child means this involves a children's rights analysis, while for Hester, an obligations approach is important because it emphasizes the unique nature of both the parental and the medical role. While their background reasoning differs, both authors conclude that demands for harmful treatment cannot be honored in this case. The chapter concludes by considering what value this sort of ethical consultation has, and whether, and how, it might be augmented by using more procedural approaches such as mediation or legal adjudication. Because procedural approaches themselves have limitations and weaknesses, we conclude that using both ethical analysis and robust procedure is the best way to overcome some of the weaknesses and problems inherent when each is used separately.

Keywords Harmful treatments · Futility · Reasonableness in Medicine · Benefit · harm

Giles Birchley's work on this chapter was part of the Balancing Best interest in Medical Ethics and Law project (BABEL) generously funded by the Wellcome Trust (Grant Number 209841/Z/17/Z).

G. Birchley (✉)
Centre for Ethics in Medicine, University of Bristol, Bristol, UK
e-mail: Giles.Birchley@bristol.ac.uk

D. M. Hester
Department of Medical Humanities & Bioethics (UAMS), Clinical Ethicist (UAMS & Arkansas Children's Hospital), Little Rock, USA
e-mail: DMHester@uams.edu

© Springer Nature Switzerland AG 2022
N. Nortjé and J. C. Bester (eds.), *Pediatric Ethics: Theory and Practice*,
The International Library of Bioethics 89,
https://doi.org/10.1007/978-3-030-86182-7_11

11.1 Introduction

In this chapter we consider the ethical implications of demands for potentially harmful treatments. We approach the problem by considering a case where an older child and her mother demand cancer treatment that is likely to shorten the child's life without any apparent benefits. At first face and despite apparently approaching the problem in thoughtful and reasoned ways, we, the authors, do not share the same assessments of the challenges posed by cases where the proposed interventions are considered, at least by some persons, as non-beneficial, even harmful. This lack of shared approaches to value-based assessments is a common problem, present in all areas of ethics, including clinical ethics, described by Rawls as the "fact of reasonable pluralism" (Rawls 2001, 151). While exploring the issues raised by demands for harmful treatment, this chapter also considers the significant challenge of reasonable pluralism to ethical deliberation, and explores how this challenge can be met. Our overall approach is to treat this chapter as a quasi-clinical ethics consultation, where we both give our separate assessments of problem, before coming together to reach some common ground.

After considering the case study, our discussion begins by considering four concepts that appear important to the case: futility, reasonableness, benefit and harm. Our analysis reveals the highly mutable nature of these concepts, and we suggest that further purchase on the problem may be gained by using a framework to guide our understandings. It is here we diverge, with Birchley taking a children's rights approach and Hester an obligations approach. Both approaches reach similar conclusions about the case. We conclude by considering what ethical (dis)agreement may mean, what ethical analysis brings to disputed cases, and how it might sit within a more strictly procedural approach to the ethical challenges inherent in clinical cases.

11.2 Conceptual Clarification

In this chapter, we want to analyze and apply four important concepts in medical care—namely, futility, reasonableness, harm and benefit. We believe that in order to understand best these concepts, it is useful to consider them through the lens of a specific case. The clinical facts of are loosely informed by a case[1] heard by the courts in the United Kingdom (although the motivations of the protagonists are significantly altered).

---

[1] *An NHS Trust v BK and others* (2016) EWHC 2860.

11.3 Case Study

At eight years old, Amy was diagnosed with a high-grade osteosarcoma (bone cancer) in her upper femur. Scans also showed more than a dozen lung metastases (secondary cancers in the lungs). The osteosarcoma and metastases were treated with chemotherapy but responded poorly. Definitive surgery was performed, with a significant mass of viable tumor requiring excision from the femur. Imaging during routine follow-up over the next 18 months twice revealed a recurrence of lung metastases, each requiring more chemotherapy before surgical excision. At two years, investigations revealed a recurrence of the cancer in the femur, which had now invaded the hip joint, necessitating further chemotherapy and amputation, followed by reconstruction. In year three the cancer returns again in the pelvis, with multiple metastatic sites including the lungs and kidneys. Amy's doctors believe Amy's treatment is futile because Amy has a no reasonable chance of long-term survival. They assert that further attempts at cure, which will involve increasingly radical surgery that risks shortening Amy's life, would amount to harm with no chance of benefit. Despite this, Amy—now eleven years old—and her mother, are adamant they want further curative treatment.

11.4 Some Key Concepts

Futility

By the time we arrive at the end of the description of Amy's case, we learn that "doctors believe Amy's treatment is futile because she has a no reasonable chance of long term survival." Often when cases like these arise in the hospital, providers can be heard suggesting that further curative interventions would be "futile." This concept has a great deal of rhetorical power as it suggests that providers would be wasting their (valuable) time providing intensive therapies. But claims of futility around such cases need a good deal of unpacking before accepting their implications for next steps. For more on the concept of futility please refer to Chap. 16.

Futility can be analyzed at least three ways: physiological futility, quantitative futility, and qualitative futility. In fact in the 1990s, there were a good many attempts to gain traction in defining futility in line with one of these three qualifiers. Physiologic futility is the narrowest version where futility is determined only when no physiological effects leading to specific physiologic goals can be obtained (Bosslet et al. 2015). Quantitative futility relies on evidence and experience by setting a lower numeric limit for futility—for instance, if the procedure has not had a positive effect more than once in the last 100 times it was tried, then it is futile (Schneiderman et al. 1990). Qualitative futility is the broadest of the concepts where futility is determined by whether or not the outcomes are of adequate quality, are good enough, or are "worth it" (Fost 2011).

What clinical ethics has come to in light of these different characterizations of futility is that quantitative and qualitative are really two sides of the same "is it worth it?" coin. And further, determining whether an intervention is worth it is precisely where ethical debates arise. Thus, neither qualifier does much to eliminate, or even just mitigate, ethical challenges regarding so-called futile care. As such, in a recent paper five professional medical societies came to consensus that "futile" should be reserved only for cases of physiologic futility, and for all other cases that push on the concept of "futility," the concept of "potentially inappropriate treatment" should be used (Bosslet et al. 2015). However, what makes treatment "inappropriate," then leads to another concern—namely, the scope of reasonable medicine.

Reasonableness in Medicine

To provide effective health care takes a certain expertise in the field of medicine. Medicine is a social science based on natural sciences and human considerations. The practice of medicine is a profession which, over its long history, has delimited its scope, focused on a person's ability to live healthily through the caring application of evidenced-based science in light of an individual's (and even the public's) interests and values. On one hand, then, medicine is not simply the basic sciences nor is it solely psychology. It is not what is good in the lab but what is good for actual people. On the other hand, it is not about all things that people find to be good for them. It is not religious practice; it is not entertainment. It differs in recognizable ways from a great many other human practices and achievements. That is, it has its own space in our society.

As such, medicine is to be practiced by those trained in the field, expert in what lies within its scope. There will be reasonable responses that those practitioners qua professionals in medicine should make towards patients who seek care, and thus, there will be some requests by patients which will simply be unreasonable for medicine to attend to. Unfortunately, there is no clear way cleanly and universally to distinguish reasonable and unreasonable requests. Of course, some are obvious—it is reasonable to ask a physician to attempt to repair a simple fracture of a finger; it is unreasonable to ask a physician to call out demons from a person's soul. What makes these cases obvious? It seems likely that it is based on the degree patients, society, and professionals agree about both material conditions and beliefs in both cases. A body of medical opinion, society, and the patient themselves both accept the material basis of a fractured finger and hold the belief that medicine has the tools to fix the fracture. In the case of demonic possession, neither is the material basis demonstrated in a way that is generally accepted to all parties, nor is the belief that medicine is a way to fix the soul mutually held. But cases like Amy's are less straight forward to categorize.

In this case, while the material basis of cancer is accepted, beliefs about the suitability of medicine to treat the cancer are not. At one end of the spectrum, the health care team might try to claim that the request for further cancer treatment is simply futile, and thus would be unreasonable medical care. But if we take the position of the multi-society consensus paper, can we claim it is obviously physiological futile? Amy and her mother may claim that attacking cancer and staving off death as a result of a disease process is precisely what medicine is all about and it is completely reasonable to continue. How might such a conflict be resolved, then?

In cases of request for *potentially* inappropriate medical care, a persistent suggestion is to work through a process either of adjudicating the claims of each side in order to determine which side should be supported (Bosselt et al. 2015) or of mediating between the two parties to come to a resolution between them (Fiester 2015; Dubler and Liebman 2011). The former can be more contentious, and typically leads to outcomes in line with medical providers' interests. The latter may mitigate contention but can ultimately be dissatisfying for both parties—indeed, it may be questionable whether diametrically opposed courses of action where there is no obvious middle ground are even amenable to mediation.

Maybe, then, analyzing cases like that of Amy's using other concepts and norms can help break through the ethical deadlock. Amy's case also contains competing claims of harmfulness and benefit, and these are similarly important concepts to scrutinize. While to some extent harm and benefit are two sides of the same coin, there are important differences.

Benefit

Utilitarianism argues the goal of morality is the maximum benefit for the greatest number. While Kantian deontology is primarily concerned with self-legislation of the good, it nevertheless allows for beneficial corrections of inaccurate precepts that give rise to unreasonable beliefs. Requiring professionals to benefit patients, acts as a counterbalance to human impulses of self-interest and convenience that may otherwise cause professionals to find reasons to ignore or deprioritize the needs of patients. Seeking to benefit is thus an important characteristic of good health care and, as the principle of beneficence, a principle of clinical ethics.

Benefit may not always guide short term actions, permitting some harms for the greater good. However, unlike harm, benefit—however defined—will tend to form part of an overarching plan, so has a unique place as a guiding principle. Unfortunately, any rule that one should create benefits is rather vague and unaccountable, because understandings of benefit vary. They include, for instance, physical, emotional and metaphysical benefits: indeed what counts as a harm or a benefit at any particular time may vary depending on the context and the goal that is sought. This said, the situation does not appear completely unsalvageable, for like reasonable and unreasonable requests for treatment, there do seem to be cases that are more or less obvious. There are few persons or contexts who would consider dining on cyanide as a benefit. This has led to efforts to give certain types of benefit and harm a positive grounding, for example attempting empirically to define health and disease on a statistical basis (Boorse 2014). However, no matter how successfully we identify benefits in functional terms, it is hard to uncouple functional definitions from a person's perception of the benefit.

The fact that benefits are hugely diverse means they are difficult to demarcate successfully. To some extent the status of a putative benefit will depend on who is receiving it, but the insights of both patient and a variety of experts are also important. The inherent fallibilities of our perceptions and reasoning processes will mean that we should resist strictly demarcation of who has a say about what. Any analysis of benefit in any particular case will therefore be improved by involving a range of actors with a variety of perspectives and expertise, for example including psychologists, clinical ethicists and veteran patients and families. To avoid excessively busy or muddled communication, these many perspectives are delegated to relatively few participants. In Amy's case our response will therefore depend on both the medically informed benefits and harms that health care professionals identify, as well as Amy and her mother's perceptions, but will be finessed with the insights of a range of other actors.

Harm

An idealized aim of medicine seeks medical benefit while avoiding undue harm. However, because harms are both varied and often unavoidable, the impulse to benefit must also, sometimes, be restrained. With benefit, harm speaks to a conception of the best interest principle where benefits and burdens must be balanced. Importantly, considerations of harm as a principle of medical ethics tend to be versed as negative instructions to avoid doing harm, rather than positive instructions to prevent harm. Thus, harm is importantly distinguished from benefit in Beauchamp and Childress' (2009) ethical principlism, because to benefit demands positive action, whereas avoiding harm demands refraining from action. This follows the importance of avoiding harm in most schools of moral philosophy. Avoiding harm is an essential threshold in JS Mill's utilitarian conception of liberty. In his deontological analysis, Rawls argued that avoiding harm was among the natural duties that apply to us irrespective of any other commitments or contracts (Rawls 1999).

Harm may manifest in different ways, including physically, psychologically or metaphysically. For example, depriving a person of liberty could be physically harmless, but cause metaphysical or psychological harm. Because of this mutability, harms may be only prima facie prohibited, and thus allowable in one domain to avoid harms in another. Diekema (2004) has proposed that a harm threshold should demarcate parental freedoms (refer to Chap. 4). This threshold prohibits State interference with parental choices that do not cause imminent and significant harm. The sorts of parental choices that Diekema argues should be respected include making choices for cancer treatments with fewer side effects but less likelihood of success, and Diekema has recently extended his arguments to parental requests for treatment (Shah et al. 2017). Nevertheless in Amy's case, her age and experience of illness may mean the case is not a *parental* decision at all, but a question of whether society should respect a *patient's* requests for a harmful treatment. In either case the response of anyone involved in the case will depend on how they interpret harms. They might feel that the physical harms arising from surgery are prohibitive, the metaphysical harms arising from interference with Amy's and/or her Mother's choices outweigh those in the physical domain, or that harms are whatever the patient themselves determines as such. Any perspective will be influenced by roles as well as reasoning, meaning that both no perspective is inherently correct or faulty, and any interpretation is open to the challenge of others where a comprehensive account is sought.

Given the mutability of the concepts we have discussed, it is clear that concepts alone cannot offer us a determinative framework, and we may seek to impose one by, for example, demarcating authority over specific areas to particular parties. Alternatively, we could stipulate a series of objective considerations that should guide decisions. We now consider how these different approaches might operationalize our concepts.

11.5   Discussion

Practical ethics means a constant interplay between the theory and practice, a process Beauchamp and Childress (2009) characterize as 'reflective equilibrium'. In Rawls's (Rawls 1999) description, reflective equilibrium is where the thinker strikes a balance between the demands of theory and the necessities of context, unavoidably guided by the fact of personal experience and intuition. The thinker's conclusion is developed in line with the scientific principles of coherence, simplicity and consilience. From the standpoint of a co-authored chapter—particularly one that aims to give an international perspective—it is therefore necessary to note that differences between nation states will affect context, and thus the fit of different theoretical approaches. While the jurisdictions of both authors are liberal democracies, the USA and the UK are different in ways that are important to this case, including different approaches to health risks (manifested in the dominance of either socialized or private health care), different levels of social welfare provision and differences in the social and legal status of children. These ultimately create different priorities among the liberties enjoyed by citizens. It is not necessary to belabor these differences, other than to observe that similar cases may be approached in different ways in the two jurisdictions. We therefore offer two distinct approaches. First Birchley details a rights approach to the case, before Hester offers an account based on fiduciary obligations.

11.5.1   Birchley: A Rights Approach

As we concluded above, the concepts of reasonableness, futility, benefit and harm do not, by themselves, show us how to resolve cases. It may be helpful to follow a determinative framework, and one potential framework is that which identifies specific rights. A rights approach is, on most accounts, based on a premise that rights holders have fundamental entitlements. These entitlements may amount to personal privileges to act or refrain or may impose obligations on others to act or refrain.

Internationally, all jurisdictions subscribe to the view that children have rights, inasmuch as every nation state on Earth is a signatory of the United Nations Convention on the Rights of the Child (UNCRC), and only the United States of America has failed to subsequently ratify the treaty. Nevertheless, many prominent thinkers have voiced skepticism about rights, memorably encapsulated in Bentham's description of natural rights as "nonsense upon stilts" (Waldron 1987). Without rehearsing the many nuanced arguments for and against rights, we can cut through some of the resulting noise by following the pragmatist formulation of Oliver Wendel Holmes: "a right is only the hypostasis of a prophecy—the imagination of a substance supporting the fact that the public force will be brought to bear upon those who do things to contravene it" (Holmes 1995, 447). In other words, so long as rights are commonly supported and commonly enforced, they represent a potent power to which arguments for equality can appeal. Stipulating rights is thus an important remedy to attempts to place any person "outside ... the body politic of citizens" (Arendt 1998, 107), as seen in the tendency of some pediatric bioethicists to see children as incomplete persons and moral agents (Engelhardt 2010).

An immediate difficulty for any analysis of children's rights is that there are competing theories of rights. One of these ("will" theory) grants rights only to those who can, and do, claim them (Feinberg 1970). The other ("interest" theory) ascribes rights independently of choices (Raz 1986). If rights are not chosen but are ascribed it does not prevent protections of important choices (because respecting choices may itself be a right), whereas if rights are choices, all persons who cannot exercise choices—including young children—cannot have rights. While parents may have moral rights related to their own interests (Birchley 2018), no family is guaranteed unlimited freedoms or absolute privacy. Although the threshold of intervention is contested, there is widespread agreement that parents are not free to cause their children serious harms or deny life defining benefits without challenge. Ascribing interest-based rights to children therefore concurs with intuitions that there are obligations to nurture children when parents cannot or do not. This is not to say children's rights should be indifferent to parents, despite concerns in some quarters that children's rights undermine family life and parental authority (Ross & Swota 2017). Importantly, to ascribe children rights based on their interests must in many cases imply a duty toward other persons besides the child. For example, children's rights to adequate health care will—occasionally—imply that parents "have a duty to manipulate and coerce" (Brighouse and Swift 2014, 70) a child who is reluctant to have treatment. Society therefore has a duty to allow parents to exercise some discretion in this regard. Moreover, children have a right to family life, which usually implies that the state must guarantee levels of support to parents to furnish such a life (also refer to Chap. 9).

In Amy's case, because her rights focus on interests, she has no simple right to treatment just because she desires it. Instead we must appeal to a specific account of rights. Although we will not list all 54 articles of the UNCRC, at least 9 of its articles may be germane to Amy's case for treatment. While the convention considers only parental responsibilities (not rights), a further 4 articles seem germane to Amy's mother (we list these in Table 11.1).

Before giving our attention to these rights and responsibilities it is important to recognize what rights can, and cannot, do. The rights in the UNCRC are incompletely theorized, that is, they do not contain fully prescriptive rules that give an answer to any question (Tobin 2013). This is in common with most principled approaches, which set

Table 11.1 Summary of UNCRC rights and responsibilities germane to Amy's case

Summary of UNCRC rights and responsibilities germane to Amy's case	
Rights germane to Amy	Responsibilities germane to Amy's mother
A right to life, such that the state must ensure survival and development where possible (Art. 6)	A responsibility to provide direction and guidance to her child in the exercise of these rights (Art. 5)
A right to express her view (Art. 12)	Responsibilities must be guided by a basic concern for her child's best interest (Art. 18)
Freedom of expression (Art. 13)	Primary responsibility to secure moral and social development (etc.) of her child (Art. 27)
Freedom of thought, conscience and religion (Art. 14)	Education should build respect for a child's parents and culture (Art. 29)
Privacy of family, home and correspondence (Art. 16)	
Protection from maltreatment neglect and abuse (Art. 19)	
Enjoyment of highest attainable standard of health, access to health facilities and information (Art. 24)	
Prevention of inhumane or degrading treatment or punishment (Art. 37)	

out a broad direction of travel and leave the details to be worked out by those closest to a decision. There are good reasons for this. No legislator (moral or legal), no matter how wise, can conceive of all potential situations and combinations of situations. The very essence of rights is that they focus attention on the interests of the individuals concerned, so blanket prescriptions of specific actions are a contradiction in terms.

What we can see from the breakdown of rights in Table 11.1 is that Amy has rights to: life and health, where this is attainable; expression of thought and beliefs; protection from maltreatment or degradation. Her mother is recognized as the person primarily responsible for guiding Amy's decisions and moral development. Respect for the authoritative nature of her role is reinforced through Amy's education. In this case, as in many others, much hinges on whether treatment is in Amy's best interest, that is, what is best for Amy, all possible harms and benefits considered. Best interest must guide her mother's responsibilities and be the 'primary' consideration (refer to Chap. 2) in guiding the actions of any institutional decision-maker.[2] Nevertheless, best interest is a vague principle to follow. While acknowledging that the relationship between rights and interests is much debated, Snelling (2016) argues that rights and best interest are complimentary in the sense that best interest tells us to seek what is

---

[2] Per UNCRC Art. 3,1—There is some debate over whether Art. 3,1 represents a right as such, thus we exclude it from our summary table. See: Archard (2013).

best overall,[3] while rights provide a powerful restraint from devolving into a narrow and unnuanced utilitarian approach where we simply seek to maximize benefits and minimize harms irrespective of what that entails.

The focus of rights on the individual to express themselves and their viewpoint rules out all but the broadest notions of futility. Physiologic and Quantitative futility is inadequate as it denies the enormous complexity of human feelings and beliefs. Only qualitative futility can account for the phenomenological qualities of being human: treatment is therefore not futile because Amy feels it is worth it. Because a rights approach indicates that the bond between Amy and her mother is one we respect, it is important that Amy's views in this case concord with those of her mother. However Article 12 and 13 rights assert Amy 's views should be accorded importance irrespective of her age and her mother's viewpoint. Any assessment of harms and benefits must give full weight to Amy's views and experiences. Amy and her mother know and have experienced the trauma of treatment, although not the terminal phase of Amy's illness. While Amy's Article 6 and 24 rights suggest an obligation toward attaining the highest possible standard of health, Amy has received treatment up to the point that it ceases to provide more physical benefit than harm. Lack of functional benefit does not mean that Amy and her views are irrelevant, a rights approach recognizes that she is owed the time and effort of maintaining a therapeutic relationship where both she and her clinicians can have full and frank conversations about her health and choices.

Nevertheless, in seeking further treatment the professionals involved, informed by expert understanding, are sure Amy and her mother are making an error. As we discussed above, for Health professionals to agree that a patient's request is reasonable requires that they share their patient's understanding of material conditions of illness and beliefs in what medicine can achieve. Amy and her mother's demand are making direct assumptions about what medicine can achieve that are unsupported by expert assessment. Amy's choice may be founded on hope, but it is not founded on what is achievable *in Amy's situation*. The medical cure Amy seeks cannot be delivered—any benefits she attains by following her choice may easily topple into disappointment and despair when she comes to realize the cure she desires cannot be realized. On a balance of harms and benefits, treatment is therefore unsupportable. There is no actionable right to trump this conclusion: Amy's right to self-expression and attainable standards of health cannot change facts about the world. Her choices in this case cannot oblige professionals to provide cancer treatments. Importantly, this does not remove her right to health care choices that can improve her health, including access palliative care and making choices about how her death will be managed.

---

[3] Since our focus is on what is best for Amy overall, we shall focus on Amy's best interest and park the issue of whether there is a minimal "threshold" account of interests that would leave space for separate parental rights.

11.5.2   Hester: An Obligations Approach

While not universally agreed upon, it is commonly held that rights come along with corresponding obligations, and this relationship between rights and obligations (aka, responsibilities) can be seen in Table 11.1. However, rights are arguably not the only way that obligations arise, and this is important because recognizing other sources of obligations may prove instructive for how we are to weigh and respond to claims about obligations we hold. This account does not rely on any particular claims about rights but holds an expectation that obligations should be fulfilled (James 1891; Dewey 1932).

When obligations arise from sources other than as the corresponding result of a claim of someone's right, it is typically because there exist certain expectations from two primary sources: the roles we fulfil and the relationships we have (Dewey 1932). Roles have scope and set (if not always well-defined) parameters of conduct while relationships establish connections and assumptions about how parties will act towards each other. We might call the obligations that arise from roles and relationships "fiduciary," as they are grounded, not on the moral framework of rights, but on trust and confidence that comes from an understanding and expectation of what those roles fulfil, or those relationships promise.

Medical situations bring this out into stark relief. For example, whether or not one has a right to health care, physicians have myriad obligations to their patients and to society. The above-discussed concepts, like harm and benefit, translated into principles like non-maleficence and beneficence, are themselves norms of responsibility that providers have (cf. Beauchamp and Childress 2009). Further, there simply are expectations of what physicians and other providers owe to patients, based on what it means to provide good care in the exercise of medicine (Pellegrino 1979; Zaner 1988).

Of course, family members also acquire obligations to each other based on their relationships to each other and the roles they perform within the family. Children may have rights, it is true, but even if no account of children's rights were available, parenting as a role in society comes with expectations to which parents are held accountable. It is a set of activities in light of the "nature" of children as developing humans that can establish the responsibilities to which society holds parents (Blustein 1982). As such, it may not be the case that an analysis based on rights of parents and children is either useful or needed in order to analyze all of the obligations all the parties have to each other and to other parties affected by medical cases.

In Amy's case, the request for further "curative" treatment is perfectly understandable. Amy is struggling with a fatal disease, and her mother, as caretaker and protector, is motivated by those responsibilities. To ask a parent if it is ok for them to let their child die is to ask something otherwise incomprehensible, at least in everyday, common conditions. It is the relationship between parents and child that warrants such a response. And yet, remitting cancer is not a common condition, and it is this new set of conditions that forces a reconsideration of the obligations parents have to children and demands serious consideration of the question—is it acceptable to let your child die?

Further, while curative measures are often the expected medical practice in most confrontations with illness and injury, physicians themselves fail in their professional duties if they continue to offer and provide treatment qua curative if the conditions are such that the interventions available to Amy are not, in fact, curative. Physicians are responsible to provide quality care, unto death, but that does not entail providing any and all technically available interventions. In fact, it may *require* that they specifically do not offer some interventions as they would create false hope and set up the conditions for harm, rather than benefit. Such interventions would be inappropriate—not futile, necessarily, but inappropriate as they would violate obligations to care, of beneficence and of non-maleficence. Again, these obligations are fiduciary, not following from the rights of physicians or patients or parents, but from the natures of the medical relationship and practice of medicine themselves.

11.6   Practical Conclusion

While approaching the case using different frameworks, both authors have reached the conclusion that Amy and her Mother's request for more treatment cannot be justified. However, like many cases that reach clinical ethics consultants and committees, Amy's case was not one that many people would find finely balanced. Moral psychology appears to show that moral reasoning involves constructing justifications for immediate intuitions (Haidt 2013). The function of an ethicist on this account is at best to make sure that all aspects of a decision are thoroughly considered, even if the eventual conclusion is likely to be a post hoc justification of their first reaction to the case.

Moreover, ethical concepts are vague and socially contingent. There are clear difficulties in operationalizing such factors as futility, reasonableness, harm and benefit. This reflects the wide scope of discretion allowed to patients, their proxies/parent, and their health professionals in bioethics. There is evidence that such principles are thoroughly mutable given changes in circumstances. For example, evidence from social work practice, where interventions in child protection are guided by the harm threshold, shows that the way the threshold was interpreted was influenced by economic austerity (Devaney 2018). Such problems seem unlikely to be solved by strict definitions, given that "…every interpretation hangs in the air together with what it interprets, and cannot give it any support. Interpretations by themselves do not determine meaning" (Wittgenstein 1958, ss198). Such a view suggests, although we may strengthen our approaches by falling back on specified lists (as we have done with reference to the UNCRC), this may not always be sufficient to overcome the failings of language. In such circumstances many have suggested that procedural approaches can overcome definitional confusion (Wilkinson and Savulescu 2019), and specifically in medical cases like Amy's, professional societies have suggested procedural consensus-building over hard-liner policies and regulations (Bosslet et al.

2015). Procedural approaches proposed include mediation, ethical review and the courts.

Nevertheless, while there is strong warrant for procedural approaches, they do come loaded with their own baggage which can lead to inadequacies. We note for example that mediation is most useful when each party has interests that can be traded. It may therefore be better used in only certain situations—for example, in marital breakdown when the house is traded against the car, a weekend against a summer holiday with the kids, and so on. It may be less useful in zero sum situations where life or death is at stake. Similarly, while the courts are designed to adjudicate differences, they do so on an adversarial platform and can be expensive. Further, legal venues may not be well equipped to navigate the significant ethical issues at stake (Huxtable 2018).

Considering all aspects of a situation through the operationalization of ethical concepts, norms, and principles seems indispensable. This notwithstanding, philosophy is at heart an aporetic approach, better at identifying problems than agreeing solutions. Our own approach, where each author has taken a consciously different framework to consider the problem, may be an exemplar of the ability of clinical ethics consultation to use this philosophical tendency to broaden the considerations that make up a decision. This can result in agreement (as here) but, we might view such agreement as lacking probity, given the psychological tendency for the agreement to be based on first intuitions rather than 'working through' the problem from first principles. Moreover, in some very finely balanced situations there are difficulties in reaching good faith agreement despite a broad analysis of the problem taking place and all parties being attentive to one another's concerns.

Given all these challenges, though, we still believe that, like Rawls's appeal to "overlapping consensus," parties can hold on to many of their personal and philosophical commitments while finding common ground. Procedural approaches to ethical challenges (like we see in Amy's case) that decidedly focus on addressing transparency concerns and building agreement can avoid the entanglements of theoretical disagreements and prior political and metaphysical commitments. Cutting through such entanglements can come to a better understanding for every one of the ethical challenges and opportunities of the case and, possibly, even agreement on which path to take.

Guiding Principles When Dealing with Demands for Harmful Treatment

Harm is undue medical interference:
-Avoiding harm to patients is an injunction against undue medical interference
-Where treatment cannot achieve the intended end, a concern that it may be harmful should be taken seriously

Harms require specification:
-Where a treatment is demanded that is believed to be harmful it is important to recognize that what is perceived as a harm by one person may not be perceived as a harm by another
-Harms can be clarified by engaging with all parties
-Wherever possible, it is important to listen to the child, who will have insights into their experience, and how they believe they will benefit from the treatment they demand

(continued)

(continued)

There are limits to specification and a place for proceduralism:
-Frameworks and approaches may further clarify arguments against harmful treatment
-The limits of language mean that these may most effectively aid clinical decision-making in combination with procedural approaches

References

Archard, D. 2013. Children, adults, best interest and rights. *Medical Law International* 13 (1): 55–74.

Arendt, H. 1998. *The human condition*, 2nd ed. London: University of Chicago Press.

Beauchamp, T.L., and J.F. Childress. 2009. *Principles of biomedical ethics*, 7th ed. Oxford: Oxford University Press.

Birchley, G. 2018. Charlie Gard and the weight of parental rights to seek experimental treatment. *Journal of Medical Ethics* 44 (7): 448–452. https://doi.org/10.1136/medethics-2017-104718.

Blustein, J. 1982. *Parents and children: The ethics of the family*. Oxford University Press.

Boorse, C. 2014. A second rebuttal on health. *Journal of Medicine and Philosophy* 39 (6): 683–724.

Bosslet, G.T., et al. 2015. An official ATS/AACN/ACCP/ESICM/SCCM policy statement: Responding to requests for potentially inappropriate treatments in intensive care units. *American Journal of Respiratory and Critical Care Medicine* 191 (11): 1318–1330.

Brighouse, H., and A. Swift. 2014. *Family values: The ethics of parent-child relationships*. Oxford: Princeton University Press.

Devaney, J. 2018. The trouble with thresholds: Rationing as a rational choice in child and family social work. *Child & Family Social Work* 24 (4): 458–466.

Dewey, J. 1932. *Ethics*, 2nd ed. New York: Henry Holt and Co.

Diekema, D.S. 2004. Parental refusals of medical treatment: The harm principle as threshold for state intervention. *Theoretical Medicine and Bioethics* 25 (4): 243–264.

Dubler, N.N., C.B. Liebman. 2011, *Bioethics mediation: A guide to shaping shared solutions*, 2nd edn. Vanderbilt University Press.

Engelhardt, H.T., Jr. 2010. Beyond the best interest of children: four views of the family and of foundational disagreements regarding pediatric decision-making. *Journal of Medicine and Philosophy* 35 (5): 499–517.

Feinberg, J. 1970. The nature and value of rights. *Journal of Value Inquiry* 4 (4): 243–257.

Fost, N. 2011. Futility. In *Clinical ethics in pediatrics: A case-based textbook*, ed. D. Diekema, M. Mercurio, and M. Adam, 106–111. Cambridge: Cambridge University Press.

Haidt, J. 2013. *The righteous mind*. London: Penguin.

Holmes, W. 1995. *Collected works of justice holmes*, vol. 3. Chicago: Chicago University Press.

Huxtable, R. 2018. Clinic, courtroom or (specialist) committee: In the best interest of the critically Ill child? *Journal of Medical Ethics* 44 (7): 471–475.

James, W. 1891. The moral philosopher and the moral life. *International Journal of Ethics* 1: 330–354.

Pellegrino, E.D. 1979. The Anatomy of Clinical Judgments. In *Clinical judgment: A critical appraisal. Philosophy and medicine*, vol 6, ed. Engelhardt H.T., S.F. Spicker, B. Towers B. Springer, Dordrecht.

Rawls, J. 2001. *Justice as Fairness: A Restatement*. London: Belknap Press.

Rawls, J. 1999. *A theory of justice (Revised Edition)*. Cambridge: Harvard University Press.

Raz, J. 1986. *The morality of freedom*. Oxford: Oxford University Press.

Ross, L.F., and A.H. Swota. 2017. The best interest standard: Same name but different roles in pediatric bioethics and child rights frameworks. *Perspectives in Biology and Medicine* 60 (2): 186–197.

Schneiderman, L.J., N.S. Jecker, and A.R. Jonsen. 1990. Medical futility: Its meaning and ethical implications. *Annals of Internal Medicine* 112 (12): 949–954.

Shah, S.K., A.R. Rosenberg, and D.S. Diekema. 2017. Charlie gard and the limits of best interest. *JAMA Pediatrics* 171 (10): 937–938.

Snelling, J. 2016. Minors and contested medical-surgical treatment. *Cambridge Quarterly of Health Care Ethics* 25 (1): 50–62.

Tobin, J. 2013. Justifying children's rights. *The International Journal of Children's Rights* 21 (3): 395–441.

Waldron, J. 1987. *Nonsense upon stilts*. London: Methuen.

Wilkinson, D., and J. Savulescu. 2019. *Ethics, conflict and medical treatment for children: From disagreement to dissensus*. London: Elsevier.

Wittgenstein, L. 1958. *Philosophical investigations. Translated by G.E.M. Anscombe*, 3rd ed. Oxford: Blackwell.

Zaner, R.M. 1988. *Ethics and the clinical encounter*. Englewood Cliffs, NJ: Prentice Hall.

Further Reading

Huxtable, R. 2012. *Law, ethics and compromise at the limits of life: to treat or not to treat*. London: Routledge.

Moreno, J. D. 1995. *Deciding together: Bioethics and moral consensus*. New York: Oxford University Press.

Tobin, J. 2019. *The UN convention on the rights of the child: A commentary*. Oxford: Oxford University Press.

Wilkinson, D., and J. Savulescu. 2011. Knowing when to stop: futility in the ICU. *Current Opinion in Anesthesiology* 24 (2): 160-165.

Chapter 12
Family or Community Belief, Culture, and Religion: Implications for Health Care

T. Rossouw, P. Foster, and M. Kruger

Abstract This chapter explores the implications of differences in belief, religion, and culture, both among different family members and between families and health care providers, and how these manifest in medical practice. The authors principally aim to provide a practical approach to ethical pediatric and cross-cultural care. Acknowledging the complexity of the clinician-parent-patient triad, it provides examples of cases involving seemingly irreconcilable differences, as well as an approach to managing them, with emphasis on the need for mediation and continued communication. It further provides a practical approach, centered on respect for persons that culminates in truly informed consent and assent, to managing conflicts between families and health care workers. It delineates three principles by which to approach disagreements about care decisions: the best interest standard; the family-centered approach; and the harm principle. The chapter concludes with a discussion of cultural competence, viewed as health services that are culturally and linguistically sensitive as well as respectful of and responsive to the health beliefs of diverse patients. It explores the notions of cultural safety and cultural humility as two fundamental components needed to realize respectful and culturally competent care.

Keywords Cultural competence · Respectful care · Ethical pediatric care

T. Rossouw (✉)
Department of Immunology, Faculty of Health Sciences, University of Pretoria, Pretoria, South Africa
e-mail: theresa.rossouw@up.ac.za

P. Foster
Cook Children's Hospital, Fort Worth, TX, USA
e-mail: Pam.Foster@cookchildrens.org

M. Kruger
Department of Paediatrics and Child Health, Faculty of Medicine and Health Sciences, Stellenbosch University, Stellenbosch, South Africa
e-mail: marianakruger@sun.ac.za

© Springer Nature Switzerland AG 2022
N. Nortjé and J. C. Bester (eds.), *Pediatric Ethics: Theory and Practice*,
The International Library of Bioethics 89,
https://doi.org/10.1007/978-3-030-86182-7_12

12.1 Introduction

What ethical implications are involved when two parents of the same child have significant differences in their religious preferences and practice? Is it ethical to give preference to one parent over another? What should be done if the health care worker, parents, and child patient experience conflict regarding the health care management of the child? How does a health care practitioner from one culture effectively communicate with a patient from a different culture? These are the kinds of ethical challenge explored in this chapter. The focus is on the implications of differences in belief, religion, and culture, both among different family members and between families and health care providers. The chapter examines various arguments and theoretical perspectives and concludes with an exploration of and a practical approach to ethical pediatric and cross-cultural health care.

12.2 Definition of Concepts

A chapter dealing with value-laden concepts such as belief, religion, and culture would not be complete without first defining the terms and exploring the meanings embedded in them. 'Belief' can be regarded as a state of mind according to which one considers something to be true even though it cannot be proven to be so. Beliefs permeate our ideas about life and the world, and mutually supportive beliefs may come together to form philosophical, ideological, or religious belief systems (Council of Europe, n.d.).

'Religion' can be considered as a belief system that relates humanity to spirituality, in other words to that which people "regard holy, sacred, absolute, spiritual, divine, or worthy of especial reverence" (Encyclopædia Britannica, n.d.). Religions are often built on narratives, symbols, traditions, and sacred histories and are organized around specific practices that create a community of people with a shared morality, ethics, and often a preferred lifestyle. The belief systems surrounding people influence their identity, regardless of whether the person identifies as religious or spiritual or not (Council of Europe, n.d.).

Despite its common use, the concept of 'culture' is difficult to define. The American anthropologists Kroeber and Kluckhohn compiled a list of 164 different definitions of culture. While this diversity of meaning lies at the heart of the problem of defining the concept, it extends beyond mere conceptual and semantic differences and points to "different political or ideological agendas" that are often ingrained, yet subliminal (Avruch 1998, p. 6). The following definition by the British anthropologist and founder of cultural evolutionism, Sir Edward Burnett Tyler, posited in 1870 has been influential:

> Culture ... is that complex whole which includes knowledge, belief, art, morals, law, custom, and any other capabilities and habits acquired by man as a member of society (cited by Avruch 1998, p. 6).

While this definition speaks to the complex and acquired nature of culture, it does not acknowledge the individual experiences and manifestations of culture as well as its transitory nature. Matsumoto (1996, p. 16) defines culture as "... the set of attitudes, values, beliefs, and behaviors shared by a group of people, but different for each individual, communicated from one generation to the next."

The expression of culture is not homogeneous among people and should be seen not only as a social construct but also as an individual one: people adopt and engage with their culture to different degrees, and this impacts on their attitudes, values, beliefs, and behaviors. For this reason, Spencer-Oatey (2008, p. 3) expands on Tyler's definition by incorporating an element of individual agency:

> Culture is a fuzzy set of basic assumptions and values, orientations to life, beliefs, policies, procedures and behavioral conventions that are shared by a group of people, and that influence (but do not determine) each member's behavior and his/her interpretations of the 'meaning' of other people's behavior.

This differential expression of culture is also partly determined by the fact that people belong to different groups and categories simultaneously which can be seen to form several conceptual layers within a person. These layers could be national (country); regional, ethnic, religious, or linguistic; gender; generational; role (e.g. parent or child); social class (influenced by educational opportunities and occupation or profession); and organizational or corporate (for those who are employed) (Hofstede 1991). Many factors therefore affect a person's cultural identity, as illustrated in Fig. 12.1. Hence, while all people from a specific culture are expected to hold similar norms, these norms will be relevant to different degrees and expressed in different ways by people within that culture (Matsumoto 1996).

12.3 Ethical Challenges Emanating from Differences in Belief, Religion, and Culture

Cultural distinctions are easily identifiable when multiple families from the same culture are gathered in one place. Health care providers assume that cultural similarities such as dress, language, religion, ethnic background, and customs will be present (Avruch 1998). Expectations prevail that members of the same family share values, ethics, and morals that are lived out in ways distinctive to the particular community to which they belong and may be observed during communication and decision-making. Health care providers may be aware of how a particular community views its narrative including meaning making, storytelling, and fulfilling one's destiny. Identifying cultural differences within a single-family system, however, proves more challenging.

The typical underlying assumption is that the members of a single family are homogenous, meaning that the family members share the same thoughts and feelings about matters such as hope, trust, love, and belonging. In addition, there is a strong presupposition that each member of a family knows how the emotional

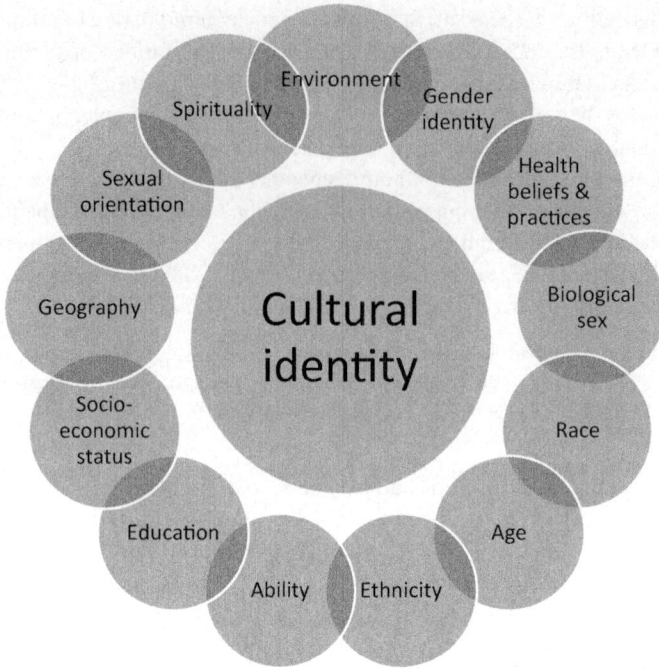

Fig. 12.1 Factors affecting a person's cultural identity (2001 adapted from U.S. Department of Health and Human Services, Office of Minority Health)

system operates in the family, workplace, and social system (Kerr 2000). However, experience shows this to be a faulty assumption.

The members of a nuclear family emulate patterns of behavior, speech, work, and social interactions from watching others in their family in similar situations. When a child is ill, however, there may be no pattern from which to operate. Each factor at play brings its own presuppositions, values, and behaviors to a setting that is already emotionally charged by the patient's illness. The parents of the patient may or may not have been acquainted for any length of time, may or may not be married, and may or may not be in a stable relationship. The extended family may include grandparents, aunts, uncles, siblings, or even great-grandparents. Each member brings their own values and presuppositions with them.

A dilemma seen typically in a pediatric care setting involves religious preferences. What are the ethical implications when the parents of a child have significant differences in their religious preferences and practice? For example, if the patient is diagnosed with leukemia and the mother professes Jehovah's Witness as her faith, she may reject any intervention of blood or blood product as treatment for her child's illness even though best practices indicate the need for such. The patient's father may profess Catholicism as his faith, not adhering to the same tenets as the mother regarding treatment options. These differing philosophies create tension in the family and for the treatment team. Is it ethical to give preference to the beliefs of one parent over another? Giving blood products, according to the mother's understanding of her faith system, may imperil the child's spiritual eternity. The medical care team is charged with providing evidence-based care that provides the greatest opportunity for improvement in the timeliest fashion. Yet, the patient's mother prefers treatment options that align with her beliefs but are less effective. The father prefers more aggressive treatment and expresses his displeasure with the mother's choices to the team. Does the care team side with the parent whose views align with the current standard for medical treatment? Sensitivity and dedicated clear communication are needed to navigate this case and others where ongoing treatment decisions are impacted while making certain that all persons needed for decision-making are present.

Another dilemma often encountered in pediatrics involves faith healing. Belief in miracles and the use of miracle language is prevalent in many faith traditions. *Cambridge Dictionary* defines a miracle as an unusual and mysterious event that is thought to have been caused by a god because it does not follow the usual laws of nature. Families embracing miracles as fact may rely heavily on their faith when making decisions about all aspects of the members' personal lives. In one instance, a family chooses prayer as intervention in difficult situations involving work, school and relationships, both inside and outside the family, as well as the members' physical wellbeing. However, other families utilizing similar faith language may use miracle language to avoid making hard decisions regarding care planning for their member. Understanding this particular faith stance is challenging to health care providers. Communication may feel non-existent because the family seemingly refuses to acknowledge the medical information shared. Another family professing their faith, in contrast, acknowledges that faith is an important part of the family narrative but embraces the notion that faith is informed by science, particularly medical science. When making decisions for a pediatric patient, paternal and maternal relations may disagree about the place of faith and miracles in treatments and outcomes. What responsibility does the care team have in navigating faith language with families? Ideally, the health care team will include a professional chaplain who can join the conversation advocating for calm, clear conversation between providers and family. When needed, the chaplain can access faith-based community resources to facilitate these challenging conversations.

As individuals and families embrace unique perspectives that signify their personal faith, morals, and values, health care providers are increasingly challenged to communicate effectively with each person involved. Taking the time to discover the values that are important to each patient and each family places a burden on the health care team. What is the ethical obligation in each case? When and to what degree is one compelled to preserve the dignity of another person and to hear the case for autonomy? These are questions that must be considered and will be increasingly important for ethical medical care in the future.

12.4   Practical Approach to Respectful Pediatric Care

Caring for the child patient is ethically complex, as indicated in the previous section. The parents, as custodians of their children, are included in any decision-making involving the child patient, which is important for maintaining the integrity of the family relationship (Downie & Randall, 1997). This creates a clinician-patient-parent triad, which can be difficult to navigate. The assumption is that parents are the best situated persons to be the decision-makers for their child, except in situations where harm is expected due to parental decisions (Katz and Webb 2016). Respect for persons (autonomy) is critically important in health care, and the key elements of informed consent include the sharing of comprehensive information, assessing of decision-making capacity and obtaining voluntary consent without coercion. In the context of children, there is a need to acknowledge the minor's potential lack of capacity which, by default, makes parents the primary decision-makers.

According to the Convention on the Rights of the Child (1990) (also refer to Chap. 5), the child should be involved in medical decision-making regarding her/his own health care to the extent that she/he can participate in the decision-making process. In practice, children should provide assent for medical treatment. Assent is different from consent as it is elicited in minors who are still developing their decision-making capacity. Assent relies on informing the minor about health care options in a developmentally appropriate way, coupled with assessment of their understanding of such options before an indication of agreement or disagreement is obtained from the minor (also refer to Chap. 13). Leikin (1983) made the key conceptual move form assent as mere acquiescence or deference to authority, to an active engagement with the pediatric patient in a climate that encourages the development of the capabilities required for competency to consent, such as understanding, reasoning and voluntariness.

Determination of pediatric capacity for medical decision-making is challenging. Usually, the 'rule of sevens' (refer to Chap. 2), as defined in the Tennessee Supreme Court case of Cardwell versus Bechtol (1987), is used. Accordingly, it is accepted that there is a lack of decision-making capacity for children under 7 years of age, while children from 7 to 14 years have evolving capacity, and children 14 years and older may have decision-making capacity equal to that of adults in matters pertaining to health care. Assent should therefore be elicited from 7 years of age. Local policies in different countries provide guidance regarding consent or assent of children in medical care. In contrast to consent, assent is not binding and is subject to the minor's level of understanding.

In standard health care practice, it is therefore advisable to obtain both parental consent and minor assent to health care treatment. It is important to determine whether the child is able to understand the information provided, possesses a stable value system, and can deliberate the potential outcome of the treatment plan (Ackerman, 2001; Katz and Webb 2016) (see Table 12.1).

In many countries, adolescents are allowed to consent to medical treatment independently if they are deemed mature enough to understand the information (refer to

Table 12.1 Determining the ability to provide assent

Cognitive ability	Determine whether the child is able to understand the information provided in an age-appropriate manner
	Determine whether the child is able to provide voluntary permission
Key attributes	Determine whether the child is able to reflect thoughtfully on the information provided
	Determine whether the child is able to voice an opinion
	Determine whether the child is able to take action
Consequences	Determine whether the child can deliberate the consequences of the decision
Assent	Solicit acceptance of the proposed health care from the child

Chap. 10), can deliberate the potential outcome, and can make decisions that reflect their own value system (Katz and Webb 2016; Weithorn and Campbell 1982). Major professional bodies such as the American Academy of Pediatrics and the Confederation of European Specialists in Pediatrics provide recommendations regarding parental consent and the child's assent in health care decision-making (De Lourdes Levy et al, 2003; Katz and Webb 2016). For a detailed discussion on this please refer to Chap. 7.

In the context of treatment refusal, the reasons for refusal need to be examined and discussed carefully. As mentioned above, parents can differ about the management of their child. In the case of minors not yet able to participate in decision-making, the assumption is that the parents and the health care workers will use the 'best interest' standard to guide them (refer to Chap. 4). The concept involves assessing how to maximize benefits while minimizing harm or risks. Best interest is regarded as the highest achievable benefit after evaluation of all available treatments. In this context, the expected benefits of a treatment should be significantly more than the potential harm or risks.

According to Kopelman (1997), the best interest standard determines three aspects of care: the threshold for intervention (e.g. custody issues), the *prima facie* duties (e.g. right to health care), and the standard for reasonableness (e.g. resource allocation). However, the interpretation of the best interest standard is often subjective and variable. For example, parents' decision for one child may influence the wellbeing of their other children negatively. This may create a dilemma for them, leading to an inability to make a decision. The best interest standard is also applied differently in pediatric bioethics, where it is used as a guiding principle, compared to the human rights community, where it is used as both a guiding and an intervention principle (Ross and Swota 2017).

There is currently a move towards a family-centered approach whereby the best interest of each child in the family is considered in the decision-making process to ensure that the whole family's needs are considered (Committee on Hospital Care, American Academy of Pediatrics 2003). The family's values and belief systems, as mentioned above, may influence their perspective on disease and influence their consent to essential treatment, crucial for the cure of their child. If consensus cannot

Table 12.2 Parental or surrogate decision-making for children

Best interest	Maximize the benefits and minimize the risks or harm for the child patient
Family-centered decision-making	Parents, family members (e.g. siblings), and health care workers cooperate in decision-making
Harm principle	Determine a threshold for harm where parental decisions will be overruled. Leads to constrained parental decision-making

be reached among the health care workers, the child patient, and the parents, there should be a thorough deliberation, focused on the best interest standard, whereby benefits are maximized, and risks and harm are minimized (Table 12.2).

An alternative is to use the harm principle to determine the potential harm of either the disease process or the treatment and to compare that to alternative treatment options that are more acceptable to the family (Diekema 2004). When using the harm principle, it is important to first determine the wishes or decision of the parents as well as the child, if old enough to voice an opinion, and to assess the effects of these decisions on the health or wellbeing of the child. If the potential effect will pose significant harm to the child, it may be necessary to overrule the parents' decision. It is, however, also important to assess what the effect will be on the child if the parents' decision is overruled and whether this may constitute greater harm than the parents' original decision. If this is the case, the parents' decision should be allowed. However, if the potential harm of the parents' decision is greater than that of the proposed health care intervention, the parents' decision should be overruled.

Another alternative is to recognize that there is the potential for constrained parental autonomy, although parents have the ethical right to make medical decisions on behalf of their children in general. An example is the universal vaccination program to protect children against harmful infectious diseases; however, some parents may not support vaccinations. In this scenario, the state may enforce vaccination as vaccinations both protect the individual child and the public at large. There is therefore a limit to or constraint on parental authority, which should be overruled if parents' decision poses significant harm to the child and/or the public (Ross, 1998). In conclusion, there is an ongoing debate with regards to the specific roles of the best interest standard , the harm principle and constrained parental autonomy (refer to Chap. 4). According to Bester (2018) any medical decision regarding the child patient's best interest should be substantiated by a reasonable argument, while the decision should not pose a risk of harm to the child's welfare (Bester 2018).

12.5   Practical Approach to Managing Conflicts Between Families and Health Care Workers

Treatment refusal is a major issue in health care and occurs when there is a conflict of opinions among the health care worker, the parents, and the minor, with a tension between beneficence and autonomy. Three scenarios are possible:

1)   Both the parents and the child refuse the proposed treatment.
2)   The parents refuse the proposed treatment, but the child assents to the treatment.
3)   The parents accept the proposed treatment, but the child refuses the treatment.

In the context of treatment refusal, it is important to determine the reasons for this refusal, which can be due to a different or incomplete understanding of the disease or even a disbelief in the existence of the disease. These differences often emanate from different belief or value systems, especially in a multicultural context. Fear of stigmatization could also be brought about by a diagnosis or the treatment. Standard recommended practice in pediatric medicine usually involves parental and pediatric counselling prior to testing and diagnosis. This is the ideal time to elicit any potential for treatment refusal by also obtaining information about the family's understanding of diseases, their belief systems, and their cultural background. After the diagnosis of a disease has been made, parental and child post-counselling takes place. Depending on the complexity of the disease, there could be several consultations with the doctor or multidisciplinary team before and during initiation of a treatment plan.

If both the parents and the child refuse treatment, the best interest standard and assessment of harm should guide the health care worker. The health care worker should also assess how to achieve maximum benefit while at the same time assessing the potential harm to the child if the wishes of the parents and child are disregarded. An example is treatment refusal on the grounds of religious beliefs such as refusal of an emergency blood transfusion to a minor if the parents are Jehovah's Witnesses. The clinician or health care worker is advised to consult with peers or higher authorities to ensure that the proposed treatment will maximize benefit while minimizing harm and to ensure an objective rather than a subjective decision.

The same process is necessary if either the parents or the child refuse treatment, except that there is then a need to facilitate between the parents and the child with regard to the differences in opinion. An example is a child who refuses cancer treatment due to the fear of loss of hair while the parents agree to the treatment with curative intent. The child's fear of body image should be addressed, but the treatment refusal cannot be upheld as the alternative may be death. Another scenario is a child who refuses life-sustaining treatment while the parents insist on treatment. Here it is important to determine the child's understanding of the consequences of the decision, whether the treatment or intervention poses significant risks, and whether the potential for cure is small. If the child has adequate understanding in the face of risky treatment with a small possibility for cure, the child's refusal should be respected while the parents' decision is respectfully overruled.

Health care workers dealing with minor patients are challenged by the complexity of the clinician-parent-patient triad. In this section, the importance of eliciting assent from minors in medical decision-making as well as treatment refusal by either minors or their parents has been discussed. In dealing with the child patient, clinicians should promote ethically sound participation in decision-making by parents and the minor patient.

12.6   Culture and Cross-Cultural Health Care

Cross-cultural care is the art of "learning how to transcend one's own culture in order to form a positive therapeutic alliance with patients from other cultures" (Deagle 1986, p. 1315). Cross-cultural care represents both an opportunity to provide and receive care across cultural boundaries as well as the reality of modern medical practice, given the mosaic of cultures, races, and ethnic groups that makes up our societies.

Care across cultural boundaries is complicated by diverse beliefs about health, practices, values, and communication (Tavallali et al. 2017). Health care can be fraught with special risks and hazards that are often based on cultural stereotypes and perceptions about normality and even superiority. This is especially true in the context of illness where the experience of illness varies among cultures and often manifests in communication difficulties and tension in the therapeutic relationship. These challenges are exacerbated by the unfamiliarity of health care workers with cultural diversity and of patients with the health care system, especially in the context of a foreign country (Nkulu Kalengayi et al. 2012). A first step and prerequisite to providing effective cross-cultural care is therefore to understand the concept of culture and then to reflect critically on one's own culture (Deagle 1986).

Scholars of culture believe that there is no unique or absolute set of characteristics that can differentiate definitively among different cultures (Spencer-Oatey, 2012). Cultural divides are also blurred by the process known as cultural diffusion, which is defined as "the spreading of cultural items from one culture to another " (Spencer-Oatey 2012, p. 13). An understanding of culture as "acquired through the process of learning" and hence being more fluid and contextual than a fixed attribute can create greater tolerance for cultural differences, something that is essential for effective intercultural communication (Ferraro 1998, p. 19).

Most scholars refer to culture as a learned human behavior. However, behavior can be difficult to decipher and requires an exploration of the deeper values that underpin actions. Values, though important, are also limited by the fact that they only represent the "espoused values of a culture" (Spencer-Oatey 2012, p. 3), in other words, that which people assert as or would like to be the reasons for their behavior. The *truth* lies much deeper, in the concealed or unconscious underlying assumptions "about how things really are", which truly determine how people perceive, think, and act (Spencer-Oatey 2012, p. 3). These assumptions are powerful and difficult to confront since they are regarded as the 'truth' and, hence, not open to debate.

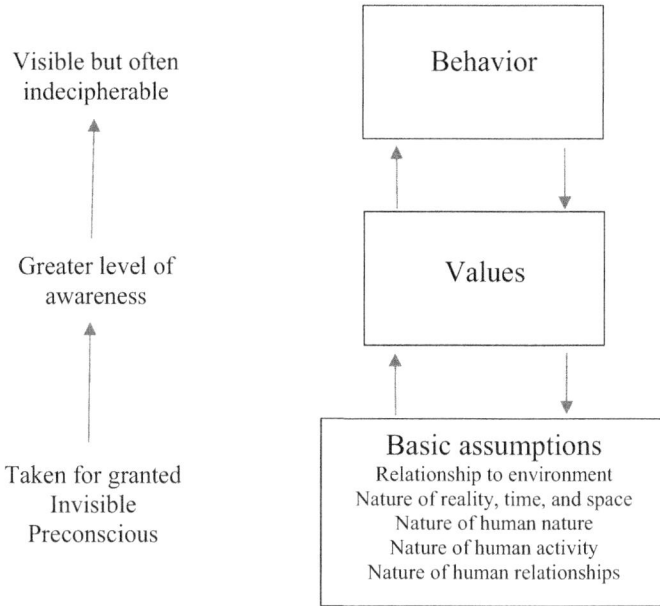

Fig. 12.2 The levels of culture and their interaction

Figure 12.2, adapted from Spencer-Oatey (2012, p. 4), represents culture as an interplay among assumptions, values, and behavior. To truly understand one's own and other cultures, it is therefore essential to critically evaluate and reflect on one's underlying assumptions.

12.7 The Role of Culture in the Therapeutic Relationship

Culture is something that is present in all patients and care providers and plays a role in the many interactions encountered in the health care setting, for instance among patients, their families, health care providers, and health care systems (Canadian Paediatric Society 2018). The illness experience is learned through culture, often through certain 'explanatory models' that each person in a culture possesses, often at a subconscious level (Kleinman 1980). It is therefore not unexpected that health care workers and their patients may have very different explanatory models or beliefs about illness and hence different expectations of the therapeutic encounter (Deagle 1986). Failing to understand these differences could lead to miscommunication, mistrust and misunderstanding, and ultimately, on the side of the patient, nonadherence to the health care workers' treatment plan.

12.8 Practical Approaches to Respectful Cross-Cultural Health Care

Respectful cross-cultural care is based on cultural competence. While no single definition of cultural competence exists, the following definition is helpful: "… a set of congruent behaviors, attitudes, and policies that come together in a system, agency, or among professionals that enables effective work in cross-cultural situations" (U.S. Department of Health and Human Services, Office of Minority Health 2001). Essentially, cultural competence refers to services that are culturally and linguistically appropriate as well as respectful of and responsive to the health beliefs of diverse patients (U.S. Department of Health and Human Services, Office of Minority Health, 2001). This is underscored by the fact that patients experience nurses' cultural competence through the expression of respect and their regard for people as individuals rather than merely members of a specific culture (Tavallali et al. 2017).

Two aspects are integral to cultural competence, namely cultural safety and cultural humility. Cultural safety entails creating an environment where the cultural identities of all persons are recognized and respected. Importantly, it does not pretend that cultural differences do not exist (so-called 'cultural blindness') so as not to minimize real and important cultural differences that could affect the therapeutic encounter. Instead, it acknowledges and respects differences while preventing practices that belittle, humiliate, or disempower the cultural identity of others (Nursing Council of New Zealand, 2002). Cultural safety embodies an ethos of shared respect, meaning, and knowledge and values the experience of learning together (Williams 1999). Cultural humility is one of the scaffolds of cultural safety. It has an inward focus and emphasizes a commitment to lifelong self-evaluation and -critique together with an awareness of one's own implicit or explicit biases, with the goal of managing inherent power imbalances related to culture (Canadian Paediatric Society 2018).

The LEARN model is a practical framework for teaching cultural competence and contains the following actions (Berlin and Fowkes 1983):

- **Listen** with sympathy and understanding to the patient's perception of a problem.
- **Explain** your perceptions of a problem.
- and **discuss** differences and similarities.
- **Recommend** treatment.
- **Negotiate** agreement.

The following guide, adapted from the Canadian Paediatric Society (2018), presents a useful approach to practicing cultural safety and cultural humility:

- Focus on the patient during the consultation; specifically listen to the way in which people talk about themselves.
- Use interactive communication that signals to people that their perspectives are valued. Avoid talking in a controlling and authoritarian style.
- It is not necessary to understand every culture or group's health beliefs; understanding the extent to which individual people view their culture to influence an issue is sufficient.

Table 12.3 Guiding principles for ethical pediatric care

Respect	Recognize and respect the wishes of children as they develop and mature
	Determine how they view their illness: cause, how it changed bodily functions, personal significance, etc
	Consider dissent seriously and explore underlying causes
Emerging capacity	Communicate with minors in terms they can understand
	Recognise the wide variation in competence among individual children
	Determine each child's level of understanding and reasoning ability
	Evaluate individual decision-making capacity considering the patient's age, developmental level, and ability to understand risks, benefits and consequences of a decision and its alternative
	As competencies develop, children's participation should increase
Privacy and confidentiality	Recognize children's need for privacy and confidentiality
	Offer older minors the opportunity for a private and confidential consultation to discuss concerns, needs, fears, etc
	Deliver confidential health services for issues involving mental health, reproductive and sexual health, and substance abuse to consenting adolescents
Collaboration	Create conditions that are supportive of a collaborative effort among parents, the minor, and health care team

- Be aware of the cultural differences between yourself and others; this awareness helps to avoid conflict during diagnosis or while developing a treatment plan.
- Be aware of and critically examine your own culturally influenced beliefs and biases, especially as they are present in clinical encounters.
- Allow patients to take on the role of experts and become true partners in determining their care.
- As needed, guide patients and their families on local social and cultural norms and laws. If a specific tradition conflicts with local laws or human rights, it is the duty of the health care worker to ensure that the person is informed (Table 12.3).

12.9   Conclusion

Dealing with differences in family or community belief, culture, and religion in the health care setting is not easy. Given the inter- and intracultural diversity of people, it has become imperative for health care workers to develop an improved understanding of the impact of these difference on the care they provide. Health care workers dealing with minor patients are especially challenged by the complexity of the clinician-parent-patient triad. This chapter argues for the importance of respectful care that is exemplified by ongoing communication, eliciting assent from minors where appropriate, and ethically sound participation in decision-making by parents and the minor patient. It proposes a framework for developing cultural humility and competence in order to realize respectful cross-cultural care. By providing a practical approach to ethical pediatric and cross-cultural care, the authors hope to empower health care workers across various disciplines to practice sound decision-making in ethically complex situations.

References

Ackerman, T.F. 2001. The ethics of drug research in children. *Paediatric Drugs* 3: 29–41.
Avruch, K. 1998. *Culture and conflict resolution.* Washington, D.C.: United States Institute of Peace Press.
Berlin, E., and W. A. Fowkes. (1983). A teaching framework for cross-cultural health care. *Western Journal of Medicine* 139: 934–938. Available from: http://www.pubmedcentral.nih.gov/picren der.fcgi?artid=1011028&blobtype=pdf. Accessed 28 March 2020.
Bester, J. C. 2018. The harm principle cannot replace the best interest standard: Problems with using the harm principle for medical decision-making for children. *American Journal of Bioethics* 18 (8): 9–19. [Taylor & Francis Online], [Google Scholar]
Canadian Paediatric Society. (2018). *Cultural competence for child and youth health professionals.* Retrieved from https://www.kidsnewtocanada.ca/culture/competence
Cardwell v. Bechtol. (1987). Retrieved from https://law.justia.com/cases/tennessee/supreme-court/ 1987/724-s-w-2d-739-2.html. Accessed 10 March 2020.
Committee on Hospital Care, American Academy of Pediatrics. (2003). Family-centered care and the pediatrician's role. *Pediatrics* 112 (3 pt 1): 691–697.
Convention on the Rights of the Child. (1990). Retrieved from https://www.ohchr.org/en/professio nalinterest/pages/crc.aspx. Accessed 15 February 2020.
Council of Europe. (n.d.). *COMPASS: Manual for Human Rights Education with Young People.* Retrieved from https://www.coe.int/en/web/compass/religion-and-belief. Accessed 2 February 2020.
Deagle, G.L. 1986. The art of cross-cultural care. *Canadian Family Physician* 32: 1315–1318.
Diekema, D.S. 2004. Parental refusals of medical treatment: The harm principle as threshold for state intervention. *Theoretical Medicine and Bioethics* 25 (4): 243–264.
Downie, R.S., and F. Randall. 1997. Parenting and the best interest of minors. *The Journal of Medicine and Philosophy* 22: 253–270.
Encyclopædia Britannica. (n.d.). *Religion.* Retrieved from https://www.britannica.com/search? query=religion+. Accessed 2 March 2020.
Ferraro, G. 1998. *The cultural dimension of international business,* 3rd ed. New Jersey (NJ): Prentice Hall.
Hofstede, G. 1991. *Cultures and organizations: Software of the mind.* London, England: Harper Collins Business.
Katz, A. L., and S. A. Webb. 2016. Committee on Bioethics. Informed consent in decision-making in pediatric practice. *Pediatrics* 138 (2): e20161485. https://doi.org/10.1542/peds.2016-1485
Kerr, M. E. 2000. One family's story: A primer on Bowen Theory. *The Bowen Center for the Study of the Family.* http://222.thebowencenter.org
Kleinman, A. 1980. *Patients and healers in the context of culture: An exploration of the borderland between anthropology, medicine, and psychiatry.* Berkeley, CA: University of California Press.
Kon, A.A., E.K. Shepard, N.O. Sederstrom, S.M. Swoboda, M.F. Marshal, B. Birriel, and F. Rincon. 2016. Defining futile and potentially inappropriate interventions: A policy statement from the Society of Critical Care Medicine Ethics Committee. *Critical Care Medicine* 44 (9): 1769–1774.
Kopelman, L. 1997. The best interest threshold, ideal, and standard of reasonableness. *The Journal of Medicine and Philosophy* 22: 271–289.

Leikin, S.L. 1983. Minors' assent or dissent to medical treatment. *The Journal of Pediatrics* 102 (2): 169–176. https://doi.org/10.1016/s0022-3476(83)80514-9.

De Lourdes Levy, M., V. Larcher, and R. Kurz. 2003. Ethics Working Group of the Confederation of European Specialists in Paediatrics (CESP). Informed consent/assent in children. Statement of the Ethics Working Group of the Confederation of European Specialists in Paediatrics (CESP).*European Journal of Pediatrics* 162 (9): 629–633.

Matsumoto, D. (1996). *Culture and psychology.* Pacific Grove, CA: Brooks/Cole.

Nkulu Kalengayi, F.K., A.K. Hurtig, C. Ahlm, and M. Ahlberg. (2012). 'It is a challenge to do it the right way': An interpretive description of caregivers' experiences in caring for migrant patients in Northern Sweden. *BMC Health Services Research* 12: 433. Retrieved from http://www.biomed central.com/1472-6963/12/433. Accessed 20 March 2020.

Nursing Council of New Zealand. 2002. *Guidelines for cultural safety, the treaty of Waitangi, and Maori health in nursing and midwifery education and practice.* Wellington, New Zealand: Nursing Council of New Zealand.

Ross, L. 1998. *Children, families, and health care decision-making.* New York, NY: Oxford University Press.

Ross, L.F., and A.H. Swota. 2017. The best interest standard: Same name but different roles in pediatric bioethics and child rights frameworks. *Perspectives in Biology and Medicine* 60 (2): 186–197.

Spencer-Oatey, H. 2008. *Culturally speaking. Culture, communication and politeness theory,* 2nd ed. London, England: Continuum.

Spencer-Oatey, H. (2012). *What is culture? A compilation of quotations.* Retrieved from http://www.warwick.ac.uk/globalpadintercultural. Accessed 8 March 2020.

Tavallali, A.G., M. Jirwe, and Z.N. Kabir. 2017. Cross-cultural care encounters in paediatric care: Minority ethnic parents' experiences. *Scandinavian Journal of Caring Sciences* 31: 54–62.

U.S. Department of Health and Human Services, Office of Minority Health. 2001. *National standards for culturally and linguistically appropriate services in health care.* Washington, D.C.: U.S. Department of Health and Human Services.

Weithorn, L.A., and S.B. Campbell. 1982. The competency of children and adolescents to make informed treatment decisions. *Child Development* 53 (6): 1589–1598.

Williams, R. 1999. Cultural safety—what does it mean for our work practice? *Australian and New Zealand Journal of Public Health* 23 (2): 213–214.

Further Reading

Derrington,k S.F., Paquette, E., Johnson, K.A.K. (2018). Cross-cultural interactions and shared decision-making. *Pediatrics* 142 (3): S187-S192. https://doi.org/10.1542/peds.2018-0516J

Fadiman, A. (1998). *The Spirit Catches You and You Fall Down: A Hmong Child, Her American Doctors, and the Collision of Two Cultures.* New York: Farrar, Straus and Giroux.

Saha, S., Beach, M.C., Cooper, L.A. (2008). Patient Centeredness, Cultural Competence and Health care Quality. *Journal of the National Medical Association* 100 (11): 1275–1285. https://doi.org/10.1016/s0027-9684(15)31505-4

United Nations (UN) (2014). Inter-agency support group on indigenous issues. Thematic paper on the health of indigenous peoples. UN, New York.

Chapter 13
Children Requiring Emergency Health Care

I. Mitchell and J. Guichon

Abstract Children present to physician's offices, urgent care centers or emergency departments for many reasons, not just with presumed severe and/or life-threatening illness. Children might be accompanied by the parents/guardians, neighbors or relatives, babysitters, sometimes bystanders or paramedics. Whatever the severity, an initial assessment is essential. After that assessment, consent is required for all other interventions, unless the patient's situation is immediately life or limb threatening. Consent for medical treatment is usually given by a parent/guardian. In the case of a minor patient, generally those 14 years of age and older, there must be an assessment of the patient's capacity to consent for him or herself. In all situations, the child should be informed about what is being done and why it is being done. Where appropriate, his/her assent should be sought. Some situations require a special approach, such as teenagers coming to the emergency department by themselves for care involving mental health, including addiction, and reproductive issues. Another special situation is the medical assessment and care of refugee children. Some children presenting for urgent/emergent care might have suffered from maltreatment. This fact might be obvious at first or it might become obvious during the initial assessment; in such cases, the help of child protection authorities should be sought and must be sought under the law of most jurisdictions. Severely ill and injured children should have the benefit of evidence-based intervention. Hence, research, with appropriate safeguards and after consent, is needed in emergency situations.

Keywords Focus on the child · Informed consent · Challenges of urgency · Mature minor · Importance of context · Child protection

I. Mitchell (✉) · J. Guichon
Cumming School of Medicine, University of Calgary, Calgary, Canada
e-mail: imitche@ucalgary.ca

J. Guichon
e-mail: guichon@ucalgary.ca

© Springer Nature Switzerland AG 2022
N. Nortjé and J. C. Bester (eds.), *Pediatric Ethics: Theory and Practice*,
The International Library of Bioethics 89,
https://doi.org/10.1007/978-3-030-86182-7_13

13.1 Introduction: Consent in the Urgent or Emergent Pediatric Setting

Informed consent is a keystone of modern pediatric practice. Yet in TV dramas—
the source of much public information about pediatric practice—seeking informed
consent for medical treatment is depicted as unimportant, or even irrelevant. These
dramatic events—medical crises—are usually set in an Emergency Department (ED),
or sometimes in an Intensive Care Unit (ICU). The characters are depicted to believe,
with virtuous certainty, that the ordered treatment is the only possible one, so they
must hurry, hurry! Why provide information or choices to the decision-makers? The
process of seeking formal consent seems an obviously unnecessary delay in a drama
in which the child's very life can turn on medical decisions that must be taken quickly.

Actual pediatric practice is different. Consent for all medical investigations and
treatment of the child must be received at every stage (with one very important
exception, which we shall come to). The need for specific consent is not based on the
physician's assessment of the importance of any one intervention; it is an ethical and
usually legal requirement for any and all interventions being conducted on the child.
For example, informed consent must be received prior to a blood test, something
seen as routine and trivial by most health care professionals. Having said that, there
is a limit to the granularity of information that must be given to the decision-maker.
A request for consent for "routine blood work" need not encompass details about the
test—hemoglobin, white cell counts and differential, electrolytes and liver function
tests—though the physician must be open to giving more detail on request. Some
tests should be mentioned explicitly and may require their own consent; an example
is testing for HIV.

Time might be of the essence but receiving informed consent before proceeding is
mandatory. The exception to the need to seek informed consent is when the situation
is an extreme emergency. For example, if a child arrives moribund, then there is no
time, and therefore no need, to seek permission for the potentially lifesaving insertion
of IV lines, endotracheal intubation and so on.

The requirement for consent for medical treatment is based on respect for persons.
In pediatrics, the respect is primarily for the child patient, though documents are
usually signed by a parent/guardian. In other words, in most pediatric situations, it
is assumed the parents are the ones who should be "authorizing" the intervention.
Indeed, the word "authorizing" is more appropriate than the word "consent", because
the parent is agreeing that something should happen to and for the child (refer to
Chap. 7). Nevertheless, there are many situations when the minor patients themselves
may consent, irrespective of whether they are accompanied by their parents. Three
distinct types of consent were described in a paper titled "Minors' assent or dissent
to medical treatment" published almost 40 years ago. (Leikin 1983). The underlying
concepts remain relevant, and some of the controversial aspects remain unresolved.
Firstly, some minors who seek particular forms of treatment, e.g. termination of
pregnancy and treatment for sexually transmitted infections, should be regarded as
being sufficiently competent to give consent "independent of parental consent or

knowledge". Secondly, should minors have the power to dissent when parents have given consent? Can minors overrule parents? Leikin did not answer this question; indeed, it cannot be answered in any absolute sense, but at the very least, minors' objections to a course of treatment should be acknowledged and questions answered in detail, even if not acted upon. The third type of consent, "assent" follows from the second. The professional must ensure that the minor knows and participates in discussion about treatment, without implying a veto. In other words, even when consent is appropriately sought from a parent/guardian, the child must be involved, at a level suitable to the age and stage of development. In this chapter, we make no overarching assumption regarding who might consent; we will describe ways to determine whether the decision-maker should be the parent/guardian or the minor patient, and the limitations involved.

All medical interactions involve attention to consent, confidentiality (information protection) and privacy (freedom from being observed or disturbed by other people). While these are separate concepts, they are related. For example, the fact that consent, or authorization, requires the decision-maker to have full information implies that the decision maker requires full access to all confidential information. Yet when minors present for urgent care, even if the parent/guardian is to be the one authorizing an intervention, there are limits regarding what the parent/guardian may be told. Specifically, information about the mental health of the minor, including drug abuse issues and reproductive information should be kept confidential. Health care professionals must take care to protect the minor's confidentiality interests, without adversely affecting the parent/guardian's ability to give a valid consent (please see Chap. 8). Privacy might also be essential, for example in physical examination.

13.2 Conceptual Qualification and Definition

Why should health care providers need to seek consent before giving medical treatment? The answer is based on the fundamental principle that all persons are equal in their human dignity and that they are, therefore, entitled to respect for their personhood. Respect for persons means that no one should interfere with another's physical person without that person's permission. Most people are capable of making decisions for themselves and, therefore, consenting to medical treatment. In these cases, we say that persons have "autonomy"; they are self-governing. But in the case of people who have not become self-governing (e.g., young children, adults with significant intellectual disabilities) or who have temporarily lost that ability (e.g., due to illness or injury) we still respect their personhood by requiring that another person (typically a loved one) take medical decisions in that person's best interest.

Respect for persons has replaced the older value (paternalism) that justified encouraging the patient to do what the physician thinks is best for the patient, or the physician conducting the treatment without permission. The change from paternalism to respect for persons is supported by legal precedent, as well as ethical reasoning. Dreisinger and Zapolski (2018) quote a US legal case from 1914 and describe increasing acceptance of the concept throughout the 20th Century so that now "little is done without a careful discussion between the doctor, patient, and the patient's family." Given that the legal precedent was a 1914 ruling, and changes in medical practice to align with the ruling came many decades later, it is likely that there was a gradual change in practice in keeping with societal norms. Mention of this delay is not to disparage legal precedents; they remain important. For example, the UK Medical Defence Union (MDU) (an organization that gives legal advice and support to physicians) uses a legal precedent rendered in 1999 after litigation followed a medical mishap to emphasize to physicians that for patient consent to be valid, full information must be given (MDU 2017a). The historical background of ethicists' shift in thinking from paternalism to respect for persons is described in detail by Beauchamp and Childress (2013).

Almost all health care practitioners respect the right of patients to accept or to decline treatment; they also aim to do good, to be beneficent.[1] Beneficence is an important value even in the current era when paternalism is no longer dominant (refer to Chap. 11). In other words, the concepts of respect for persons and beneficence are both relevant, and the balance tips in favor of respect for persons. In many situations of urgent and emergent care for children, the balance tilts toward beneficence, but without veering towards paternalism or ignoring the profound need to respect the child and the family. Today, there is a degree of tension between the concepts of respect for persons and beneficence; both are important (Pellegrino and Thomasma 1987).

Respect for persons requires that the patient or someone who can act in the patient's interests decides whether to give permission for the proposed treatment. A person is in a position to decide whether to consent, when the person:

1. Is competent, that is, has the ability to reason and to reflect on information received;
2. Has clear information from the health care provider about the treatment, the serious risks and common minor risks and any consequences of refusal;
3. Understands and appreciates that giving consent means that he/she has a responsibility to ensure he/she understands the information;
4. Is free from coercion, that is, the person may make the decision completely voluntarily.

When all four conditions obtain, then we say the person has "capacity to consent". Giving or refusing consent is a process; but in many situations, the requirement for formal documentation means that giving consent appears to occur at a specific moment in time (Mitchell and Guichon 2019).

Children should always be involved in this process in, of course, a manner consistent with the developmental stage, emotional and cognitive level, and medical condition. Consent might be implicit, for example if a child is brought to the ED by a

---

[1] Beneficence must be distinguished from paternalism. Beneficence is attempting to achieve the best outcome taking into account all patient factors. By contrast, paternalism is premised on the understanding that 'doctor knows best' and focuses on the physician and his or her understandings.

caregiver, then there is an assumption that initial assessment is being requested. Consent may be given verbally, but for procedural reasons, consent for all surgical interventions should be documented in writing. In most cases, institutions will have rules about which procedures require written consent. Written consent by itself is no guarantee that consent is ethically valid (for example, the doctor might not have described the possible risks), but the process of seeking and receiving written consent creates a moment when all parties concerned are reminded of the proposed intervention's importance. A detailed description of the discussion preceding the signing of the consent document is an essential part of the health record.

When a parent/guardian must make a medical decision for a child, that person needs some help in identifying how to proceed. This situation is different from when adults make medical decisions for themselves. When adults are presented with a medical decision for themselves, they may follow the principle of "pure autonomy". In other words, because they are self-governing, their decision may be based on their personal choice alone. (For example, they might choose not to take prescribed blood pressure medication.) But the situation is otherwise when the proposed medical treatment will take place on the body of another person. When parents/guardians are making decisions for their minor children, they are constrained to consider the child. Thus, in the case of parents/guardians and children, principles other than autonomy (self-governance) are required to help parents. In making the decision, the parents are not acting for themselves but for and on behalf of the child. They are aided by being encouraged to think of the child's "best interest". This principle means that the focus should be on the child, not what the parents themselves wish (please see Chap. 9). This principle is supported with appropriate language to describe the process. The use of the word "permission" to describe parental authorization, is closer to what should be happening, rather than using the phrase "parental consent". In addition, the parent should involve the child in the process, even when the child is young. The child should know what is to happen, and why, with explanations that are age appropriate and sensitive.

"Best interest" is described in detail in Chap. 4. In essence "best interest" is a way of maximizing benefits and minimizing harms for the child. Yet there can be problems with the concept of "best interest". At its the most basic, a problem arises with the word "best", a word that might imply there is only one solution. There might be more than one legitimate solution. Indeed, there might be many or, at least, some variations. Moreover, the information provided by the health care professional will, of course, be specific, detailed, health information. But when making a decision, parents/guardian might include other considerations such as the family's values, the impact of any action for one child on his/her siblings and so on. In other words, parents and health care professionals might each think they are dealing with "best interest", but can arrive at quite contrary decisions.

Another way that parents might reach acceptable health care decisions, is use of the "not unreasonable standard". (Rhodes and Holzman 2004) This standard accepts that there is no ideal choice in most health care situations, and allows parents/guardians to consider some items mentioned above, such as the interest of siblings. Using this standard, parents may choose options with which health care professionals might not fully agree. Such choices can be considered legitimate provided the interest of the child patient remains the dominant concern.

13.3   Issues to Be Discussed

13.3.1   The Circumstances and the Setting

In the scheduled medical encounter, such as where the child visits a general practitioner or a pediatrician, the parents usually take the child to a physician known to them. In a pediatric medical emergency, the situation is different. The physician in this setting might be a stranger to the family. This is almost certainly the case when care is sought in a hospital or clinic emergency department (ED). The accompanying adults might be different, too. Often children arrive with their parents/guardians. But minors might present for emergency treatment without a legal guardian. Children might be accompanied by relatives such as grandparents, a baby-sitter, school official, some other health care provider, home nurse or aide (in the case of children with medical complexity), or a bystander when there has been roadside trauma. The arrival might be by ambulance or helicopter, with only the health care providers accompanying the child.

The frequency with which children appear in ED without their legal guardians is not clear; it might be common. One study of children with "non-trivial blunt head trauma", reported that nearly one half might not have a guardian available during the initial evaluation (Holmes et al. 2009). The chance of a child at home becoming injured or sick with a caregiver who is not a parent/guardian was increasing before the COVID-19 epidemic as both parents were then likely to be in the work force. For example, Canadian employment data report that "between 1976 and 2015, the employment rate of women aged between 25 and 54 years increased from 48.7% to 77.5% (Statistics Canada 2017). A high rate of women in the workforce will likely be true in many countries when the pandemic ends. So caregivers might be among those who accompany minors to the ED. It is also possible that a minor will appear in an ED with no one at all. Adolescents might choose to attend an ED because they don't want their family to know and they wish to **avoid** a physician familiar to them, such as when seeking health care related to reproductive or drug abuse issues.

Given the importance of trust in any medical encounter, all staff in an ED must approach children and families with a friendly, non-judgmental welcome. In whatever fashion the child arrives, the child must receive support even as the triage process continues. The support might be a volunteer, or, for example, a child life specialist. The support person helps in comforting the child, and can decrease some of the distress, which, if unaddressed by a support person, will interfere with good decision-making. When a child arrives without a parent/guardian, a delegated health care provider should, in most situations, make strenuous efforts to find the parents/guardian, and to achieve reunification, which is essential (Dudley et al. 2015).

Of course, the opposite is true when a teenager seeks care in an ED to avoid the parent/guardian knowing about the need for health care advice. In such cases, the physician's job is to guide the adolescent to sound decision-making given that the patient is young and unsupported by a loved one.

This chapter assumes that most medical visits for urgent or emergent care will be in an ED, rather than a private medical office. The ED might not be in a pediatric hospital and so might serve mostly adults, not children. In this setting, administrators should ensure that all staff understand the requirements for good pediatric care. Staff must recognize the need to ensure the child is comfortable and supported, whatever the circumstances, and that whoever accompanies the child is helped in the role of supporting the child. Even where a child arrives in a pediatric ED, the team might need to be reminded to adjust their professional behavior depending on the age and developmental status of the child. A newborn might present, followed by a 15-year-old, who in turn might be followed by a 7-year-old with developmental challenges. Physicians must adjust accordingly. According to JM Baren, "The overall approach to the emergency care of a minor must take into account the physical, mental, behavioral, and emotional differences that exist between children and adults and between children of different ages, intellect, and developmental stages" (Baren 2006).

The ED must also accommodate the child or adolescent to permit privacy for confidential and emotional discussions (refer to Chap. 10). This need is rarely addressed in the literature. In Iran, Schrabi and Ali (2007) surveyed EDs in Tehran and showed deficiencies in providing for privacy. In a review of how ready EDs in Europe were for pediatrics, equipment and training was discussed but there was no mention of private space (Mintegi et al. 2020).

As discussed above, children attending an ED might be met and cared for by strangers. This chapter is being written amidst the COVID-19 pandemic, which has exacerbated such challenges for children. In this era, the health care professionals involved are not only strangers, they can be hidden behind personal protective equipment (PPE). The child, if older than 2 years, and parents will be asked to wear masks. The precautions are necessary but fear inducing. Moreover, all parties, professionals and children and parents alike, lose some ability to see and hear, and to read faces and body language. If major decisions are required, a situation in which family communication is paramount, then it is more than likely that only one parent will be in the ED. The other parent, or family supporter, might have a virtual presence via electronic communications. Even after the pandemic is over, there will be lessons in communication learned in this extreme situation that might help us to improve our communication in more normal times.

13.3.2   The Initial Assessment

In medical emergencies, there is always a process of "triage". This involves a rapid review of the severity of the illness or injury and assigning a priority (Christian 2019). This process is essential to managing the limited resources for all patients and does not require specific consent. The next stage, a more detailed assessment, will not require specific consent, at least initially. For example, that assessment might include such issues as measurement of pulse, temperature, blood pressure, oxygen saturation and a screening physical examination. The justification for still proceeding without formal consent while continuing the initial assessment is that this stage of medical work aims to identify the needs of the patient (rather than the first stage which was about managing the resources required). During this assessment, situations might be identified requiring very urgent interventions. In the case of minor patients, some assessment of their ability to consent to their own treatment can be incorporated into the assessment. To quote the *American Academy of Pediatrics* (AAP), "an MSE [Medical Screening Examination] and any medical care necessary and likely to prevent imminent and significant harm to the pediatric patient with an EMC [Emergency Medical Condition] should never be withheld or delayed because of problems with obtaining consent." (Sirbaugh et al. 2011). This AAP statement moves quickly from discussing considerations about the initial assessment, through subsequent more detailed assessment to initiating treatment. It is reasonable to consider these together, given the pace of action required in many emergencies. In addition, in a modern emergency department with a severely sick or injured child, triage is very complicated. Again, to quote the AAP this assessment includes "laboratory testing, radiographic imaging, and subspecialty consultation."

13.3.3 *Initiating Interventions When the Child has Been Brought by Someone Other Than the Parent/guardian*

Sometimes it is appropriate to treat the child even when the parent has not arrived in the ED. The reason for so doing is similar to the rationale for initiating assessment in any injured/sick child. Treating the child before the parent has arrived can be justified on the grounds of beneficence; it is helping a child survive. It might also be said that the physician has a duty to act in this way. Legal advice supports such action, with qualifications. For example, the Canadian Medical Protective Association (CMPA), (the organization that provides medico-legal help to most Canadian physicians), states "In cases of medical emergency when the patient (or substitute decision maker) is unable to consent, a physician has the duty to do what is immediately necessary without consent" (CMPA 2016). The CMPA qualifies this statement by noting that a physician may proceed without parental or guardian consent only when action is needed to respond to "demonstrable severe suffering or an imminent threat to the life or health of the patient". That organization cautions physicians that moving ahead without consent can never be because of physician preference or convenience, and that the treatments used must be limited to those that "prevent prolonged suffering or deal with imminent threats to life, limb or health".

After the initial assessment, physicians may conduct or order necessary investigations and life-saving treatment. For important components of treatment that are not life or limb saving, consent is required—hence the urgent need to trace and contact parents/guardians. Absent parents/guardians, the advice of institutional legal counsel should be sought, as a judicial order might be required in some circumstances, but not always. The UK Medical Defence Union tells the story of a 13-year-old who quite properly had surgery without parental permission or a judicial order (MDU 2017b).

> "Thirteen-year-old E is attending boarding school while his parents are in Africa working. One evening his house master, Mr. G, brings him to the Emergency Department (ED); E has a raised temperature and is complaining of severe abdominal pain. A diagnosis of acute appendicitis is quickly made, and arrangements to take E down to theatre are put in motion. The doctor asks Mr. G for E's parents' contact details but is told that they are unavailable as they are currently in a locality with no communication infrastructure.
>
> Mr. G offers to sign the consent form on their behalf, explaining that he is acting *in loco parentis*[2] while Mr. and Mrs. S are away. The doctor, however, is unsure about this and contacts the trust's solicitor. She tells him that it is likely that Mr. G could consent on behalf of E's parents, provided they assigned such rights to him, but suggests that the doctor first assess E's capacity.
>
> *Even if E had been deemed not competent to make a decision, the treatment could still have gone ahead as it was in E's best interest*
>
> When his condition is explained to him in terms he can understand, E readily grasps the situation, his need for urgent surgery and the consequences of delay. He is therefore competent to consent to treatment on his own behalf, so parental consent is not necessary. However, even if E had been deemed not competent to make a decision, the treatment could still have gone ahead as it was in E's best interest. The consequences of not operating in this case would be profound."

13.3.4 Initiating Interventions When a Parent/Guardian is Present

Given that there may be differences between health care professionals and parents regarding the best course of action for the child, there is the possibility of conflict. However, conflict can usually be avoided by the health care professional taking the time to understand fully the parents' background and previous experience of health care, and the reasons for them preferring an alternate course of action, or no action. It is a reality of medical practice today that parents are exposed to a variety of online sources of information and disinformation. Hence there must be an exploration of where parents receive information that informs their decision-making. Aside from any other action to be taken, if the parents/guardians do not have the same first language as a health care professional, translators must be used. With that proviso aside, unhurried explanation and persuasion on the part of the health care professional is extremely important. Persuasion is the correct course of action; the health care

---

[2] *In loco parentis* is a Latin term meaning, "In the place of the parent." The legal concept is derived from common law, and means that in the absence of a parent, another person (or organization) may assume some of the functions and responsibilities of a parent.

professional must not cross the border into coercion. An alternative course of action is to use what are called "nudges" (Levy 2017). Levy uses the term, "nudges to reason". One very helpful way to "nudge" is to present positive information first with a straightforward statement, then follow with much more detailed information. For example, a 2-week-old infant might present with fever. Words that convey the reality that meningitis is a possibility and that a lumbar puncture is always needed in this situation, will "nudge" parents towards giving permission. Of course, the initial words must be followed up by detailed explanation of what is involved, risk as well as benefits, and with time to answer all of the parents' questions. Nudges might sometimes regrettably regress into paternalistic behavior, which is to be avoided. At their best, nudges are a compassionate use and understanding of psychology to help parents make decisions that will help the child.

The situation in critically ill and injured children is almost always time sensitive. Health care professionals might be strongly tempted to present only positive arguments in favor of a treatment, and not risks. Such a presentation is never appropriate, even for parents who have been misinformed by family, friends, or any form of media. Strong presentations of what the physicians perceives as facts might increase resistance, the so-called "backfire effect" (Peter and Koch 2016). Time taken to understand parents' or guardians' existing beliefs and sources of information, followed by careful explanation focusing on the child, is likely to be the most effective strategy.

Generally, agreement can be reached on a course of action, perhaps with some accommodations to the standard plan that meets the parents' concerns. When there is an impasse, of course, all parties experience distress. At this stage, the health care professional must review all the information available on the child, to determine whether any delay in treatment is likely to lead to severe damage to the child or is a clear threat to life. If the situation is less dire, then discussion should continue with the family and other resources should be brought to bear, such as help from social worker, family members, religious figures and so on. If treatment must be carried out immediately, then there are legal processes available in all jurisdictions via child protection services. The contact information is available in all emergency departments. While specifics will vary from one legal jurisdiction to another, the generality is that help is available around the clock for medical emergencies.

It is the physician's duty to bring the matter of a child needing necessary medical treatment to the state's attention through its child protection officers so that the issue can be resolved legally. "Necessary medical treatment" is generally understood to mean legitimate life or limb threatening crises. The state will use its coercive power to grant a temporary guardianship order and a medical treatment order rarely; but when necessary, the state will act through its child protection officers and lawyers, and judges, rapidly and decisively. This matter is discussed in more detail below.

13.3.5 Who Gives Consent When the Physician Suspects Abuse by the Caregiver?

Children who attend for urgent/emergency care after an episode of maltreatment are not usually identified as such by the caregiver. However, child maltreatment/abuse is an important cause of child morbidity and such children do present in in the ED. In a Canadian Survey, 1997–2011 of 18 EDs, 200 child maltreatment cases were identified. "Physical abuse was the most prevalent type (57%), followed by sexual assault (31%), unspecified maltreatment (7%), injury as the result of exposure to family violence (3%) and neglect (2%)." Recognition matters. King et al. (2006) reviewed information on 44 child abuse fatalities and showed that 19% had been medically evaluated in the month before death, most often in an ED. The complaints ranged "from fussiness to vomiting to poor feeding".

In terms of consent for treatment, there is no difference in the ethical issues relevant in the early stages of what might be child abuse. If the child is unconscious or severely ill or injured, then treatment will continue apace, without specific consent being sought or even needed. Part of the initial action will include obtaining a history from the immediate caregiver, whether or not this is a parent/guardian, and examining the child. There might be clues in a history that does not correspond with the type or extent of the findings. There might be other clues on examination, such as the pattern of bruising or bites. However, care must be taken not to misinterpret a pattern of bruising. Bruising is common in all children; most have an innocent explanation. The physician should adopt an attitude of "trust but verify".

Issues of consent become very important in the next stages of management of the severely injured/sick child. It is assumed, by this stage, there is at least a suspicion of maltreatment. The physician will have to decide whether to inform the parents of the suspicion and inform child protection services at this stage, or to seek more information. Pediatric institutions will have specialists in the area who can be consulted, and who can help guide the treating physician.

Whatever is decided, it is in the child's interest that assessment and treatment continue. If the parents have capacity, and the minor him/herself is not so deemed, then the physician should proceed as described earlier in this chapter. The child should be spoken to in a private area away from the parents. There might be a need for specific tests—for example, coagulation if multiple bruises and an X-ray examination on almost all children in whom maltreatment is suspected. At this stage, the conversation should continue with the provision of full information, responding to questions from the parents, and using persuasion and nudges.

If sexual abuse is suspected, then the physician in ED should request specialist help. Almost all facilities have teams trained and skilled in investigating sex abuse and have specific skills in taking a full and appropriate history, examining the child and collecting specimens. This team will inform child protection services, who may in turn inform the police. There has been tension in the past between child protection workers and police (Cross et al. 2005). Now many jurisdictions are developing ways of working together that respect the roles and skills of each professional involved.

In all case of suspected abuses, if the parents block conversations with the child, or oppose essential investigations, an application should be made for consent for medical treatment from the court, via child protection services. This is a very big step for most pediatricians, but hesitation is dangerous for the child, as outlined earlier. If need be, physicians can seed assistance from colleagues as they perform their duty. In whatever fashion permission for assessment and treatment of the child is given, even when by a court, the parents should be informed of results of the assessment and progress of treatment and the implications for their child's medical management.

It is worth stating, though it might be obvious to readers, that parents or guardians who mistreat their child almost always love their child. Parents who mistreat their children are over-represented among "the poor, unemployed, socially and economically disadvantaged families" (Lewis and Creighton 1999). Sometimes, the offending parent or parents have mental health disorders or a temporary mental health challenge. An underappreciated advantage of involving child protectives services in jurisdictions (especially where health care is publicly-funded) is that some family hardships that are causing distress in the parent might be addressed and a sick adult or adults will likely also receive needed medical attention, strengthening the family with the goal of making it possible for the child to return home and to be safe there.

13.3.6   Initiating Interventions When the Child is a Refugee, Whether or not Accompanied by the Parent/Guardian

The situation where a child is a refugee is obviously complex. This chapter will solely address issues of consent for urgent/emergency care in refugee children. The United Nations (United Nations Convention 1951) define a refugee thus:

> A refugee is a person who, owing to a well-founded fear of being persecuted for reasons of race, religion, nationality, membership of a particular social group or political opinion, is outside the country of his nationality and is unable, or owing to such fear, is unwilling to avail himself of the protection of that country; or who, not having a nationality and being outside the country of his former habitual residence, as a result of such events, is unable to or, owing to such fear, is unwilling to return to it.

Children as refugees fall into a number of important sub-categories. These include Unaccompanied Asylum-Seeking Child (UASC) defined as "a child or young person seeking asylum without the presence of a legal guardian". Other terms used include separated children or Unaccompanied Minors (UAM). There might also be a dispute over age: is the person seeking care really a minor? Sadly, some minors seen in emergency will be victims of trafficking.[3]

Refugees have been shown to use ED for routine health care as well as emergencies in USA (DeShaw 2006; Semere et al. 2018), Australia (Mahmoud and Hou 2013),

---

[3] Guidance to address such a situation might be found here: Refugee-and-unaccompanied-asylum-seeking-children-and-young-people—guidance-for-paediatricians.pdf. In summary, the child is

Canada (Saunders et al. 2018) and Afghanistan (Higgins-Steele et al. 2017). General concerns include: lack of access to a primary care physician, hence use of ED for non-urgent care: need for translators; unclear status in terms of insurance to pay for health care and necessary drugs; difficulty/impossibility of accessing information on previous health interventions; unknown parental relatedness and immunization status; and limited ability to provide follow up. While not directly relevant to the issue of consent in an urgent/emergent situation, the words of Semere et al. should be heeded: "Designing systems to ensure timely evaluations of newly arrived refugees may reduce frequent acute care utilization."

In general, in terms of life or limb threatening situations, there will be no difference in the strictly medical components of proceeding with an assessment, including necessary investigations. Of course, support of the child patient and family, if present, is even more important than with non-refugee families; such support must be sensitive to cultural issues recognizing previous extraordinary suffering, and availability of translation services. The next steps will be even more time consuming than usual, even in the face of a crisis. Assessing the minor's capacity to consent will be difficult with cultural and language barriers; everything mentioned in previous sections about providing information, and avoiding coercion remain relevant. The only specific caution in dealing with refugee children is to be very hesitant before involving child protection services. Refugees come from many parts of the world including oppressive regimes, and might have significant fear of any government authority. Some refugees might come from parts of the world where parents control all aspects of a child's life, and will have difficulty in understanding situations where external agencies may make decisions for their child. Nevertheless, physicians must meet their statutory obligations to report children who need state intervention. Especially in cases where the child is a refugee, it is important to review the statutory obligation against the medical facts with legal advice.

13.3.7 Initiating Interventions When the Minor Comes for Urgent/Emergency Care Without an Adult

Whenever an adolescent comes for urgent/emergency care without an adult, questions are immediately raised about who will give consent. The adolescent? Should the staff immediately contact and find a parent? Some other legal authority? Previous sections have dealt with those severely sick or injured and brought by, for example, ambulance. This section addresses adolescents who choose to attend for medical care by themselves. This might be a common occurrence, or it might be rare. It might happen simply because illness manifesting as pain or an injury occurs when the teenager is with friends. As mentioned above, it might occur because the teenager

---

entitled to confidential health care; interpreters should be used as needed; consent for examination and procedures follows the same principles outlined earlier in the chapter: and attention should be a mental and emotional health.

deliberately seeks health care in an urgent care/emergency department with the goal of having privacy and confidential care in relative anonymity (Dreisinger and Zapolsky 2018).

Adolescents use pediatric and general EDs for a variety of reasons. Given the wide variations in how health care services are organized within and between countries, it is difficult to know its frequency in any comprehensive fashion. In one US ED, a study concluded that of the adolescents seen "Eighteen percent had emergency, 60% urgent, and 21% non-urgent conditions" (Melzer-Lange and Lye 1996). These numbers suggest some of the teens use the ED for routine medical care. The study also showed that the majority came in the evening or overnight. Injuries were the most common reason for male visits (47%), and were also frequent in females (42%). Gynecologic problems were the commonest reason for adolescent females to visit the ED. Exacerbations of chronic illnesses accounted for more than a quarter (27%).

We have suggested that teens might choose to access care in an ED because of privacy and confidentiality concerns. When the teen is seen in the ED, the focus must be on the teen's best interest. Ethical norms will require an assessment of capacity, and the physician will proceed with a detailed history, a physical examination, appropriate investigations followed by treatment (Michaud et al. 2010). However, the law, often with a desire to protect teens, might not support proceeding without parental consent. A report from the United Nations Agency leading the global effort against AIDS, reviewed international laws regarding adolescent consent. (UNAIDS) This agency noted that "68 of 108 reporting countries required parental consent to access sexual and reproductive health services." The report concluded that laws often discouraged teens from accessing medical services. Practitioners must be aware of the laws in their jurisdiction about teenage consent. Even when parental permission is the norm, there may be exceptions for reproductive and substance use issues. For example, in the US, minor adolescents may provide consent only for a range of treatment decisions such as reproductive issues and mental health concerns, while parental permission is needed for all other treatment.

In other western countries, the law is often less restrictive than that in the US. In Canada adolescents over 14 years may generally provide consent for treatment if judged by the clinician to have the necessary capacity to do so. In the UK, when a teen is determined to have capacity to give consent for medical procedures, the teen is said to have "Gillick competence". A ruling of the UK House of Lords in 1985 (Academy of Medical Royal Colleges 2020):

> "permits under-16s to consent to medical treatment if the child shows sufficient understanding and intelligence to enable them to understand fully what is proposed. This includes the purpose, nature of treatment, likely effects and risks, chances of success and the availability of other options."

Dreisinger and Zapolsky (2018) describe mature minors as those who may give consent themselves because they display "sufficient maturity and intelligence to understand and appreciate the benefits, risks, and alternatives of a proposed treatment to make a voluntary and reasonable choice."

There is a specific subgroup of adolescents who will almost always be declared "mature minors". These are teens attending for issues of mental health including drug, alcohol and vaping addiction, and reproductive issues, whether pregnancy related, need for contraception or treatment of sexually transmitted diseases, confidentiality is essential. In such situations, the teenager might not give an accurate history without an assurance of privacy and confidentiality. Once this assurance is given, then the ability to give consent surely follows (Dreisinger and Zapolsky 2018). If the issue relates to sexual assault, then the matter is both medical and criminal. This situation must be handled, as always, with respect for the adolescent; the first priority is the safety of the teenage victim. Is there a risk of continuing harm to the teen? Or to others in her or his circle? As mentioned above, this situation needs specialist medical and social service help. In terms of involving the police, typically the specialists in the area are aware of whom and when to contact.

Respect is due to adolescents in the assessment of whether or not they should give their own consent. Even when it is concluded that a guardian should give consent, respect is still required. In that situation, there is still a need for privacy and confidentiality. Health care professionals will recognize that any decision on consent in an adolescent is taken in their specific context. Adolescence is a time when the major health risks are not biomedical, they are behavioral and social (Ziv et al. 1998). An apparent visit for a "minor" health concern might be a cry for help; the leading causes of mortality in teenagers are injury, suicide and homicide. A mental health assessment might be indicated that can be conducted privately and confidentially.

13.4 Research on Children in an Emergency

As in clinical care, so it is in research; obtaining parental consent is essential. For many years there were prohibitions on children being included in research studies (Fisher et al. 2011). It was the 1990s before legislation in US and Europe addressed and removed legal barriers to research in children. But pediatric research does matter because children deserve the benefits of evidence-based investigations and treatments. Children suffer from different conditions than adults, and even when they have the same conditions, frequently drug metabolism, idiosyncrasies and responses are very different. Hence children must be studied. Parents will usually consent to the research, but children should always be involved, in an age and developmental stage appropriate manner. Such assent by the child is often formalized with a brief description of the study that the child signs, if developmentally capable. As an aside, this section will not address issues of minors giving their own consent because such an event is unusual in research. When such research does occur, researchers will find the principles described in earlier sections helpful.

Before there is an invitation to participate in research, researchers require the "consent" of an oversight body. These bodies have various titles, from Health Ethics Committees (HECs), through Institutional Review Boards (IRBs) to Research Ethics Board (REBs). While there are detailed processes and regulations to guide these bodies, their goal "parallels that of the researcher; both are charged with ensuring that human subjects research is conducted ethically, with sound scientific rationale, to maximize benefits and minimize risks." (Rice 2008).

The research oversight board will consider all aspects of research, paying particularly close attention to issues of recruitment and consent. In addition, in the context of research, members of the board will see children as "vulnerable", requiring even more careful consideration. In terms of research in an emergency, the situation itself might be considered as, at least to some extent, coercive. This is not to say that the researchers are themselves coercive in their approach to parents, but as in all aspects of medical activities in an emergency, time is of the essence. That having been said, it is important to reiterate that the challenges of conducting research on pediatric emergency medical treatment are not reasons not to do research, but rather reasons to be scrupulous regarding the details of recruitment and consent.

The usual expectation is that a clinician involved in the child's care will request permission from the child before the researchers speak to the parents about consent. The clinician who has this initial conversation will want to ensure that the parent who will be giving consent is, in fact, competent to do so. If the parents agree to be approached, then the researchers will explain the study in detail, including the risks and possible benefits, and will answer all questions. Within the constraints of an urgent or emergency situation, time will be given to the parents to reflect upon the issues presented by the researcher, and to seek advice from another parent/guardian, relatives, or knowledgeable friends. The oversight body will have ensured that the written document, required to be signed as evidence of consent,—the consent form—contains all the information required, and is written in an easily understandable fashion. There might also be an assent form for the child to sign. Once the consent form has been signed, then the research may begin. The research in this situation might take many forms, including clinical trials, observational and qualitative studies.

It is obvious that the process described above, while appropriate and in keeping with ethics norms, poses great difficulties in all emergency situations. For this reason, research ethics boards now accept methodology in adult and pediatric research whereby research is permitted without individual patient consent, or research may begin provided consent is sought later (Jansen et al. 2009). In such cases, the "consent" is provided by a research ethics oversight body. The ethical justification for such an approach is that it allows research to be conducted that usually has a strong possibility of benefit to the subject and that is essential in advancing understanding and treatment of pediatric conditions. Research oversight bodies might additionally require community consultation before they provide approval (Fordyce et al. 2017). Such consultation might be with specific communities affected, such as those who live in the same locality in which the research will happen. Consultation might also occur with patient populations affected, such as families of children with seizure disorders.

One example of deferred consent is a study of catheter-related infection. Two different catheters were both in use, one impregnated with an antibiotic and one not (Mok and Gilbert 2011). In a study of acceptance of methodology, 275 parents completed the questionnaire, and 20 families participated in an interview. Some parents were shocked or angered that the child was entered in a study without their consent. Yet once the reason for deferred consent was explained, and the parents understood that both interventions were already in use, most of the parents were satisfied (Woolfall et al. 2015). Explanation of the reason for deferred consent is clearly important, though the optimal timing for this discussion is not yet clear. Another example of research being conducted without consent is in the management of seizures, with comparisons of two approved drugs (Chamberlain et al. 2014). A third example is examination of different techniques of cardiac resuscitation (Carroll et al. 2012).

Overall, research is extremely important in all aspects of pediatrics, and particularly in emergency situations. Consent will usually be sought from the parent or guardian. Rarely, and only in exceptional, unusual circumstances, the research oversight body will permit initiation of the research without consent, always with a proviso that parents must be informed.

13.5 Parent/guardians Want to Leave ED Against Medical Advice (AMA)

One might think that in urgent and emergency medical situations, all parties would agree on the need for the child to be assessed and treated. This is not always the case. Sometimes, parents might be unwilling to wait for proper assessment of their child in the ED; they might leave before the child has been seen by a physician.

How the events unfold will depend on how the facility is organized, and how busy and how crowded it is at any one moment in time. Typically, a clerk will take basic information about the child, including who are the family physician and regular pediatrician. If that particular family and child leave without telling anyone, then it might be some time before the fact is recognized. It is difficult to see how anything can be done in this situation. In another situation, basic information might have been taken by the clerk, and then a nurse might have collected more information to help with triage, assessed that the situation was less urgent, and decided that the child could wait. If the family leave before being seen, then that fact is likely to become known to the staff when the name is called for more extensive physician assessment. If it becomes clear that the family has indeed left, then the triage information should be reviewed to determine whether, on a second look, the child's care needs are more urgent than initially assessed. The family physician or pediatrician named by the family, should be telephoned to inform him or her of the situation.

A more difficult situation arises when parents wish to leave after the child has been seen and diagnostic tests and/or medical interventions have been proposed to the parents, but no action has yet been taken. In effect, the parents are refusing the recommended treatment in that ED. Parents might make it clear that they are going to another medical institution or going home. If the former, a brief assessment that transfer is safe is needed, and after the child has left, a telephone call to the other institution to confirm the child has turned up and has been seen. If the parents intend to take the child home, then health care providers must determine whether the child's life would be threatened by such action or whether the need for immediate further action is not critical in terms of threat to the child's life or limb. In both situations, the child must be protected.

In the case where tests and treatment can wait, most institutions have specific procedures to deal with this eventuality. Institutions usually have a form to be signed in which the parent or the mature minor acknowledges personal responsibility for the refusal to stay in the facility for further investigation or treatment. Before reaching the stage of signing a specific form, the treating physician must spend time with the parent or minor, and always with respect, in reviewing all the options, the reasons for them, and the risks and benefits of moving ahead with medical recommendations as soon as possible. The possible consequences of not proceeding with assessment and treatment must be described, but in non-coercive terms. An attempt should be made to identify why there is refusal and an attempt should be made to negotiate, perhaps by altering the treatment plan in one way or another.

On the other hand, if immediate action (investigation or treatment) is essential, then further discussion and explanation, focusing on parental anxieties is now a priority. If the parents are adamant, then the next steps are clear but unpleasant. In Sect. 1.3.5, a specific course of action is described, culminating, when all else fails, with the involvement of child protection organizations.

These situations do occur, but fortunately are uncommon. They must be addressed very thoughtfully. The health care practitioner might receive some personal protection from complaints or lawsuits by obtaining other medical opinions specifically on the risks of the child not being treated at that moment and in that specific facility.

13.6   After Emergency Treatment

The need for rapid and correct decision-making to help a child survive a medical crisis is the rationale for the very existence of emergency departments and pediatric intensive care units. Other hospital units are also involved in the course of the child's care that must be delivered in a number of different settings, with timely transfer between settings, and always with great skill and attention to detail. We hope that the critically ill child will be received in a "health care system [that] would identify sick children early in their illness, and provide treatment that is safe, effective, patient centered, efficient, timely and equitable" (Hodkinson et al. 2016) (also refer to Chap. 16).

Even in the COVID-19 pandemic, severely ill and injured children must be evaluated promptly, in the same manner and time frame as if there had been no pandemic. Unfortunately, some children have appeared late to the ED during the pandemic because of parental concern that their child might be exposed to SARS-CoV-2 in the hospital. In one specific childhood emergency, acute appendicitis, Snapiri and colleagues described seven children with late diagnosis of appendicitis because the parents feared contracting the virus by attending hospital during the pandemic (Snapiri et al. 2020). While this chapter focused on issues of guardians authorizing a health care practitioner to treat a minor, the first recognition that assessment and treatment will be required is the responsibility of whoever is the immediate caregiver of the child. That person must appreciate that their duty is not to diagnose an ailment but to recognize in a timely manner that the child needs to be assessed by a physician and to deliver the child to a physician. Parents and other caregivers usually do take a sick child to a physician or an emergency department. Media engagement by physicians occurred during the pandemic to encourage people to attend an ED if they are unwell (for example, Snowdon 2020).

Sadly, some children die in the emergency department, or shortly after transfer to a pediatric intensive unit. Their families require considerable support, and great sensitivity in all of our communication. In situations of sudden death, statutes will require mandatory reporting to legal authority such as medical examiners or coroners. Parental consent is not a consideration because these authorities are empowered by those statutes to make their own decisions about whether or not a full autopsy will be performed. In other situations, there might be advantages in recommending an autopsy to the families; they should be approached carefully for their consent to this procedure. The approach should be as described above, carefully explaining the advantages of an autopsy, what it entails, some of the disadvantages, and phone information will be available. Many clinicians might see this task as particularly unpleasant, but this sentiment should not deter them from making the request. The pathologist who performs the autopsy is the child's last doctor.

The majority of children seen for urgent/emergency care survive but many will require aftercare. For example, it is appropriate to provide information on injury or poison prevention in appropriate circumstances. If a teenager is seen for a sexually transmitted infection, then issues related to whether the teen consented to sexual activity might be explored and resources or further counselling on sexuality should be provided as described earlier in the chapter. Consent need not formally be sought by the physician to make follow-up arrangements; the children and families should be informed about their importance.

Guiding Principles for the Ethical Emergency Health Care of Children

An initial assessment is always essential

Every other intervention requires consent

Consent is usually given by parents, and should be designated "authorization"

The minor child may give consent in some situations

The minor must always be involved in a developmentally appropriate manner, even if not giving consent

Privacy and confidentiality are important for the child, and must be provided even when the child is not the one giving consent

Research is important to ensure treatment is based on the best evidence, Appropriate safeguards must be in place

13.7   Conclusion

Consent is a prerequisite of all medical interventions, irrespective of their nature. Written permission is an institutional requirement in many situations, such as surgical interventions. In other situations, even in a non-acute emergency, time is required to explain all the details about the intervention, and the risks and benefits; to answer the parents' questions, and to ensure that whoever gives consent is competent to do so. In an acute emergency—where life or limb are imminently threatened—critical time cannot be used to seek consent; the child must be treated. In pediatric practice, consent will usually be given by the parent/guardian provided they are acting in the child's best interest, but there are a number of situations when consent is appropriately requested from the minor himself/herself. Even when the parent gives consent, an attempt should be made to explain what is happening to the child and to obtain his/her agreement, so-called assent, as things move along. The fundamental value that underlies the duty to seek consent for medical treatment is that we are each equal in our human dignity; the ethical principle that derives therefrom is respect for the personhood of the patient—in this case, the child.

References

Academy of Medical Royal Colleges. 2020. https://www.rcpch.ac.uk/sites/default/files/2020-06/academy_of_medical_royal_colleges_statement_on_gillick_competency_final_003.pdf. Accessed 11 Feb 2021

Baren, J.M. 2006. Ethical challenges in the care of minors in the emergency department. *Emergency Medicine Clinics* 24 (3): 619–631.

Carroll, T.G., V.V. Dimas, and T.T. Raymond. 2012. Vasopressin rescue for in-pediatric intensive care unit cardiopulmonary arrest refractory to initial epinephrine dosing: A prospective feasibility pilot trial. *Pediatric Critical Care Medicine* 13 (3): 265–272.

Chamberlain, J. M., P. Okada, M. Holsti, P. Mahajan, K. M. Brown, C. Vance, and J. Grubenhoff. (2014). Lorazepam vs diazepam for pediatric status epilepticus: a randomized clinical trial. *Jama* 311 (16): 1652–1660.

Christian, M.D. 2019. Triage. *Critical Care Clinics* 35 (4): 575.

CMPA. (2016). *Canadian Medical Practice Association, Consent: A guide for Canadian physicians.* Fourth edition. https://www.cmpa-acpm.ca/en/advice-publications/handbooks/consent-a-guide-for-canadian-physicians#Emergency%20treatment Accessed 9 Feb 2021

Cross, T.P., D. Finkelhor, and R. Ormrod. 2005. Police involvement in child protective services investigations: Literature review and secondary data analysis. *Child Maltreatment* 10 (3): 224–244.

DeShaw, P.J. 2006. Use of the emergency department by Somali immigrants and refugees. *Minnesota Medicine* 89 (8): 42.

Dreisinger, N., and N. Zapolsky. 2018. Complexities of consent: Ethics in the pediatric emergency department. *Pediatric Emergency Care* 34 (4): 288–290.

Dudley, N., A. Ackerman, K. M. Brown, S. K. Snow, and American Academy of Pediatrics Committee on Pediatric Emergency Medicine, & Emergency Nurses Association Pediatric Committee. 2015. Patient-and family-centered care of children in the emergency department. *Pediatrics* 135 (1): e255–e272.

Fisher, H.R., C. McKevitt, and A. Boaz. 2011. Why do parents enrol their children in research: A narrative synthesis. *Journal of Medical Ethics* 37 (9): 544–551.

Fordyce, C.B., M.T. Roe, and N.W. Dickert. 2017. Maximizing value and minimizing barriers: Patient-centered community consultation for research in emergency settings. *Clinical Trials* 14 (1): 88–93.

Higgins-Steele, A., D. Lai, P. Chikvaidze, K. Yousufi, Z. Anwari, R. Peeperkorn, and K. Edmond. 2017. Humanitarian and primary health care needs of refugee women and children in Afghanistan. *BMC Medicine* 15 (1): 196.

Hodkinson, P., A. Argent, L. Wallis, S. Reid, R. Perera, S. Harrison, and A. Ward. (2016). Pathways to care for critically ill or injured children: a cohort study from first presentation to health care services through to admission to intensive care or death. *PLoS one* 11(1): e0145473.

Holmes, J. F., R. Holubkov, N. Kuppermann, and Pediatric Emergency Care Applied Research Network (PECARN). 2009. Guardian availability in children evaluated in the emergency department for blunt head trauma. *Academic emergency medicine* 16 (1): 15–20

Jansen, T.C., E.J. Kompanje, and J. Bakker. 2009. Deferred proxy consent in emergency critical care research: Ethically valid and practically feasible. *Critical Care Medicine* 37 (1): S65–S68.

King, W.K., E.L. Kiesel, and H.K. Simon. 2006. Child abuse fatalities: Are we missing opportunities for intervention? *Pediatric Emergency Care* 22 (4): 211–214.

Leikin, S.L. 1983. Minors' assent or dissent to medical treatment. *The Journal of Pediatrics* 102 (2): 169–176.

Levy, N. 2017. Nudges in a post-truth world. *Journal of Medical Ethics* 43 (8): 495–500.

Lewis, V., and S.J. Creighton. 1999. Parental mental health as a child protection issue: Data from the NSPCC national child protection helpline. *Child Abuse Review: Journal of the British Association for the Study and Prevention of Child Abuse and Neglect* 8 (3): 152–163.

Mahmoud, I., and X.Y. Hou. 2013. Utilisation of hospital emergency departments among immigrants from refugee source-countries in Queensland. *Clinical Medicine and Diagnostics* 3 (4): 88–91.

MDU. 2017a. Medical protection, new judgment on patient consent, June 30, 2017. https://www.medicalprotection.org/uk/articles/new-judgment-on-patient-consent. Accessed 9 Feb 2021.

MDU. 2017b. An essential guide to consent—Cases, May 9, 2017. https://www.medicalprotection.org/uk/articles/an-essential-guide-to-consent-cases. Accessed 11 Feb 2021

Melzer-Lange, M., and P.S. Lye. 1996. Adolescent health care in a pediatric emergency department. *Annals of Emergency Medicine* 27 (5): 633–637.

Michaud, P.A., K. Berg-Kelly, A. Macfarlane, and L. Benaroyo. 2010 . Ethics and adolescent care: An international perspective. *Current Opinion in Pediatrics*. 22 (4): 418–422.

Mintegi, S., I. K. Maconochie, Y. Waisman, L. Titomanlio, J. Benito, S. Laribi, and H. Moll. 2020. Pediatric preparedness of european emergency departments: A multicenter international survey. *Pediatric Emergency Care*.

Mok, Q., and R. Gilbert. 2011. Interventions to reduce central venous catheter-associated infections in children: Which ones are beneficial? *Intensive Care Medicine* 37: 566–568.

Pellegrino, E.D., and D.C. Thomasma. 1987. The conflict between autonomy and beneficence in medical ethics: Proposal for a resolution. *The Journal of Contemporary Health Law and Policy* 3: 23.

Peter, C., and T. Koch. 2016. When debunking scientific myths fails (and when it does not) The backfire effect in the context of journalistic coverage and immediate judgments as prevention strategy. *Science Communication* 38 (1): 3–25.

Rhodes, R., and I. Holzman. 2004. The *not unreasonable* standard for assessment of surrogates and surrogate decisions. *Theoretical Medicine and Bioethics* 25: 367–386.

Rice, T.W. 2008. The historical, ethical, and legal background of human-subjects research. *Respiratory Care* 53 (10): 1325–1329.

Saunders, N.R., P.J. Gill, L. Holder, S. Vigod, P. Kurdyak, S. Gandhi, and A. Guttmann. 2018. Use of the emergency department as a first point of contact for mental health care by immigrant youth in Canada: A population-based study. *CMAJ* 190 (40): E1183–E1191.

Semere, W., P. Agrawal, K. Yun, I. Di Bartolo, A. Annamalai, and J.S. Ross. 2018. Factors associated with refugee acute health care utilization in southern Connecticut. *Journal of Immigrant and Minority Health* 20 (2): 327–333.

Sirbaugh, P. E., D. S. Diekema, K. N. Shaw, A. D. Ackerman, T. H. Chun, G. P. Conners, and S. M. Selbst. 2011. Policy statement-Consent for emergency medical services for children and adolescents. *Pediatrics* 128 (2): 427–433.

Snapiri, O., C. Rosenberg Danziger, I. Krause, D. Kravarusic, A. Yulevich, U. Balla, and E. Bilavsky. 2020. Delayed diagnosis of pediatric appendicitis during the COVID-19 pandemic. *Acta Paediatrica.*

Snowdon, W. 2020. Avoiding emergency care during pandemic a dangerous trend. Edmonton ER doctor warns: Patients are coming in quite a bit sicker than what would normally be seen'. https://www.cbc.ca/news/canada/edmonton/covid-coronavirus-emergency-rooms-1.5571469. Accessed 11 Feb 2021

Sohrabi, M. R., and M. H. Ali. 2010. Privacy, confidentiality and facility criteria in designing emergency departments of the teaching hospitals of Shahid Beheshti University of Medical Sciences in 2007.

Statistics Canada. 2017. https://www150.statcan.gc.ca/n1/pub/89-652-x/89-652-x2017001-eng.htm. Accessed 11 Feb 2021

UNAIDS. 2021. Parental consent is required in the majority of countries worldwide https://www.unaids.org/en/resources/presscentre/featurestories/2019/april/20190415_gow_parental-consent. Accessed 13 Feb 2021

United Nations Convention. 1951. https://www.unhcr.org/1951-refugee-convention.html. Accessed 11 Feb 2021

Woolfall, K., L. Frith, C. Gamble, R. Gilbert, Q. Mok, and B. Young. 2015. How parents and practitioners experience research without prior consent (deferred consent) for emergency research involving children with life threatening conditions: a mixed method study. *BMJ open* 5 (9).

Ziv, A., J.R. Boulet, and G.B. Slap. 1998. Emergency department utilization by adolescents in the United States. *Pediatrics* 101 (6): 987–994.

Further Reading

Beauchamp, C. (2013). Beauchamp TL, Childress JF Principles of biomedical ethics. Chapter 4 Respect for autonomy and Chapter 6 Beneficence.

Mitchell, I., & Guichon, J. R. (2019). *Ethics in Pediatrics.* Springer, Cham. Chapter 4. How Do I Know from Whom I Need Permission and When I Have It?

Chapter 14
Ethical Issues and Considerations for Children with Critical Care Needs

B. M. Morrow and W. Morrison

Abstract Pediatric critical care refers to the health care of children with life-threatening illness or following major surgery or severe injury. This care is offered in different contexts across the globe. In well-resourced environments, critical care may be provided in pediatric intensive care units (PICU), which provide highly complex medical care with advanced, potentially expensive technological devices aimed primarily at sustaining life; whereas in poorly resourced regions, only primary care may be available for critically ill or injured children. Even where PICU facilities are available, they are a scarce and expensive resource. The knowledge and ability to sustain a critically ill or injured child's life, potentially with effects on the child and family's quality of life and at the expense of other children's health care, leads to frequent ethical challenges. Ethical issues related to how best to ensure optimal quality of life and the fair distribution of scarce resources are common. In-depth knowledge and understanding of the individual patient's medical, scientific and psychosocial context, through interdisciplinary discussion and collaboration; as well as familiarity with the overriding, relevant ethico-legal policies and principles is essential for appropriate decision-making. Truthful communication with the family is essential, using a shared decision-making process, and the child's best interest must be central to every decision. This chapter presents an overview of the recommended approach to, and important considerations for, ethical decision-making in the PICU, using a hypothetical case example.

"What every physician wants for every one of his patients old or young, is not just the absence of death but life with a vibrant quality"—(Elkinton 1966).

B. M. Morrow (✉)
Department of Paediatrics and Child Health, University of Cape Town, Cape Town, South Africa
e-mail: Brenda.Morrow@uct.ac.za

W. Morrison
Department of Anesthesiology and Critical Care, The Children's Hospital of Philadelphia, Philadelphia, PA, USA
e-mail: MORRISONW@email.chop.edu

Perelman School of Medicine at the University of Pennsylvania, Philadelphia, PA, USA

© Springer Nature Switzerland AG 2022
N. Nortjé and J. C. Bester (eds.), *Pediatric Ethics: Theory and Practice*,
The International Library of Bioethics 89,
https://doi.org/10.1007/978-3-030-86182-7_14

Keywords Pediatric critical care · Intensive care unit · Ethics · Child rights · Life-sustaining therapy · Spinal cord injury

14.1 Introduction

Pediatric critical or intensive care can be defined as the care of children who undergo major surgery or who suffer a life-threatening injury or illness, from the moment they first present (Kissoon et al. 2009). The provision of critical care services varies according to available resources, the level of training of health care professionals and the severity of illness or injury. It may include delivery of highly technologically advanced and complex care in a well-equipped pediatric intensive care unit (PICU); but where resources are limited, appropriate primary health care of potentially life-threatening conditions also constitutes critical care. There is global inequity, with complex PICU services being scarce, expensive and standing to benefit relatively few children, whilst primary critical care is relatively inexpensive and can benefit many. Where PICU facilities are available, there needs to be an ethically sound approach to appropriate selection of patients ('who gets the bed?'), as well as rational and morally sound decision-making throughout the child's critical illness (also refer to Chap. 29). In many cases we have the knowledge and technology to treat a condition and to sustain life, but this is not always achievable or in the patient's best interest. Ethical decision-making can be accomplished through interdisciplinary discussion and collaboration, with a clear understanding of local laws, policies and ethical principles.

This chapter, using a hypothetical case study example, will provide an overview of some of the ethical issues and principles relating to clinical pediatric critical care provided in a PICU. For the purposes of this chapter, neonatal-specific considerations as well as ethical considerations for pediatric research will be excluded. Many clinical situations that raise ethical questions in the PICU, such as end-of-life care are discussed in Chap. 15 in this book.

Hypothetical clinical case

> A 10-year old boy, J.E. is referred to the pediatric intensive care unit (PICU) following an acute spinal cord injury as a result of a pedestrian motor vehicle accident. J.E. was resuscitated on the scene by a bystander, and intubated by emergency services personnel, who initiated invasive mechanical ventilation. He was transferred to a regional hospital for stabilization prior to possible transfer to a tertiary facility. The pediatrician at this center reported observing some lower limb movements. On arrival in the PICU, J.E. is awake and appears to be orientated—he can blink his eyes in response to yes/no questions. On investigation, an MRI showed the spinal injury to be a complete resection at C2/C3 level.
>
> Social circumstances: J.E. and his family live in low socio-economic circumstances, in a two-roomed informal dwelling without electricity, and with shared outside water and toilet facilities. He shares the house with his two parents and two siblings, aged 3 and 12. J.E. is currently in Grade 5 at school, performing adequately academically. He enjoys playing soccer with friends. His father is the breadwinner, working as a bricklayer; his mother is unemployed. Both parents are at the bedside.

14.2 Ethical Decision-Making

When making ethical decisions regarding clinical care in the PICU, all available options need to be considered, through extensive discussions amongst all stakeholders, and the ethical or moral dilemma/s defined. Sufficient information (scientific/medical, social, legal, ethical, financial, and rights-based) must be established to holistically and comprehensively evaluate every available option. A patient- and family centered care approach should be used, with appropriate inclusion and consideration of the family in shared multidisciplinary team discussions.

14.2.1 Defining the Dilemma

A moral dilemma is described by Beauchamp and Childress (2001) as "*circumstances in which moral obligations demand or appear to demand that a person adopt each of two or more alternative actions*", without the possibility of performing all the alternatives (Beauchamp and Childress 2001).

In the above case, after J.E.'s admission in the PICU, the ethical or moral dilemma relates to determining what the best, ethically acceptable management plan for J.E. is. For this patient, the following main therapeutic options have been identified:

(1) To perform a tracheostomy and continue long-term invasive ventilation
(2) To compassionately withdraw life-sustaining therapy and allow natural death (terminal extubation).

14.2.2 Consider the Available Information and Possible Outcomes for All Options

If the first option is considered, J.E. currently could not be cared for at home, as electricity is not available to support mechanical ventilation; he may therefore need long-term institutional care, which may not be available or appropriate in a low resourced environment. Potential provision of electricity could be pursued through local governmental processes, or alternative housing could be explored. If electricity were available, J.E.'s caregiver could be taught tracheostomy home- care and ventilator management, as well as the other necessary day-to day maintenance care required (e.g. intermittent catheterization, passive movements and stretches, pressure care). J.E.'s life would, however, likely be profoundly limited in both quality and quantity, as he would be severely limited functionally (fully dependent on others for all activities of daily living); as well as being at high risk of developing complications such as pneumonia, bed sores, urinary tract infections, and joint contractures, ultimately with an expected premature death. Although these secondary complications are considered preventable, in reality they are still highly prevalent in low resourced regions, with high associated mortality (Bickenbach et al. 2013).

If long-term care is implemented, it needs to be directed at maintaining health (preventing secondary complications) and maximizing function and participation (including return to education/schooling). Health care systems will vary substantially regarding how much clinical and financial support is available to J.E.'s family for ongoing care. Without sufficient access to appropriate rehabilitation and provision of assistive devices, there is little hope that a patient with a complete high spinal cord injury could participate meaningfully in society (Bickenbach et al. 2013). However, even with limited rehabilitation services within the poor socio-economic circumstances of J.E. and his family, providing long-term respiratory support might allow J.E. to enjoy quality-time with his family. If this option were chosen, ethical discussion should occur, ideally together with a palliative care team, regarding limitation of further life-sustaining therapy (e.g. no escalation of ventilation; "do not resuscitate"/DNR orders).

If the second therapeutic option were followed, J.E. would die quickly, as his high cervical spinal injury is too severe to allow spontaneous, unsupported breathing. Terminal extubation can be accomplished in a sensitive manner, with enough time for appropriate preparatory counselling and medication to limit unpleasant end-of-life symptoms and anxiety for J.E., as well as allowing appropriate inclusion of the family and pastoral or other support required. This approach may limit the social and financial cost on the family and health care service as well as limiting long-term suffering for J.E. In some situations, however, this choice might be considered an expedient option, and may lead to substantial "survivor guilt" for the family if not managed carefully and holistically, with appropriate follow-up bereavement counselling (Suttle et al. 2017).

14.2.3 Consider the Overriding Policies, Laws and Principles

14.2.3.1 Rights of the Child

Children's rights are human rights as they apply to children. Children need extra protections because they are dependent on adults for their care, making them vulnerable to abuse and exploitation. The following articles listed in the United Nations Convention on the Rights of the Child (refer to Chap. 5) are particularly relevant to pediatric critical care, including the long-term provision of critical care for technology-dependent children (United Nations 1989):

- **Article 2 (non-discrimination)**: *A child may not be discriminated against on the basis of race, sex, religion, language, ethnic background, social circumstances, disability* etc. This would include discrimination based on poverty and lack of amenities. In order to make consistent and fair decisions, multidisciplinary team discussions should include an honest discussion of what decision would be made if a child from a middle to upper income family were to present with a similar injury, but the family were able to offer high tech assistive devices (e.g. eye-gaze controlled electric wheelchair), round the clock at-home nursing care and advanced augmentative- communication technology, for example. The right to non-discrimination also needs to be balanced against other Child Rights, such as the right to act in an individual child's best interest (which may differ according to socio-economic circumstances, for example); the right not to be involuntarily separated from parents (in one scenario the family could care for the child at home whilst this may not be possible in the current case); and the right to the "highest attainable health", which again differ according to resource availability.

- **Article 3 (best interest' principle)**: *In deciding on all actions concerning children, the best interest of the child must be the primary consideration.* In this regard, all relevant information about the patient's condition and all available treatment options, including the resources available to initiate and maintain this care, must be considered to decide on what course to follow. This article does not, however, require unlimited invasive therapy to ensure prolongation of life, where this is not in the child and family's best interest and/or conflicts with societal or family values.
- **Article 6 (right to life)**: *Every child has the inherent right to life and* health care *professionals should ensure, "to the maximum extent possible", the survival and optimal development (physical, emotional, intellectual, social, cultural and spiritual) of the child.* In the case of J.E., it may not be possible to provide the ongoing care necessary to achieve high quality life; however, this would depend on available structures and resources.
- **Article 9 (right to parental care)**: *A child shall not be separated from his or her parents against their will.* In this case, it would potentially contravene a basic child right to remove J.E. from his parents' care purely for the purposes of therapeutically prolonging his life, if long-term ventilation were not feasible in the home.
- **Article 12 (right to be heard)**: *The views of the child should be considered and given due weight according to their age and stage of development.* This includes the right to participate in decision-making. J.E. is ten years old, and therefore lacks legal capacity to make final decisions regarding his health care. He is, however, cognitively able to understand the situation and choices available and according to this right should be allowed to express his views and wishes, and for those to be considered in decision-making. The presence of an endotracheal tube and invasive ventilation are a surmountable barrier, as augmentative and alternative communication (AAC) methods are available to facilitate non-verbal communication (Bickenbach et al. 2013).
- **Article 13 (right to information)**: *Children have the right to seek and receive accurate information* about their illness and symptoms. According to this right, J.E. should be provided with full and honest disclosure about his condition, and the options available for treatment, at a level appropriate to his age and cognitive ability.

- **Article 23 (right to a full life)**: "…[the] *physically disabled child should enjoy a full and decent life, in conditions which ensure dignity, promote self-reliance and facilitate the child's active participation in the community*". This item further recognizes the right of disabled children to special care (including education, health care and rehabilitation services, and recreation opportunities) in order to achieve the "*fullest possible social integration and individual development*". It has been reported that many individuals with severe physical disability following injury report higher long-term quality of life years than would have been anticipated at the time of injury, with the physical domain of quality of life improving with time after injury (Lude, et al. 2014). However, this article notes that this right is subject to available resources and might not be universally achievable.
- **Article 24 (right to health)**: *Children have the right to enjoy the "highest attainable" standard of health, including access to* health care *resources*. The caveat (in bold print) here implies that this standard is also not universal, and must take other considerations, including resource availability, into account.
- **Article 28 (right to education)**: *Education is part of a child's optimal development* and should be integrated in the provision of critical care, as a basic right. Different contexts would have variable ability to offer optimal educational opportunities to children with severe disabilities, and this would need to be considered as part of the decision-making process. In the case provided, it may be extremely difficult to provide optimal educational opportunities to J.E. if he were offered long-term ventilation.
- **Article 31 (right to play)**: *Children have the right to leisure, to engage in play and recreational activities.* Given the nature and severity of the spinal cord injury, J.E. would not be able to participate in traditional play activities, such as those he enjoyed previously. The potential for alternative activities and access should be explored as part of the decision-making process.
- **Article 37 (freedom from torture)**: *No child may be subjected to torture or other cruel, inhuman or degrading treatment or punishment.* In this regard, many acute and chronic critical care interventions could be considered to be cruel, especially if the ultimate outcome (in terms of quality and duration of life) is likely to be poor.

14.2.3.2   Ethical Principles

Beneficence and Non-Maleficence

The ethical principle of beneficence requires health care professionals to provide care that benefits the patient; whilst non-maleficence refers to the requirement to avoid harm (Beauchamp and Childress 2001). Although these may seem obvious, deciding on "benefit" and "harm" is not always easy or clear-cut, especially in the PICU context. In the above case, offering J.E. PICU admission, performing a tracheostomy and providing long-term ventilation might seem to offer this child the greatest benefit, in that it is the only reasonable way to save his life and may allow some valuable time with his family prior to his inevitably premature demise.

However, in the context of a family living in poor socio-economic circumstances, without easy access to basic amenities, offering those interventions without ensuring sufficient processes and resources are in place to sustain high-quality care might cause more harm than good, including prolonged physical discomfort and potential social and psychological harm to both the child and the family. Using technology to prolong life requires a balance of clear benefit to justify the associated burdens.

"Benefit" in this regard does not just refer to saving a life, it must also consider quality of life. Quality of life may be defined as that which gives a person meaning, value and pleasure and may include, but is not limited to, health; wellbeing; the ability to participate in recreational activities; financial security and material provision; and social interactions or relationships (Post 2014). Quality of life does not equate to perceived social value. In J.E.'s hypothetical case, offering long-term invasive ventilation may provide some time with his family; however, J.E. would be tetraplegic, unable to move from the neck down and unable to breathe for himself. He would be dependent on others for all his activities of daily living; dependent on the ventilator literally for every breath (with potential heightened anxiety as a result); and unable to participate in the sport which contributed substantially to his quality of life. Participating in school/educational activities may be extremely difficult in these socio-economic circumstances, although potential adaptations could be considered to allow enrollment in a school for children with special needs, with provision of transport (where available). Conversely, withdrawing life-sustaining therapy could be performed compassionately and appropriately, in order to minimize anxiety and pain for J.E.; the family could (and should) be part of the end-of-life process, and provision of counselling could facilitate appropriate grieving and enable the family to attain acceptance and closure (Suttle et al. 2017). This approach may ameliorate many long-term effects on the family, in terms of burden of care, however this option should not be considered purely for expedience.

Respect for Persons/autonomy

This principle speaks to the right of an individual to self-determination, according to their own beliefs, values and preferences (van Niekerk 2011).

Since the advent of pediatric critical care in the 1950s, there has been an appropriate shift from the traditionally paternalistic "doctor knows best" approach to a patient- and family- centered, shared care approach, where the patient and family are active participants in the decision-making process (Antoniadou et al. 2007). Autonomy means "self-rule" and refers to every individual's right to make decisions regarding themselves, including deciding what is in their own best interest. In health care this relates to people having the right to choose their treatment, including the right to refuse treatment (Beauchamp and Childress 2001). Young children, however, are not considered autonomous agents as they are dependent on adults for their care and protection and may not be cognitively competent to make fully informed decisions. Although the legal age to consent to medical care varies amongst different countries, in general young children's parents or legal guardians are required to make the final decisions regarding their child's care. However, children develop autonomy with advancing age and cognition, and therefore should be included in decision-making processes about their health care at a developmentally appropriate level, wherever possible (refer to Chap. 10). This also upholds the fundamental child rights to be heard and to be informed (United Nations 1989). Wherever possible, attempts should be made to facilitate non-verbal communication, for children unable to speak owing to the presence of an invasive airway, but who have sufficient level of conscious-ness. Where the child's wishes cannot be ascertained, similar to the role of surro-gate decision-makers for adult patients, parents or guardians can consider the *prior wishes*, if known, of a child regarding their medical care; or use *substituted judgment* to consider what the child would choose if they were able (Arnold and Kellum 2003). In pediatric critical care practice, patients and families may have different values or goals to that of the health care team, which sometimes causes further ethical chal-lenges (Kon et al. 2016). However, unless against the "best interest" principle, in general decisions should be in accordance with the family's views and values.

For appropriate decision-making, and valid permission granted for therapeutic procedures, there must be transparent, comprehensive information provision; the parent/guardian must be able to fully understand the information given; and have the ability to both make and communicate a final decision after considering the available options. The person providing permission must therefore have full decision-making capacity (refer to Chap. 13). In the context of the PICU, it has been shown that parents of critically ill children present with profound anxiety and psychological distress, which might impact their decision-making capacity (Kon et al. 2016). In this context, including extended family or other parental sources of support in clinical team discussions, and providing appropriate psychosocial and spiritual support and counselling may improve the ability to make a properly informed decision.

"Shared decision-making is a collaborative process that allows patients, or their surrogates, and clinicians to make health care decisions together, taking into account the best scientific evidence available, as well as the patient's values, goals, and preferences"—(Kon et al. 2016).

Although informed consent/permission is considered a tenet of the principle of respect for autonomy, in practice decision-making in the PICU is often (and may be ideally) a shared process, based on mutual respect between the patient/family and health care provider (Kon et al. 2016). Shared decision-making includes char-acteristics of deliberation and negotiation; flexibility in using an individualized approach; information exchange; multiple individuals; compromise; partnership; mutual respect; and patient/family participation and education. In the process of shared decision-making, all interested and appropriate parties (e.g. family, patient, extended community) come together to consider all options; discuss the individual psycho-social, spiritual and economic context, the child patient's best interest and family values; and arrive at an acceptable treatment decision for the individual patient. This model allows the health care worker to play an active role in decision-making, without being paternalistic. In some circumstances, families may even ask the health care team to make difficult decisions for them, which may be appropriate in specific circumstances. Although the parents are still the ultimate decisional authority, one of their decisions could be to rely on the medical team for guidance (Kon et al. 2016). It can also be seen that through shared decision-making, with the individual patient's context and interests central to the process, it may be ethically acceptable to make completely different clinical decisions between two patients with similar clinical presentations (but different contexts and circumstances).

Using the case of J.E. above, it may be argued that parents should not be required to make the final decision regarding withdrawal of life-sustaining therapy, where this would lead either to the child's death or to a life of severe disability, as this could contribute to feelings of guilt, psychological distress and, in the case of death, could result in complicated grief, with further major psychological impact (Boelen and Smid 2017; Kon et al. 2016). The family should, however, be part of the shared decision-making process and ideally should have at least reached a point of acceptance of the management plan prior to implementation.

The principle of respect for autonomy also includes the ethical requirement to tell the patient and family the truth, thereby also upholding their right to information, and to respect the privacy (relating to the person) and confidentiality (relating to personal information) of patients and their families. In the case of J.E., if it was deemed most appropriate to withdraw life-sustaining therapy, there may be conflict between the duty to tell the truth and not deceive the patient (autonomy), and the potential harms in terms of increased anxiety, psychological distress and fear for a child being told that they are going to die. In this case, some may argue that a compassionate approach would be to with-hold information from the child, where there is good evidence that disclosure would be against the child's best interest.

If the decision were made to withdraw life-sustaining therapy and allow J.E.'s natural death, parents and health care providers may opt to with-hold information and sedate the child, making him unaware of his circumstances (non-maleficence). However, this could be seen as being overly expedient and denying J.E. his fundamental rights, which do not fall away when dying. At the age of ten years, J.E. would have the cognitive ability to understand the concept of death, and there is great value in talking to and listening to children's thoughts, wishes and fears; and responding in a truthful and age-appropriate manner. Children need to know that they will be cared for, and they need *support through*, not *protection from* their sadness and worries (Tanchel 2003). Providing honest information, answering J.E.'s questions and addressing his concerns could allow him to participate actively in his own end-of life process, including saying the necessary good-byes, performing memory work or legacy building, having some choice in the process (e.g. who he wants with him) and receiving counselling to reach a point of acceptance and understanding (beneficence) (Levetown et al. 2004). This would include reassurance that any unpleasant symptoms arising could and would be managed through both pharmacological and non-pharmacological approaches. This process requires an extended multidisciplinary team approach.

Justice

The principle of justice relates to fairness, and encompasses respect for local, morally acceptable laws; respect for people's rights; and fair distribution of limited resources (distributive justice), without discrimination based on non-medical factors (Beauchamp and Childress 2001).

In high-income countries, most children with critical illness or injury would have access to ongoing life-sustaining therapy, high-quality rehabilitation, and assistive technologies; however these are not universally available or accessible (Bickenbach et al. 2013). In many countries patients with medical health care patients with medical health care insurance who are able to afford private health care may have access to care that is not available to poorer patients in the public health sector, for example (Norval and Gwyther 2003). This is a clear example of the fundamental injustices of poverty. Offering J.E long-term ventilation and ongoing costly supportive care, when this could not be offered to all children requiring chronic technological support, is contrary to the principle of distributive justice (the fair distribution of limited resources) (Mkhize 2008); yet it may be supported by rights-based justice (Moodley et al. 2011). Offering high-cost technology to a few patients, at the expense of basic care for many, would also violate the principles of distributive justice.

The Ethics and Law Advisory Committee of the Royal College of Paediatrics and Child Health state that decisions about who is offered technologically advanced care should not be primarily motivated by resource availability or constraints, but rather by what is appropriate for a given child at a given time (essentially the child's best interest). They do, however, acknowledge the challenges in applying this principle (Larcher et al. 2015).

"*Justice is an expression of our mutual recognition of each other's basic* [human] *dignity*" (Velasquez et al. 2014). This principle, as well as a basic child right, is violated if individuals are treated unequally or discriminated against because of arbitrary or irrelevant characteristics, including ethnicity, age, and education levels. However, differential treatment may be ethically justified if patients have differences that are directly relevant to the clinical scenario. In this regard, it may be considered fair or just to make different decisions for J.E. than, for example, a child living in different circumstances, with easy access to appropriate rehabilitation services and technological aids. In this case poverty is directly relevant to the ability to provide long-term care to J.E. and must therefore be factored into the decision-making process. Societal policies can attempt to decrease such disparities; however, applying them in a fair and just manner is not always feasible.

Utilitarianism versus Deontology—Moral Theory

Ethical considerations can also be discussed using other moral theories, including the theories of utilitarianism/consequentialism and deontology. Utilitarian followers endeavor to do the greatest good for the greatest number, speaking to distributive justice, with the outcome or consequence justifying the action (the end justifies the means). Should high intensity and high- cost critical care be offered to a few individuals, when most children in lower socio-economic circumstances cannot access a PICU at all? Should resources not be better used by focusing on lower-cost, lower technological care for all children?

In the context of J.E.'s case above, a utilitarian argument in support of performing the tracheostomy could focus on J.E.'s potential future contributions to society or argue that providing him the advanced care does not disadvantage other individuals. A utilitarian argument for withdrawing life-sustaining therapy could also be made if this action allowed for the "greatest good for the greatest number".

Deontology or Kantianism on the other hand considers the morality of the action itself (the means justifies the end). This theory upholds the duty of care, which is familiar to health care professionals—the requirement to do the "right" thing for the "right" reason. Of course, what is deemed "right" by one person, might not be considered the same by another. In our hypothetical case, if with-holding long-term ventilation was judged to be in the child's best interest (therefore the "right" thing to do), then J. E's death after withdrawal of life-sustaining treatment could be considered justifiable. On the other hand, any action causing death (withdrawal of care) may be considered immoral using this theory as well. In this context it might be argued that as long as the intent of the intervention was morally sound, the outcome or consequence could be justified.

14.2.3.3 Criteria for With-Holding or Withdrawing Life-Sustaining Therapy

The updated Royal College of Paediatrics and Child Health (United Kingdom) guidelines present updated categories, which might support a decision to withdraw or with-hold life- sustaining treatment (Larcher et al. 2015). These replace the outdated categories of brain death, permanent vegetative state and 'No Chance', 'No purpose' and 'Unbearable' situations (Royal College of Paediatrics and Child Health 2004).

All of these categories may be applicable in the case of J.E., as discussed throughout this chapter (Larcher et al. 2015):

1. **Life is limited in quantity**: treatment is unable or unlikely to significantly prolong life and/or it may not be in the child's best interest to provide this treatment.
2. **Life is limited in quality**: treatment may be able to prolong life significantly but will not alleviate associated burdens of either or both the condition and treatment.
3. **Informed competent refusal of treatment**: based on a model of shared knowledge and mutual respect.

Summary Table: Guiding principles for the ethical care of children with critical careneeds

Patient- and family-centered care
– The family and, wherever possible, the child should be included as active participants in clinical decision-making and treatment planning
Rights of the child
– Children need extra protection as they are vulnerable to abuse and exploitation owing to their dependent status
– The rights and interests of the child must be considered when making clinical decisions. The United Nations Convention on the Rights of the Child is one source of important guidance
Ethical principals
– The ethical principles of beneficence (doing good)/non-maleficence (avoiding harm), justice and respect for persons must be considered and upheld in all decisions related to critically ill children

(continued)

(continued)

Beneficence/non- maleficence
- Care provided should aim to optimally benefit to the patient, whilst minimising or avoiding harm
- Quality of life should be considered independently of social value

Respect for persons
- Individuals have the right to self-determination
- Decisions about medical care for young children, particularly those who are critically ill, are usually taken by the parent or legal guardian, in the best interests of the child
- Children develop autonomy with advancing age and cognition, and therefore should be included in decision-making processes about their health care at a developmentally appropriate level, wherever possible
- Where children are unable to speak owing to medical intervention (e.g. mechanical ventilation), attempts should be made to facilitate non-verbal communication
- Decision-making in the paediatric intensive care unit should be a shared process, based on mutual respect between the patient/family and health care provider
- Truth-telling is an ethical imperative
- Health care professionals have a duty to protect patients' privacy of person and confidentiality of information

Justice
- This principle relates to fairness, including the fair distribution of limited resources and non-discrimination
- No child should be unfairly discriminated against, based on arbitrary or unrelated, non-medical factors (e.g. ethnicity, language, gender etc.)
- Non-medical factors that may impact on patient outcome may be factored into medical decision-making
- Utilitarian principles can be used to ensure scare resources are optimally used to benefit the greatest number of children (distributive justice)
- Offering high-cost technological intervention to selected patients, at the expense of basic care for many, violates the principle of distributive justice but may uphold rights-based justice

14.3 Conclusion

Ethical argument is, unfortunately, often circular or conflicting, and easy answers to complex questions arising in the PICU are difficult to find. The highly advanced medical care and technologies provided in the PICU make considerations of quality of life and the fair distribution of scarce resources (e.g. ventilators, transplanted organs, specialized expertise) common dilemmas faced by the PICU team. Approaching such dilemmas with knowledge of the relevant ethical and legal background and in a spirit of shared decision-making with the patient and family can help clinicians navigate difficult situations.

For the hypothetical case presented in this chapter, either clinical decision may be considered ethically sound if it is reached through in-depth discussions with the patient's family (including the child if appropriate), other appropriate stakeholders such as community representatives, and the multidisciplinary clinical team (Gwyther 2008); changed goals of therapy are understood and agreed to; J.E. and his family's human rights are not infringed; and with J.E.'s best interest being the primary motivating factor, with the likely benefits of the decision outweighing the potential physical, psychosocial and spiritual risks or harms (Larcher et al. 2015).

References

Antoniadou, A., F. Kontopidou, G. Poulakou, E. Koratzanis, I. Galani, E. Papadomichelakis, and H. Giamarellou. 2007. Colistin-resistant isolates of Klebsiella pneumoniae emerging in intensive care unit patients: First report of a multiclonal cluster. *The Journal of Antimicrobial Chemotherapy* 59 (4): 786–790. https://doi.org/10.1093/jac/dkl562.

Arnold, R.M., and J. Kellum. 2003. Moral justifications for surrogate decision-making in the intensive care unit: Implications and limitations. *Critical Care Medicine* 31 (5Suppl): S347-353. https://doi.org/10.1097/01.CCM.0000065123.23736.12.

Beauchamp, T.L., and J.F. Childress. 2001. *Principles of biomedical ethics*, 5th ed. New York: Oxford University Press.

Bickenbach, J., C. Bodine, D. L. Brown, A. Burns, R. Campbell, D. Cardenas, and X. Xiong. 2013. International perspectives on spinal cord injury. Switzerland: World Health Organisation and International Spinal Cord Association. Retrieved from https://www.who.int/publications/i/item/international-perspectives-on-spinal-cord-injury. Accessed 8 Aug 2020.

Boelen, P.A., and G.E. Smid. 2017. Disturbed grief: Prolonged grief disorder and persistent complex bereavement disorder. *BMJ* 357: j2016. https://doi.org/10.1136/bmj.j2016.

Elkinton, J.R. 1966. Medicine and the quality of life. *Annals of Internal Medicine* 64 (3): 711–714. https://doi.org/10.7326/0003-4819-64-3-711.

Gwyther, E. 2008. Withholding and withdrawing treatment: Practical applications of ethical principles in end-of-life care. *South African Journal of Bioethics and Law* 1: 24–26. https://doi.org/10.7196/SAJBL.7.

Kissoon, N., A. Argent, D. Devictor, M.A. Madden, S. Singhi, E. van der Voort, and J.M. Latour. 2009. World federation of pediatric intensive and critical care societies—its global agenda. *Pediatr Crit Care Med* 10 (5): 597–600. https://doi.org/10.1097/PCC.0b013e3181a704c6.

Kon, A. A., J. E. Davidson, W. Morrison, M. Danis, and D. B. White. 2016. Shared decision-making in intensive care units. Executive summary of the American College of Critical Care Medicine and American Thoracic Society Policy Statement. *American Journal of Respiratory and Critical Care Medicine* 193 (12): 1334–1336. https://doi.org/10.1164/rccm.201602-0269ED

Larcher, V., F. Craig, K. Bhogal, D. Wilkinson, and J. Brierley. 2015. Making decisions to limit treatment in life-limiting and life-threatening conditions in children: a framework for practice. *Archives of Diseases in Childhood* 100 (Sup 2): S1–S23.

Levetown, M., S. Liben, and M. Audet. 2004. Palliative care in the pediatric intensive care unit. In *Palliative care for infants, children, and adolescents: A practical handbook*, ed. B.S. Carter and M. Levetown, 273–291. Baltimore (MD): John Hopkins University Press.

Lude, P., P. Kennedy, M.L. Elfstrom, and C.S. Ballert. 2014. Quality of life in and after spinal cord injury rehabilitation: A longitudinal multicenter study. *Top Spinal Cord Inj Rehabil* 20 (3): 197–207. https://doi.org/10.1310/sci2003-197.

Mkhize, N. 2008. Communal personhood and the principle of autonomy: The ethical challenges. *Continuing Medical Education* 24 (1): 26–29.

Moodley, K., R. Moosa, and S. Kling. 2011. Justice. In M*edical ethics, law and human rights. A South African perspective*, ed. K. Moodley, 73–85. Pretoria, South Africa: Van Schaik Publishers.

van Niekerk, A. A. 2011. Ethics theories and the principilist approach in bioethics. In *Medical ethics, law and human rights. A South African Perspective*, ed. K. Moodley, 19–40. Pretoria, South Africa: Van Schaik Publishers.

Norval, D., and E. Gwyther. 2003. Ethical decisions in end-of life care. *Continuing Medical Education* 21 (5): 267–272.

Post, M.W. 2014. Definitions of quality of life: What has happened and how to move on. *Top Spinal Cord Inj Rehabil* 20 (3): 167–180. https://doi.org/10.1310/sci2003-167.

Royal College of Paediatrics and Child Health. 2004. *Withholding and withdrawing life sustaining treatment in children: A framework for practice*, 2nd ed. London: RCPCH.

Suttle, M.L., T.L. Jenkins, and R.F. Tamburro. 2017. End-of-life and bereavement care in pediatric intensive care units. *Pediatric Clinics of North America* 64 (5): 1167–1183. https://doi.org/10.1016/j.pcl.2017.06.012.

Tanchel, I. 2003. Psychosocial issues in palliative care. *Continuing Medical Education*, 21 (5).

United Nations. 1989. Convention on the rights of the child. Retrieved from https://www.ohchr.org/en/professionalinterest/pages/crc.aspx. Accessed 9 Aug 2020

Velasquez, M., C. Andre, T. Shanks, and M. J. Meyer. 2014. Justice and Fairness. Retrieved from https://www.scu.edu/ethics/ethics-resources/ethical-decision-making/justice-and-fairness/. Accessed 10 Aug 2020.

Further Reading

Burns, J. P., and C. H. Rushton. 2004. End-of-life care in the pediatric intensive care unit: research review and recommendations. *Critical Care Clinics* 20 (3): 467–485.

Davidson, J. E., R. A. Aslakson, and A. C. Long et al. 2017. Guidelines for Family-Centered Care in the Neonatal, Pediatric, and Adult ICU. *Critical Care Medicine* 45 (1): 103–128.

Kon, A. A., J. E. Davidson, W. Morrison, M. Danis, and D. B. White. 2016. Shared decision-making in intensive care units. Executive summary of the american college of Critical Care medicine and American thoracic society Policy statement. *American Journal of Respiratory and Critical Care Medicine* 193 (12): 1334–1336.

Orioles. A., and W. E. Morrison. 2013. Medical ethics in pediatric critical care. *Critical Care Clinics* 29(2):359–375.

Chapter 15
End of Life: Resuscitation, Fluids and Feeding, and 'Palliative Sedation'

R. Hain and F. Craig

Abstract In this chapter, we consider how a commitment to acting in a child's interests can be brought to bear on three specific ethical quandaries that face those caring for children at the end of life, and how such a commitment might seem to cohere or be in tension with other principles such as autonomy and justice. We examine the status of 'do not resuscitate' orders in children and argue that they cannot exist in children in the same form as in adults. Since the standard of ethical permissibility is set by the interests of the child, rather than the preferences of parent or doctor, there is no single person with authority to agree an adult-style DNR. Looking at clinically-assisted feeding or hydration, we argue that withholding, withdrawing, commencing or continuing it each constitutes an intervention and needs to be carefully justified by considering its burdens and harms in the broadest sense reasonable. Finally, we consider 'palliative sedation' and suggest that, while there are circumstances under which sedation is an inevitable result of adequate symptom control, impaired consciousness does not of itself represent a way to control symptoms and should not be deliberately induced.

Keywords Palliative care · Symptom control · End of life care · Resuscitation · Artificial nutrition and hydration · Palliative sedation · Triage ethics · Interests · DNR/DNAR · Parental authority

R. Hain (✉)
Paediatric Palliative Medicine, College of Human and Health Sciences, All-Wales Paediatric Palliative Care Network, University of Swansea, Swansea, Wales
e-mail: richard.hain@southwales.ac.uk

F. Craig
Paediatric Palliative Medicine, The Louis Dundas Centre for Paediatric Palliative Care, Great Ormond Street Hospital NHS Foundation Trust, London, UK
e-mail: Finella.Craig@gosh.nhs.uk

15.1   Introduction

While Beauchamp and Childress' 'four principles' approach (Beauchamp and Childress 2009) does not represent a single coherent moral theory, it remains the best known and perhaps the most useful framework for analyzing ethical quandaries to which clinicians turn in practice. In end of life decisions about a child, the emphasis principlism places on interests, autonomy and justice needs to be specified to children and to the circumstances a child and family are facing as the child's death approaches.

Interests (the balance of benefit and harm) represent the gold standard in decision-making over a child (United Nations General Assembly, 1989). All those making decisions about a child, whether parent, health care professional or Court, must put their own preferences aside and consider what, on the basis of their own knowledge, is likely to be best for the child. That requires assessment of the potential benefits of an intervention and ensuring, as far as possible, that they outweigh the risks and burdens. Consideration must be given not only to the disease process but to the child's quality of life, as well as her psychological, spiritual, cultural and social values. Where there is no perceived benefit, or the benefit is outweighed by risks and burdens, an intervention should be withheld or discontinued (also refer to Chap. 14). Where there is uncertainty, it may be started but should be discontinued if the desired aims are not achieved.

Autonomy recognizes the right of the patient herself, rather than her parents or the doctor, to make a personal choice. It needs to be based on a full and appropriate understanding of the situation, consequences of the intervention and consequences of refusing the intervention. The ability to weigh up the benefits and harms requires a child or young person to be adequately informed, and to possess the cognitive skills necessary to weigh them rationally. On the modern medical understanding, it is that capacity that defines 'autonomy' (Beauchamp and Childress 2009). The degree of autonomy necessary to engage in a medical decision depends on the gravity of the decision and its consequences. A three-year-old with a small cut to the knee will usually have sufficient autonomy to be allowed to decide whether or not to have a sticking plaster. When it comes to making decisions at the end of life, however, deliberation requires more complex comprehension. Human comprehension of one's own death evolves slowly, reaching full understanding only in the mid-twenties (Steinberg 2010). In relation to decisions that might hasten death, most humans probably lack meaningful autonomy until well beyond the legal age of majority.

The principle of justice requires that all children have equitable access to the medical interventions available. They should not be excluded on any grounds, including pre-existing illness or disability or because their care is considered too burdensome to their family or society. Decisions to commence, withhold or discontinue an intervention must be based on the principle of overall benefit, with each child considered individually and holistically.

In this chapter, we bring those principles to bear on three specific ethical quandaries that face those caring for children at the end of life: the status of 'do not resuscitate' orders in children and who should make them, the challenge of commencing or continuing clinically-assisted feeding or hydration, and the contentious therapeutic intervention often called 'palliative sedation'.

15.2 'Do Not Resuscitate' Orders

Cardiopulmonary resuscitation can sometimes restore individuals who have experienced a cardiorespiratory arrest to their baseline level of health. The process of resuscitation, and the interventions needed to support a patient afterwards are unpleasant, such that they need to be carefully weighed against the likelihood that resuscitation will achieve its objective. Advanced age, failure of several organ systems, the nature of the underlying condition and any co-morbidities may all conspire to reduce the likelihood that resuscitation will succeed (Siriphuwanun et al. 2014) In the past, it was acceptable practice for the responsible doctor to conclude that the benefits of resuscitation were so small that they did not justify the inevitable harms it would cause. A doctor would record that conclusion in the patient's medical notes using the abbreviation 'DNR' ('do not resuscitate') or, 'DNAR' ('do not attempt active resuscitation' or 'do not attempt resuscitation'). In most Western cultures, DNR decisions today are made in discussion with the patient herself and, in many countries, adult patients even have a legal right to refuse resuscitation in advance (Hooper et al. 2020).

The DNR order is consistent with the philosophy of palliative care. It provides clear instructions for health care staff at the patient's bedside, expressing the important idea that some interventions cause more harm than benefit and that there are circumstances under which attending to the comfort of the dying patient is more appropriate than summoning the arrest team to institute an unpleasant and futile intervention. The decision to refrain from instituting resuscitation is a heavy responsibility to take, and without a DNR it is probable that resuscitation would be inflicted on patients who are, in reality, highly unlikely to benefit from it and might otherwise have died a peaceful death.

A DNR is also important because it symbolizes the important conversations that need to take place around the end of a patient's life.

15.3 Resuscitation in Children: Who Can Say 'no'?

The adult patient has the authority to refuse resuscitation because resuscitation is an intervention, and it is wrong to inflict an intervention on a patient who does not want it. It remains wrong whether or not the intervention would, as a matter of fact, be of benefit. As a form of consent to non-treatment, DNR orders are effective because the person who gives that consent is entitled to do so.

There is, however, often no such entitled consent-giver in children, as many children lack the capacity to make such complex decisions for themselves. The authority of parents over medical decisions in their children is considerably more restricted than the authority of an adult to refuse or request treatment for herself. Nor are doctors entitled unilaterally to decide that resuscitation should not be carried out. They too are constrained in their decision-making by the harm and benefit to the child that are likely to result. So, for children, unless the child has capacity to do so, there is no single individual with the moral authority to decide that resuscitation should not be attempted.

Yet the problems that the DNR was designed to solve in adults are also problems in children. Professionals attending the child still need clear guidance if inappropriate invasive interventions are to be avoided, and it is as important in children as it is in adults to begin explorations in a timely manner about what might happen at the end of life.

There are three voices that might hypothetically contribute to a decision not to proceed with resuscitation in a child. The first is that of the child or young person herself who, given the opportunity, might choose to decline the intervention at the time or in advance. The second is that of her parents, and the third is that of the doctors who would be called on to perform the intervention.

15.4   Child or Young Person

The fact that ideas of mortality are not yet developed in children and young adults does not mean that a young patient's views are of no importance. Whether or not they constitute a legally binding decision, the preferences of a child or young person should always be solicited. Cognitively normal children with life-limiting conditions are much more aware of the implications of their illness than adults often imagine (Bluebond-Langner et al. 2010), and often value being involved as a gesture of respect and compassion in itself. Children and young people with cancer and those with Duchenne Muscular Dystrophy[1] are examples of patients who are not adults, but who usually possess a degree of autonomy that is always meaningful, and should sometimes be determinative, in making decisions about resuscitation.

In practice, however, many children with life-limiting conditions do not have the cognitive skills to engage in an exploration of their mortality, or the communication skills to articulate them, or both.

15.5   Parents

Most societies acknowledge the family to be the key unit in caring for children (United Nations General Assembly 1989) and the nature of family relationships makes parents an obvious choice to speak for their child (refer to Chap. 5). Parents are usually the people who know an individual child best and so are most likely to be

---

[1] Progressive muscle weakness.

correct in their estimations of what the child might want, or how medical decisions will impact on the child's quality of life, both in the immediate and long term.

The authority given to parents by society is, however, more restricted than many parents assume. The concept of parental rights does not represent an extension of parents' own autonomy over themselves, nor does it imply that the child is a possession whose interests are always those of her parents. The rights that society accords to parents (Guggenheim 2005) are correlative to the duty they have to care for their child.[2] In relation to medical decision-making, parental authority is essentially limited to giving consent to interventions that are likely to confer more benefit that harm. Parents' rights, however, do not extend to forcing doctors to intervene in a way that would harm their child, or to withhold an intervention that is certainly in the interest of their child.

Again, that does not mean that parents' views are unimportant. Parental preferences are highly relevant and should be solicited. Most parents should be considered 'experts in their own child' as well as in some aspects of the specific condition their child has. Furthermore, other than the child, it is parents who will bear the greatest impact of any decision that is made. Parental preferences cannot, however, be the only authority for medical decision-making because the preferences of parents, if acted on, would sometimes be against the interests of their child and so are ethically impermissible. That is not usually because parents choose to enact deliberate harm on their child. It can, however, be difficult for parents living through the death of their child to distinguish between their own anguish and the suffering of their child. For some, the pain of losing their child is so great that there is no amount of suffering the child could endure that is not justified if it delays that death. Inadvertent preferences for harming a child can also arise from an incorrect understanding of the facts and probable outcomes. Such strongly held misbeliefs may have their origin in miscommunication, in emotional coping mechanisms, including denial or a sense of powerlessness, or in cultural or religious world-views. Each of these can masquerade as any or all of the others.

It is inevitable and understandable that parents faced agonizingly with the death of their child sometimes express preferences that, if acted on, would act against the child's interests. Their preferences do not usually reflect any intention to harm the child. Nevertheless, it would not be permissible for doctors to act on the basis of them. Parental preferences are always important, but they are only determinative if they are for a course of action that is in the child's interests (refer to Chap. 4).

---

[2] As adults in their own right, parents have certain legal rights which might conflict with their duty to care for their child. For example, a parent is entitled to believe on religious grounds that, on balance, measles vaccination is likely to do her child more harm than good. That belief is, however, incorrect for most children, and if she acts on it she is harming the child and so is acting immorally. In the name of adult freedom the law offers some protection (albeit limited) to the right of parents to choose to harm their child. But even that *legal* permission does not alter the *moral* obligation on parents not to do so.

15.6   Doctors

On the face of it, it seems that doctors are perfectly placed to authorize a DNR in a child. A medical opinion is often clear and unequivocal, and it gives a voice to those who are most expert in the medical condition that a child has and in the health status of the child at the time resuscitation is considered.

But doctors alone are not usually in a position to decide what actions are in the interests of the child. The authority that society gives to doctors to make medical decisions over children has the same source as the authority it gives to parents. A doctor is not entitled to harm a child, even if parents request it, but neither is she entitled arbitrarily to set aside the preferences of a child or her parents. What it comes down to is that the doctor is entitled to endorse preferences of parents when they are for an action she believes are in the interest of the child, but is required to set aside those preferences if she believes they are for an action that will harm the child. Some of the facts that are material to determining harm and benefit are not the sort of objective medical fact that doctors know, but rather relate to the child's own subjective experience of life and its quality. It is unlikely that the doctor will know those facts as well as the parents do.

It is also true that doctors often feel they need to consider the interests of other children as well as the patient in front of them. In some circumstances, such as during the recent coronavirus pandemic, the need for an intervention can exceed the resources available to provide it. Under such 'triage' conditions (and only under those conditions), a new factor enters into ethical decision-making; prioritization. The issue is no longer simply whether a certain intervention will offer a patient more benefit than harm, but whether the excess of benefit over harm is as great in this patient as it might be in other patients. When triage ethics obtain, it may be reasonable for doctors to withhold resuscitation from a child, not because the child herself will not benefit from it, but because another child will benefit more.

In resource-poor societies where the need for health care outstrips the resources available, it can be argued that circumstances almost always permit such 'triage ethics' (refer to Chap. 29). Most of the time that is probably not true in most high- and middle-income countries. There can nevertheless be a perception that triage ethics apply, because the public *demand* for health care (as distinct from the *need* for it) usually outstrips the finite resources that are available. One result is that doctors can be vulnerable to prioritizing the needs of hypothetical children inappropriately over those of an actual child.

Although doctors are better placed than parents to know some of the facts that are relevant to establishing the benefits and harms of resuscitation, there are other facts that are equally relevant that parents are more likely to know. Furthermore, doctors are vulnerable to making decisions about an individual patient that are shaped partly by considerations other than the interests of that patient. That can allow doctors to feel justified in allowing harms to a child that are in fact ethically impermissible. Doctors alone are not in a position to authorize a DNR in a child.

15.7   Who Can Authorize a DNR?

Compassionate care demands that a decision to withhold invasive resuscitation must sometimes be taken. The child with a life limiting condition may not be able to make that decision for him- or herself.

The right of parents, as well as that of doctors, to make medical decisions over a child is restricted to a responsibility to ensure that actions are only taken if they are likely to be in the child's interests. The inconvenient truth is that in children there is no single voice with the authority to permit, still less to instruct, that resuscitation be withheld.

Nevertheless, there is a point in the trajectory of most life-limiting conditions at which the benefits of resuscitation to a child are so small or so unlikely that they cannot justify the harms. Under those circumstances, resuscitation would be ethically impermissible and should not be carried out. That point is not always clear and must be established in dialogue between the various 'experts' involved. At the very least, that should be the parents and the doctors, and of course the child if that is practical. It might also include wider members of the health care team and family.

The purpose of dialogue is to establish–as closely as can be known–what decision is likely to be in the interests of the child. It is not in order for parents or doctors to give each other instructions about what must happen, nor is it intended to negotiate a compromise acceptable to both parties. Its purpose is to set out what course is most likely to be in the interest of the child. When it comes to making end of life decisions in the child, such dialogue is not an optional extra, but is always necessary.

Of course, without shamanic powers of prophecy, parents and doctors cannot know for certain what the outcome of their decision will be. At best, the result of dialogue will be a reasonable approximation of the child's true interests, based on the combined expertise of those who know the child and the condition best. It is also important to acknowledge that it is impractical to hold to a standard in which no harm is permitted to a child at all. Some harms are imposed even by such an obviously correct decision as starting intravenous antibiotics in an otherwise healthy child with pneumonia. The child experiences, for example, the pain of a cannula, and is exposed to the risk of anaphylaxis. The question is whether those certain harms are justified by the magnitude and/or probability of benefit.

One consequence is that it may be that more than one course of action is reasonable (Gillam 2016) (refer to Chap. 11). Doctors in particular need to distinguish carefully between a decision they feel is not the one they themselves would prefer, and a decision they feel will offer so much more harm than benefit that it cannot be in the child's interests. The latter are impermissible, and doctors must refuse to act on them. But the former are permissible and, since it is the parents who will bear the brunt of the consequences, where parents and doctors disagree about two courses of action that are, in reality, both reasonable given the information available, the preferences of parents should be given priority over those of doctors.

There is no doubt that a DNR will impose some harm in a child, even as the end of life approaches. If resuscitation is not offered, then the child will certainly die whereas the chance of restoring a child to baseline health through resuscitation, even where it is very small, is rarely zero. But resuscitation is highly unpleasant, as is any subsequent admission to intensive care unit with all its attendant painful and distressing interventions. Against the speculative possibility that a DNR might cause harm must be weighed the known certainty that a child will significantly benefit from avoiding the discomfort of the resuscitation.

In many institutions, including our own services, the process of advanced care planning is supported by formal policy and documentation that supports clinicians in having those necessary conversations in a timely manner, records what discussions have taken place, and disseminates them so that all those who need to know are kept fully informed. That includes health care staff in primary care and at entry points into hospital care such as the emergency department or assessment unit, as well as ambulance services.

Parents often assume it is entirely up to them how their child is treated; a prospect which many find terrifying. Doctors can inadvertently reinforce that misunderstanding by offering parents the 'choice' whether or not a child should be resuscitated, even when the doctors have already decided that would not be the right thing to do. The gamble is that if parents make the 'right' decision, the doctor has been relieved of the responsibility for making it. If resuscitation is not in the child's interest, however, the choice parents are being offered is a false one because in reality they are not entitled to make it. If the gamble fails to pay off, the risk is that parents are confused and infuriated by doctors' disingenuousness (Meert et al. 2008).

The decision not to resuscitate a child is always a decision made together by all those who work as colleagues in that child's care, and who bring their own expertise to establishing what will, as a matter of fact, be best for the child. It is important that doctors are honest with parents from the outset about the extent and limitations of their authority to decide what happens to their child at the end of life.

15.8   Feeding

Hydration and nutrition are basic requirements to survival. When oral intake cannot provide adequate fluids and nutrition, due to mechanical limitations or illness, clinically assisted nutrition and hydration (CANH) must be considered. This is the provision of nutrition or fluids in any other form other than through the mouth and includes nasogastric, nasojejunal, gastrostomy, gastrojejunostomy as well as subcutaneous or intravenous routes. Whilst it is technically a medical intervention, withholding or withdrawing CANH often carries much greater emotional and ethical significance than other, more 'aggressive', interventions such as mechanical ventilation or dialysis. Many consider providing calories and fluids to children to be basic, rather than medicalized, care. Like those other interventions, however, CANH is not without burdens to the child, requiring procedures that can be unpleasant and may be associated with side effects and complications.

As with other interventions, it is the best interest standard that should determine whether CANH should be commenced, withheld or withdrawn (refer to Chap. 4). CANH is not permissible if the chance of harm outweighs the likelihood of benefit. Although the context of benefits, risks and burdens to the family as a whole is important, it is the welfare of the child herself that must be the primary focus.

15.9 Clinically Assisted Nutrition and Hydration at the End of Life

As death approaches, oral intake of food and fluids can be significantly reduced for a variety of reasons, including reduced appetite, weakness, impaired swallow, reduced consciousness or sedation.

The importance of balancing harm and benefit remains as great in the last few weeks of life as at other points in the child's illness trajectory, but the potential burdens and benefits of CANH are likely to be different. The potential benefits of CANH are not limited to the extent that it can preserve or prolong life or health. CANH can also preserve or enhance the *quality* of life through improved symptom control. CANH can, for example, help maintain a child's comfort by reducing her experience of hunger or thirst.

Where there are no such clearly defined and achievable goals, however, there is a risk that CANH serves at best to prolong the dying process. Under those circumstances, the intervention would be burdensome with no perceived benefit. Burdens include the harm of potentially prolonging the dying process and all the symptoms that accompany it, and also the burdens of the intervention itself. Inserting a nasogastric tube, subcutaneous or intravenous line are unpleasant. Enteral feeds or fluids may cause vomiting and risk aspiration. Intravenous fluids are likely to require further blood tests to monitor electrolytes, increase the risk of sepsis and fluid overload and may mean that care cannot be provided outside a hospital environment. Even where CANH offers some benefits, therefore, the intervention could, on balance, be more harmful than beneficial.

Furthermore, there are often other, potentially less burdensome, interventions that can relieve symptoms of thirst or hunger without the need for CANH. For example, dryness of the mouth and lips, increasing the sensation of thirst, can be relieved with good oral hygiene and lip care and, where appropriate, small quantities of fluid. Some children may manage limited amounts of nutrition and fluid but insufficient to sustain growth or to sustain life. Under those circumstances it may be in the child's best interest, on the balance of burdens and benefits, to continue with limited enteral feeding rather than introduce tube or gastrostomy feeds or escalate to parenteral fluids or nutrition.

Finally, there is some evidence from adult studies, that reducing calorie intake during terminal illness may be beneficial through production of ketones and endorphins that can paradoxically reduce hunger and enhance a feeling of wellbeing (Bruera et al. 2013; Good et al. 2014; Winter 2000).

15.10 Can Different Decisions About CANH Be Equally Permissible for Patients in Apparently Similar Situations?

The burdens and benefits of CANH must be assessed for each patient individually, taking into account not just the medical situation but all other relevant factors, including family value systems, social, spiritual and cultural factors. That broad assessment of burdens and benefits means that different decisions can be equally reasonable in seemingly similar situations.

So, for example, a baby with severe hypoxic ischemic encephalopathy may feed enough orally to satisfy hunger and comfort, but CANH might be necessary to avoid malnutrition. If she also has severe respiratory problems such that, irrespective of her nutritional state, she is thought unlikely to survive more than a few weeks, it is improbable that she will survive long enough to derive benefit from CANH and its harms might be difficult to justify. Under those circumstances it would be reasonable to decide to withhold or to withdraw CANH.

But that decision depends on a judgement about how long she is likely to survive that is inherently uncertain. It might also be reasonable to judge that she will survive longer than that and, on that basis, to decide to continue CANH because she will live to derive benefit. Under those conditions, the preferences of parents and the wider social, emotional and spiritual context which shapes those preferences, can be the final arbiters of the decision. Those elements of judgement, uncertainty and complexity mean that, even in the context of seemingly similar medical situations, opposing decisions can both be ethically sound.

15.11 The Emotional Burden of Withholding or Withdrawing CANH

Withholding or withdrawing feeds from a child can carry a high emotional burden for family members and professional caregivers. There is a natural instinct to care for our young, with the provision of nutrition and fluids being one of the most basic and natural forms of care. Where other forms of medical interventions may be viewed as intrusive or aggressive, maintaining nutrition and hydration is often seen as nurturing. Without CANH, a child may be perceived to be 'starving to death' and suffering. Where there is visible weight loss this can be especially difficult.

Good symptom management is essential, to reduce the burden of suffering for the child, family and professional caregivers. Good mouth and lip care are essential. Skin care, with attention to pressure areas, requires careful attention to prevent skin breakdown, a particular risk if there is weight loss. It can at times be helpful for teams and family members to revisit the best interest's principle of decision-making, remembering that CANH is not a burdenless intervention.

15.12   Palliative Sedation

A discussion about sedation at the end of a child's life is complicated by the fact that it means different things to different people. For some, the term 'palliative sedation' describes a clinician's active intervention to induce unconsciousness in a patient when she considers that the end of life is imminent. For others, the term merely denotes acceptance that, as a patient's death approaches, reduced consciousness may be an inevitable side effect of medications necessary to manage her physical and existential distress. The issue of palliative sedation is highly contentious because it illustrates two controversial ethical principles at the end of life in children.

The first is the principle of double effect (PDE). In practice, every action a doctor might choose to take will have more than one consequence. The PDE articulates that, since the doctor can intend only one of those consequences (even if she is aware that others are possible), it is inevitable that the result of her action will not always be what the doctor hoped or intended. Even if the consequence is undesirable, the decision that led to it might be morally correct providing the clinician did not intend it, and providing that the outcome she did intend is both sufficiently good and sufficiently likely to justify the risk of the undesirable consequence.

Faced with a patient whose symptoms are difficult to manage, the PDE shows that there is a morally significant difference between the decision a clinician might take to induce unconsciousness in her patient, and a decision to optimize her symptom control while recognizing that reduced consciousness is a possible result.

The second contentious issue associated with conscious sedation is the idea that the quality of an individual's existence can be equated with the absence of suffering. Since suffering requires consciousness, there is one sense in which rendering someone unconscious represents a way to alleviate their suffering. Since one goal of palliative care is precisely to alleviate a patient's suffering, abolishing consciousness appears, then, to be ethically permissible in the context of end of life care.

Properly expressed, however, the purpose of palliative care is not to alleviate suffering in some abstract sense, but to facilitate an actual person's experience of a good quality existence. Since suffering usually jeopardizes that kind of flourishing, it is true that the goal of palliative care is usually to alleviate it. But the goal of facilitating a patient's enjoyment of his or her life is not furthered by deliberately inducing unconsciousness, since an unconscious patient is just as unable to enjoy life as she is unable to suffer from it.

Palliative sedation also illustrates the less contentious idea that there is an obligation on the part of a clinician to make decisions that will, on balance, tend more to help her patient than to harm her. The objective of all interventions in pediatric palliative care should be to facilitate as far as possible the child's flourishing in the last days and weeks of her life.

The harms of actively inducing unconsciousness are clear. While she is asleep, a child is not able to engage with those around her, nor is she able to process or understand what is happening to her. As well as to the child, there are harms to her family, who effectively lose their child earlier than is necessary. Against those harms must be set some important benefits. The presence of uncontrolled physical symptoms, too, would make it difficult for a conscious child to engage with others. The sight of their child in unmanaged pain is a significant source of distress for families. There is no doubt that avoiding those symptoms represents an important benefit to families that palliative care should aspire to offer.

It is not clear, however, that inducing unconsciousness is the only, or even the best, way of achieving that. It could be argued that inducing unconsciousness deliberately is acceptable if there is no other way to alleviate the child's suffering to a point where consciousness is preferable to unconsciousness. In practice, however, it is not clear that such a situation ever obtains, *providing the child has adequate access to appropriately skilled pain and palliative care expertise*. It is surely hard to argue that inducing unconsciousness is a justifiable *alternative* to such expertise. In that case, palliative sedation in the sense of actively inducing unconsciousness in a patient as the primary goal must be ethically impermissible.

But many symptom control modalities, particularly those for treatment of anxiety, pain and nausea and vomiting, carry the possibility of causing drowsiness. That drowsiness is usually seen as an adverse effect or harm that should be minimized through careful therapeutic decision-making. For some children, however, those symptoms are so severe that in controlling them, drowsiness becomes likely or indeed inevitable. Under those circumstances, the benefits of adequate doses to control symptoms are so great, and the harms of drowsiness so relatively small, that it becomes ethically permissible to tolerate or even welcome a reduction in consciousness. If palliative sedation refers to the clinician's willingness to accept that a patient will become unconscious as a result of adequate management of her symptoms, then it is ethically permissible.

15.13 Conclusion

In this chapter, we have outlined the way in which the broad principles of medical ethics can be specified to three decisions about medical care that commonly need to be made as a child approaches the end of life.

The first is the DNR or DNAR order. In pediatric palliative care, a resuscitation decision is made through a process of dialogue, often referred to as advance care planning, in which all those who have significant expertise in this child with this condition, particularly (but not only), the child, the doctor and the parents set out together to establish what actions will be in the interests of the child.

The second is whether or not to continue artificial feeding or hydration. CANH is a medical procedure that merits the same gravity of decision-making as other seemingly more aggressive interventions. It may be of benefit where it relieves symptoms or maintains or prolongs a life of acceptable quality for that particular child. It may not be an acceptable intervention if it serves only to prolong suffering or to increase suffering as a consequence of its introduction.

The final question is around 'terminal sedation'. Induction of unconsciousness is not a therapeutic intervention in palliative care and should never be its primary aim. In an individual patient, however, it can sometimes be an inevitable consequence of the interventions needed to achieve good symptom control. Palliative care clinicians should not deliberately induce unconsciousness in their patients, but they should recognize that a reduction in conscious state is sometimes inescapable as a result of adequate symptom control, and that under those circumstances, and those alone, it need not only be tolerated but sometimes welcomed.

These are not, of course, the only ethically challenging questions that face those who care for dying children and their families. But they illustrate both how the principles that govern acute medical decision-making in adults are the same as, and how they are different from, decision-making in pediatric medicine.

15.14 Guiding Principles

Balancing benefit and harm
At the end of life, as at other times, a medical intervention is ethically permissible only if its harms are justified by the likelihood of meaningful benefit
Considering benefits and harms broadly
Benefits and harms can be physical, emotional, psychological or spiritual (existential). Sometimes benefits and harms of one sort need to be weighed against those of another
Working with parents as colleagues
All decisions regarding children require dialogue between professionals, parents and (where possible) the child herself. Parents' moral authority to make medical decisions about their children is not unrestricted but, because they know their child well, parents typically possess expertise that others do not have about the child's likes and dislikes and the quality of the child's usual life. Those are relevant to a decision based on benefits and harms that are considered appropriately broadly (see above) and the health care professional must actively solicit them
Accommodating reasonable alternatives
Where more than one course of action will probably offer more benefit than harm, health care professionals should actively support the preferences of a child's family. Where parents have a preference for a course of action that in reality is likely to cause the child more meaningful harm than meaningful benefit, health care professionals must not accede to it

(continued)

(continued)

Being honest
From the outset of their relationship with a family, health care professionals must be open about what lies ahead, including acknowledgement of uncertainty and candid clarification about the extent and limitations of parents' authority over interventions in their child

References

Beauchamp, T., and J. Childress. 2009. *Principles of biomedical ethics.* New York: Oxford University Press.

Bluebond-Langner, M., J.B. Belasco, and M. Demesquita Wander. 2010. I want to live, until i don't want to live anymore: Involving children with life-threatening and life-shortening illnesses in decision-making about care and treatment. *Nursing Clinics of North America* 45: 329–343.

Bruera, E., D. Hui, S. Dalal, I. Torres-Vigil, J. Trumble, J. Roosth, S. Krauter, C. Strickland, K. Unger, J.L. Palmer, J. Allo, S. Frisbee-Hume, and K. Tarleton. 2013. Parenteral hydration in patients with advanced cancer: A multicenter, double-blind, placebo-controlled randomized trial. *Journal of Clinical Oncology* 31: 111–118.

Gillam, L. 2016. The Zone of parental discretion: An ethical tool for dealing with disagreement between parents and doctors about medical treatment for a child. *Clinical Ethics* 11: 1–8.

Good, P., R. Richard, W. Syrmis, S. Jenkins-Marsh, and J. Stephens. 2014. Medically assisted hydration for adult palliative care patients. *Cochrane Database Syst Rev* Cd006273.

Guggenheim, M. 2005. *What's wrong with children's rights, Cambridge.* Mass: And London, England, Harvard University Press (Kindle).

Hooper, S., C. P. Sabatino, and R. L. Sudore. 2020. Improving medical-legal advance care planning. *J Pain Symptom Manage.*

Meert, K.L., S. Eggly, M. Pollack, K.J. Anand, J. Zimmerman, J. Carcillo, C.J. Newth, J.M. Dean, D.F. Willson, and C. Nicholson. 2008. Parents' perspectives on physician-parent communication near the time of a child's death in the pediatric intensive care unit. *Pediatr Crit Care Med* 9: 2–7.

Siriphuwanun, V., Y. Punjasawadwong, W. Lapisatepun, S. Charuluxananan, K. Uerpairojkit, and J. Patumanond. 2014. The initial success rate of cardiopulmonary resuscitation and its associated factors in patients with cardiac arrest within 24 hours after anesthesia for an emergency surgery. *Risk Manag Healthc Policy* 7: 65–76.

Steinberg, L. 2010. A dual systems model of adolescent risk-taking. *Developmental Psychobiology* 52: 216–224.

United Nations General Assembly 1989. *Convention on the rights of the child: Adopted and opened for signature, ratification and accession by general assembly resolution 44/25.* Geneva.

Winter, S.M. 2000. Terminal nutrition: Framing the debate for the withdrawal of nutritional support in terminally ill patients. *American Journal of Medicine* 109: 723–726.

Further Reading

Beauchamp, Thomas and Childress, James. 2009. *Principles of Biomedical Ethics* 6 edn. New York: Oxford University Press.

Wilkinson, Dominic. 2013. *'Best interest and the Carmentis machine', Death or disability ?: the 'Carmentis machine' and decision-making for critically ill children.* Oxford: Oxford University Press.

Chapter 16
Medical Futility in Pediatrics: Goal-Dissonance and Proportionality

I. D. Wolfe and A. A. Kon

Abstract Futility is a concept long fraught with issues over its definition and application. Its normative use rose alongside advances in life-sustaining technology and innovations in medicine. Over the decades, futility discourse has accelerated in the pursuit for definitional clarity. However, multiple cross-national organizations between North America and Europe have come out against the use of the term "futility" for any treatment other than those that are strictly physiologically futile (in contrast to value-based judgements). This chapter explores the history of the normative conceptions of futility, its history, evolution, cross-national discourse, and then presents current complexities specifically relevant to this concept in pediatrics. Finally, we propose a move away from using the term futility, supported by multiple organizations, towards the antecedent problem of goal dissonance and a focus on the proportionality of benefit to burden.

Keywords Futility · Goals of care · Value judgment · Conflict · End-of-life · Proportionality

16.1 Introduction

...[a futility dispute] is not a debate about the permissibility of futile intervention but, rather, struggles over differing conceptions of a 'good enough life,'...Framing these struggles in terms of a futility discourse disregards this inherent strife over trust, respect and power; indeed, futility talk seems to perpetuate these problems further. (Carnevale 1998, p. 515)

Very few words in health care hold both a moral weight and an ambiguity than "futility." Often, when clinicians state that an intervention is "futile" they do not

I. D. Wolfe
Children's Minnesota, Minneapolis, MN, USA
e-mail: ian.wolfe@childrensmn.org

A. A. Kon (✉)
University of California San Diego School of Medicine, San Diego, CA, USA
e-mail: aakon@uw.edu

© Springer Nature Switzerland AG 2022
N. Nortjé and J. C. Bester (eds.), *Pediatric Ethics: Theory and Practice*,
The International Library of Bioethics 89,
https://doi.org/10.1007/978-3-030-86182-7_16

mean that it is literally, or physiologically, futile. Rather, in such cases the intended goals of the intervention fail to comport with what clinicians' feel are the proper ends of medicine. This conception of futility as a disagreement over the value, the proportion of benefit to burden, of the stated goals of treatment has been termed "normative futility" by Youngner (1988) and reaffirmed by Veatch (2013), while Schneiderman, Jecker, and Jonsen coined the term "qualitative futility" to discuss such value-based disagreements (Schneiderman et al. 1990, 1996).

In contrast, physiologically futile medical interventions are ones that simply cannot produce the intended physiologic result (e.g., performing CPR on a patient in rigor mortis is physiologically futile because there is no possibility that this action will restore spontaneous circulation, which is the intended physiologic goal of CPR). This "literal" definition is often referred to as "strict" or "physiologic futility" to conceptually differentiate it from normative futility.

Interventions that are physiologically futile are so clearly incongruous with standard medical treatment that they almost never lead to conflict. Indeed, when disagreements arise regarding the provision of physiologically futile interventions, such disagreements generally center on medical facts rather than on value judgements (i.e., one party believes that the intervention can bring about the intended physiological outcome whereas the other party does not) (Veatch 2013). The overwhelming majority of discourse surrounding the provision of futile care focuses on what clinicians believe to be normatively futile interventions rather than physiologically futile interventions (Truog et al. 1992; Danis et al. 1997; Bagheri 2013a). This is particularly true in pediatrics and suggests that when there is agreement on the proper course of action, there is no conflict, no claims of futility (Carnevale 1998; Wolfe 2019a).

Unfortunately, health care professionals often use the term "futility" when discussing either normative futility or physiologic futility. Such vernacular usage leads to confusion and a misunderstanding of the ethical underpinnings of denying "futile" care when patients or family members request it. The ethics of denying physiologically futile interventions is fundamentally different than the ethics of denying normatively futile interventions.

In this chapter we review relevant history and discourse around normative futility in health care, and specifically in pediatrics. We review international literature around the concept of futility and its use. We then discuss complexities that exist in "futility" disputes, and provide guidance for addressing "futility" in practice. We conclude that using the term "futility" when describing anything other than its literal, physiologic definition has been widely discouraged and argue that the terms "normative futility", "qualitative futility" and "futility" (in the normative sense) are misleading and should be abandoned. Clinicians and scholars should instead focus on the tensions that exist prior to these disputes which center around proportionality of burden to benefit and a dissonance in goals between parents and clinicians. We propose a move further upstream from where normative futility emerges, towards earlier and intentional complex decision-making that more properly balances autonomy and responsibility, interests and harms, hopes and goals.

16.2   History and Discourse

16.2.1   The Beginnings of Medicine

Futility as a term in medicine is often attributed to Hippocrates. He compared the attempt to employ a futile treatment as something akin to madness (Hippocratic Corpus 1977). In this comparison, the central focus is on the act (treatment) in relation to its ability to achieve some result or the powerlessness of medicine to cure a patient overtaken by their disease (Hippocratic Corpus 1977). The powerlessness to see an effect in the context of overwhelming disease often weighs the likely utility with the harms, the proportionality of benefit to burden.

16.2.2   Catholic Contributions

Catholic doctrine from the sixteenth century first began discussions focused on quality of outcome in proportion to the harms of the treatment (Kelly 1958). This is the first known consideration of treatments as ordinary or extraordinary, proportionate or disproportionate (Kelly 1958; Danis et al. 1997). This consideration of treatments in their relativeness to ordinariness and proportionality continues to inform Catholic clinicians and ethicists through the Ethical and Religious Directives on issues that often invoke normative futility (Catholic Church 2018). It is not simply a question of proportional burdens to benefit but also the extraordinary nature of the treatment weighed against the likelihood and impact of the benefit.

16.2.3   The Era of Innovation

The post-WWII era saw significant advances in medical technology. Surgical advancements grew exponentially with the advent of the mechanical ventilator, developed in large part by Dr. Forrest Bird, based on his experiences with military aviation breathing aparati. Pilots were limited in the altitude at which they could fly due to atmospheric pressure. Dr. Bird, an army pilot himself, designed a respirator that would help push air into a pilot's lungs allowing a strategic military advantage. He would later develop this technology into the first positive pressure ventilator, which has saved countless lives and is still used in every ICU today.

   With the ability to control breathing, the ventilator allowed for longer and more invasive surgeries through the ability to use stronger anesthetics. Eventually, patients could even stay on the ventilator after surgery. The innovation spread to non-surgical indications and medical ICUs began sprouting up with the need for higher level and specialized nursing care. The benefits of new technologies like the ventilator were remarkable; so too were the advent of newer drugs and treatments. But these advances began to push the limits of what we could do and clinicians began to voice discomfort with emerging applications for these new technologies that pushed past the usual conception of meaningful recovery and survival allowing instead merely adequate biological survival.

By the late 1950s, there was already international debate about the extent to which we ought to keep patients alive. In 1959, the French doctors Mollaret and Goulon described "coma depasse," a state beyond coma in which patients appeared to be between life and death (Mollaret and Goulon 1959). Their work formed the basis for the later Harvard Medical School ad hoc Committee definition of brain death (Beecher 1968). These milestones marked the progression of the international debate regarding futility and appropriate limits to life-prolonging interventions.

With the new concepts of "coma" and "brain death" patients who did not fit either criteria but were neurologically injured and dependent on technology to breathe and eat fell in an undefined space. Long-term use of technology and medications to keep patients alive increased but not without significant questions and controversy. Clinician discomfort became expressed through the invocation of the word futility.

Discussion of medical futility in academic literature reached ubiquity in the 1980's (Helft et al. 2000). The number of publications with the keyword "futility" rose exponentially through the 1990's (Carnevale 1998). The discourse began to contain an added element of conflict over demands for treatment as the pendulum moved away from medical paternalism to shared decision-making. Attempts to adequately define the concept of futility shifted towards achieving clarity for conflict resolution. No work could be more foundational in this area then that of Lawrence Schneiderman, Nancy Jecker, and Albert Jonsen.

16.2.4   Defining Futility

Schneiderman, Jecker, and Jonsen set a benchmark in medical futility discourse (1990, 1996). The authors set a definitional distinction between quantitative and qualitative futility with the former similar to statistical evaluations where if a medical treatment has not been successful in the last 100 cases then it can be considered futile. The latter, qualitative futility, is specifically focused on medical treatments where the only goal is either to prolong permanent unconsciousness or dependence on intensive medical treatments, such as a ventilator (Schneiderman et al. 1990, 2017). Qualitative futility focused on the moral questions clinicians at the bedside were being confronted with, whether a treatment is likely to achieve "...an effect that the *patient* has the capacity to appreciate as a *benefit*... (Schneiderman 2011, p. 125)."

Quantitative futility was rare at the bedside because the power of its clarity enabled clinicians to say "no." An increase in innovative therapies blurred these boundaries and thus quantitative futility became better described as malpractice or experimentation (Veatch 2013). However, as clearly quantitatively futile treatments typically lack clinical indication, most everything that cannot be identified as such requires subjective assessment. The majority of futility conflicts for this reason center around if and how a person will benefit from a specific treatment. Qualitative futility, as Schneiderman, Jecker, and Jonsen conceptualize it, is more frequently encountered in clinical practice; most notably when there are differing conceptions of what a "good life" is, where exactly that falls on the spectrum of "unconsciousness," and what is an "appreciated benefit."

Despite the value of their contribution to the futility discourse, the utility of Schneiderman, Jecker, and Jonsen's practical definitions have now been contested due to reliance on controversial value judgments and an unobtainable degree of prognostic accuracy (Bosslet et al. 2015). There is a growing realization that defining futility into a technical term with specific meaning might be impossible, or at least not helpful (Bosslet et al. 2015; Kon et al. 2016). The difficulty with defining futility objectively and the desire for practical guidance when and where these situations arise led several professional organizations to publish statements around these issues.

16.3 The Era of Societal Statements

Cases where futility is invoked are relatively rare, accounting for a small number of patients, however, the cases themselves often account for a significant amount of patient-days and resources, mainly in ICUs (Huynh et al. 2013). Futility determinations are complicated and value-laden, often emerging as conflicts. Conflicts involving futility become intractable and the focus has been on mitigation and resolution, rather than prevention (Wolfe 2019a). Clinicians and ethicists began looking into protocols and guidelines to help clinicians, patients, and families in these situations. Guidance about clinical issues related to conceptions of futility emerged most notably from critical care professional organizations (American Thoracic Society 1991; Danis et al. 1997; Bossaert et al. 2015; Bosslet et al. 2015; Kon et al. 2016). Many of these statements emerged generally, not specific to any particular age group. Later, pediatric and neonatal guidance began to emerge.

16.3.1 Adult/General

One of the first policy statements came from the American Thoracic Society (ATS) in 1991 with the purpose of defining acceptable standards of medical practice and making recommendations around withholding and withdrawing life-sustaining treatments for adults with or without decision-making capacity (it was noted that it might be useful in some children but neonates were excluded) (American Thoracic Society 1991, p. 478). The objectives of this statement were to (1) enhance understanding about the issues involved with withholding and withdrawing treatments, (2) promote medically and ethically sound decision-making related to life-sustaining care, and (3) assist in the development of hospital and public policies (American Thoracic Society 1991, p. 478).

The ATS statement's definition of futility is similar to Schneiderman, Jecker, and Jonsen's quantitative and qualitative futility; lack of medical efficacy and lack of meaningful survival, respectively (American Thoracic Society 1991). The statement invokes beneficence and nonmaleficence in concluding that "… the purpose of a life-sustaining intervention should be to restore or maintain a patient's wellbeing… not as its *sole* goal the unqualified prolongation of a patient's biological life (American Thoracic Society 1991, p. 481)." The authors go on to state that a "… life-sustaining intervention is *futile* if reasoning and experience indicate that the intervention would be highly unlikely to result in a *meaningful survival* for that patient (American Thoracic Society 1991, p. 481)."

The next published consensus statement specifically addressing futility was from the Society of Critical Care Medicine (SCCM) Ethics Committee in 1997. It called for retiring the word "futility" from clinical decisions and conversations (Danis et al. 1997). SCCM stated that the concept of futility is not useful because strictly futile treatments are rare, not usually offered, and not disputed. The committee concluded that treatments that are extremely unlikely to be beneficial, extremely costly, or have questionable benefit should be considered "inappropriate" and hence "inadvisable" (Danis et al. 1997).

In 2015, the five leading critical care organizations in North America and Europe came together to publish a consensus statement regarding the issue of futility (Bosslet et al. 2015). SCCM, ATS, the American Association of Critical Care Nurses (AACN), the American College of Chest Physicians (ACCP), and the European Society of Intensive Care Medicine (ESICM) published this statement in the American Journal of Respiratory and Critical Care Medicine. This statement (herein referred to as the "multiorganization statement") reiterated the sentiment from the 1997 SCCM statement, the term futility should be used to describe only those interventions that cannot accomplish the intended physiologic goal. They called for the term "potentially inappropriate" to be used to describe treatments that have at least some chance of accomplishing the effect sought, but clinicians believe that competing ethical considerations justify not providing them (Bosslet et al. 2015). The statement provides guidance regarding due process and an ethically supportable and robust dispute resolution process.

The 2015 European Resuscitation Council Guidelines for Resuscitation put forward a section on the ethics of resuscitation and end-of-life decisions (Bossaert et al. 2015). In it, they use the definition of medical futility from the World Medical Association (WMA) which defines futile treatments as ones without reasonable hope of recovery or improvement, and the patient is in a permanent state where they cannot experience any benefit from it (Bossaert et al. 2015, p. 303). The authors point out the harm to the family that might come from starting a futile treatment by instigating false hope and undermining autonomy.

Shortly after the multiorganization statement, the Ethics Committee of SCCM published further guidance (Kon et al. 2016). Because the multiorganization statement provided no guidance regarding what should, and what should not, be considered inappropriate treatment, and because the dispute resolution process could be significantly truncated (or entirely abandoned) in emergency situations (Bosslet et al.

2015), SCCM published guidance detailing general principles of what may be considered inappropriate treatment to ensure some measure of consistency in patient care particularly in emergency situations. The 2016 SCCM statement drew heavily from the earlier 1997 SCCM statement as well as from the work of Schneiderman, Jecker and Jonsen. SCCM stated that "ICU interventions should generally be considered inappropriate when there is no reasonable expectation that the patient will improve sufficiently to survive outside the acute care setting, or when there is no reasonable expectation that the patient's neurologic function will improve sufficiently to allow the patient to perceive the benefits of treatment" (Kon et al. 2016, p. 1771). SCCM continued, stating that this definition was neither obligatory (i.e., clinicians may provide ICU treatment is such circumstances if they believe it to be appropriate) nor exhaustive (i.e., clinicians may deem treatments to be inappropriate when the definition is not met; however, they should complete the full dispute resolution process in such cases).

These guidelines were written with clinicians from a multitude of disciplines and specialties as well as academic scholars, ethicists, and philosophers, and were presented towards a more general patient age population, but there are some very important nuances that must be considered when discussing issues of futility in children and neonates.

16.3.2   Pediatric and Neonates

An added complexity in futility disputes in pediatrics and neonatology is the state's role and obligation of *parens patriae* (see Chap. 26). The state has a paternalistic interest over children, separate from parents. Generally, this is meted out through limits of parental authority as protection from harms and/or promotion of interests (depending on the country). Many of the issues dealing directly with medical futility and authority in pediatrics actually came out of an influential chapter in neonatology.

In the U.S., the Baby Doe case led to significant changes in end-of-life care and the concept of futility in the care of newborns, and this legacy persists today (Lantos and Meadows 2006). Baby Doe was born in 1982 and was diagnosed with trisomy-21 and tracheoesophageal fistula (TEF). The pediatrician advised that TEF is easily repairable and many individuals with trisomy-21 lead fulfilling lives, therefore the child should undergo TEF repair surgery. The parents opted to follow the advice of the obstetrician, declined surgery due to the diagnosis of trisomy-21. The case was brought to court, and the court ruled that because there was no clear medical consensus, the parents have the authority to choose. Baby Doe died when he was one week old.

News of Baby Doe's death became widespread and the Reagan administration promulgated new "Baby Doe Regulations." The Baby Doe Regulations significantly limited physician and parent decision-making regarding limiting or withdrawing life-prolonging interventions in newborns. Further, the regulations set up a telephone line, manned at all times, where people who suspected that infants were being mistreated could call a "Baby Doe Squad" to the hospital to assess the baby's care. Baby Doe Squads had significant power including the ability to cut federal funding to hospitals and prosecute physicians when they believed the rights of disabled infants were being violated.

In a series of court cases in 1983 and 1984 the Baby Doe regulations were struck down. The Reagan administration revised the regulations; however, in 1986 the U.S. Supreme Court again ruled against it and permanently shut down the regulations. In lieu of the federal regulations, the U.S. Congress amended the Child Abuse Protection and Treatment Act (CAPTA) to state that physicians must provide treatment (including "appropriate nutrition, hydration, and medication") unless the infant is irreversibly comatose, the treatment would merely prolong dying, the treatment would not be effective in treating all of the infant's life-threatening conditions, the treatment would be futile in terms of survival, or the provision of such treatment would be virtually futile in terms of survival and the treatment itself would be inhumane.

This amended section of CAPTA is interpreted conservatively at many facilities (i.e., life-prolonging interventions are withheld from neonates only when the specific criteria are met). However, many other facilities use a more liberal interpretation (i.e., any intervention, including nutrition, hydration, and medications, is not "appropriate" when the physician and parents agree that it is not appropriate; therefore, physicians and parents may limit or withdraw any life-prolonging interventions when they believe that doing so is appropriate), and this liberal interpretation is consistent with the position of the American Academy of Pediatrics (Committee on Bioethics 1996) and has been supported by law-enforcement and child protective services (Paris and O'Connell 1991).

Many neonatal ICUs continue to limit or withdraw life-prolonging interventions only in cases that meet the strict CAPTA criteria. In contrast, because pediatric critical care medicine is a much newer field than neonatology, few pediatric intensivists currently practicing were working at the time of the Baby Doe Squads. Further, because the CAPTA restrictions are limited to newborns, most pediatric intensivists are not curtailed by CAPTA. For this reason, in the U.S. there remains significant variability in how newborns are treated at different facilities, and there are significant differences between the care of newborns in neonatal ICUs and the care of infants and children in pediatric ICUs (PICU).

Guidelines on end-of-life decisions in neonatology issued in the 1990's by the Dutch Association of Pediatrics and the Dutch Medical Association put forward two categories to determine if life-prolonging treatment is inappropriate (Moratti 2010). The first category is "impossible," meaning there is no chance of survival. The second category is if the treatment is futile (Moratti 2010). Here futile is defined as the "… expectations for the baby's future are so poor that treatment would be pointless (Moratti 2010, p. 4). The current and predictable future condition should be taken into account (Moratti 2010).

In the "Ethical charter of the Union of European Neonatal and Perinatal Societies," a futile treatment is one that would be "highly unlikely to result in a meaningful survival for the patient (Guimaraes et al. 2011, p. 857). This European charter draws heavily from the 1991 ATS statement's focus on futility defined through the chances of "meaningful survival" in navigating the difficult spaces around resuscitation, withholding and withdrawing, and litigation concerns. The European charter holds futility as an acceptable criterion in forgoing intensive life-sustaining care (Guimaraes et al. 2011, p. 857). They make a distinction that their focus is on the future possibility of the infant's consciousness, not on disability, as a reason to withhold intensive therapy.

The phenomenon of medical futility, or perception and discussion of it, is not limited to the U.S. and Europe. The normative conception of futility is generally relative to the level of resources a country has, however, discussion of it in what are typically considered "resource-poor" countries has been growing. There are likely a number of factors other than resources that influence how futility is conceptualized, defined, and viewed within each country.

16.4 Futility Discourse Around the World

Discourse is often difficult to discern as arising from within or without countries and regions. Language and context certainly influence this, as does the permeation of Western medicine and culture around the world. Much of the futility discourse likely comes from Western bioethics literature which rose in voice along with the rise in advanced life-sustaining technologies. Many of the differences around the phenomenon of futility between countries and cultures are due to the way autonomy and authority are distributed between patients and their families, the state, and the level of paternalism in medicine.

Many countries, notably the U.S. and Canada, were founded on ideas of individualism, liberty, and the pioneering spirit. From this there is a strong sense of autonomy over one's body and one's children's bodies. This authority is not unlimited though, and as in many countries the state has an interest in protecting children and vulnerable people. Futility discourse in pediatrics in this sense has progressed often as a conflict between parental autonomy and the integrity of clinical practice.

The 2013 book "Medical Futility: A Cross-national Study" provides a collection of perspectives regarding futility from many countries (Bagheri 2013a). Prior to and since that book's publication, there has been a growth in the literature regarding futility from around the world, present in every region and continent, save Antarctica.

16.4.1 Canada, Europe, and Australia

A review of the policies and attitudes around end-of-life decisions in Europe and North America found two differences. North American clinicians utilize more standard and formal procedures and use relatives as rightful decisional surrogates versus a more paternalistic physician decision-making approach in Europe (Moselli et al.

2006). In the U.S., autonomy is given very strong force, based on the uniquely American ethic, or interpretation of, individualism and self-governance. In Europe, social relationships and obligations hold far more sway than individualism. Doctors are given far more authority to make decisions (Carlet et al. 2003).

Canada is aligned through the Commonwealth and thus is generally more similar in social discourse to the U.K. and Europe, but its proximity to the US, and likely colonial and pioneering foundations, renders it susceptible to the diffusion of U.S. ideas that inform how futility is viewed. Much of the Canadian discourse has developed around similar court cases, societal statements, definitional clarity (quantitative or qualitative, futile or inappropriate) and debates around CPR and other end-of-life issues (College of Physicians and Surgeons of Ontario 2015; Downar et al. 2016; Kyriakopoulos et al. 2017; Downar et al. 2019; Vivas & Carpenter 2020).

The most recent and influential writing on futility in the U.K. comes from Wilkinson and Savulescu (2018). The authors frame their most recent discussion on the Charlie Gard case paying attention to the language the courts used and the antecedents of futility, differing views of benefit and burden. Their analysis is similar to the multiorganizational statement (which included European and American organizations), that the use of the word futility doesn't apply and is not helpful (Wilkinson and Savulescu 2018). While the Justices in the Gard case referenced both the medical and legal definition of futility, Wilkinson and Savulescu argue that the only clear definition of futility is one that is based on physiologic possibility (2018, p. 29). Furthermore, these authors argue that physiologic futility would be the only helpful definition of the concept, and that this was not present in the U.K. cases that have come in front of the courts, such as Gard (Wilkinson and Savulescu 2018, p. 30).

In Europe, the issue of futility had a different trajectory than in the U.S. One reason for this is the difference in the legal standardization of surrogate decision-making. The U.S. sees surrogate decision-making as a continuance of patient individualism and autonomy when the patient is incapacitated. In many European countries surrogate decision-making is less formalized. In Britain, France, and Sweden when a patient is no longer competent or able to make decisions for themselves, the physician claims decisional authority, not the surrogate (Moselli et al. 2006). No disagreement, no futility disputes.

An Australian qualitative survey of adult physicians found that all of the respondents relied on qualitative approaches to defining futility with patient benefit at the center (White et al. 2016). The study participants engaged familiar concepts of futility supported by the Australian Medical Association that defines futile treatment as having no, or only a small chance of prolonging meaningful survival or benefit (White et al. 2016). However, other work on futility in Australia finds similar issues that are confronted in the U.S. (Martin 2013), driven by a technological optimism that drives the hopes of patients and families that there is always something more that *can* be done.

In a study interviewing Dutch neonatologists around what considerations were important to them when considering a treatment "medically futile," the three top criteria were communication (ability to), mental condition, and motor condition

(Moratti 2010, p. 8). This indicates a more normative qualitative conception of futility.

Belgium and other European countries with legal physician-assisted suicide are unique because of their ability to essentially accept something as futile in the normative sense, with agreement even from the patient (Bernheim et al. 2013). In Switzerland, medical futility is thought of at a social and political level which incorporates accepted moral standards of the country's medical academy and nursing association (Krones and Monteverde 2013). These European countries are less concerned with specificity and objectivity in their definitions of futility then they are with balancing professional practice and appropriate treatment with societal resources and patient benefit.

16.4.2 Middle East

In Turkey, physicians have historically taken a more paternalistic approach to declaring when something was medically futile (Kalkan and Mirici 2018). Patient and family involvement in decision-making is reportedly growing. It is considered unacceptable to withdraw any treatment or intervention that will result in the patient's death, even if they request it (Arda and Aciduman 2013). However, a physician's determination of medical futility is sufficient for unilateral do-not-resuscitate decisions (Kalkan and Mirici 2018). There is no clear definition, guidance, nor consensus around futility, nor are there laws or guidelines about nor condoning withdrawal or refusal of treatment in children, even if the patient or family expresses wishes to stop (Arda and Aciduman 2013).

A review article from two Saudi Arabian authors reference the 2015 multiorganization statement in discussing and examining futility (Ali and Hassan 2018). The authors discuss the overestimation of survival and unwillingness to declare imminent death by Saudi physicians as partly due to a lack of experience in distinguishing end of life issues. They note that in Saudi Arabia, futile treatments are often demanded by patients and family members which creates conflict. They note that Islamic views on futile care are in great dispute among Islamic Scholars (Ali and Hassan 2018). They note that despite Islamic teachings that have been used by Muslim jurists to allow the withdrawal of futile treatment, the belief in enduring suffering and God's power to heal and cure despite feelings of futility continues to inspire conflict (Ali and Hassan 2018, p. 15).

Controversy around defining futility in Iran appears similar to that of the U.S. and Europe. The balance of power between physicians and patients, paternalism and autonomy, plays a similar role in Iran (Bagheri 2013b). Other Middle Eastern countries, such as the United Arab Emirates, follow Saudi Arabia and generally adhere to Islamic rules when it comes to end-of-life issues (Abuhasna and Al Obaidli 2013).

16.4.3   Asia

Medical futility as a term is reportedly absent from health care in the Russian Federation (Kubar et al. 2013). Issues that invoke futility in other parts of the world, namely end-of-life issues are exclusively conceptualized through palliative medicine incorporating "… all ethical, social, legal and medical aspects associated with end-of-life treatment … (Kubar et al. 2013, p. 101)."

In a 2006 survey of Japanese bioethicists, 67.6% of respondents felt that refusal to continue treatment based on a futility judgment was never justifiable (Bagheri et al. 2006). The study concluded that despite Japanese health care being considered medically paternalistic and while medical futility is a growing issue in Japan, unilateral decision-making was not justified (Bagheri et al. 2006). Modern Japanese bioethics emerged in the 1980s triggered by the issues of brain death and organ transplantation and was largely influenced by the growing ethical debates in Western medicine (Akabayashi and Slingsby 2003). In fact, most Japanese health care workers when asked to identify medical futility did so in cases that involved aggressive or life-sustaining treatments either in around end-of-life or in the presence of neurological impairment (Kadooka and Asai 2013).

In a more recent survey of current practices and barriers around end-of-life care in twenty-seven Japanese PICUs, physicians reported fear of lawsuits related to withholding or withdrawing which the authors relate to one particularly bad case that went public as well as confusion among physicians regarding end-of-life practices (Seino et al. 2019). The authors suggest that due to these factors "… life-sustaining support for the critically ill is still routinely applied in Japan, even when it is medically futile (Seino et al. 2019, p. 863)."

In Korea, end-of-life issues were traditionally taboo and ethical duties of both physicians and families were to preserve life in any condition (Kwon 2013, p. 189). With the rise in the population of elderly and advances in medical technology unnecessary ICU treatment increased, partly instigated by families due to the unfamiliarity of physicians with the concept of patient autonomy (Kwon 2013). In 2009, guidelines were published by the Korean Association of Hospitals which advised that futile medical treatment as determined by medical judgment or the patient, could be withdrawn (Kwon 2013).

In China, medical futility is dealt with primarily under hospice care, similar to the Russian Federation (Shi et al. 2013). A 2014 survey of health care providers around "terminal care," including questions about medical futility, found that most of the respondents thought these issues should be joint discussions between patients and families, with only 5.6% suggesting unilateral decision-making (Pazooki 2014). Many participants thought there should be more central policy and guidance on issues of medical futility.

16.4.4 Africa

A 2019 original research paper by Onyeka et al. published in the University of Toronto Medical Journal, reports a phenomenological study from Nigeria on resident physician perceptions of futility in a "low-resource ICU." They found that participants commonly viewed care to be futile in the same way other countries did, though there was more concern about financial hardships on surviving families of patients (Onyeka et al. 2019).

In South Africa, the 2008 Health Professions Council guidelines on withdrawing and withholding treatment, futile treatment is mentioned but not defined. The guidelines suggest that families may be given the option to transfer the patient to another facility, but that the treatment can be withdrawn legally and ethically against family wishes by the health care team as long as futility is confirmed by an independent practitioner (Health Professions Council of South Africa 2008; McQuoid-Mason and Naidoo 2019).

16.4.5 South America

One Argentinian study about forgoing life-sustaining treatment in PICUs found that most children received CPR prior to death and that family involvement in decisions around withholding and withdrawing interventions was low, suggesting a more paternalistic approach (Althabe et al. 2003). There is little published scholarship about futility in South America but there are also large inequities and disparities in the level of resources available, and access to those resources (Piva et al. 2005). (We could only find an article from Central America which was about "therapeutic futility" in regard to medications, not normative futility).

Two authors in Brazil have noted that the term dysthanasia is used over "futile treatment" which is more common in Anglo-Saxon countries (Pessini and Hossne 2013). These authors discuss the obsession with pursuing life at all costs leading to a therapeutic obstinacy, which has led to the popularization of "orthothanasia" which translates to a "good," "natural," or "correct" death in Brazil (Pessini and Hossne 2013, p. 36).

Writing from the Venezuelan perspective, d'Empaire notes that the meaning of the Spanish word for futility, "futilidad," translates more accurately to mean something of little importance (2013, p. 87). Venezuela addresses physician obligations through several codes directing medicine that focus more on the relief of suffering over prolongation of life. The most recent of these codes sets this relief of human suffering as the primary obligation for physicians and in this theme advises that terminally ill patients should not be subjected to technologic life support measures (d'Empaire 2013, p. 94).

While much ink in the West has been spilled trying to define futility into something pragmatic, many other countries such as Brazil, China, and the Russian Federation have naturally evolved their thought around normative futility into something much more practical. While thought evolves naturally out of culture we might examine these examples as guiding lights amidst the fog of the complexity current high-profile cases have set us in.

16.5   Approaching Futility in Practice

Several recent high-profile U.S. and U.K. court cases have brought the discourse around medical futility to the international stage. U.K. courts are more willing to override parental authority (Miller-Smith et al. 2019; Vizcarrondo 2019), as was seen in the cases of Charlie Gard in 2017 and Alfie Evans in 2018. In the U.S., courts often attempt to appease both parents and providers as was seen in the Jesse Koochin case in 2004 and the Jahi McMath case in 2014. When unable to appease both sides, U.S. courts have at times sided with parents (e.g., the Allen Callaway case in 2016) and at other times with providers (e.g., Hunt v DFS in Delaware, original ruling in the Tinslee Lewis case). In general, the custom and practice in the U.S. is to transfer such patients if there is another facility able and willing to provide the treatment desired by the parents (Bosslet et al. 2015).

Legal protection strategies have been attempted in the U.S., most notably in Texas, California, and Virginia (Lantos 2018a). Many of these policies follow a deliberative process that gives notice, uses the institutions ethics committee, and allows time for parents to find an institution to transfer to if they disagree. While the U.K. generally lets these cases go to court where judges often side with the hospital, U.S. hospitals, especially pediatric ones, are hesitant to let these cases slip into the public domain (Miller-Smith et al. 2019, p. 79).

In many countries there is a strong, rebuttable assumption that parents should have authority to make medical decisions for their children, even if those decisions are sub-optimal but not harmful. While there is concern around decisional burdens put on parents, particularly when asked to make life and death decisions (Clark and Dudzinski 2013), the few studies available suggest parents still want to have the authority. One study found that even when making these decisions have left parents with significant long-term distress, parents still feel they are the ones that should make them (Botti et al. 2009). In contrast, other societies in Europe, Middle East, and historically in Japan, there is an expectation that clinicians make decisions around end-of-life (refer to Chap. 1).

Clearly differentiating between physiologic futility and disagreements regarding appropriate goals of care (i.e., potentially inappropriate treatments) is essential when considering the ethical issues surrounding futility in patient care. Unfortunately, many clinicians use the term "futile" to refer to both categories. This puts the term "futility" front and center, distracting clinicians and parents from finding consensus.

Many of these conflicts center around differences of opinion regarding appropriate goals of treatment. Parental goals and clinical goals often coalesce. When they don't, conflicts emerge that, if not dealt with carefully, can become intractable

(Lantos 2018b; Wolfe 2019b). Goal dissonance may happen suddenly or gradually with changes in a child's condition. Often, conflicts develop due to suboptimal communication. There are several strategies that can be useful in avoiding such conflicts and resolving conflicts that have already developed.

16.5.1 Family Meetings

SCCM recommends regularly scheduled family meetings utilizing a standardized approach to communication (Davidson et al. 2017). Many clinicians may require education and training in communication techniques in order to successfully lead such discussions. Meeting with parents early in the child's hospitalization, within the first 48 h, and then at least twice weekly is recommended by SCCM and can help keep parents and clinicians on the same track with common goals (Davidson et al. 2007).

16.5.2 Re-goaling

Parents of ill or injured children who progress to a state where cure is no longer possible often have difficulty transitioning their goals of care and maintaining hope. Clinicians need to help parents make such transitions. Concordance around medical problems and hopes between parents and clinicians has been shown to be lower than between parents (Hill et al. 2015). This suggests that purposefully attending to hope at the outset might improve goal setting through more open and meaningful communication (Feudtner 2010; Feudtner 2014; Rosenburg and Feudtner 2016).

Hill et al. (2014) present the concept of "re-goaling" to describe a process of transitioning to different goals of care in these situations. Re-goaling contains the process of disengagement of old goals and reengagement of new ones throughout changes in a child's care (Hill et al. 2014).

16.5.3 Innovation in Choice Architecture

Another strategy concerns how choices are presented to parents. There is a growing appreciation for the complexity of certain decisions and how those factors affect where they fall on the spectrum between autonomy and paternalism.

Morrison et al. (2018) specifically focuses on decision-making in pediatrics with their approach called "titrated clinician directiveness." Their well-developed approach offers a spectrum of directiveness depending on the context, from low to near-complete (Morrison et al. 2018, p. S180). Low degrees of directiveness are appropriate when there are reasonable and medically appropriate choices for parents to make. Near-complete directiveness are appropriate when there are no medically sound decisions to be made and the medical team is unanimous that a treatment, though technically possible, is non-beneficial and potentially harmful (Morrison et al. 2018).

Clinicians are under no obligation to present medically inappropriate options to parents. Parents shouldn't have to make those choices which might negatively impact their sense of hope and duty. Maintaining common goals is an important part of Morrison et al.'s approach (2018). We believe that the purposeful attention to shifting goals through re-goaling, and the use of titrated clinician directiveness can help avoid conflicts that invoke futility. These approaches serve the purpose of maintaining communication and shared goals with parents while preserving clinical integrity to appropriate and ethical practice.

16.6   Guiding Principles

Approaching Futile and Inappropriate Treatment
- The term "futility" should only be applied to treatments that cannot strictly meet its physiologic goal
- Where a treatment provides no benefits, or benefits that are not outweighed by the burdens, clinicians should not use the term "futile" but instead focus on communication
- Inappropriate treatments should be differentiated from physiologically futile treatments and discussion with parents and the care team should be pursued to deliberate goals of care and what appropriate treatments might be

Communication, Goals, and Decision-Making
- Communication is an essential skill for clinicians and is an important part of care particularly in complex pediatric illness
- Frequent and regular communication should occur with families
- Careful and explicit focus should be taken in partnering on goals of care, and re-goaling should occur as a child's conditions or treatments change or shift
- Shared decision-making methods and choice architecture are extremely important skills for clinicians but are deemphasized when the concept of futility is invoked and resolutions are sought through policy and procedure

16.7   Conclusion

There is broad agreement that physiologically futile interventions should not be performed even when parents demand them (Bosslet et al. 2015). Disagreements regarding such interventions are rare and generally center around disagreements of medical facts rather than values. How to handle disagreements regarding what constitutes appropriate of goals of treatments is more complex and varies widely. Some scholars argue that defining futility is not possible (Brody and Halevy 1995;

Halevy and Brody 1996; Wilkinson and Savulescu 2018, p, 29). Many scholars and professional organizations have reframed the debate and published clear and useful definitions and guidelines for use in clinical practice (American Thoracic Society 1991; Danis et al. 1997; Bossaert et al. 2015; Bosslet et al. 2015; Kon et al. 2016).

Global differences in the conceptualization of futility are most pronounced when comparing China and the Russian Federation with non-Asian countries. In China and the Russian Federation, futility is conceptualized as part of end-of-life care rather than as a separate entity (Kubar et al. 2013; Shi et al. 2013). Though further comparative analysis would be needed, it appears that the focus in China and the Russian Federation is on what can be done, palliative and hospice care, rather than disagreement over what ought to be done as in most other countries. Brazil also takes a unique approach in their use of terms such as "orthonasia" that balances the limits of treatments, interests and harms, with an acceptance and consideration around how one dies (Pessini and Hossne 2013, p. 36).

However, the way futility is dealt with through palliative care in the Russian Federation, and through hospice in China, suggests that it is more than simply clinical paternalism, rather a cultural concordance of goals with consideration to the proportionality of benefit to burden.

When parents and health care professionals disagree about appropriate goals of treatment, there are significant differences in how such cases are handled across the globe. In some countries, parental wishes are generally followed. In others, physicians simply do what they believe is best. In others, courts often are involved. In still others, disagreements are generally handled at the institutional level with a dispute resolution process. Approaches to futility disagreements vary considerably but most focus on conflict resolution rather than prevention. There is a need to focus on issues of trust, respect, and power that are inherent to these difficult situations in the care of critically or terminally ill children. There is wide agreement that early and excellent and consistent communication between clinicians and parents in the clinical setting is essential in order to avoid such conflicts before they arise.

References

Abuhasna, S., and A.A. Al Obaidli. 2013. Medical futility in the United Arab Emirates. In *Medical futility: A cross-national study*, 247–261. World Scientific.

Akabayashi, A., and B.T. Slingsby. 2003. Biomedical ethics in Japan: The second stage. *Cambridge Quarterly of Health Care Ethics* 12: 261.

Ali, A.M., and C.P. Hassan. 2018. Futility of medical treatment. *International Journal of Human and Health Sciences* 2 (1): 13–17.

Althabe, M., G. Cardigni, J.C. Vassallo, D. Allende, M. Berrueta, M. Codermatz, et al. 2003. Dying in the intensive care unit: Collaborative multicenter study about forgoing life-sustaining treatment in Argentine pediatric intensive care units. *Pediatric Critical Care Medicine* 4 (2): 164–169.

American Thoracic Society. 1991. Withholding and withdrawing life-sustaining therapy. *Annals of Internal Medicine* 115 (6): 478–485. https://doi.org/10.7326/0003-4819-115-6-478.

Arda, B., and A. Aciduman. 2013. Medical futility in Turkey. In *Medical futility: A cross-national study*, 227–245. World Scientific.

Bagheri, A., A. Asai, and R. Ida. 2006. Experts' attitudes towards medical futility: An empirical survey from Japan. *BMC Medical Ethics* 7 (1): 8.

Bagheri, A. 2013a. *Medical futility: A cross-national study.* World Scientific.

Bagheri, A. 2013b. Medical futility in Iran. In *Medical futility: A cross-national study.* World Scientific.

Beecher, H.K. 1968. A definition of irreversible coma: Report of the Ad Hoc Committee of the Harvard Medical School to examine the definition of brain death. *Journal of the American Medical Association* 205 (6): 337–340.

Bernheim, J.L., T. Vansweevelt, and L. Annemans. 2013. Medical futility and end-of-life issues in Belgium. In *Medical futility: A cross-national study*, 59–83. World Scientific.

Bossaert, L.L., G.D. Perkins, H. Askitopoulou, V.I. Raffay, R. Greif, K.L. Haywood, et al. 2015. European Resuscitation Council Guidelines for Resuscitation 2015: Section 11. The ethics of resuscitation and end-of-life decisions.

Bosslet, G.T., T.M. Pope, G.D. Rubenfeld, B. Lo, R.D. Truog, C.H. Rushton, et al. 2015. An official ATS/AACN/ACCP/ESICM/SCCM policy statement: Responding to requests for potentially inappropriate treatments in intensive care units. *American Journal of Respiratory Critical Care Medicine* 191 (11): 1318–1330. https://doi.org/10.1164/rccm.201505-0924ST.

Botti, S., K. Orfali, and S.S. Iyengar. 2009. Tragic choices: Autonomy and emotional responses to medical decisions. *Journal of Consumer Research* 36 (3): 337–352.

Brody, B.A., and A. Halevy. 1995. Is futility a futile concept? *Journal of Medicine and Philosophy* 20 (2): 123–144. https://doi.org/10.1093/jmp/20.2.123.

Carlet, J., L.F. Thijs, and M. Antonelli. 2003. Statement of the fifth international consensus conference in critical care: Challenges in end-of-life care in the ICU. *Intensive Care Medicine* 30: 770–784.

Carnevale, F.A. 1998. The utility of futility: The construction of bioethical problems. *Nursing Ethics* 5 (6): 509–517. https://doi.org/10.1191/096973398670048722.

Catholic Church. 2018. *Ethical and religious directives for Catholic* health care *services*, sixth edition. Washington, D.C.: United States Conference of Catholic Bishops. https://www.usccb.org/about/doctrine/ethical-and-religious-directives/upload/ethical-religious-directives-catholic-health-service-sixth-edition-2016-06.pdf.

Clark, J.D., and D.M. Dudzinski. 2013. The culture of dysthanasia: Attempting CPR in terminally ill children. *Pediatrics* 131 (3): 572–580.

College of Physicians and Surgeons of Ontario. 2015. Planning for and providing Quality End of Life Care, Policy Statement #4–15, September 2015. http://www.thaddeuspope.com/images/ONT_CPSO_End-of-Life.pdf.

Committee on Bioethics. 1996. Ethics and the care of critically ill infants and children. *Pediatrics* 98 (1): 149–152.

Hippocratic Corpus, The Art. (1977). In *Ethics in Medicine Historical Perspectives and Contemporary Concerns*, ed. Reiser S.J., A.J. Dyck, and W.J. Curran, 6–7. Cambridge: MIT Press.

d'Empaire, G. 2013. The concept of medical futility in Venezuela. In *Medical futility: A cross-national study*, 85–97. World Scientific.

Danis, M., M. Devita, M.A. Baily, D. Beyda, D. Chalfin, F. Dagi, et al. 1997. Consensus statement of the Society of Critical Care Medicine's Ethics Committee regarding futile and other possibly inadvisable treatments. *Critical Care Medicine* 25 (5): 887–891.

Davidson, J.E., R.A. Aslakson, A.C. Long, K.A. Puntillo, E.K. Kross, J. Hart, et al. 2017. Guidelines for family-centered care in the neonatal, pediatric, and adult ICU. *Critical Care Medicine* 45 (1): 103–128. https://doi.org/10.1097/CCM.0000000000002169.

Davidson, J.E., K. Powers, K.M. Hedayat, M. Tieszen, A.A. Kon, E. Shepard, et al., American College of Critical Care Medicine Task Force 2004–2005, Society of Critical Care Medicine. 2007. Clinical practice guidelines for support of the family in the patient-centered intensive care unit: American College of Critical Care Medicine Task Force 2004–2005. *Critical Care Medicine* 35(2): 605–622.https://doi.org/10.1097/01.CCM.0000254067.14607.EB.

Downar, J., M. Warner, and R. Sibbald. 2016. Mandate to obtain consent for withholding nonbeneficial cardiopulmonary resuscitation is misguided. *Canadian Medical Association Journal* 188 (4): 245–246.

Downar, J., E. Close, and R. Sibbald. 2019. Do physicians require consent to withhold CPR that they determine to be nonbeneficial? *Canadian Medical Association Journal* 191(47): E1289 LP–E1290. https://doi.org/10.1503/cmaj.191196.

Feudtner, C. 2010. Taking care of hope. *AJOB* 10 (5): 26–27.

Feudtner, C. 2014. Responses from palliative care: Hope is like water. *Perspectives in Biology and Medicine* 57 (4): 555–557.

Guimaraes, H., M. Sanchez-Luna, C.V. Bellieni, G. Buonocore, and Uenps. 2011. Ethical charter of union of European neonatal and Perinatal societies. *The Journal of Maternal-Fetal & Neonatal Medicine* 24 (6): 855–858.

Halevy, A., and B.A. Brody. 1996. A multi-institution collaborative policy on medical futility. *JAMA* 276 (7): 571–574.

Health Professions Council of South Africa. 2008. Booklet 12: Guidelines on withholding and withdrawing treatment. Pretoria: HPCSA. http://www.hpcsa.co.za/Uploads/editor/UserFiles/downloads/conduct_ethics/Booklet7.pdf.

Helft, P.R., M. Siegler, and J. Lantos. 2000. The rise and fall of the futility movement. *New England Journal of Medicine* 343 (21): 1576–1577.

Hill, D.L., V. Miller, J.K. Walter, K.W. Carroll, W.E. Morrison, D.A. Munson, et al. 2014. Regoaling: A conceptual model of how parents of children with serious illness change medical care goals. *BMC Palliative Care* 13 (9): 1–8.

Hill, D.L., V.A. Miller, K.R. Hexem, K.W. Carroll, J.A. Faerber, et al. 2015. Problems and hopes perceived by mothers, fathers and physicians of children receiving palliative care. *Health Expectations: An International Journal of Public Participation in Health Care and Health Policy* 18(5): 1052–1065.

Huynh, T.N., E.C. Kleerup, J.F. Wiley, T.D. Savitsky, D. Guse, B.J. Garber, and N.S. Wenger. 2013. The frequency and cost of treatment perceived to be futile in critical care. *JAMA Internal Medicine* 173 (20): 1887–1894.

Kadooka, Y., and A. Asai. 2013. Medical futility in Japan. In *Medical futility: A cross-national study*, 145–162. World Scientific.

Kalkan, E.A., and A. Mirici. 2018. Opinions of chest physicians about the do-not-resuscitate (DNR) orders: Respect for patient's autonomy or medical futility? *Journal of Medical and Surgical Intensive Care Medicine* 9 (2): 34–39.

Kelly, G. 1958. *Medico-moral problems*. St. Louis, Missouri: Catholic Hospital Association of the United States and Canada.

Kon, A.A., E.K. Shepard, N.O. Sederstrom, S.M. Swoboda, M.F. Marshall, B. Birriel, and F. Rincon. 2016. Defining futile and potentially inappropriate interventions: A policy statement from the society of critical care medicine ethics committee. *Critical Care Medicine.* https://doi.org/10.1097/CCM.0000000000001965.

Krones, T., and S. Monteverde. 2013. Medical futility from the Swiss perspective. In *Medical futility: A cross-national study*, 205–226. World Scientific.

Kubar, O.I., G.L. Mikirtichian, and M.I. Petrova. 2013. Medical futility in the Russian Federation. In *Medical futility: A cross-national study*, 99–117. World Scientific.

Kwon, I. 2013. Medical futility in Korea. In *Medical futility: A cross-national study*, 181–203. World Scientific.

Kyriakopoulos, P., M. Fedyk, and M. Shamy. 2017. Translating futility. *Canadian Medical Association Journal* 189 (23): E805–E806. https://doi.org/10.1503/cmaj.161354.

Lantos, J. 2018. Intractable disagreements about futility. *Perspectives in Biology and Medicine* 60 (3): 390–399.

Lantos, J.D. 2018. Ethical problems in decision-making in the neonatal ICU. *New England Journal of Medicine* 379 (19): 1851–1860.

Lantos, J.D., and W.L. Meadow. 2006. *Neonatal bioethics: The moral challenges of medical innovation.* The Johns Hopkins University Press.

Martin, D. (2013). Medical futility in Australia. In *Medical futility: A cross-national study*, 119–144. World Scientific.

McQuoid-Mason, D.J., and N. Naidoo. 2019. Palliative care ethical guidelines to assist health care practitioners in their treatment of palliative care patients. *South African Journal of Bioethics and Law* 12 (1): 14–18.

Miller-Smith, L., Á.F. Wagner, and J.D. Lantos. 2019. *Bioethics in the pediatric ICU: Ethical challenges encountered in the care of critically ill children.* Springer.

Mollaret, P., and M. Goulon. 1959. Ğ Le coma dépassé ğ. *Rev. Neural 101.*

Moratti, S. (2010). Non-treatment decisions on grounds of medical futility and quality of life: Interviews with fourteen Dutch Neonatologists. *Issues Law & Medicine* 26(3).

Morrison, W., J.D. Clark, M. Lewis-Newby, and A.A. Kon. 2018. Titrating clinician directiveness in serious pediatric illness. *Pediatrics* 142 (Supplement 3): S178–S186.

Moselli, N.M., F. Debernardi, and F. Piovano. 2006. Forgoing life sustaining treatments: Differences and similarities between North America and Europe. *Acta Anaesthesiologica Scandinavica* 50 (10): 1177–1186.

Onyeka, T.C., I. Okonkwo, U. Aniebue, I. Ugwu, F. Chukwuneke, and D. Agom. 2019. 'Wrong treatment': Doctors' take on medical futility in a low-resource ICU. *University of Toronto Medical Journal* 96 (3): 17–23.

Paris, J.J., and K.J. O'Connell. 1991. Withdrawal of nutrition and fluids from a neurologically devastated infant: The case of baby T. *Journal of Perinatology: Official Journal of the California Perinatal Association* 11 (4): 372.

Pazooki, M. 2014. Medical futility and palliative care gain momentum in China. *Asian Bioethics Review* 6 (3): 315–319. https://doi.org/10.1353/asb.2014.0018.

Pessini, L., and W.S. Hossne. 2013. The reality of medical futility (Dysthanasia) in Brazil. In *Medical futility: A cross-national study*, 35–57. World Scientific.

Piva, J.P., E. Schnitzler, P.C. Garcia, and R.G. Branco. 2005. The burden of paediatric intensive care: A South American perspective. *Paediatric Respiratory Reviews* 6 (3): 160–165.

Rosenberg, A.R., and C. Feudtner. 2016. What else are you hoping for? Fostering hope in paediatric serious illness. *Acta Paediatrica* 105 (9): 1004–1005.

Schneiderman, L.J. 2011. Defining medical futility and improving medical care. *Journal of Bioethical Inquiry* 8: 123–131. https://doi.org/10.1007/s11673-011-9293-3.

Schneiderman, L.J., N.S. Jecker, and A.R. Jonsen. 1990. Medical futility: Its meaning and ethical implications. *Annals of Internal Medicine* 112 (12): 949–954.

Schneiderman, L.J., N.S. Jecker, and A.R. Jonsen. 1996. Medical futility: Response to critiques. *Annals of Internal Medicine* 125 (8): 669–674.

Schneiderman, L.J., N.S. Jecker, and A.R. Jonsen. 2017. The abuse of futility. *Perspectives in Biology and Medicine* 60 (3): 295–313.

Seino, Y., H. Kurosawa, Y. Shiima, and T. Niitsu. 2019. End-of-life care in the pediatric intensive care unit: Survey in Japan. *Pediatrics International* 61 (9): 859–864.

Shi, Y., M. Zhao, Y. Yang, C. Mao, H. Zhu, and Q. Hu. 2013. Medical futility in China: Ethical issues and policy. In *Medical futility: A cross-national study*, 163–179. World Scientific.

Truog, R.D., A.S. Brett, and J. Frader. 1992. The problem with futility. *The New England Journal of Medicine* 326 (23): 1560–1564. https://doi.org/10.1056/NEJM199206043262310.

Veatch, R.M. 2013. So-called futile care: the experience of the United States. In *Medical futility: A cross-national study*, 9–33. World Scientific

Vivas, L., and T. Carpenter. 2020. Meaningful futility: Requests for resuscitation against medical recommendation. *Journal of Medical Ethics.*

Vizcarrondo, F.E. 2019. Medical futility in pediatric care. *The National Catholic Bioethics Quarterly* 19 (1): 105–120.

White, B., L. Willmott, E. Close, N. Shepherd, C. Gallois, M.H. Parker, et al. 2016. What does "futility" mean? An empirical study of doctors' perceptions. *Medical Journal of Australia* 204(8), 318–318.
Wilkinson, D., and J. Savulescu. 2018a. *Ethics, conflict and medical treatment for children: From disagreement to dissensus.* Elsevier.
Wolfe, I.D. 2019a. A critical analysis of futility discourse in pediatric critical care. *Journal of Pediatric Ethics.* 1 (2): 82–90.
Wolfe, I.D. 2019b. Intractable conflict in pediatric critical care: A case examination and analysis of futility. Doctoral dissertation, University of Minnesota. https://core.ac.uk/download/pdf/226939 643.pdf.
Youngner, S.J. 1988. Who defines futility? *Journal of the American Medical Association* 260 (14): 2094–2095.

Further Reading

Bagheri, A. 2013a. *Medical futility: A cross-national study.* World Scientific.
Miller-Smith, L., Á.F. Wagner, and J.D. Lantos. 2019. *Bioethics in the pediatric ICU: Ethical challenges encountered in the care of critically ill children.* Springer.
Perspectives in Biology and Medicine. 2017. *Special issue on futility*, vol. 60, Number 3. The Johns Hopkins University Press. ISSN: 1529-8795. https://muse.jhu.edu/issue/37924.
Wilkinson, D., and J. Savulescu. 2018. *Ethics, conflict and medical treatment for children: From disagreement to dissensus.* Elsevier.

Chapter 17
Newborns with Severe Disability or Impairment

M. Devereaux and K. L. Marc-Aurele

Abstract This chapter focuses on the medical and ethical issues involved in decision-making for newborns with severe disability or impairment. We begin by defining the conceptual terminology used in neonatology, distinguishing it from more general social terms such as "newborn." We then move to a brief historical analysis of the Baby Doe rulings, the case law that lays the foundation for the model of shared medical decision-making currently operating in the US. We examine how these legal rulings reflect and support a changing cultural landscape in medical decision-making and the emergence of the current model of "shared decision-making," which is presented, both as an ethical ideal, and as a practical guide. We evaluate the use of the "best interest of the child" standard in resolving goals of care disagreements. We then analyze competing ethical priorities in two complex pediatric conditions: extreme prematurity and a serious congenital heart condition. The chapter concludes with recommendations for best practices in shared decision-making.

Keywords Newborns · Disability · Impairment · Ethics · Shared decision-making

17.1 Introduction

This chapter focuses on the medical and ethical issues involved in decision-making for newborns with severe disability or impairment. While our attention is directed at the newborn period, some cases of disability or impairment are diagnosed prior to delivery, thus involving prenatal as well as postnatal decisions. It is worth noting from the start that decision-making in this very early period has special challenges.

M. Devereaux (✉)
Research Ethics Program, Department of Pathology, University of California, San Diego, CA, USA
e-mail: mdevereaux@ucsd.edu

K. L. Marc-Aurele
Divisions of Neonatology and Palliative Medicine, Department of Pediatrics, Health and Rady Children's Hospitals, University of California, San Diego, CA, USA
e-mail: kmaurele@health.ucsd.edu

© Springer Nature Switzerland AG 2022
N. Nortjé and J. C. Bester (eds.), *Pediatric Ethics: Theory and Practice*,
The International Library of Bioethics 89,
https://doi.org/10.1007/978-3-030-86182-7_17

Whether diagnosed prenatally or not, a child with serious illness confronts parents with unusual stresses and often wrenching ethical decisions, particularly in the face of grave uncertainty about long-term outcomes.

We begin by defining the conceptual terminology used in neonatology and distinguishing it from more ordinary terms such as "newborn." We then move to a brief historical analysis of the Baby Doe rulings, case law that lays the foundation for the model of shared medical decision-making currently operating in the US. We examine how these legal rulings reflect and support a changing cultural landscape in medical decision-making and the emergence of the current model of "shared decision-making." The model of shared decision-making involves parents and medical team working together to identify goals of care and how best to reach them. Shared decision-making is presented, both as an ethical ideal, and as a practical guide, using the "best interest of the child" standard in resolving disagreements. Lastly, we explore competing ethical priorities in pediatric decision-making in two complex medical conditions: extreme prematurity and a serious congenital heart condition, Hypoplastic Left Heart Syndrome. Our aim throughout is to illustrate best practices in ethical decision-making for children with severe disability or impairment, particularly in the neonate and infant period.

17.2   Conceptual Definitions and Terminology

As our topic is newborns with severe disability or impairment, it is worth beginning by noting that the word, "newborn," is an ordinary term that typically means someone "recently born" without any boundaries for the numbers of days or months of life. In contrast, the terms "neonate" and "infant" have medical definitions, specifying the first 28 days and first year of life, respectively. This chapter confines itself to roughly the first six months of life. Many tough decisions, such as whether to resuscitate or perform surgeries, including gastrostomies or tracheostomies, typically present during this period.

We refer to the WHO paper on *Early Childhood Development and Disability,* which defines disability as an "impairment in body function or structure, limitation in activity, or restriction in participation. Children with disabilities include those with cerebral palsy, spina bifida, muscular dystrophy, traumatic spinal cord injury, Down syndrome, and children with hearing, visual, physical communication and intellectual impairments" (World Health Organization and UNICEF 2012). Here we use the term "impairment" in its strict definition: the diminishment or loss of function or ability. According to the WHO paper, "a number of children can have a single impairment while others may experience multiple impairments. For example, a child with cerebral palsy may have mobility, communication, and intellectual impairments."

The available research for each condition determines the metrics for newborn survival. Most studies look at survival to hospital discharge or survival to early childhood (i.e., age 2). Survival for a given condition can vary from center to center or from country to country, reflecting different health care resources and health determinants. Survival can also change over time because of improved medical care and/or increased acceptance of medical interventions for that condition. For example, a rapid rise in survival for children with Trisomy 21 began in the 1950s. At that time, just under 50% of newborns with Down syndrome survived the first year of life and life expectancy was estimated to be 12 years. Fifty years later, the proportion of first year survivors rose to 96% and the median life expectancy increased to 60 years of age (Bittles and Glasson 2004). This change was likely the result of an increased awareness of the cognitive and social skills of those with Down syndrome and a subsequent shift in attitudes (Haslam and Milner 1992; Milner 2016). In short, it is important to recognize that survival statistics depend on the time, place, and type of care received as well as the cultural values of parents and providers.

17.3   Analysis of the Baby Doe Rulings and Case Law

Historically, authority for medical decision-making in the US was largely in the hands of physicians. Prior to the Internet and social media, most patients and families had little medical knowledge or access to clinical data, thus paternalism ruled. Physicians largely determined the medically appropriate options and made recommendations that patients mostly followed. In that setting, limitations in treatment were in the hands of the medical team and occurred primarily when death was believed imminent or treatment deemed potentially inappropriate. As therapies and intensive care for newborns improved, doctors began asking whether providing life-prolonging treatment was always appropriate (Duff and Campbell 1973).

What came to be known as the Baby Doe rulings in the US legally challenged and ultimately altered the culture of medical paternalism. In 1982, the first Baby Doe was born with a tracheoesophageal fistula and Trisomy 21. Together with physicians, parents decided not to repair the fistula. A court order was filed anonymously in an attempt to mandate performing the surgery. The case went to the Indiana State Supreme Court, which ultimately supported the family's decision. The following year, the second Baby Doe case emerged. Baby Jane Doe was born with spina bifida, hydrocephalus, kidney damage, and microcephaly. Parents were told that their daughter would likely have severe mental retardation, epilepsy, and be confined to her bed. In consultation with physicians, parents decided to forgo meningomyelocele repair and ventriculoperitoneal shunt placement. Although the surgeries would prolong Baby Jane Doe's life, they would not improve her anticipated condition. A "right to life" lawyer, having learned about Baby Jane Doe, brought the case to the New York State Supreme Court. That judge ruled that the infant needed immediate surgery to preserve her life. The case ultimately went to the U.S. Supreme Court which ruled, in *Bowen v American Hospital Assoc 1986*, that the "Hospital's withholding of treatment from a handicapped infant *when no parental consent has been given* cannot violate section 504 of the Rehabilitation Act of 1973" (italics added). The Reagan Administration had, in response to the original Baby Doe case, warned that hospitals that discriminated against handicapped persons would risk termination of Medicare and Medicaid funds, citing section 504 of the Rehabilitation Act of 1973. In Bowen, however, the court ruled that section 504 did not give the US Department of Health and Human Services authority to regulate or interfere with parents' rights in medical decision-making (Annas 1984).

In a subsequent case, that of Baby K, the courts also ruled in favor of the parental right to determine the best interest of the child. Baby K was born in 1992 with anencephaly, a severe congenital condition in which the "breathing center" in the brain stem is not well developed (Darr, 1995). Due to this underdevelopment, babies stop breathing and die from apnea. To live, they need a ventilator long-term. At Baby K's delivery, mechanical ventilation was initiated. The child's mother wanted life prolonging treatment, including a tracheostomy. The hospital filed suit on the grounds that providing Baby K with life-sustaining medical care was futile and inhumane. In 1994, the appellate and federal courts ruled that, "Physicians have an ongoing duty to provide stabilizing treatment under the Emergency Medical Treatment and Labor Act ... even when a responsible physician determines such treatment is not medically or ethically appropriate." In sum, the two Baby Doe cases demonstrate a legal precedent acknowledging a newborn's degree of disability as relevant to making treatment decisions. All three cases reflect a shift away from medical paternalism to a focus on patient- and family-centered decision-making.

17.4   The Model of Shared Decision-Making

Decision-making for newborns at risk of severe disability or impairment extends well into childhood, representing not just one decision but continuous decisions throughout the child's life. Especially during the first six months of life, parents may face an unknown or poorly understood diagnosis. Even with a clear diagnosis, parents and team may face a high degree of uncertainty regarding outcomes. Good outcomes data for a given condition nevertheless offers a range of possibilities; it cannot guarantee a good outcome for a particular child. Early uncertainty thus leaves parents and the medical team balancing hard-to-estimate risks and benefits.

In the US, more so than in some countries, parental values and choices play a major role in deciding among equally appropriate, but differing, medical approaches. The emphasis on autonomy in adult medical decision-making carries over to pediatrics, allotting parents a strong role in directing goals of care for their child. It is largely assumed that parents in a multi-cultural, heterogeneous society will display a broad range of personal, religious, and cultural perspectives. Parents facing a life-limiting or high-risk fetal diagnosis in pregnancy may respond in a variety of ways. Some families elect to terminate a high-risk pregnancy. For other families, terminating a pregnancy for any reason violates their faith or values. They may want whatever interventions increase the chances of their baby being born alive and living as long as possible, whatever the likelihood of severe disability or early death. Other families in the same circumstances elect to carry to term to confirm a diagnosis of a life-limiting or serious medical condition, thus providing assurance that allowing natural death is the best option.

In a shared decision-making model, the provider works with families to identify their values and goals for their child, making medical recommendations and helping parents understand the likely consequences of each decision. Providers need to set aside their own values and biases, but this does not mean merely presenting families with every possible option (Lantos, 2018). It is important to recommend treatment options that reflect the family's expressed hopes and fears, their values, and past experience, e.g., with the medical system. Shared decision-making requires empathy, listening, and good communication skills, with information tailored to the needs of the individual family. For some parents, their priorities, values, and experiences play a more important role in decision-making than predicted outcomes or statistics do. Belief in the healing power of prayer or divine intervention may keep hope alive despite a poor prognosis, providing comfort and direction in navigating medical decisions (refer to Chap. 12). Other parents may opt for a more passive role in decision-making, believing that their child's future is in God's hands (Uveges et al. 2019).

Compassionate health care requires that providers recognize that parents, religious or not, need to maintain hope. Parents do not want to feel that they—or the medical team—have "given up." Parents may also have a strong need to feel that they have given their child a chance (Janvier et al. 2016). Time-limited trials of therapy allow parents and team to see what a child can do, to give a child every chance to "beat the odds." In short, while values differ, nearly all parents aim to be good parents, and do the best that they can for their child.

17.5   Ethical Disagreement

Shared decision-making does not always go well. Even with the best of intentions, shared decision-making, particularly in managing severe illness in early childhood, is difficult. To be successful, shared decision-making requires regular communication between the medical team and family, and between different members of the care team. When parents and providers disagree on the goals of care or the means to achieve them, special care is needed. Many disagreements occur when parents want treatment that doctors feel is medically inappropriate or likely of little or no potential benefit. As neonatologists and pediatric intensivists report, the pivotal factor in decisions to withdraw or withhold treatment is most often severe intellectual disability or the likelihood of such (Wilkinson 2006). Highly trained professionals typically place great importance on cognitive ability and intellectual achievement (Streiner et al. 2001). Medical intensivists also tend to see the worst cases, thus biasing their sense of the grim future facing such children (Kon et al. 2004) (refer to Chap. 16). Newborn providers may also feel responsible for burdening a child and family with life-long medical needs, or not fully informing parents of the known strains of extreme caretaking.

Important to recognize is that many parents, however, willingly tolerate a higher risk of complications or long-term disability than physicians themselves may wish to accept (Lam et al. 2009). Two other points are worth noting. First, evidence suggests that physicians are poor prognosticators, often over or underestimating how long patients have to live. Many children do better (or worse) than the medical professionals predict. Second, while physicians may believe they themselves would not accept a given complication or long-term disability for their own child, they do not in fact *know* what they would do when faced with a similar decision. Nor is it easy to imagine secondhand the meaning and joy many families find in caring for their children, whatever the severity of disability or impairment.

An illustration of changing medical practice is Trisomy 13[1]. Medical schools taught until relatively recently that nearly all children with Trisomy 13 (and 18) died early, with the few who survived infancy doing so with profound disability. This dire picture not surprisingly led to the belief that starting artificial nutrition, pursuing heart surgery, or placing a tracheostomy would merely delay inevitable death and provide parents with false hope. Parents were standardly offered only comfort care. In recent decades, this medical and ethical landscape has changed. As is now understood, historically poor survival and outcomes data for both Trisomy 13 and Trisomy 18[2] patients reflect the medical practice at the time of not offering treatment. Without heart surgery and other interventions, children mostly fulfilled the prediction of early death. The doctors' predictions were self-fulfilling. In recent decades, medical practice has shifted. A rise in medical interventions has led to subsequent decreased mortality (Peterson et al. 2017). In fact, for each day that a baby with Trisomy 13 or 18 lives, survival increases (Brewer 2002). Descriptions of positive family experiences, both in the medical literature and on social media, along with changing societal perspectives of disability, have further empowered more families to advocate for newborns with Trisomy 13 and 18 (Hasegawa and Fry 2017).

From an ethics perspective, respect for shared decision-making obliges providers to acknowledge that parental values and goals may not align with what providers might choose for themselves or their children. As pediatrician John Lantos argues, "when decisions about treatment fall within the zone of parental discretion... parents' preferences ought to prevail" (Lantos 2018). Respect for parental preferences in medical decision-making rests on the view that parents (or guardians) are generally positioned to determine what is best for their child and their family. Legally, parents serve as decisional surrogates on behalf of minor children until such time as the child is old enough and has sufficient mental capacity to make decisions for themselves. In the case of newborns with severe impairment, this transition to actual self-determination may never occur, requiring parents to serve as surrogates throughout the child's life.

---

[1] Also known as Patau syndrome, is a chromosomal condition associated with severe intellectual disability and physical abnormalities.

[2] Also known as Edwards syndrome, is a condition that is caused by an error in cell division, known as meiotic disjunction.

Features of the health care and health care insurance system also unavoidably play a role in parental decision-making. In the UK, as the Charlie Gard case illustrates (refer to Chap. 16), what is offered to parents of newborns with severe disability or impairment is constrained by the National Health Service and national health policy. In the Gard case, despite parental insistence that they wanted to try a "last ditch" offer by a US doctor for an unproven nucleoside treatment, the UK courts sided with Charlie's medical providers in refusing to support the proffered treatment. The court argued that this therapy was not in Charlie's best interest and granted the hospital's petition to withdraw treatment (Wilkinson 2006). In the US, the absence of a national health care system places approval for procedures and resource allocation more directly in the hands of medical providers. However, while the model of shared decision-making affords US parents broad discretion in deciding among medically appropriate alternatives, in practice, options may be limited by insurance coverage, disparate hospital access, socioeconomic status, race and ethnicity. The medical team may play a determinative role in overcoming such barriers, but health care disparities are increasingly recognized to play a determinative role in outcomes (refer to Chaps. 28 and 29).

17.6   The Best Interest Standard

The considerable latitude accorded to parents in shared decision-making is not without limits. Refusals of consent for necessary medical care, as in emergency blood transfusions for minors, fall outside parental decision-making authority (*Prince vs. Massachusetts 1944*). Child protective services may intervene when parental behavior raises concerns about, or meets the legal definition of, child abuse or neglect. Short of such concerns, providers generally assume that parents are acting in the best interest of the child. However, physicians ought not to accede to parental requests for interventions unsupported by a correct understanding of the medical situation and accurate analysis of risks and benefits. Shared decision-making requires that physicians ensure adequate understanding of the medical facts (i.e., diagnosis and prognosis, and the risks and benefits of treatment options). Achieving this under-standing can be difficult, particularly if parents are confronted at delivery with an unanticipated medical condition or impairment. As noted above, it may take time to achieve a correct diagnosis and rare disorders may provide little data on survival and outcomes. A clear picture of possible outcomes may be especially challenging to achieve when decisions must be made within a short time frame, e.g., electing or foregoing life sustaining heart surgery.

In addition to medical interests, a child's interests may include being held, spending time at home or having contact with siblings and other family members. For many families, these are important interests, ones that may need to be weighed against life prolonging measures that require ICU care and other aggressive interventions.

Another difficulty in establishing what is in the best interest of medically complex newborns is the lack of information about who this child will be or what preferences they might develop. Decision-making is generally guided by the ethical standard of the "best interest of the child." This generic standard contrasts with the more familiar "substituted judgment" standard used when adults, who once possessed decision-making capacity, lose the ability to make decisions for themselves. The health proxy or surrogate acts on the basis of respect for the autonomy of the patient by following advanced care planning documents, e.g., an Advance Directive, or by following past wishes. The legal standard for newborns or very young children traditionally assumes a generic view of interests, seeking to weigh which course of action will balance net benefits and burdens for the child (Wilkinson 2006). The best interest standard is a familiar, common-sense guide. It works well when everyone agrees what is best for the child (refer to Chaps. 4 and 9).

In the setting of the NICU, however, it is not uncommon that parents and the medical team disagree over the burdens of interventions or the benefits of prolonging life, even when that life will be short and lived in the intensive care unit. When disagreements over fundamental values occur, the guidance of the best interest standard is less clear. One practical problem is that the concept of "best interest" is itself vague. Its indeterminacy allows different parties to import their own values and beliefs of what should be done under the flag of acting *on behalf of the child*. Parents may find it difficult or impossible to separate their own interests or that of their family from that of the sick child (Beauchamp and Childress 1989; Wilkinson 2006). Physicians and bedside staff may struggle to distinguish their own values ("if this were my child") from the values governing parental decisions about how to proceed given the available medical options. Both providers and parents see themselves as acting in the best interest of the child but may disagree over what that interest is.

The philosophical problem is that without agreement on what constitutes best interest, the concept itself does little to settle disputes. The ethical question is precisely which values should be given priority, e.g., life prolongation with aggressive care in the intensive care unit or time at home. Thus, although the best interest of the child provides a nearly unanimously agreed upon standard, it may provide limited practical guidance in deciding among competing values and goals of care. In the case of newborns with severe disability or impairment, the notion of best interest is particularly difficult to apply given the possibility that whatever outcome is achieved, the child may be unable fully to perceive or experience those benefits. Lastly, as Douglas Diekema argues, however useful in medical decision-making, the best interest standard is not appropriate in deciding when to seek state intervention, e.g., child protective services. Diekema's claim is that when physicians and ethics consultants feel the need to challenge a parent's decision-making authority in the courts, the harm principle provides a more appropriate standard (Diekema et al. 2009). However, while the harm principle provides a clear standard in settings such as child abuse, the concept of harm is itself, like the concept of best interest, open to interpretation and disagreement (Birchley 2016). For example, some might argue that a child's severe intellectual disability or inability to get out of bed is a harm worse than death. Others may view a child with these deficits as having an acceptable quality of life (Adams et al. 2020). In short, the harm principle, like the best interest standard, doesn't by itself settle disputes over fundamental values.

17.7 Tough Cases

We conclude this analysis by looking in detail at the ethical issues presented by two of the more difficult cases in neonatology: extreme prematurity and a severe congenital cardiac condition.

17.7.1 Extreme Prematurity

The most recent American Academy of Pediatrics (AAP) guideline on counseling regarding resuscitation before 25 weeks of gestation states that the lower limit of viability at birth is generally accepted as 22 weeks' gestation and below that age, comfort-focused care is recommended (Cummings and Newborn, 2015). Between 22- and 25-weeks' gestation, decision-making should be shared and family-centered because although the risk of death or significant neurodevelopmental delay is high, it is also difficult to predict. The report recommends taking into account more than a fetus' gestational age when counseling parents regarding resuscitation because the estimate of gestational age can be plus or minus 2 weeks. Furthermore, at that point in gestation, each additional day of development can significantly influence survival.

Although there are several predictive models to estimate risk for mortality and neurodevelopmental issues for extremely premature infants, models do not provide certainty (Dukhovny 2017). Although a few morbidities acquired in the first several months of life predict death or disability by 5 years of age, cognitive and motor impairments decrease over time (Schmidt et al. 2015). In addition, parental education and socio-economic status also impact neurodevelopmental outcomes. Unfortunately, predicting survival, impairment, and disability based on information from the newborn period is difficult. Because the chances of death and disability-free survival depend on extremely premature infants surviving through the first few days of life and subsequent medical events during the hospital stay, parents need ongoing and individualized counseling to make sense of how prognosis changes as time goes on and to sift through the known limitations in predicting outcomes (Cheong et al. 2018).

For these reasons, the AAP guideline states, "Although general recommendations can guide practice, each situation is unique; thus, decision-making should be individualized." The 2015 European Resuscitation Council Guidelines recommend a similar approach: "In conditions associated with uncertain prognosis, where there is borderline survival and a relatively high rate of morbidity, and where the anticipated burden to the child is high, parental desires regarding resuscitation should be supported" (Wyllie et al. 2015). Finally, the Japan Resuscitation Council guideline falls in line with the AAP and European guidelines. Despite this agreement, the practice of resuscitating infants born at 22 to 23 weeks' gestation is common in Japan. The view is that resuscitation should be offered because of the uncertainty and difficulty predicting outcomes and because medical care for sick newborns is supported by local governments (Itabashi et al. 2009).

At the time of this writing (2021), the consensus is that by 25 weeks outcomes are generally good enough that resuscitation is recommended, with parents no longer asked to make a choice. Unless there are factors known to impact survival such as growth restriction, chromosomal abnormalities, infection, or signs of fetal deterioration, forgoing resuscitation at 25 weeks' gestation or thereafter is generally against medical advice. The vast majority of cases are resolved by thoughtful discussion about parental fears and concerns. Often alignment can be achieved by informing parents that resuscitation at birth does not commit their baby to continued intensive and invasive treatment should medical events or new information alter the baby's prognosis. The concepts of time-limited trials and ongoing shared decision-making are generally well received.

17.7.2   Hypoplastic Left Heart Syndrome

Hypoplastic Left Heart Syndrome (HLHS) is a serious congenital condition resulting from under-development of the left side of the heart. In the past, newborns with this cardiac condition would have died in the first month or two of life. Parents of children born with HLHS now, however, have the option of surgical and other life prolonging interventions. The Norwood procedure, the first of a three-stage sequence of palliative surgery developed in the 1980s, offers children with HLHS much improved chances of surviving infancy and living even into adulthood. This surgery is performed in the first days of life, followed by a second surgery, the Glenn, around 4–6 months of life. The third surgery, the Fontan procedure, typically takes place between 3 and 5 years of age.

Recent data from high volume, established pediatric cardiac programs, indicate a five-year survival rate in the range of 60–70%, with survivors living meaningful lives at home (Devereaux and Kon 2017). Less established centers, or children with other medical conditions, have fewer good outcomes. However, without surgery, 90% of children born with HLHS die within the first year (Kane et al. 2016). Given these odds, pediatric medical centers now routinely offer the Norwood procedure to families or refer them to suitable centers for treatment. The choice of opting for surgery is left to parents in consultation with the medical team. HLHS, like extreme prematurity, thus confronts parents and teams with options that raise complex medical and ethical choices.

Few parents would turn down the possibility of life-prolonging surgery given the poor odds of the neonate surviving without intervention. However, each of these surgeries requires hospitalization, often including long-term ICU care. The post-op course is not easy and may include the need for a gastrostomy or feeding tube. Families may need to relocate to an area with adequate medical expertise, with disruption to work and the education of other children. Moreover, the Norwood sequence is merely palliative, not curative. Outcomes vary and not all children survive to complete the three-stage sequence. Even in the best-case scenario, the Norwood will need to be followed by heart transplant, first in late adolescence or early adulthood, and then again, every 10–15 years, with the need to take immunosuppressant drugs throughout life. Many children with HLHS face cognitive challenges and psychomotor disorders. At the very least, such children face a long, medically complex road, with repeat hospitalizations and lifelong follow-up.

Despite these hurdles, many children go on to do well, with families grateful for the chance to extend their child's life. In the US, parents increasingly elect the surgical option. Although Europe and other developed countries initially took a less interventionist approach to HLHS, rates of surgery there have also increased since the early 2000s (Murtuza and Elliott 2011).

Some physicians argue that surgical outcomes have improved sufficiently that the option of declining surgery in favor of comfort care or hospice for HLHS should no longer be offered (Wernovsky 2008). In medical terms, the claim is that surgical treatment of HLHS should be considered "standard of care" and hence the only option presented. From an ethics point of view, the question is whether outcomes are predictably good enough that foregoing surgical intervention is clearly harmful to the child. In some cases, established evidence supports a clear benefit despite risks of intervention. Parents of a child bleeding profusely from a severed artery, and at risk of death, are not offered options. They do not have "right of refusal" as refusal would result in recognized harm to the child. Treatment in emergency situations is ethically mandatory.

Many situations in neonatology however are not so clear-cut. Currently, in the US, pediatric cardiology centers offer, and may even strongly recommend, the Norwood surgery, but do not require it. The burden of intervention is still significant and the outcome uncertain. The uncertainty is not only medical, but also developmental. Determining the best interest of the child in such cases is difficult and open to differing opinions. The best interest of the child is also increasingly recognized by bioethicists to include some consideration of the values and capacities of the child's family.

As of this writing, the decision to opt for surgery remains ultimately in the hands of parents. Values and beliefs vary as do family circumstances. Confronted with a choice about the best path forward, different parents will make different decisions. Some want to avoid suffering, or to keep their child at home. Others want to take every chance at extending their child's life even if those efforts may not succeed. In some cases, parents may want surgery, but the child does not meet criteria, e.g., because of cardiac physiology that makes surgery impossible or chromosomal abnormalities that increase the risk of mortality.

17.8 Conclusions and Recommendations

The cases of HLHS and extreme prematurity both illustrate how what is medically possible changes over time. With new drugs and surgical techniques, we may move from "there's nothing medically to be done" to early experimental or "long shot" approaches, and eventually to established protocols with known risks and benefits. Only when medical knowledge and clinical skill achieves a reliably beneficial outcome, with manageable risks, do we start to see an intervention become "standard of care." Until then, parents with very sick or impaired newborns may face wrenching ethical choices. The AAP, as we've seen, gives parents wide leeway in making decisions for their minor children, taking into consideration the interests of close others such as siblings. Families confronting such choices, however, need more than a list of options. They need information presented in understandable terms, the reassurance that the medical team will be with them every step of the way, and on-going open communication, including listening on the part of attending physicians and clinical staff. Having a primary attending or single "point person" during hospitalizations often helps to coordinate care and establish on-going trust. Whatever the diagnosis, parents should be kept medically informed and supported in choosing goals of care reflective of their values, perspectives, and hopes.

It is worth stressing that medical decision-making for very sick newborns is one of the most challenging circumstances for both parents and medical providers. Early in this process, calling an ethics consult can help providers and families identify the ethical issues, clarify the competing values, and open up space for the parties to hear and understand each other. Ethics consultants also have skills in mediation and empathetic communication that can help the parties find their way forward.

In a similar way, the palliative consulting team can foster a safe, compassionate, and supportive environment to allow parents to process their emotions and discuss their values. Before exploring hopes for their baby's care and especially before making changes to the plan of care, parents often need to address their feelings, including fear, anger, and blame. Realizing the meaning of the situation in the context of their values (including possibly religious faith) is necessary before approaching decision-making about treatment options. By validating and reflecting what parents express and in turn building trust, the palliative team can support parental voice in decision-making and facilitate communication between parents and medical teams.

Chaplaincy and local spiritual leaders may also, if relevant, play a key role in maintaining the family's trust (refer to Chap. 12).

Guiding principles for the ethical care of newborns with severe disability or impairment

---

Decision-Making
- For newborns at risk of severe disability or impairment, decision-making extends well into childhood, representing not just one decision but continuous decisions throughout the child's life
- Respect for shared decision-making obliges providers to acknowledge that parental values and goals may not align with what providers might choose for themselves or their children

---

The Shared Decision-Making Model
- Set aside your own values and biases
- Work with families to identify their values and goals for their child
- Employ empathy, listening, and good communication skills, with information tailored to the needs of the individual family
- Help parents understand the likely consequences of each decision
- Recommend treatment options that reflect the family's expressed hopes and fears, their values, and past experience
- Recognize that parents, religious or not, need to maintain hope
- Consult Palliative Care and Ethics teams for support and advice

---

References

Adams, S.Y., R. Tucker, M.A. Clark, and B.E. Lechner. 2020. "Quality of life": Parent and neonatologist perspectives. *Journal of Perinatology: Official Journal of the California Perinatal Association* 40 (12): 1809–1820. https://doi.org/10.1038/s41372-020-0654-9.

Annas, G.J. 1984. The case of Baby Jane Doe: Child abuse or unlawful federal intervention? *American Journal of Public Health* 74 (7): 727–729. https://doi.org/10.2105/ajph.74.7.727.

Beauchamp, T.L., and J.F. Childress. 1989. *Principles of biomedical ethics*, 3rd ed. Oxford University Press.

Birchley, G. 2016. Harm is all you need? Best interest and disputes about parental decision-making. *Journal of Medical Ethics* 42: 111–115. https://doi.org/10.1136/medethics-2015-102893.

Bittles, A.H., and E.J. Glasson. 2004. Clinical, social, and ethical implications of changing life expectancy in Down syndrome. *Developmental Medicine & Child Neurology* 46 (4): 282–286. https://doi.org/10.1111/j.1469-8749.2004.tb00483.x.

Brewer, C.M. 2002. Survival in trisomy 13 and trisomy 18 cases ascertained from population based registers. *Journal of Medical Genetics* 39 (9): 54e–554. https://doi.org/10.1136/jmg.39.9.e54.

Cheong, J.L.Y., K.J. Lee, R.A. Boland, A.J. Spittle, G.F. Opie, A.C. Burnett, L.M. Hickey, G. Roberts, P.J. Anderson, L.W. Doyle, J.L.Y. Cheong, C. Anderson, P.J. Anderson, M. Bear, R.A. Boland, A.C. Burnett, C. Callanan, E. Carse, M.P. Charlton, et al. 2018. Changes in long-term prognosis with increasing postnatal survival and the occurrence of postnatal morbidities in extremely preterm infants offered intensive care: A prospective observational study. *The Lancet Child & Adolescent Health* 2(12): 872–879https://doi.org/10.1016/S2352-4642(18)30287-6.

Cummings, J., and Committee on Fetus and Newborn. 2015. Antenatal counseling regarding resuscitation and intensive care before 25 weeks of gestation. *Pediatrics* 136(3): 588–595.https://doi.org/10.1542/peds.2015-2336

Darr, K. 1995. In the matter of baby K: implications for hospital administration. *Hospital Topics* 73 (1): 4–6. https://doi.org/10.1080/00185868.1995.10543738.

Diekema, D.S., J.R. Botkin, and Committee on Bioethics. 2009. Forgoing medically provided nutrition and hydration in children. *Pediatrics* 124(2): 813–822. https://doi.org/10.1542/peds.2009-1299.

Devereaux, M., and A.A. Kon. 2017. May We Take Our Baby With Hypoplastic Left Heart Syndrome Home? *American Journal of Bioethics* 17 (7): 72–74.

Duff, R.S., and A.G. Campbell. 1973. Moral and ethical challenges in the special-care nursery. *The New England Journal of Medicine* 289 (17): 890–894. https://doi.org/10.1056/NEJM19731025 2891705.

Dukhovny, D. 2017. 'Does _ predict neurodevelopmental impairment in former preterm infants?' Is this the right question to be asked? *Journal of Perinatology* 37 (5): 467–468. https://doi.org/10.1038/jp.2017.19.

Hasegawa, S.L., and J.T. Fry. 2017. Moving toward a shared process: The impact of parent experiences on perinatal palliative care. *Seminars in Perinatology* 41 (2): 95–100. https://doi.org/10.1053/j.semperi.2016.11.002.

Haslam, R.H.A.H. and R. Milner. 1992. The physician and down syndrome: Are attitudes changing? *Journal of Child Neurology* 7: 304–310. https://journals.sagepub.com/doi/abs/10.1177/088307 389200700312.

Itabashi, K., T. Horiuchi, S. Kusuda, K. Kabe, Y. Itani, T. Nakamura, M. Fujimura, and M. Matsuo. 2009. Mortality rates for extremely low birth weight infants born in Japan in 2005. *Pediatrics* 123 (2): 445–450. https://doi.org/10.1542/peds.2008-0763.

Janvier, A., B. Farlow, and K.J. Barrington. 2016. Parental hopes, interventions, and survival of neonates with trisomy 13 and trisomy 18. *American Journal of Medical Genetics Part c: Seminars in Medical Genetics* 172 (3): 279–287. https://doi.org/10.1002/ajmg.c.31526.

Kane, J.M., J. Canar, V. Kalinowski, T.J. Johnson, and K. Sarah Hoehn. 2016. Management options and outcomes for neonatal hypoplastic left heart syndrome in the early twenty-first century. *Pediatric Cardiology* 37 (2): 419–425. https://doi.org/10.1007/s00246-015-1294-2.

Kon, A.A., L. Ackerson, and B. Lo. 2004. How pediatricians counsel parents when no "best-choice" management exists: Lessons to be learned from hypoplastic left heart syndrome. *Archives of Pediatrics and Adolescent Medicine* 158 (5): 436–441. https://doi.org/10.1001/archpedi.158.5.436.

Lam, H.S., S.P.S. Wong, F.Y.B. Liu, H.L. Wong, T.F. Fok, and P.C. Ng. 2009. Attitudes toward neonatal intensive care treatment of preterm infants with a high risk of developing long-term disabilities. *Pediatrics* 123 (6): 1501–1508. https://doi.org/10.1542/peds.2008-2061.

Lantos, J.D. 2018. Ethical problems in decision-making in the Neonatal ICU. *The New England Journal of Medicine* 379 (19): 1851–1860. https://doi.org/10.1056/NEJMra1801063.

Murtuza, B., and M.J. Elliott. 2011. Changing attitudes to the management of hypoplastic left heart syndrome: A European perspective. *Cardiology in the Young* 21 (S2): 148–158. https://doi.org/10.1017/S1047951111001739.

Peterson, J.K., L.K. Kochilas, K.G. Catton, J.H. Moller, and S.P. Setty. 2017. Long term outcomes of children with Trisomy 13 and 18 after congenital heart disease interventions. *The Annals of Thoracic Surgery* 103 (6): 1941–1949. https://doi.org/10.1016/j.athoracsur.2017.02.068.

Schmidt, B., R.S. Roberts, P.G. Davis, L.W. Doyle, E.V. Asztalos, G. Opie, A. Bairam, A. Solimano, S. Arnon, R.S. Sauve, B. Schmidt, J. D'Ilario, J. Cairnie, J. Dix, B.A. Adams, E. Warriner, M.-H.M. Kim, P. Anderson, P. Davis, et al. 2015. Prediction of late death or disability at age 5 years using a count of 3 neonatal morbidities in very low birth weight infants. *The Journal of Pediatrics* 167(5): 982–986.e2. https://doi.org/10.1016/j.jpeds.2015.07.067.

Streiner, D.L., S. Saigal, E. Burrows, B. Stoskopf, and P. Rosenbaum. 2001. Attitudes of parents and health care professionals toward active treatment of extremely premature infants. *Pediatrics* 108: 152–157. https://doi.org/10.1542/peds.108.1.152.

Uveges, M.K., J.B. Hamilton, K. DePriest, R. Boss, P.S. Hinds, and M.T. Nolan. 2019. The influence of parents' religiosity or spirituality on decision-making for their critically Ill child: An integrative review. *Journal of Palliative Medicine* 22 (11): 1455–1467. https://doi.org/10.1089/jpm.2019.0154.

Wernovsky, G. 2008. The Paradigm Shift Toward Surgical Intervention for Neonates with Hypoplastic Left Heart Syndrome. *Archives of Pediatrics & Adolescent Medicine* 162 (9): 849. https://doi.org/10.1001/archpedi.162.9.849.

Wilkinson, D. 2006. Is it in the best interest of an intellectually disabled infant to die? *Journal of Medical Ethics* 32 (8): 454–459. https://doi.org/10.1136/jme.2005.013508.

World Health Organization, and UNICEF. 2012. *Early childhood development and disability: A discussion paper.* https://apps.who.int/iris/bitstream/handle/10665/75355/9789241504065_eng.pdf;jsessionid=E80CB5FA43B76CB8365356C9B5BFC3F5?sequence=1.

Wyllie, J., J. Bruinenberg, C.C. Roehr, M. Rüdiger, D. Trevisanuto, and B. Urlesberger. 2015. European resuscitation council guidelines for resuscitation 2015. *Resuscitation* 95: 249–263. https://doi.org/10.1016/j.resuscitation.2015.07.029.

Further Reading

Adams R.C., S.E. Levy, COUNCIL ON CHILDREN WITH DISABILITIES. 2017. Shared decision-making and children with disabilities: pathways to consensus. *Pediatrics* 139(6): e20170956.https://doi.org/10.1542/peds.2017-0956.

Hester D.M., C.D. Lew, A. Swota. 2016. When rights just won't do: Ethical considerations when making decisions for severely disabled newborns. *Perspectives in Biology and Medicine* 58(3): 322–327. https://doi.org/10.1353/pbm.2016.0004.

Zablotsky B., L.I. Black, M.J. Maenner, L.A. Schieve, M.L. Danielson, R.H. Bitsko, S.J. Blumberg, M.D. Kogan, C.A. Boyle. 2019. Prevalence and trends of developmental disabilities among children in the United States: 2009–2017. *Pediatrics*, 144. https://doi.org/10.1542/peds.2019-0811.

Chapter 18
Neonatal Euthanasia and the Groningen Protocol

Jacob J. Kon, A. A. Eduard Verhagen, and Alexander A. Kon

Abstract Neonatal euthanasia has been legal in the Netherlands since 2005. Data indicate that neonatal euthanasia is practiced sub rosa by some clinicians in other countries as well; however, the true extent of neonatal euthanasia practice remains unknown. In this chapter, we review end-of-life options to describe the ethical background in the adult setting and how these translate into the neonatal setting. Further, the ethical arguments in favor and opposed to allowing euthanasia of infants, and those in favor and opposed to the use of paralytics in neonatal euthanasia, are presented.

Keywords Euthanasia · End-of-life · Neonatal ethics · Physician assisted dying · Life sustaining treatment

In 2002, experts in neonatology and bioethics from the University Medical Center Groningen, in collaboration with the Groningen district attorney's office, developed the Groningen Protocol that provides a systematic approach to decision-making regarding euthanizing infants (Verhagen and Sauer 2005a, b), which was adopted as a national guideline in 2005. The Protocol was spurred by data indicating that neonatal euthanasia was not uncommonly performed in The Netherlands (van der Heide et al. 1997); however, there were no standards for the practice. These findings raised significant concerns that some neonatologists may be euthanizing at least some infants without adequate oversight and without appropriate standards of practice.

J. J. Kon
University of California Santa Cruz, Santa Cruz, CA, USA
e-mail: jkon@ucsc.edu

A. A. E. Verhagen
University Medical Center Groningen, University of Groningen, Groningen, The Netherlands
e-mail: a.a.e.verhagen@umcg.nl

A. A. Kon (✉)
University of California San Diego School of Medicine, San Diego, CA, USA
e-mail: aakon@uw.edu

University of Washington School of Medicine, Seattle, WA, USA

© Springer Nature Switzerland AG 2022
N. Nortjé and J. C. Bester (eds.), *Pediatric Ethics: Theory and Practice*,
The International Library of Bioethics 89,
https://doi.org/10.1007/978-3-030-86182-7_18

The goal of the Protocol is to allow parents and doctors to end the life of infants in cases where the infant is suffering unbearably[1] with no hope for improvement, but is neither actively dying nor dependent on medical technology (e.g., a ventilator) for life. Under the Protocol, physicians may administer substances to the infant to rapidly and painlessly end the infant's life[2] when specific criteria are met (Table 18.1).

In this chapter, we will review end-of-life options to describe the ethical background in the adult setting and how these translate into the neonatal setting. We will then present the ethical arguments in favor and opposed to allowing euthanasia of infants.[3] Finally, we will discuss the arguments in favor and opposed to the use of paralytics in neonatal euthanasia.

18.1 End-of-Life Options in Adult Medicine

In order to appreciate the ethical issues surrounding end-of-life (EOL) care in infants, it is imperative to understand these issues in the adult setting. Clearly, options that are not ethically permissible in the care of a competent adult patient would not be ethically permissible in the care of an infant. Although some have argued that infants are not

---

[1] NB: There is variability in the understanding of the term "unbearable suffering." In this chapter, we take it to mean subjective suffering to the extent that the patient herself feels that she can no longer bear it, and she believes that being dead would be better than being alive in her current state. That is, a degree of suffering that to the patient constitutes a fate worse than death.

[2] Note on terminology: Many clinicians, bioethicists, and authors use various terminology for similar acts. We have chosen to use terminology that is as unbiased and non-inflammatory as possible. Some authors, particularly those with strong moral objections to euthanasia, may use terminology such as "killing patients," "killing babies", "executing children," etc. We find such terminology to be overly biased and unhelpful when deep consideration of the ethical issues is appropriate.

[3] It should be noted that the Groningen Protocol has not been widely accepted. Indeed, neonatal euthanasia remains illegal in all countries except the Netherlands; however, as noted in this chapter, neonatal euthanasia is practiced to some extent widely. It should further be noted that two of the authors of this chapter have written extensively on this topic. E. Verhagen was the primary author of the protocol and has written extensively on the ethical justification of neonatal euthanasia (Brouwer et al. 2018; de Vries and Verhagen 2008; Dorscheidt et al. 2013; Sauer and Verhagen 2009; Verhagen 2013; Verhagen and Sauer 2005a, b, 2008; Verhagen et al. 2005; Verhagen 2006). A. Kon has written arguing that the protocol is not ethically justifiable and should be abandoned (Kon 2007, 2008, 2009). It should also be noted that many authors have argued strongly against the protocol based on ethical concerns and/or moral objections. We believe that this is an evolving area in health care. When withdrawal of life-sustaining treatment was first considered, many believed such an act to be immoral and equivalent to killing patients; however, such practice is now accepted in nearly all societies. When the discussion around allowing physicians to prescribe life-ending substances to terminally ill competent adults was first discussed, again many authors raised serious ethical and moral objections and stated that such acts are merely killing patients; however, Physician Assisted Dying is now widely accepted. Based on these historical occurrences, we believe that when considering novel end-of-life options, ethicists should attempt to approach topics in an unbiased and open fashion. As such, the current chapter is written to present a balanced consideration of the ethical arguments on both sides of this important issue.

Table 18.1 The Groningen Protocol for Euthanasia in newborns

Requirements that must be fulfilled
The diagnosis and prognosis must be certain
Hopeless and unbearable suffering must be present
The diagnosis, prognosis, and unbearable suffering must be confirmed by at least one independent doctor
Both parents must give informed consent
The procedure must be performed in accordance with the accepted medical standard

Information needed to support and clarify the decision about euthanasia
Diagnosis and prognosis
Describe all relevant medical data and the results of diagnostic investigations used to establish the diagnosis
List all the participants in the decision-making process, all opinions expressed, and the final consensus
Describe how the prognosis regarding long-term health was assessed
Describe how the degree of suffering and life expectancy were assessed
Describe the availability of alternative treatments, alternative means of alleviating suffering, or both
Describe treatments and the results of treatment preceding the decision about euthanasia

Euthanasia decision
Describe who initiated the discussion about possible euthanasia and at what moment
List the considerations that prompted the decision
List all the participants in the decision-making process, all opinions expressed, and the final consensus
Describe the way in which the parents were informed and their opinions

Consultation
Describe the physician or physicians who gave a second opinion (name and qualifications)
List the results of the examinations and the recommendations made by the consulting physician or physicians

Implementation
Describe the actual euthanasia procedure (time, place, participants, and administration of drugs)
Describe the reasons for the chosen method of euthanasia

Steps taken after death
Describe the findings of the coroner
Describe how the euthanasia was reported to the prosecuting authority
Describe how the parents are being supported and counseled
Describe planned follow-up, including case review, postmortem examination, and genetic counseling

Adapted from Verhagen and Sauer (2005b)

persons, and therefore are not entitled to protection (Giubilini and Minerva 2013), this view is shared by few and is not supported by major professional organizations nor statutes.

It is widely agreed that because they lack decision-making capacity and because they are entirely dependent on others for even basic care (e.g., feeding), infants enjoy special protections. Indeed, most jurisdictions globally afford special protections to infants and children, and when parents fail to meet these obligations, states have an obligation to provide for these minors. The doctrine of *parens patriae* requires states to assume responsibility for infants and children when they are subject to abuse or neglect; however, there is significant variability in the definition of abuse and neglect globally. Because infants and children enjoy special protections, EOL options that are not ethically permissible in competent adults would clearly be impermissible in the care of infants. At the same time, one could argue that denying infants EOL options that are ethically and legally accepted for competent adults in selected countries, seems unjust as well. We therefore present the range of EOL options in the adult setting, listed roughly in order of most widely accepted to least permissible, with brief discussion of each.

18.1.1 Limiting Life-Prolonging Interventions

In many cases, patients, families, and the care team decide that some life-prolonging interventions are not appropriate. For example, in some cases the patient, family, and care team determine that cardiopulmonary resuscitation is not appropriate should the patient suffer a cardiac arrest, and in such cases the doctor may write a Do Not Resuscitate (DNR) order (NB: Some use other terminology such as Do Not Attempt Resuscitation or Allow Natural Death; however, the ethical arguments are unchanged regardless of the terminology employed). Similarly, a decision may be made to not intubate the trachea of a patient should she develop respiratory failure. Such decisions to not initiate specific life-prolonging interventions are widely viewed as ethically permissible. Indeed, in some cultures (e.g., the United States) it is considered unethical and illegal to perform such interventions over the patient's objection even when failure to perform the interventions will lead to the patient's death.

18.1.2 Withdrawal of Life-Prolonging Interventions

In some cases, patients may already be receiving life-prolonging interventions and the patient, family, and care team determine that removal of such interventions is appropriate even when the team understands that removal of such interventions will most likely lead to the patient's death. In the famous 1976 American case of Karen Ann Quinlan, the parents of Ms. Quinlan, a 21-year-old woman in persistent vegetative state (PVS), wished to remove the breathing tube keeping their daughter alive

("In Re Quinlan, 355 A.2d 647" 1976). There was significant debate in the media and academic journals regarding the ethical appropriateness of what was then termed "passive euthanasia."[4] Ultimately, the New Jersey Supreme Court ruled that patients have a right to decline any medical intervention, including life-saving/prolonging interventions, based on a right to privacy; when patients lack decision-making capacity, their agents can make such decisions on their behalf; in such cases, physicians must remove the intervention(s) and they are not liable for such actions; and legal review is not required for subsequent cases ("In Re Quinlan, 355 A.2d 647" 1976). While withdrawal of life-prolonging interventions is not universally viewed as ethically permissible (e.g., in Japanese culture, such acts are widely considered a form of murder (Asai et al. 1997)), it is accepted as standard practice in most societies.

18.1.3 Voluntary Stopping of Eating and Drinking

In some cases, patients choose to stop eating and drinking as a means to hasten their death when they view their life as unbearable. There remains some debate as to the ethical permissibility of such action; however, because this is an individual choice that does not require the participation of the health care team, most agree that the voluntary stopping of eating and drinking (VSED) does not violate health care ethics norms, although in many cultures it does violate social norms. In some cases, however, VSED can itself be very distressing to patients, particularly as they suffer severe hunger and thirst. In some cases, clinicians may use palliative sedation (sometimes called terminal sedation) to ease the suffering of such patients. Research shows that palliative sedation does not hasten death (Maltoni et al. 2009; Mercadante et al. 2009); however, there remains ethical debate as to the appropriateness of palliative sedation as many view such action as participating in suicide (Quill et al. 1997, 2009). In the United States, palliative sedation in the care of patients who opt for VSED is generally considered ethically supportable and is practiced openly; however, in many other countries it remains taboo.

18.1.4 Withholding Medically Provided Nutrition and Hydration

In some cases, patients are not dependent on a ventilator or other medical technology to maintain basic physiologic functions, therefore withdrawal of life-prolonging

---

[4] As noted above, many authors use different terminology. In the Quinlan case, many referred to the option of removing her breathing tube as "killing" her. It is important to note that some continue to view withdrawal of life-sustaining interventions as killing patients and therefore morally unacceptable.

interventions is not an option; however, the burdens of treatment may be viewed as outweighing the benefits. In some such cases, patients, families, and care teams may determine that medically provided nutrition and hydration (MPNH) is not indicated. In the United States, removing MPNH was deemed permissible in the 1990 case of Nancy Cruzan ("Cruzan v. Director, Mo. Dept. of Health, 497 U.S. 261" 1990), Ms. Cruzan was a 25-year-old woman in PVS. Her parents wished to remove her gastrostomy tube and stop all MPNH. The case was adjudicated by the United States Supreme Court, and ultimately it was determined that stopping MPNH was both legally supported and ethically justifiable. Since that case, stopping MPNH has been widely accepted in the United States. Such practices are more controversial in other countries. Indeed, in many jurisdictions (Japan (Aita et al. 2008), many Islamic countries (Alsolamy 2014), Israel (Shalev 2009), etc.) stopping MPNH is generally considered illegal or unethical.

18.1.5 Physician-Assisted Dying

In some jurisdictions, physicians are allowed to prescribe substances to patients that will allow the patient to end her life quickly and painlessly. Some use the term "physician-assisted suicide"; however, because the term "suicide" has significant negative connotations, most prefer one of the more neutral terms: "death with dignity," "physician aid in dying," or "physician-assisted dying" (PAD). PAD is legal in The Netherlands, Belgium, Luxembourg, Colombia, Canada, Australia, and parts of the United States (in 2020, approximately one-third of Americans live in jurisdictions that allow PAD). PAD remains controversial, and many argue that it is unethical and should be illegal. Indeed, in most countries PAD is illegal.

18.1.6 Voluntary Active Euthanasia

Euthanasia (from the Greek "good death") is the deliberate termination of life, generally understood to be an act of mercy, with a goal of painlessly ending a person's life and suffering. Euthanasia may also be referred to as "mercy killing." The term "voluntary" indicates that the patient herself chooses life termination. Further, the term "active" indicates that there is an affirmative act (e.g., administration of life-ending substances) to terminate the patient's life, in contrast to "passive" (described above) in which life-prolonging interventions are removed and the patient is allowed to die from her underlying disease process.

Some jurisdictions allow not only PAD, but also voluntary active euthanasia (VAE). In such cases, patients wish to have their life ended quickly and painlessly in order to relieve their suffering. VAE is less accepted than PAD because physicians play an active role in administering the life-ending substances (Quill et al. 1997). It has been argued, however, that if we allow PAD, then we must allow VAE because some patients lack the physical ability to self-administer the life-ending substances, and to deny such individuals the option of PAD while allowing others who are more physically able to access this option is not ethically supportable (Brock 1992). Currently, VAE is practiced legally in only a few jurisdictions (The Netherlands, Belgium, Luxembourg, Colombia, and Canada). In all of these jurisdictions, patients must be actively involved in the choice to end their life. There is evidence, however, that VAE is practiced sub rosa in the United States, Australia, the United Kingdom, and elsewhere (Back et al. 1996; Brahams 1992; Maitra et al. 2005; Meier et al. 1998).

18.1.7 Nonvoluntary Active Euthanasia

VAE requires the voluntary choice of the patient herself. Clearly, some patients (e.g., infants) lack the decision-making capacity to make such a choice. In very few jurisdictions, families and care providers may choose to end the patient's life without the express consent of the patient herself. Such practice is termed "nonvoluntary active euthanasia" (NVAE) or "active life ending without consent."[5] Because others choose on behalf of the patient, active life ending without consent is legally permissible only in The Netherlands; however, there is some evidence that NVAE may be administered to adult patients clandestinely in Australia (Stevens and Hassan 1994), the United States (Meier et al. 1998), and potentially elsewhere.

18.1.8 Involuntary Active Euthanasia

In some cases, providers may end the life of persons against that person's expressed wishes; so-called "involuntary active euthanasia" (IVAE). Most notably, IVAE was carried out in Nazi Germany as part of the eugenics movement. The atrocities of the Nazis lead to the eugenics movement being widely discredited and abandoned, and IVAE is illegal universally and globally considered inconsistent with bioethical principles. Indeed, IVAE is ethically and legally simply murder.

---

[5] Those who are strongly opposed to such practices often refer to NVAE as "killing patients." Here, we use less pejorative terminology to allow the reader to weigh the arguments of both sides of this controversy; however, it is important to note that many clinicians and ethicists believe that such acts are merely killing and have strong moral objections to use of any other terminology.

18.2   End-of-Life Options in Neonates

Because neonates cannot participate in decision-making, any EOL decisions are necessarily nonvoluntary. In general, parents and care providers work collaboratively to determine the option or options that are in the infant's best interest. At times, it may be appropriate to consider the interest of others in decision-making for neonates and children; however, in general, the interests of the infant herself must be of primary concern (Katz et al. 2016).

Several EOL options are well-accepted in neonatal care. For example, limiting or withdrawing life-prolonging interventions is widely accepted and practiced when doing so is deemed consistent with the infant's best interest. There is significant variability, however, regarding when and in which cases life-prolonging interventions can or should be limited or withdrawn. Further, there is variability regarding who holds the authority to make such decisions (Lago et al. 2008; Liu et al. 2020; McHaffie et al. 1999; National Institute for Health and Care Excellence 2016 (updated 2019); Weise et al. 2017).

Clearly, infants cannot voluntarily stop eating and drinking, therefore VSED is not relevant to neonatal medicine; however, withholding MPNH is relevant and raises special ethical issues. Unlike most adults, neonates are incapable of feeding themselves. Indeed, feeding a baby is widely viewed as a primary obligation of parents and care providers, and withholding oral feeds from a baby who can bottle or breast feed and who is hungry is widely considered unethical and cruel. Because all infant feeding is provided by others, withholding MPNH raises special ethical issues that do not exist in the care of adult patients for whom being fed is not "normal." Because many view the obligation to feed an infant as paramount, and because MPNH (through a nasogastric tube, percutaneous gastrostomy tube, or other enteral tube feeding devices) may be viewed as feeding the baby, some argue that withholding MPNH is never ethically permissible in infants. Withholding MPNH in the neonatal setting is practiced openly in the United States and Europe (Bucher et al. 2018; Diekema et al. 2009; Kuhn et al. 2017; Moreno Villares 2015; National Institute for Health and Care Excellence 2016 (updated 2019); Weise et al. 2017); however, it is unclear the extent to which this practice is accepted elsewhere, and there remains significant debate regarding the ethical permissibility of withholding MPNH in the neonatal setting.

Due to the infant's inability to participate in decision-making, PAD, VAE, and IVAE are also irrelevant in the neonatal setting. Any hastening of death in the neonatal period would necessarily be a form of NVAE. In such cases, the ethical implications of NVAE in an infant would be similar to those in an adult patient who lacks the ability to participate in decision-making or the ability to have or express her wishes and preferences.

Research shows that neonatal euthanasia is practiced openly or sub rosa[6] in many countries. In 2000, the EURONIC Study Group reported that among survey respondents, 73% of French neonatologist, 47% of Dutch neonatologists, and several German, British, Swedish, Italian, and Spanish neonatologists reported personally administering drugs with the purpose of ending the life of an infant (Cuttini et al. 2000). Further, a 2004 study showed that a significant proportion of physicians in Lithuania had been personally involved in at least one case of neonatal euthanasia (Cuttini et al. 2004). In a 2020 study of Greek neonatologists, one subject indicated that they had performed neonatal euthanasia in one case (Dagla et al. 2020). Further, in a study of French doctors, nurses, and lay public, the overwhelming majority in each group favored neonatal euthanasia at least in some cases (Teisseyre et al. 2010). A similar attitude among nurse and doctors working in Flanders, Belgium was reported in a nationwide study in 2020 (Dombrecht et al. 2020). While the above references demonstrate health care professionals have reported euthanizing infants in many countries, it is likely that neonatal euthanasia is practiced even more widely than has been reported. Due to the criminal nature of such acts in many jurisdictions, and wide condemnation of such practices, however, it is likely that those who have euthanized infants would not report such activity.

18.3 Neonatal Euthanasia: Pro and Con Arguments

18.3.1 *Arguments in Favor of Allowing Neonatal Euthanasia*

The arguments supporting acceptability of neonatal euthanasia, and ultimately the endorsement of the Groningen Protocol in the Netherlands can probably best be understood in the context of the developments in the adult setting.

The majority of the population in the Netherlands has always been in favor of euthanasia, defined as deliberate medical life ending on the patients' own request, since 1966 in public opinion polls (Griffiths et al. 1998). Surveys among doctors have consistently shown their willingness to perform euthanasia for patients with unbearable suffering. Reports in the 1980s and 1990s from professional organizations and governmental institutions examined Dutch EOL care and made suggestions for practical ways to allow euthanasia and monitor and review these cases. During these years, several doctors were prosecuted after reporting they had ended the life of a patient who suffered unbearably, at the patient's request.

---

[6] The term "sub rosa" is defined by the Merriam-Webster dictionary as: in confidence; secretly. However, the term has a connotation of something carried out undercover, in confidence, privately, or with discretion whereas the term "secretly" has a connotation of being specifically designed to escape notice. See https://www.merriam-webster.com/dictionary/sub%20rosa, https://www.urbandiction ary.com/define.php?term=sub%20rosa, https://www.dictionary.com/browse/sub-rosa, and https://www.dictionary.com/browse/secret?s=t.

In 1984, in a landmark Dutch Supreme Court decision, the concept of necessity resulting from a conflict of duties was formulated as a potential legal justification for not prosecuting doctors who perform euthanasia. The conflict was between the duty to alleviate the patient's hopeless suffering and the duties to obey the law and to preserve the patient's life. Ultimately, this court's verdict resulted in the enactment of the Dutch Termination of Life on Request and Assisted Suicide Act ('Euthanasia Law') in 2002. This law stipulated that physicians may perform euthanasia if they are convinced that the patient: (a) made a voluntary and well-considered request; (b) suffers unbearably with no prospect of improvement, and there's no reasonable alternative to relieve the suffering; (c) understands the situation and prognosis. Additionally, an independent physician is consulted who must visit the patient and the physician performs the act with due care and reports the case to the regional review committee. Patients over the age of 16 years may request and consent to euthanasia. Patients aged 12–15 years may request euthanasia; however, parental participation in the decision-making process is also required.

The ethical justification of VAE for the Dutch originates in the principles: self-determination, beneficence, responsibility, and compassion (Widdershoven 2002). Politically, legalization of VAE as an option (not an obligation) in end-of-life care was justified by the broad acceptance of euthanasia in the population.

Similar to the developments for adults, a national debate started in the 1980's about end-of-life decision-making for severely ill newborns (Griffiths et al. 2008; Verhagen and Sauer 2005a, b). Influential reports from various professional organizations on the medical and ethical acceptability of EOL decisions in newborns, including neonatal euthanasia were published (Nederlandse Vereniging voor Kindergeneeskunde 1992). In two landmark cases against physicians who ended the life of a sick newborn in the 1990's, the high courts accepted necessity resulting from a conflict of interests as defense, identical to the justification in the rulings on the adult cases mentioned above.

Based on these verdicts, the 2005 Groningen Protocol was created and accepted by Dutch pediatricians (Verhagen and Sauer 2005a, b). Two years later, the Protocol became a legal governmental regulation which included the establishment of a multidisciplinary advisory committee that publicly reviews all neonatal euthanasia cases.

Clearly, the ethical justification for neonatal euthanasia differs in part from adult euthanasia especially where it concerns self-determination. The Groningen Protocol requires the agreement of both parents, which provides specific extension of the notion of self-determination that Brouwer et al. have called 'parental determination' (Brouwer et al. 2018). This parental determination is a bridge between self-determination and beneficence, which is another justifying principle. This view is based on the presumption that parents are the appropriate surrogate decision makers, and that parents give primacy to the best interests of their child. For a doctor to act beneficently, he needs to have sufficient understanding of the child's suffering. The parents provide a specific and necessary perspective on the child's suffering, informed by family values, intimate knowledge of the child, and their view on the child's quality of life. This parental determination prevents euthanasia for incompetent children from becoming an out-of-balance decision based only on beneficence.

From this brief comparative overview, it becomes apparent that the main arguments in favor of allowing neonatal euthanasia in the Netherlands are closely tied to the rationale employed in justification of adult euthanasia. Specifically, adult euthanasia was deemed permissible in order to improve the quality of dying for patients who were suffering unbearably. From the Dutch perspective, to deny loving parents the possibility to end unbearable and hopeless suffering for the newborn for whom they are responsible and whom they love feels unjust. Further, the majority of the Dutch medical community has always supported this argument. At the same time, physicians have underlined the need for procedural safeguards to prevent misuse and to protect the responsible physicians from unjust prosecution for murder. Hence the formal regulation with an obligation to report each case for review by a multidisciplinary advisory committee. In the view of most Dutch stakeholders in the debate, this set of provisions is the best way to make a complex medical practice transparent and open to review. The committee's annual reports are published as open-access documents (Committee Late Termination of Pregnancy and Termination of Life in Newborns (Commissie Late Zwangerschapsafbreking en levensbeeindiging bij pasgeborenen) 2020). Between 2007 and the writing of this chapter (2020), only 3 cases (2 with epidermolysis bullosa, 1 with progressive neurodegenerative disease) were reported and reviewed, and none was prosecuted.

18.3.2   Arguments Against Allowing Neonatal Euthanasia

Many authors have raised serious moral objection to any form of neonatal euthanasia (Jotkowitz and Glick 2006; Kodish 2008). Such moral objections are similar to those raised by opponents of VAE. Specifically, they argue that actively ending the life of another human is morally corrupt in all cases and is contrary to the obligations of a physician. Further, because infants are a vulnerable class, unable to express their wishes or advocate for their interests, actively killing an infant is even more abhorrent than killing an adult. Such an argument is persuasive to those who agree with the fundamental moral position that physicians should not kill patients; however, it is unpersuasive to those who hold a different moral belief. For those who hold this moral belief, no further argument is necessary. However, for those whose moral beliefs allow for euthanasia when doing so is consistent with bioethical principles, the patient's interests, and good medical practice, more scrutiny is required. As such, the following arguments are based on an assumption that VAE is ethically permissible; not because the acceptance of VAE is universal, but rather because neonatal euthanasia could be justified only if VAE is considered justifiable. However, as described below and as discussed by multiple authors, even if one posits that VAE is justifiable and consistent with good medical practice (in some cases), neonatal euthanasia is not justifiable

(Chervenak et al. 2006, 2008, 2009; Jotkowitz et al. 2008; Jotkowitz and Glick 2006; Kon 2007, 2008, 2009).[7]

In general, the ethical arguments against allowing neonatal euthanasia are the same as the arguments against all forms of NVAE. To best understand these arguments, a clear understand of the arguments in favor of PAD and VAE are necessary in order to then understand why they do not apply in the NVAE scenario.

The primary ethical support for PAD and VAE are two-fold: (1) The principle of respect for patient autonomy, and (2) The principle of beneficence. Many have argued that patients have a right to determine their own destiny. If a patient believes that her life is unbearable, or if she has significant concern regarding the progression of her disease with impending suffering and potentially loss of decision-making capacity, then, it is argued, she has a right to end her suffering (Quill 1991). Under this logic, if the patient chooses to end her life (either by taking a substance herself (PAD) or by having the physician administer a substance (VAE)), then the principle of respect for patient autonomy could be viewed as allowing the physician to participate. Alternatively, however, it can be argued that although the patient may have a right to request PAD or VAE, unless there is a medical indication for this intervention (e.g., the goal of treatment is to alleviate the patient's suffering, all reasonable interventions have failed to alleviate her suffering, and the patient and physician agree that the only reasonable option to end her suffering is the end her life; then it can be argued that PAD or VAE is medically indicated) the physician should not participate. It has been argued by many that PAD and VAE are contrary to the physician's duty to do no harm; however, others have argued that if the patient and physician agree that PAD or VAE is consistent with the patient's best interests, then the physician is not harming the patient by participating.

Alternatively, some point to the principle of beneficence as the primary support for PAD and VAE. If a patient is suffering unbearably, has exhausted all reasonable medical options without alleviation of her suffering, and believes that her suffering is a fate worse than death, then the patient's death may be viewed as therapeutic (Kon and Ablin 2010). That is, if the goal of treatment is to end the patient's suffering, and if all other treatments have failed to achieve this goal, and if the patient's suffering will end upon her death, then PAD or VAE may be seen as therapeutic, consistent with the principle of beneficence, and consistent with good medical practice.

In order to be ethically supportable under the principle of respect for patient autonomy, the patient must be able to make her own decisions. Clearly, in the case of infants, this is not possible. There is some measure of parental autonomy; however,

---

[7] NB: Several authors have written about ethical problems with the Groningen Protocol specific to how it has been implemented and the infants who have been euthanized under the protocol (Barry 2010; Callahan 2008; Van Der Maas et al. 1991). Here, we discuss only the broader ethical concerns with any form of nonvoluntary euthanasia, including neonatal euthanasia. We do this to focus on any form of such practice beyond the Groningen Protocol and its use. Readers should be aware, however, that even if they find the ethical arguments in favor or neonatal euthanasia compelling, there are deeper concerns with the clinical applications of the Groningen Protocol specifically as it was connected to infants with myelomeningocele.

there are significant limits to this autonomy and the state has an obligation to supersede parental authority when parents make choices that are contrary to the infant's best interests. As such, the principle of respect for patient autonomy cannot be the ethical basis for any form of NVAE, including neonatal euthanasia.

The principle of beneficence is more conducive for consideration as the ethical basis for NVAE. If the patient is suffering unbearably, and if death is the only therapeutic option that will alleviate that suffering, then NVAE could be ethically supportable. The problem here, however, is the judgement of whether the infant's suffering is *unbearable*. Through caring for adults and older children, it is clear that there is significant variability regarding what patients view as *unbearable* suffering. While many patients suffer, only the patient herself can judge whether that suffering is unbearable and whether living in her condition is worse than death. Clearly, we can judge whether a patient, even an infant, is suffering by regarding their face, listening for crying, seeing how they react to stimuli, and looking at their condition over time. We cannot, however, accurately judge whether the patient's suffering is unbearable. Many patients with severe unrelenting pain judge their suffering to be unbearable; however, other patients with the same degree of pain judge their suffering to be horrible but still better than being dead (i.e., not unbearable). The judgement of whether suffering is *unbearable* is wholly subjective and can be determined only by the patient herself (Jotkowitz and Glick 2006; Kodish 2008; Kon 2007).

Further, data suggest that health care providers, and even parents, are poor judges of the extent of children's suffering. Data suggest that when asked to self-assess their own quality of life, children with disabilities and "normal" children generally provide similar assessments (Saigal et al. 1996). Unfortunately, however, physicians generally rate the subjective quality of life of children with disabilities significantly lower than those children rate their own subjective quality of life (Janse et al. 2004). Further, in general, significant others of patients also generally underestimate patients' subjective quality of life (Sprangers and Aaronson 1992). These findings are critical because they demonstrate that physicians and parents are highly likely to overestimate the burdens to an infant and overestimate the infant's suffering (see Chapter DEVEREAUX). As such, empirical research suggests that many infants who suffer, but whose suffering is not unbearable, will be judged to have unbearable suffering by physicians and parents. This would necessarily lead to euthanizing infants who are not suffering unbearably.

Based on the above discussion, if we forbid NVAE, there will be some infants who suffer unbearably who are kept alive with unbearable suffering. Alternatively, if we allow NVAE, there will certainly be infants whose suffering is significant but not unbearable, and for whom being alive would be preferable to being dead, who will be euthanized. The question we must answer, therefore, is: Is it better to keep some patients with unbearable suffering alive, or to euthanize some patients who are not suffering unbearably?

One of the foundational tenets of medicine is *do no harm* (Hippocrates 400 B.C.E.). Clearly, there are times when physicians must harm patients in order to further their best interests. For example, cutting a person open is harming them; however, if doing so is the most appropriate treatment for their perforated appendix, then such harm is justified. Similarly, it may be argued that euthanizing a patient is harming them; however, if doing so is the most appropriate treatment for their unbearable suffering, then such harm may be ethically permissible. Because we cannot accurately judge the *unbearableness* of an infant's suffering, and because research shows that doctors and parents are highly likely to overestimate the burdens of disease and disability, it is clear that allowing neonatal euthanasia would lead to the killing of some infants whose suffering is not unbearable. We use the term "killing" here because in such cases, life-termination is not euthanasia, it is simply killing a baby. Because killing some babies whose suffering is not unbearable is widely considered worse than keeping some babies alive who are suffering unbearably, the only conclusion can be that neonatal euthanasia is unsupportable.

18.4 Use of Paralytics and Neonatal Euthanasia: Pro and Con Arguments

18.4.1 Arguments in Favor of Using Paralytics

In the last 15 years, several studies were carried out in the Netherlands to monitor end-of-life practice in the Dutch Neonatal Intensive Care Units (NICUs) (Verhagen et al. 2009, 2010, 2007). In up to 16% of NICU deaths, providers administered paralytic agents (also referred to as neuromuscular blockade) at the time of withdrawal of mechanical ventilation. The main argument from these studies was that the patients already received these agents to support respiratory treatments at the time life-sustaining treatment was withheld or withdrawn, and discontinuing these medications would likely contribute to suffering. A second argument was to stop or prevent gasping in the final phase of dying babies on parental request. Paralytic agents were always combined with the administration of opioids and/or sedatives as comfort providing medication. The main reason for the use of paralytics instead of other medications was that they simply work well for both indications. Interestingly, physicians had different rationales for using paralytics in dying newborns: some viewed it as palliative care, while others viewed it as deliberate hastening of death, which needed to be reported as neonatal euthanasia and reviewed.

When it became clear that health care providers used different definitions of newborn euthanasia and palliative care, a multidisciplinary group of experts was created to address this controversy (Willems et al. 2014). According to this expert group, administration of paralytics is permitted if the aim is to stop prolonged gasping during ventilator withdrawal and to end a dying process presumed to take several hours or more, which only adds to the suffering of the parents. This uncommon situation may occur when even state-of-the-art palliative sedation is insufficient to relief pain and suffering and despite the medical team's careful preparation of the parents. The experts ultimately concluded that administering paralytics in these circumstances should be regarded as "good medical practice," but they recommend that, in view of the ongoing debate about its legality, all cases must be reported for review to maintain full transparency and accountability.

Most of the group's recommendations were adopted in the evidence-based Guidelines for Pediatric Palliative Care that was issued by the Dutch Pediatric Association in 2013 and updated a few years later. Interestingly, the debate about use of paralytic agents in newborns has faded since 2018, and our 2020 informal survey confirmed that paralytics are no longer used in end-of-life care and that most units have removed paralytics from their EOL care protocols and palliative plans.

So, thanks to the thorough and repetitive studies of EOL care in the Dutch NICU's we know exactly how and why paralytics are administered in some severely ill babies. These data helped health care professionals and parents to reflect on this issue, rethink EOL care strategies, and find ways to control/regulate paralytic use to prevent misuse as covert euthanasia. Experts have indicated that probably as a result of data analysis and multidisciplinary debate, paralytics are no longer part of Dutch NICU care.

18.4.2 Arguments Against Using Paralytics

The primary reason to use paralytics in euthanasia is for the comfort of those seeing the patient die. Paralytics mask any signs of pain or discomfort, air hunger, anxiety, or other evidence of suffering. Indeed, one survey demonstrated that when physicians use paralytics during neonatal euthanasia, they generally do so to mask gasping, a sign of air hunger and suffering (Dorscheidt et al. 2013). To be clear, paralytics do not decrease air hunger or the patient's suffering, they merely mask the signs of such suffering so that others are less uncomfortable with the infant's demise.

Although seeing a patient suffer air hunger, anxiety, etc. while she is being euthanized can be very unsettling, it is imperative that care teams not mask such symptoms. If a patient exhibits signs of suffering while being euthanized, then providers have an obligation to aggressively treat that suffering. If all suffering is fully treated, then the patient will not show signs of anxiety, air hunger, pain, etc. and therefore paralytics would not be required. If the goal of NVAE is to relieve the patient's suffering, then masking her suffering during dying is contraindicated and ethically unsupportable.

Further, the use of paralytics has a high likelihood of harming the patient. Because paralyzed patients cannot be assessed for distress, air hunger, pain, etc., there is a high likelihood that such suffering will go untreated in the dying infant. Paralytics harm the infant, provide no benefit to the infant, and are used solely so that parents and members of the care team do not see signs of suffering during the dying process. As such, use of paralytics in any form of euthanasia is unethical and inconsistent with good medical practice.

18.5   Conclusion

Neonatal euthanasia has been practiced legally in the Netherlands since 2005. Outside of the Netherlands, neonatal euthanasia remains illegal; however, data suggest that it is practiced sub rosa by some clinicians in other countries. The true extent of neonatal euthanasia practice remains unknown. The ethical support for neonatal euthanasia stems from the duty of health care professionals to treat patient suffering. If the only intervention that can alleviate the patient's suffering is death, then death may be seen as therapeutic and hastening death may be ethically appropriate. In contrast, the ethical arguments against neonatal euthanasia stem from the ethical principle of do no harm and the belief that only the patient herself can judge whether her suffering is truly unbearable. Further discourse on this subject should illuminate these and other ethical arguments for and against NVAE.

Guiding Principles in Neonatal Euthanasia

• Neonatal euthanasia is illegal in all countries except The Netherlands
• Data suggest that neonatal euthanasia occurs occasionally in many countries in which it is illegal. The true extent of clandestine neonatal euthanasia is unknown
• In The Netherlands, neonatal euthanasia is performed legally using a detailed protocol (the Groningen Protocol). The protocol aims to ensure that only infants who face a life of unbearable suffering are euthanized. All cases undergo thorough post hoc review to ensure they were conducted appropriately
• Critics of the protocol argue that the safeguards are not sufficient to ensure that infants who suffer, but whose suffering is not unbearable, will not be euthanized. Many critics point to the use of the protocol in patients with myelomeningocele as evidence that the protocol is not ethically supportable
• Critics of neonatal euthanasia further argue that no protocol could be devised that would provide adequate protection for neonates and therefore neonatal euthanasia in all forms in not ethically supportable
• Like all areas of end-of-life care, this is an evolving field

References

Aita, K., H. Miyata, M. Takahashi, and I. Kai. 2008. Japanese physicians' practice of withholding and withdrawing mechanical ventilation and artificial nutrition and hydration from older adults with very severe stroke. *Archives of Gerontology and Geriatrics* 46(3): 263–272. https://doi.org/10.1016/j.archger.2007.04.006.

Alsolamy, S. 2014. Islamic views on artificial nutrition and hydration in terminally ill patients. *Bioethics* 28(2): 96–99. https://doi.org/10.1111/j.1467-8519.2012.01996.x.

Asai, A., S. Fukuhara, O. Inoshita, Y. Miura, N. Tanabe, and K. Kurokawa. 1997. Medical decisions concerning the end of life: A discussion with Japanese physicians. *Journal of Medical Ethics* 23(5): 323–327. https://doi.org/10.1136/jme.23.5.323.

Back, A.L., J.I. Wallace, H.E. Starks, and R.A. Pearlman. 1996. Physician-assisted suicide and euthanasia in Washington State. Patient requests and physician responses. *JAMA* 275(12): 919–925. https://www.ncbi.nlm.nih.gov/pubmed/8598619.

Barry, S. 2010. Quality of life and myelomeningocele: An ethical and evidence-based analysis of the Groningen Protocol. *Pediatric Neurosurgery* 46 (6): 409–414. https://doi.org/10.1159/000322895.

Brahams, D. 1992. Euthanasia: Doctor convicted of attempted murder. *Lancet* 340(8822): 782–783. https://doi.org/10.1016/0140-6736(92)92314-6.

Brock, D.W. 1992. Voluntary active euthanasia. *The Hastings Center Report* 22(2): 10–22. https://www.ncbi.nlm.nih.gov/pubmed/1587719.

Brouwer, M., C. Kaczor, M.P. Battin, E. Maeckelberghe, J.D. Lantos, and E. Verhagen. 2018. Should pediatric euthanasia be legalized? *Pediatrics* 141(2). https://doi.org/10.1542/peds.2017-1343.

Bucher, H.U., S.D. Klein, M.J. Hendriks, R. Baumann-Holzle, T.M. Berger, J.C. Streuli, J.C. Fauchere, and Swiss Neonatal End-of-life Study Group. 2018. Decision-making at the limit of viability: Differing perceptions and opinions between neonatal physicians and nurses. *BMC Pediatr* 18(1): 81. https://doi.org/10.1186/s12887-018-1040-z.

Callahan, D. 2008. "Are their babies different from ours?" Dutch culture and the Groningen Protocol. *The Hastings Center Report* 38(4): 4–6; author reply 7–8. https://www.ncbi.nlm.nih.gov/pubmed/18711810.

Chervenak, F.A., L.B. McCullough, and B. Arabin. 2009. The Groningen Protocol: Is it necessary? Is it scientific? Is it ethical? *Journal of Perinatal Medicine* 37 (3): 199–205. https://doi.org/10.1515/JPM.2009.058.

Chervenak, F.A., L.B. McCullough, and B. Arabin. 2006. Why the Groningen Protocol should be rejected. *The Hastings Center Report* 36(5): 30–33. https://doi.org/10.1353/hcr.2006.0073.

Chervenak, F.A., L.B. McCullough, and B. Arabin. 2008. "Are their babies different from ours?" Dutch culture and the Groningen Protocol. *The Hastings Center Report* 38(4): 6; author reply 7–8. https://www.ncbi.nlm.nih.gov/pubmed/18709907.

Committee Late Termination of Pregnancy and Termination of Life in Newborns (Commissie Late Zwangerschapsafbreking en levensbeeindiging bij pasgeborenen). (2020). *Annual Reports*. https://www.lzalp.nl/publicaties/jaarverslagen. Accessed 12 Sep 2020.

Cruzan v. Director, Mo. Dept. of Health, 497 U.S. 261, (Supreme Court of the United States 1990). https://www.courtlistener.com/opinion/112478/Cruzan-v-director-mo-dept-of-health/.

Cuttini, M., M. Nadai, M. Kaminski, G. Hansen, R. de Leeuw, S. Lenoir, J. Persson, M. Rebagliato, M. Reid, U. de Vonderweid, H.G. Lenard, M. Orzalesi, and R. Saracci. 2000. End-of-life decisions in neonatal intensive care: Physicians' self-reported practices in seven European countries. EURONIC Study Group. *Lancet* 355(9221): 2112–2118. https://doi.org/10.1016/s0140-6736(00)02378-3.

Cuttini, M., V. Casotto, M. Kaminski, I. de Beaufort, I. Berbik, G. Hansen, L. Kollee, A. Kucinskas, S. Lenoir, A. Levin, M. Orzalesi, J. Persson, M. Rebagliato, M. Reid, and R. Saracci. 2004. Should euthanasia be legal? An international survey of neonatal intensive care units staff. *Archives of Disease in Childhood-Fetal and Neonatal Edition* 89(1): F19–F24. https://doi.org/10.1136/fn.89.1.f19.

Dagla, M., V. Petousi, and A. Poulios. 2020. Bioethical decisions in neonatal intensive care: Neonatologists' self-reported practices in Greek NICUs. *International Journal of Environmental Research and Public Health* 17(10). https://doi.org/10.3390/ijerph17103465.

Diekema, D.S., J.R. Botkin, and Committee on Bioethics. 2009. Clinical report—Forgoing medically provided nutrition and hydration in children. *Pediatrics* 124(2): 813–822. https://doi.org/10.1542/peds.2009-1299.

Dombrecht, L., L. Deliens, K. Chambaere, S. Baes, F. Cools, L. Goossens, G. Naulaers, E. Roets, V. Piette, J. Cohen, K. Beernaert, and Consortium N. 2020. Neonatologists and neonatal nurses have positive attitudes towards perinatal end-of-life decisions, a nationwide survey. *Acta Paediatrica* 109(3): 494–504. https://doi.org/10.1111/apa.14797.

Dorscheidt, J.H., E. Verhagen, P.J. Sauer, and J.H. Hubben. 2013. Medication regimes in the context of end-of-life decisions in neonatology: Legal considerations with regard to Dutch NICU-practice. *Medical Law* 32(2): 215–229. https://www.ncbi.nlm.nih.gov/pubmed/23967795.

Giubilini, A., and F. Minerva. 2013. After-birth abortion: Why should the baby live? *Journal of Medical Ethics* 39(5): 261–263. https://doi.org/10.1136/medethics-2011-100411.

Griffiths, J., H. Weyers, and M. Adams. 2008. Termination of life in neonatology. In *Euthanasia and Law in Europe*, 217–255. Hart Publishing.

Griffiths, J., A. Bood, and H. Weyers. 1998. *Euthanasia and law in the Netherlands*. Amsterdam University Press.

van der Heide, A., P.J. van der Maas, G. van der Wal, C.L. de Graaff, J.G. Kester, L.A. Kollee, R. de Leeuw, and R.A. Holl. 1997. Medical end-of-life decisions made for neonates and infants in the Netherlands. *Lancet* 350(9073): 251–255. https://doi.org/10.1016/S0140-6736(97)02315-5.

Hippocrates. (400 B.C.E.). *Of the epidemics* (F. Adams, Trans.). http://classics.mit.edu/Hippocrates/ epidemics.1.i.html.

In Re Quinlan, 355 A.2d 647. (Supreme Court of New Jersey. 1976). https://www.courtlistener. com/opinion/1537678/in-re-Quinlan/.

Janse, A.J., R.J. Gemke, C.S. Uiterwaal, I. van der Tweel, J.L. Kimpen, and G. Sinnema. 2004. Quality of life: Patients and doctors don't always agree: A meta-analysis. *Journal of Clinical Epidemiology* 57(7): 653–661. https://doi.org/10.1016/j.jclinepi.2003.11.013.

Jotkowitz, A.B., and S. Glick. 2006. The Groningen Protocol: Another perspective. *Journal of Medical Ethics* 32(3): 157–158. https://doi.org/10.1136/jme.2005.012476.

Jotkowitz, A., S. Glick, and B. Gesundheit. 2008. A case against justified non-voluntary active euthanasia (the Groningen Protocol). *American Journal of Bioethics* 8(11): 23–26. https://doi. org/10.1080/15265160802513085.

Katz, A.L., S.A. Webb, and Committee on Bioethics. 2016. Informed consent in decision-making in pediatric practice. *Pediatrics* 138(2). https://doi.org/10.1542/peds.2016-1485.

Kodish, E. 2008. Paediatric ethics: A repudiation of the Groningen Protocol. *Lancet* 371(9616): 892–893. https://doi.org/10.1016/s0140-6736(08)60402-x.

Kon, A.A. 2007. Neonatal euthanasia is unsupportable: The Groningen Protocol should be abandoned. *Theoretical Medicine and Bioethics* 28 (5): 453–463. https://doi.org/10.1007/s11017-007-9047-8.

Kon, A.A., and A.R. Ablin. 2010. Palliative treatment: Redefining interventions to treat suffering near the end of life. *Journal of Palliative Medicine* 13(6): 643–646. https://doi.org/10.1089/jpm. 2009.0410.

Kon, A.A. 2008. We cannot accurately predict the extent of an infant's future suffering: The Groningen Protocol is too dangerous to support. *The American Journal of Bioethics* 8(11): 27–29. https://doi.org/10.1080/15265160802513150.

Kon, A.A. 2009. Neonatal euthanasia. *Seminars in Perinatology* 33(6): 377–383. https://doi.org/ 10.1053/j.semperi.2009.07.005.

P. Kuhn, L. Dillenseger, N. Cojean, B. Escande, C. Zores, and D. Astruc. 2017. Palliative care after neonatal intensive care: Contributions of Leonetti Law and remaining challenges. *Archives of Pediatrics* 24(2): 155–159. https://doi.org/10.1016/j.arcped.2016.11.012. (Soins palliatifs au decours d'une reanimation neonatale: apports de la loi Leonetti et defis persistants).

Lago, P.M., J. Piva, P.C. Garcia, E. Troster, A. Bousso, M.O. Sarno, L. Torreao, R. Sapolnik, and Brazilian Pediatric Center of Studies on E. 2008. End-of-life practices in seven Brazilian pediatric intensive care units. *Pediatric Critical Care Medicine* 9(1): 26–31. https://doi.org/10. 1097/01.PCC.0000298654.92048.BD.

Liu, H., D. Su, X. Guo, Y. Dai, X. Dong, Q. Zhu, Z. Bai, Y. Li, and S. Wu. 2020. Withdrawal of treatment in a pediatric intensive care unit at a Children's Hospital in China: A 10-year retrospective study. *BMC Medical Ethics* 21(1): 71. https://doi.org/10.1186/s12910-020-00517-y.

Van Der Maas, P.J., J.J. Van Delden, L. Pijnenborg, and C.W. Looman. 1991. Euthanasia and other medical decisions concerning the end of life. *Lancet* 338(8768): 669–674. https://doi.org/10. 1016/0140-6736(91)91241-l.

Maitra, R.T., A. Harfst, L.M. Bjerre, M.M. Kochen, and A. Becker. 2005. Do German general practitioners support euthanasia? Results of a nation-wide questionnaire survey. *The European Journal of General Practice* 11(3–4): 94–100. https://doi.org/10.3109/13814780509178247.

Maltoni, M., C. Pittureri, E. Scarpi, L. Piccinini, F. Martini, P. Turci, L. Montanari, O. Nanni, and D. Amadori. 2009. Palliative sedation therapy does not hasten death: Results from a prospective multicenter study. *Annals of Oncology* 20(7): 1163–1169. https://doi.org/10.1093/annonc/mdp 048.

McHaffie, H.E., M. Cuttini, G. Brolz-Voit, L. Randag, R. Mousty, A.M. Duguet, B. Wennergren, and P. Benciolini. 1999. Withholding/withdrawing treatment from neonates: Legislation and official guidelines across Europe. *Journal of Medical Ethics* 25(6): 440–446. https://doi.org/10.1136/jme.25.6.440.

Meier, D.E., C.A. Emmons, S. Wallenstein, T. Quill, R.S. Morrison, and C.K. Cassel. 1998. A national survey of physician-assisted suicide and euthanasia in the United States. *New England Journal of Medicine* 338(17): 1193–1201. https://doi.org/10.1056/NEJM199804233381706.

Mercadante, S., G. Intravaia, P. Villari, P. Ferrera, F. David, and A. Casuccio. 2009. Controlled sedation for refractory symptoms in dying patients. *Journal of Pain and Symptom Management* 37(5): 771–779. https://doi.org/10.1016/j.jpainsymman.2008.04.020.

Moreno Villares, J.M. 2015. Nutrition and hydration in newborns: Limiting treatment decisions. *Cuad Bioet* 26(87): 241–249. https://www.ncbi.nlm.nih.gov/pubmed/26378597. (Hidratacion y Alimentacion en los Recien Nacidos: Adecuacion del Esfuerzo Terapeutico).

National Institute for Health and Care Excellence. 2016 (updated 2019). *End of life care for infants, children and young people with life-limiting conditions: planning and management. NICE guideline [NG61].* https://www.nice.org.uk/guidance/NG61/chapter/Recommendations#managing-hydration.

Nederlandse Vereniging voor Kindergeneeskunde. 1992. *Doen of laten. Grenzen van het medisch handelen in de neonatologie [To treat or not to treat? Limits for life-sustaining treatment in neonatology].* Den Daas.

Quill, T.E., B. Lo, and D.W. Brock. 1997. Palliative options of last resort: a comparison of voluntarily stopping eating and drinking, terminal sedation, physician-assisted suicide, and voluntary active euthanasia. *JAMA* 278(23): 2099–2104. https://doi.org/10.1001/jama.278.23.2099.

Quill, T.E., B. Lo, D.W. Brock, and A. Meisel. 2009. Last-resort options for palliative sedation. *Annals of Internal Medicine* 151(6): 421–424. https://doi.org/10.7326/0003-4819-151-6-200909 150-00007.

Quill, T.E. 1991. Death and dignity. A case of individualized decision-making. *New England Journal of Medicine* 324(10): 691–694. https://doi.org/10.1056/NEJM199103073241010.

Saigal, S., D. Feeny, P. Rosenbaum, W. Furlong, E. Burrows, and B. Stoskopf. 1996. Self-perceived health status and health-related quality of life of extremely low-birth-weight infants at adolescence. *JAMA* 276(6): 453–459. https://www.ncbi.nlm.nih.gov/pubmed/8691552.

Sauer, P.J., and A.A. Verhagen. 2009. The Groningen Protocol, unfortunately misunderstood. Commentary on Gesundheit et al.: The Groningen Protocol—The Jewish perspective (Neonatology 2009;96:6–10). *Neonatology* 96(1): 11–12. https://doi.org/10.1159/000196883.

Seminars in Perinatology. Volume 33, Issue 6. 2009. Issue entitled "Ethical Issues in the Perinatal Period."

Shalev, C. 2009. End-of-life care in Israel—The dying patient law. *Israel Law Review* 42(2): 279–305. https://doi.org/10.1017/S0021223700000583.

Sprangers, M.A., and N.K. Aaronson. 1992. The role of health care providers and significant others in evaluating the quality of life of patients with chronic disease: A review. *Journal of Clinical Epidemiology* 45(7): 743–760. https://doi.org/10.1016/0895-4356(92)90052-o.

C.A. Stevens, and R. Hassan, R. (1994, Mar). Management of death, dying and euthanasia: attitudes and practices of medical practitioners in South Australia. *J Med Ethics,* 20(1), 41–46. https://doi.org/10.1136/jme.20.1.41.

Teisseyre, N., C. Vanraet, P.C. Sorum, and E. Mullet. 2010. The acceptability among lay persons and health professionals of actively ending the lives of damaged newborns. *Monash Bioethics Review* 20(2): 14 11–24. https://doi.org/10.1007/BF03351524.

Verhagen, E. 2006. End of life decisions in newborns in the Netherlands: Medical and legal aspects of the Groningen Protocol. *Medical Law* 25(2): 399–407. https://www.ncbi.nlm.nih.gov/pubmed/16929815.

Verhagen, A.A. 2013. The Groningen Protocol for newborn euthanasia: Which way did the slippery slope tilt? *Journal of Medical Ethics* 39(5): 293–295. https://doi.org/10.1136/medethics-2013-101402.

Verhagen, A.A., and P.J. Sauer. 2005a. End-of-life decisions in newborns: An approach from The Netherlands. *Pediatrics* 116(3): 736–739. https://doi.org/10.1542/peds.2005-0014. http://www.ncbi.nlm.nih.gov/entrez/query.fcgi?cmd=Retrieve&db=PubMed&dopt=Citation&list_uids=16140716.

Verhagen, E., and P.J. Sauer. 2005b. The Groningen protocol—Euthanasia in severely ill newborns. *New England Journal of Medicine* 352(10): 959–962. https://doi.org/10.1056/NEJMp058026.

Verhagen, A.A., and P.J. Sauer. 2008. "Are their babies different from ours?" Dutch culture and the Groningen Protocol. *The Hastings Center Report* 38(4): 4; author reply 7–8. https://www.ncbi.nlm.nih.gov/pubmed/18709906.

Verhagen, A.A., J.J. Sol, O.F. Brouwer, and P.J. Sauer. 2005. Deliberate termination of life in newborns in The Netherlands; Review of all 22 reported cases between 1997 and 2004. *Ned Tijdschr Geneeskd* 149(4): 183–188. https://www.ncbi.nlm.nih.gov/pubmed/15702738 (Actieve levensbeeindiging bij pasgeborenen in Nederland; analyse van alle 22 meldingen uit 1997/04).

Verhagen, A.A., M.A. van der Hoeven, R.C. van Meerveld, and P.J. Sauer. 2007. Physician medical decision-making at the end of life in newborns: Insight into implementation at 2 Dutch centers. *Pediatrics* 120(1): e20–e28. https://doi.org/10.1542/peds.2006-2555.

Verhagen, A.A., J.H. Dorscheidt, B. Engels, J.H. Hubben, and P.J. Sauer. 2009 (Epub ahead of print). Analgesics, sedatives and neuromuscular blockers as part of end-of-life decisions in Dutch NICU's. *Archives of Disease in Childhood-Fetal and Neonatal Edition* 94: F434–F438. https://doi.org/10.1136/adc.2008.149260.

Verhagen, A.A., A. Janvier, S.R. Leuthner, B. Andrews, J. Lagatta, A.F. Bos, and W. Meadow. 2010. Categorizing neonatal deaths: A cross-cultural study in the United States, Canada, and the Netherlands. *The Journal of Pediatrics* 156(1): 33–37. https://doi.org/10.1016/j.jpeds.2009.07.019.

de Vries, M.C., and A.A. Verhagen. 2008. A case against something that is not the case: The Groningen Protocol and the moral principle of non-maleficence. *American Journal of Bioethics* 8(11): 29–31. https://doi.org/10.1080/15265160802521039.

Weise, K.L., A.L. Okun, B.S. Carter, C.W. Christian, Committee on Bioethics, Section on Hospice and Palliative Medicine, Committee on Child Abuse and Neglect. 2017. Guidance on forgoing life-sustaining medical treatment. *Pediatrics* 140(3). https://doi.org/10.1542/peds.2017-1905.

Widdershoven, G.A. 2002. Beyond autonomy and beneficence: The moral basis of euthanasia in the Netherlands. *Ethical Perspect* 9(2–3): 96–102. https://www.ncbi.nlm.nih.gov/pubmed/15712440.

Willems, D.L., A.A. Verhagen, E. van Wijlick, and on behalf of the Committee End-of-life Decisions in Severely Ill Newborns of the Royal Dutch Medical, A. 2014. Infants' best interests in end-of-life care for newborns. *Pediatrics*https://doi.org/10.1542/peds.2014-0780.

Further Reading

Verhagen, E., and P.J. Sauer. 2005. The Groningen protocol—Euthanasia in severely ill newborns. *New England Journal of Medicine* 352 (10): 959–962. https://doi.org/10.1056/NEJMp058026.

The American Journal of Bioethics. Volume 8, Issue 11. 2008. Target article discussing the ethics of the Groningen Protocol with four commentaries.

Brouwer, M., C. Kaczor, M.P. Battin, E. Maeckelberghe, J.D. Lantos, E. Verhagen. 2018. Should pediatric euthanasia be legalized? *Pediatrics* 141(2): e20171343. https://doi.org/10.1542/peds.2017-1343. Epub 2018 Jan 9.

Theoretical Medicine and Bioethics. Volume 28, Issue 5. 2007. Issue entitled "Ethics in the Nursery."

Chapter 19
Genetic Testing and Screening of Children

M. B. Menzel and V. N. Madrigal

Abstract Scientific advancements in the genetic testing and screening of children have provided answers for some and afforded therapies and preventive guidance for others. These benefits have the potential to revolutionize preventive medicine and categorically change outcomes in specific diseases. Ethical challenges emerge, however, when the benefits of testing come with a price related to its inherent ambiguities and uncertainties. Testing a child at risk for a condition of adult onset, for example, has generated tremendous debate and though generally discouraged, continues to plague clinicians dealing with the nuanced narrative at the bedside. In this chapter we unpack some of the arguments for genetic testing and screening in children. We use the best interest standard to explore these arguments and acknowledge when it falls short of helping to answer the question "what is ethically permissible?". We explore the risks and potential harms done by performing or not performing a genetic test in a child, including psychological effects such as guilt. We highlight the importance of the child's voice with such concepts as assent, informed consent, capacity and disclosure. We explore prenatal and newborn screening, and we address the increasing complexity of the patient as consumer of knowledge and manager of health choices. Finally, we aim to give the clinician a practical guide for determining what is ethically permissible and how to navigate decisions regarding genetic testing and screening of children. We include these convictions in the context of the North American perspective and offer areas of international discrepancy.

Keywords Genetic · Newborn · Prenatal · Child · Adolescent

M. B. Menzel (✉)
Pediatric Ethics Program, Children's National Hospital, George Washington University, Washington DC, USA
e-mail: MMenzel@childrensnational.org

V. N. Madrigal
Division Critical Care Medicine, Children's National Hospital, George Washington University, Washington DC, USA
e-mail: VMadriga@childrensnational.org

© Springer Nature Switzerland AG 2022
N. Nortjé and J. C. Bester (eds.), *Pediatric Ethics: Theory and Practice*,
The International Library of Bioethics 89,
https://doi.org/10.1007/978-3-030-86182-7_19

19.1 Introduction

Questions about genetic testing and screening arise frequently within the clinical sphere and deserve examination through the lens of clinical ethics. In North America, examination via the four classic principles of bioethics (autonomy, beneficence, maleficence and justice) is critical and can often help to illustrate the dilemma (Jonsen et al. 2015). When two or more of these principles are in conflict, we turn to additional tools such as the best interest standard or the harm principle (refer to Chap. 4), along with shared decision-making, transparent communication, and recognition of bias, to elicit and examine the relevant issues. When the questions are specific to children, the questions deserve further care and attention: exploring informed consent vs. assent, and capacity and respect for persons. Outside the United Sates, many countries accept the United Nations Convention on the rights of the Child (refer to Chap. 5), recognizing that children have fundamental human rights and possess inherent dignity (Powell 2019).

Genetic testing and screening are uniquely poised to create dilemmas in clinical medicine. The identification of genetic variants and their relevance to disease processes has challenged the precision of counseling, prognosis and treatment. As uncertainty expands regarding how a particular genetic mutation may or may not affect an individual, what therapies may or may not be effective, and whether preventions may be burdensome, ethical challenges balloon and are often met with a broader range of ethically permissible actions. In cases where only seemingly bad options exist, an ethics understanding helps elicit the underlying issues, frame the important questions at stake, and dive deep into the individual narrative to emerge with viable options.

Each scientific discovery, whether in diagnosis, therapy, or prevention in this rapidly evolving field, adds an additional layer to the questions and the answers, and must constantly be re-examined. What was ethically permissible five years ago may be unacceptable five years from now. These issues are complicated by a population obsessed with knowledge-consumption met by a medical system invested in attracting and keeping patients. Restraint from testing, though ethically appropriate in certain circumstances, might be met with extraordinary resistance.

And while much of the field is science-based, the role of emotion must be a recognized contributor. Guilt, fear, anger and anxiety drive much of the parent and patient decision-making on whether or not to test and screen. An ideal approach creates a safe space and builds trust in order to explore some of these motivations.

In this chapter, we will identify many of the ethical issues that arise for children and families in the context of genetic testing and screening. We identify important risks and benefits, in the context of best interest standard and harm principle (refer to Chap. 4). We explore appropriate issues for each age group including disclosure and capacity, assent and consent. We acknowledge emotion. We give special attention to prenatal and newborn screening and testing and highlight important considerations.

19.2   Historical Context

Routine genetic testing became an option for patients in the mid-1950s, after the formal introduction by scientists James D. Watson and Francis H. C. Crick of the double-helix structure of DNA, the molecule containing human genes. It was by the late 1950s that chromosome analysis (karyotyping) became clinically available, and in the late 1960s that prenatal karyotyping via amniocentesis was introduced (Durmaz et al. 2015). Since then, the testing options and information that can be obtained from a genetic test have increased significantly. One analogy used to describe the differences between genetic tests is to compare a karyotype (counting the number of chromosomes) to counting the number of rooms in a mansion, whereas more detailed tests such as a microarray or single gene test, allow one to identify the present or missing furniture in those rooms, or sometimes even whether or not there is a scratch or stain on that furniture.

The distinction between genetic testing and screening is an important one. Genetic testing involves looking for changes in chromosomes, genes or proteins of an individual, usually via a blood draw or tissue sample. Genetic screening, often referenced during pregnancy, has the goal of identifying whether or not an individual may be at an increased risk of having or developing a particular condition. For example, The NIPS (noninvasive prenatal screening) blood test offered to pregnant women, can identify whether or not they are at increased risk of carrying a fetus with certain chromosome abnormalities, but only diagnostic testing such as an amniocentesis with karyotype or microarray can diagnose or rule out that chromosome abnormality (Schonberg and Menzel 2019). Another example is newborn screening which can identify whether or not a newborn is at increased risk of having a particular genetic condition but also requires confirmatory genetic testing to make a diagnosis. Over 6,000 genetic disorders are now known, and it is estimated that there are over 75,000 genetic tests on the market (Phillips et al. 2018). Navigating the complexity of basic genetic testing and screening is challenging for both patients and providers, and inherent in this complexity are a myriad of ethical challenges, particularly when the questions affect children.

19.3   Genetic Testing and Screening Using the Best Interest Standard

Genetic testing of infants and children can offer many benefits. For those children with developmental delays and health concerns, genetic testing may reveal a diagnosis, expanding awareness of and access to medical information and guidance. Most genetic conditions affect multiple systems, and information for families on what medical screening they need to do can be lifesaving. For example, a diagnosis of 22q11 deletion syndrome,[1] would warrant a cardiac, immunologic, endocrine and, and developmental workup along with other studies (Bassett et al. 2011). The diagnosis may relieve the parents (often the mother) of guilt that the child's symptoms were caused by something preventable in pregnancy or early infancy. The knowledge of a diagnosis may also provide an instant community and support system for children and families who may have previously felt isolated. Families with the same diagnosis may connect via social media, support groups and conferences. This can be incredibly normalizing and empowering for families and patients who can now share common experiences and resources.

For these reasons, targeted testing for a condition with a pediatric onset yields fewer ethical questions. Most clinicians and parents would agree that exploring the etiology of a constellation of symptoms is in the child's best interest, even if no therapies or prevention are available. And if those do exist for a particular condition, urgency might be recognized to implement those measures as soon as possible.

19.3.1 Clarifying Goals and Motives of Testing for Young Children

A genetic test must be offered to the child/family for the right reasons, and it is important to clarify the goals of genetic testing. Reasonable goals include the possibility that a diagnosis would affect medical management or help with obtaining local support services. Questionable goals include: the doctor feeling pressured to "do something" (coercion from parents) to figure out a diagnosis, even when genetic testing is not indicated, or offering genetic testing as part of a research study in order to obtain more participants.

19.4 Risks and Harms

19.4.1 Unintended Consequences and Implications

Depending on the type of genetic test being offered, there may be inherent psychosocial risks in learning about results for both children and parents. One example of this is when a test that is ordered to look for one genetic change instead finds another, unexpected genetic change. That unexpected change may indicate that the child is at increased risk for a genetic condition that they would not have chosen to know about, such as adolescent onset blindness, or adult onset Alzheimer's disease. For some

---

[1] 22q11 Deletion syndrome has a prevalence of approximately 1 in 4000 live births. Clinical features of 22q11 deletion syndrome vary markedly and include but are not limited to cardiac defects, developmental delays, immunodeficiencies, palatal defects, and psychiatric issues. Patients with this syndrome have a 50% chance of passing along the deletion for each pregnancy.

families, this information could be considered useful in that they can do research and prepare for these conditions, and for other families, this new unexpected information can feel debilitating. How to prepare parents and families for these possibilities, and how to include children in these complex decisions are important questions with moral implications, thus require caution and debate. Similar psychological risks are present in testing for inherited cancer syndromes. One study examined the rates of adverse psychological impact in patients undergoing genetic testing for Li-Fraumeni syndrome,[2] a rare genetic cancer syndrome characterized by early onset of a wide variety of cancers. Of patients who were tested for the gene changes associated with the syndrome; 23% reported clinically relevant psychological distress prior to testing regardless of their result status (Lammens et al. 2010).

More difficult questions surround testing children for adult-onset conditions. While most experts and organizations agree that testing for adult-onset conditions should generally not be performed until the child has reached age of consent, some notable exceptions exist (Clayton et al. 2014). First, the lines between childhood and adult onset disease may not be as clear or may change with advances in the science of therapeutics and screening. One example is Familial Adenomatous Polyposis (FAP), a familial cancer syndrome which can present in the adolescents and therefore warrants early clinical screening for affected patients.[3] (Clayton et al. 2014). Second, the psychological burden to the patient (child) of not knowing about a possible disease and the related diagnostic uncertainty may be more harmful than the implication of a possible disease. Third, further lack of clarity exists (with conflicting guidance from organizations) on the appropriateness of testing and reporting variants (including those related to adult onset conditions) when the child is being tested to address another clinical issue (Clayton et al. 2014). For example, if a patient with FAP also has an APC mutation, the penetrance of the mutation is high with 100% of patients developing colon cancer by their 40s if a colectomy is not performed. In contrast not all patients with a BRCA gene mutation will develop cancer in a lifetime, although their adult cancer risk is certainly increased. All of these examples highlight the importance of an individualistic approach to counseling and testing.

---

[2] Li Fraumeni syndrome is a heritable predisposition cancer syndrome caused by a mutation of the tumor suppressor gene TP53. Patients with a mutation in this gene are at increased risk for certain cancers (breast cancer, brain and CNS tumors, soft tissue and osteosarcoma, adrenocortical carcinoma and acute leukemia, among others) beginning in childhood and throughout their lives.

[3] Familial Adenoid Polyposis (FAP) is a an autosomal dominant disease caused by a mutation in the APC gene. Patients with FAP are at significant risk for developing colon cancer in their lifetime due to the development of 100 s of adenomatous colorectal polyps. Screening for classic FAP for those patients at risk begins at 10 years of age. Colectomy is recommended for most affected patients eventually, given the high number of polyps that will develop.

19.4.2   Genetic Testing for Adult Onset Conditions in Minors

The National Society of Genetic Counselor's (NSGC) position statement regarding
Genetic Testing of adult onset conditions of minors is as follows: *NSGC encourages
deferring predictive genetic testing of minors for adult-onset conditions when results
will not impact childhood medical management or significantly benefit the child.
Predictive testing should optimally be deferred until the individual has the capacity to
weigh the associated risks, benefits, and limitations of this information, taking his/her
circumstances, preferences, and beliefs into account to preserve his/her autonomy
and right to an open future.*

The NSGC also recognizes that an individual approach is important, stating: *The
decision for a minor to undergo genetic testing that could identify variants for adult-
onset conditions either specifically or secondarily (e.g. through genomic sequencing)
should be made cautiously, and whenever possible, with appropriate assent of the
minor. If a minor undergoes genetic testing and results are not disclosed to the
child, the* health care *provider should discuss strategies with the parents/guardian
for sharing the results as he/she develops capacity, or by the age of majority.*

*NSGC strongly recommends that families facing decisions to test minors meet with
a certified genetic counselor or other* health care *provider with genetics expertise
to review the clinical and personal implications of testing (Adopted 2017; replaces
2012 version).*

Arguments against testing minors include (Mand et al. 2012):

(1)   Psychological harm for the child/family: A positive test result may have a
      negative impact upon self-image and esteem, induce feelings of guilt and/or
      blame, lead to anxiety/depression, or result in stigmatization by immediate and
      extended family members, thus affecting family relationships.
(2)   Concerns about discrimination: Having the knowledge that an individual is at
      increased risk for developing an adult onset condition may result in discrimina-
      tion. Although the Genetics and Information Nondiscrimination Act (GINA)
      which took effect in 2009, made big strides in protecting against medical and
      employment discrimination, there continue to be barriers to equal access, such
      as obtaining life insurance, to those with a genetic diagnosis. There is also the
      possibility of social discrimination depending on the condition.
(3)   Autonomy: Many would argue that testing minors fails to respect their future
      autonomy and violates their confidentiality since results are disclosed to their
      parents. According to Fenwick and colleagues, the future autonomy principle
      which highlights the importance of protecting one's future autonomy, is often
      referenced in theory (refer to Chap. 3 for a discussion on Right to Open Future)
      and policy making, but is less often used in practice (Fenwick et al. 2017)

Arguments for testing minors include (Mand et al. 2012):

(1)   Psychological benefit for child and family: decreased uncertainty and therefore
      anxiety including the possibility of a positive effect on identity, self-image and

self-esteem. This could also translate to the potential for positive effects on family relationships.

(2)   Future planning: Results will allow discussions regarding reproductive decisions and planning for the future.

(3)   Autonomy: Adolescents are often capable of making informed decisions about their health and testing may even promote their autonomy and empower them.

19.4.3   Obligation to Other Family Members

Inherent in a genetic test result is the fact that the information received will not only be relevant to the patient but may be relevant to other family members as well. A classic example of this is testing for genetic changes in breast cancer related genes such as BRCA1 or BRCA2.[4] Female patients with a BRCA1 or 2 gene change (mutation) have up to an 80% lifetime risk of developing breast cancer and up to a 45% lifetime risk of developing ovarian cancer (Chan et al. 2017). Patients with BRCA 1 or 2 mutations also may have an increased risk of developing other cancers such as prostate, melanoma or pancreatic cancer (Mersch et al. 2015). Because BRCA1 mutations have autosomal dominant inheritance,[5] any person identified as having a BRCA1 change must assume that it was inherited from either their mother or father, and that there is a 50% chance of passing it down for each offspring they have. Therefore, arises the ethical question of whether or not knowledge about this change confers an ethical obligation to share this information with family members who may be at risk of having the same change. It appears to us that Ethicists would have good reason to argue that a patient has an ethical obligation to inform other family members of a condition that could be screened for or prevented if known, such as breast cancer. If a patient declines doing so, however, the ethical principles of duty to warn versus duty to maintain patient confidentiality come into conflict for the health care provider. (Shah et al. 2013). To complicate matters, the guidelines by professional organizations on duty to inform at risk relatives of possible genetic conditions differs, as does the legal precedent, often leaving providers alone to make these decisions (Callier and Simpson 2012).

---

[4] BRCA1 and BRCA2 are the two genes associated with Hereditary Breast and Ovarian Cancer (HBOC) Syndrome. Individuals with mutations in either of these genes are at increased risk for developing breast and/or ovarian cancer.

[5] Autosomal dominant inheritance is a pattern of inheritance wherein a person carries one copy of a change in a gene on one of their autosomes (non sex chromosomes). In autosomal dominant inheritance, all that is needed to cause the disease is one changed copy (unlike in autosomal recessive inheritance where the disease presentation relies on two changed copies). Individuals with an autosomal dominant disease have a 50% chance of passing the disease gene down in each pregnancy/offspring.

19.5   Informed Consent and Assent

Informed consent includes three main elements: (1) disclosure of information to patients and their surrogates (2) assessment of patient and surrogate understanding and capacity for medical decision-making (refer to Chap. 2), (3) obtaining informed consent before treatments and interventions (Katz and Webb 2016). For parents, the informed consent process for genetic testing of their children should include information about what the genetic test may or may not reveal, including accuracy of results, possible results of unclear significance, and the possibility of incidental findings if relevant. Further detail should include information regarding insurance authorization, coverage and possible out of pocket costs, as well as any possible effects a genetic diagnosis may have on the ability to obtain life insurance or other types of coverage. For example, will a genetic diagnosis put the patient at an advantage or disadvantage in regard to services they may receive at a local level such as early intervention? Could being diagnosed with a genetic condition disqualify a patient from being eligible for a particular surgery or study protocol?

Equally important is an extensive discussion about potential psychological and/or psychosocial consequences of a genetic diagnosis for both the child and family including how, if and when the family intends to share this information with the child, other family members, and their health care providers/school community. It is important to talk through the possible guilt feelings parents may have if they learn that they have passed down a particular gene change leading to the condition. Unintended results such as nonpaternity or the finding of an inherited genetic change for an adult onset condition that could have health consequences for the parents and other family members also need to be discussed.

For young children, the concept of informed consent is even more complicated. A child's present and future autonomy and their best interest are two ethical principles at stake when discussing genetic testing of minors. AMA Code of Medical Ethics opinion guidelines encourage genetic testing for a child when "a child is at risk for a condition for which effective measures to prevent, treat, or ameliorate it are available" (American Medical Association 2016). In such cases, they encourage informed consent of a parent or guardian with age and developmentally appropriate inclusion of the child. However, the specifics on how to make those determinations are lacking, and clinicians are often left to make the assessments on what age to begin inclusion, and how to approach testing depending on level of development. Further, these methods require an individual and contextual approach.

Because a young child does not have the capacity to consent, the goal for young children is to obtain assent, the expression of understanding and willingness to participate. The role of the parent and health care provider in facilitating this process cannot be underestimated, and begins with the assessment of understanding, reasoning and volunteerism. Astute awareness of a child's vulnerability given their inherent dependency and degree of illness is also essential to take into account while facilitating communication (Leikin 1983). Assent from children as young as 7, when children are starting to develop logical thought processes and the possibility for reasoned decision-making can be beneficial in foster moral growth and development of autonomy (Katz and Webb 2016) (also refer to Chap. 7).

For older children and adolescents, the capacity for decision-making, and thus informed consent should be addressed on an individual basis. There is no international consensus on adolescents and decision-making practice. In some countries, such as Portugal or Denmark, children as young as 15 can make medical decisions, whereas in many others including Switzerland it is determined on a case by case basis. In the United States, the legal age for making medical decisions is determined by state law, with some states allowing minors to make some decisions as young as 14, and others requiring them to be 18 (Unguru 2017).

19.5.1 Results Disclosure

Results disclosure and understanding is another important aspect of genetic testing. There are situations where parents may request that a younger or older child not be informed of their genetic testing results. In this scenario, it is important that the parents are informed and engaged in ongoing discussions of the potential harms of nondisclosure. It is equally important that parents are educated about how and when results disclosure is appropriate. Most would argue that a request for results of a genetic test by a mature adolescent should be given greater weight than a parent's opposition, and this is supported by both the American Academy of Pediatrics (AAP) and the American College of Medical Genetics (ACMG) (Ross et al. 2013).

19.6 Genetic Counselors

Genetic counselors are Master's degree board certified health professionals with expertise in discussing the complexities of genetic screening and testing with families. The genetic counseling profession was born 50 years ago (1969) and as of 2020 there are approximately 5,000 genetic counselors in the United States, making up over 1/2 of the entire world's genetic counseling professionals (Abacan et al. 2019). The fact that there are so few genetic counselors poses a real problem when it comes to informed consent for genetic counseling and screening. The vast majority of non-Genetics health care providers offering genetic screening and testing do not have the time or expertise needed, nor are they often aware of their own lack of knowledge. This ultimately leads to a lack of appropriate information or misinformation for patients, and the absence of referrals to the appropriate genetic health professionals. Studies have shown that these are global problems and can lead to confusion, misinformation, and in some cases, serious medical mistakes (Baars et al. 2005).

Creative solutions addressing the lack of genetics health professionals are needed. Primary care providers who acknowledge their lack of genetics knowledge are eager to further their education. Resources such as the Genetics in Primary Cares Institute (GPCI) in the United States, and the Gen-Equip program in Europe (www.primar cycaregenetics.org) provide education modules and tools for providers throughout Europe in multiple languages (Paneque et al. 2017). Some providers rely on genetic testing laboratories, many of whom employ genetic counselors, to interpret results and even counsel patients over the phone or telemedicine. This is common in Obstetrics where an Obstetrician is required to offer carrier screening for certain genetic conditions to all of her pregnant patients but may not have access to a genetic counselor. The usefulness of genetic testing in the Oncology space has grown exponentially in recent years, and many Oncology nurses have taken it upon themselves to pursue education and even certification (Advanced Genetics Nursing Certification AGN-BC) in Genetics. Interest in and options for genetic testing in all populations will continue to increase. We must continue to be innovative in our approach to provide access to genetic counseling for all patients.

19.7   Direct to Consumer Testing

Direct to consumer testing (DTC) is another example of the potential harm testing may have when patients aren't counseled appropriately. In DTC, genetic testing is offered directly to patients, usually via the internet, often without any requirement for genetic counseling or informed consent by any type of health care provider. The advantage to this type of testing is accessibility, but the disadvantages and ethical concerns include the lack of professional guidance about the testing, inadequacy of regulation of clinical validity and utility, and lack of transparency regarding confidentiality, privacy, and secondary use of the genetic samples and data collected (Laestadius et al. 2017) This is a global problem: Some, but not all US states now require physician involvement in health-related genetic testing. In the European Union (EU) there is no legislation regulating genetic testing or the use of genetic information or testing, nor is there regulation for DTC genetic testing. Only a few countries address DTC genetic testing directly in national laws (Paor 2018).

19.8   Costs Burdens

Genetic testing and screening are expensive, and increasingly are not covered by public insurance and even some private insurances. Genetic counselors, physicians and staff increasingly spend time advocating for their patients to receive preauthorization, or else risk having their patients slapped with hefty testing fees they did not agree to, eroding trust in their health care team, and those organizations. The process of obtaining these authorizations is time-consuming and contributes to moral distress in clinicians. One example where there is often an insurance or cost barrier to testing is testing for a suspected genetic disease in the newborn period when the newborn still appears to be unaffected. Testing is often denied in cases like these where time-sensitive knowledge and actionable therapeutics or avoidance therapy could have been instituted and would translate into significant differences in development for the future child.

The barriers to receiving equitable testing and screening places vulnerable patients and families at increased risk for not being able to have access to information even when appropriate and desired. Policy at division, hospital, state and federal levels should promote algorithms that streamline the process for efficiency and allow for fair and objective decision-making.

19.9   Newborn Screening

The purpose of newborn screening is to diagnose treatable disorders early enough to provide an intervention that will improve outcome. The screening itself varies widely between US states and between countries and is a combination of various physical/hearing exams in addition to the blood sampling that dry on filter paper spots. As with other types of genetic screening or testing, there are ethical issues with newborn screening surrounding education and informed consent, ambiguity of results or unintentional results, and obligation to other family members (Fabie et al. 2019).

While screening is mandatory in many states, education about newborn screening to parents is not. Many parents do not understand the benefits, risks and limitations of the testing. Newborn screening results can be ambiguous, requiring follow up testing, and can also reveal information about a genetic condition such as that the child is a carrier for a gene change for a particular condition rather than affected. Any of these scenarios has the potential to cause significant confusion and stress for new sleep deprived parents. With this in mind, several have recommended that education and counseling about newborn screening should be offered in the prenatal period rather than right after birth or not at all (Fabie et al. 2019).

19.10   Genetic Carrier Status

There is general consensus internationally that performing carrier testing on a child for an autosomal recessive condition for the sole purpose of identifying their carrier state is not warranted. Most health care professionals agree that unless knowing one's carrier status is something that is medically actionable, there is no need for a child to pursue this information prior to adulthood. Potential positive consequences for children knowing their carrier status for a particular condition include the opportunity for parents to normalize and educate children about it before they are of reproductive age. Potential negative effects include a misunderstanding of the results and/or negative self-esteem/increased anxiety related to a positive result.

There are scenarios where children do get carrier screening results, either intentionally or unintentionally. One *intentional example* includes when a sibling has an autosomal recessive condition and parents want to know the carrier status of their brother of sister. Some laboratories even encourage this practice by offering free carrier testing to a sibling of an affected child. Another example is population-based carrier screening offered in some high schools for select populations of individuals who are at higher risk of being carriers for certain autosomal recessive conditions. Ashkenazi Jewish individuals, for example, are at increased risk of being carriers for several autosomal recessive diseases and The American College of Obstetrics and Gynecology recommends preconception carrier screening to individuals of Ashkenazi Jewish descent for at least Tay Sachs, Cystic Fibrosis, Familial Dysautonomia and Canavan disease. (ACOG) Individuals of French-Canadian Ancestry are at increased risk to be carriers for certain genetic conditions based on their region of origin as well. Conditions more common in this population include Tay Sachs, Tyrosinemia type I, congenital lactic acidosis, spastic ataxia and/or agenesis of the corpus collosum with peripheral neuropathy (Wilson et al. 2016).

The purpose of carrier screening in these populations is to provide an easy and minimal cost opportunity to know their carrier status before they are adults of reproductive age when they are presumably less likely to spend the time or money to do so. This approach varies by country: The American Society of Human Genetics 2015 guideline does not support population carrier screening in minors. In Australia, however, carrier screening in high schools has been offered and parental consent only required if a child is under 16 years of age (Vears and Metcalfe 2015).

Carrier status for a fetus or child may be identified *incidentally* through newborn screening, following diagnostic testing or during prenatal testing. Although the benefits of these particular tests may outweigh the potential anxiety of self-esteem issues related to finding out one's carrier testing as a child, the importance of informed consent should be highlighted here. When parents or children are informed of potential incidental findings prior to having testing done or receiving results, they are much more able to process the results in a productive and healthy way.

19.11   Prenatal Screening and Testing

Almost all the genetic testing that can be offered to children can also be offered to the fetus. Pregnant women in the United States have the legal right to pursue prenatal screening and testing for anything a laboratory will offer them. For the child who has had a genetic test prenatally, the decision to opt in or out of genetic testing has been taken away, and it is up to the parents to decide if/when to disclose those results.

19.11.1 Case Example

Mia is 40 years old and pregnant with her first child. Given that her risk to have a child with a chromosome abnormality is somewhat higher because of her age, she opts to pursue an amniocentesis, a diagnostic genetic test of the fetal cells obtained by sampling of the amniotic fluid. Her doctor recommends a genetic test on those fetal cells called a microarray, allowing for the detection of small deletions of duplications of DNA. Mia thinks this is a good idea and proceeds with the testing. Ten days later she gets a call from her doctor saying that the microarray came back with a small change (duplication) that hasn't been reported in the literature before, so its significance is unclear. He recommends that Mia and her husband have their blood drawn to see if it is inherited (and therefore more likely benign) or not. They do this and learn it is not inherited. Mia and her husband are at a loss as to what to do next.

Ethical issues:

– Informed consent: Were Mia and her husband informed about the possibility of this type of "uncertain" result? If so, would they have opted to pursue this testing?
– Do no harm: Is the provider obligated to disclose the result, even if it is meaningless? Is it ethical to offer a prenatal test that may reveal ambiguous results?
– Future autonomy: If the family continues the pregnancy, what ethical obligation do the parents have to reveal this information to the child, and when?

Mia and her husband ask the laboratory for more information about the genes that are duplicated. The laboratory comes back with a list of the duplicated genes and tells them that among them there is a gene associated with schizophrenia, but that it is impossible to know whether or not the duplication will put the child at increased risk for the disease or not.

Ethical issues:

– Will this information change the way the parents feel about their child from day 1?
– Will the child be treated differently by the parents now that they know this?
– Will the parents be more anxious, different parents than they would have been without this information?
– When/how should the parents inform the child of this information or should it be kept
– quiet given that the significance is unknown?

Ethics committees are often called upon to weigh in on controversial decisions such as the above and a retrospective review of ethics committee minutes suggests that opinions and answers to questions such as these vary by hospital, country, and committee (Muggli et al. 2019).

19.11.2 Fetal and Maternal Autonomy

One of the most obvious ethical challenges in prenatal diagnosis is the right of the unborn child versus the autonomy of the mother. In his viewpoint on the moral status of the fetus, Isaacs offers three ways of looking at fetal rights: that a fetus has the same rights as a live child, that a fetus has no rights, or that a fetus has increasing moral status with advancing gestation (Isaacs 2003). Issacs argues that "if significant differences arise in the interests of the mother and fetus, the mother has a responsibility to consider the interests of both and make an informed decision for both of them. If conflicts arise, the competent mother's rights to personal autonomy should prevail over the lesser rights of the fetus early in gestation, but as the fetus matures and acquires greater moral status, the situation becomes less clear-cut". The ethic and legal discussions surrounding this topic vary significantly amongst different countries around the world, and even amongst states within countries such as the United States.

19.12   Guiding Principles for an Ethical Approach to Genetic Testing or Screening of Minors

General principles
- Appreciate that genetic counselors are essential in helping families and children to understand appropriate screening and testing
- Assess and explain the goals of testing, along with the benefits, risk and limitations, including the possibility of incidental or unintended findings such as nonpaternity, a genetic change of uncertain significance, disclosure of carrier status, or the finding of a genetic change associated with an adult onset condition
- Understand and counsel on the consequences of genetic testing on other family members
- Acknowledge and discuss direct to consumer testing benefits and risks including: insurance barriers, lack of coverage and potential discrimination based on genetic testing results

Genetic testing or screening for minors
- Adjust the process of consent or assent for the individual patient, to appropriate age, development, and maturity level
- Assess understanding, reasoning and volunteerism of the child
- Recognize Child's vulnerability given inherent dependency and degree of illness
- Discuss and develop a plan regarding disclosure of results: how, to whom and when
- Recognize the importance of a health provider's role in establishing trust and facilitating open communication

19.13 Conclusion

New discoveries of therapeutics and effective prevention will quiet the ethical challenges for some issues we currently face, and yet, the acceleration of diagnostic interpretation will yield new questions and debate. The bedside clinician across the globe faces the challenges of keeping up with the science and managing the very real humans and their emotions sitting in the clinic chairs or hospital beds. We have outlined some of the current ethics questions and approaches when considering genetic testing and screening in children. We highlighted tools and pitfalls to help recognize and begin addressing those issues. Clinicians should enlist education resources, access to genetic counselors, ethics consultants and psychology referrals at the ready. Such preparedness will give clinicians the confidence to address the questions and the skillset to work with patients and families to find solutions.

References

Abacan, M., L. Alsubaie, K. Barlowe-Stewart, B. Caanen, C. Cordier, E. Courtney, et al. 2019. The global state of the genetic counseling profession. *European Journal of Human Genetics* 183–197.

ACOG. n.d. Preconception and prenatal carrier screening for genetic diseases in individuals of Eastern European Jewish descent. *Obstetrics and Gynecology* 114 (4).

Adam, M., D. Diekema, and M. Mercurio. 2017. *AAP bioethics resident curriculum: Case-based teaching guidelines.*

American Medical Association. 2016. *Genetic testing of children code of ethics medical opinion.* Retrieved from www.AMA-ASSN.org.

Baars, M., L. Hennemen, and L. Ten Kate 2005. Deficiency of knowledge of genetics and genetic tests among general practitioners, gynecologist and pediatricians: A global problem. *Genetics in Medicine* 605–610.

Bassett, A., D. McDonald-McGinn, K. Devriendt, M. Digilio, P. Goldenberg, A. Habel, et al. 2011. Practical guidelines for managing patients with 22q11 deletion syndrome. *Journal of Pediatrics* 332–339.

Callier, S., and R. Simpson. 2012. Genetic diseases and the duty to disclose. *American Medical Association Journal of Ethics* 640–644.

Chan, J., L. Johnson, M. Sammel, L. DiGiovanni, C. Voong, S. Domcheck, et al. 2017. Reproductive decision-making in patients with BRCA 1/2 mutations. *Journal of Genetic Counseling* 594–603.

Clayton, E., L. Mcullough, L. Biesecker, S. Joffe, L. Ross, and S. Wolf. 2014. Addressing the ethical challenges in genetic testing and sequencing of children. *American Journal of Bioethics* 3–9.

Durmaz, A., E. Kuraca, U. Demkow, G. Toruner, J. Schounams, and O. Cogulu. 2015. Evolution of genetic techniques: Past, present, and beyond. *Biomed Research International.*

Fabie, N., K. Pappas, and G. Feldman. 2019. The current state of newborn screening in the United States. *Pediatric Clinics of North America* 368–386.

Fenwick, A., M. Plantinga, and S.A. Dheensa. 2017. Predictive genetic testing of children for adult onset conditions: Negotiating parents requests. *Journal of Genetic Counseling.*

Isaacs, D. 2003. Moral status of the fetus: Fetal rights or matneral autonomy. *Journal of Paediatrics and Child Health* 58–59.

Jonsen, A., M. Siegler, and W. Winslade. 2015. *Clinical ethics: A practical approach to ethical decisions in clinical medicine,* 8th edn. McGraw-Hill.

Katz, A., and S. Webb. 2016. AAP committee on bioethics: Informed consent in decision-making in pediatric practice. *Pediatrics* 138 (2).

Laestadius, L., J. Rich, and P. Auer. 2017. All your data (effectively) belong to us: data practices among direct to consumer genetic testing firms. *Genetics in Medicine* 513–519.

Lammens, C., N. Aaronson, A. Wagner, R. Sijmons, M. Ausems, and A. Vriends. 2010. Genetic testing in Li-Fraumeni syndrome: Uptake and psychosocial consequences. *Journal of Clinical Oncology* 3008–3014.

Leikin, S. 1983. Minors' assent or dissent to medical treatment. *The Journal of Pediatrics* 169–176.

Mand, C., L. Gillam, M. Delatycki, and R. Duncan. 2012. Predictive genetic testing in minors for late-onset conditions: a chronological and analytical review of Ethical arguments. *Journal of Medical Ethics* 519–524.

Mersch, J., M. Jackson, M. Park, D. Nebgen, S. Peterson, C. Singletary, et al. 2015. Cancers associated with BRCA1 and BRCA2 mutations other than breast and ovarian. *Cancer* 269–275.

Muggli, M., C. Geiter, and S. Reiter-Theill. 2019. Shall parent/patient wishes be fulfilled in any case? A series of 32 ethics consultations: from reproductive medicine to neonatology. *BMC Medical Ethics*.

Paneque, M., M. Cornel, V. Curtisova, E. Houwink, L. Jackson, and A. Kent. 2017. Implementing genetic education in primary care. *Journal of Community Genetics* 147–150.

Paor, A. 2018. Direct to consumer testing-law and policy concerns in Ireland. *Irish Journal of Medical Science* 575–584.

Phillips, K., P. Deberka, G. Hooker, and M. Douglas. 2018. Genetic test availability and spending: Where are we now? Where are we going? *Health Affairs (Millwood)* 710–716.

Powell, C. 2019. United Nations convention on the rights of a child in acute paediatrics. *Arch Dis Child* 971.

Ross, L., H. Saal, R. Anderson, and K. David. 2013. Technical report: Ethical and policy issues in genetic screening and testing of children. *Genetics in Medicine* 234–235.

Schonberg, R., and M. Menzel. 2019. Birth defects and prenatal diagnosis chapter. In *Children with disabilities: Birth defects and prenatal diagnosis*, ed. R. A. Batshaw, 37–49. Baltimore: Paul H. Brookes.

Shah, S., S. Hull, M. Spinner, B. Berkman, L. Sanchez, and R. Abdul-Karim, et al. 2013. What does the duty to warn require. *American Journal of Bioethics* 62–63.

Unguru, Y. 2017. Informed consent and assent in clinical pediatrics. In: *American Academy of Pediatrics Bioethics Resident Curriculum: Case Based Teaching Guides*.

Vears, D., & S. Metcalfe. 2015. Carrier testing in children and adolescents. *European Journal of Medical Genetics* 659–667.

Wilson, R., I. De Bie, C. Armour, R. Brown, C. Campagnolo, J. Carroll, et al. 2016. Joint SOGC-CCMG opinion for reproductive genetic carrier screening: An update for all Canadian providers of maternity and reproductive health care in the era of direct to consumer testing. *Journal of Obstetrics and Gynaecology Canada* 742–762.

Further Readings

Baig, S.S., M. Strong, E. Rosser, et al. 2016. 22 Years of predictive testing for Huntington's disease: The experience of the UK Huntington's Prediction Consortium. *European Journal of Human Genetics* 24: 1396–1402.

Caga-anan E.C.F., L. Smith, R.R. Sharp, and J.D. Lantos. 2012. Testing children for adult onset genetic diseases. *Pediatrics* 129 (1): 163–167. https://doi.org/10.1542/peds.2010-3743

Clayton, E.W., L.B. McCullough, L.G. Biesecker, et al. 2014. Addressing the ethical challenges in genetic testing and sequencing of children. *American Journal of Bioethics* 14 (3): 3–9.

Wilfond, B., and L.F. Ross. 2009. From genetics to genomics: Ethics, policy, and parental decision-making. *Journal of Pediatric Psychology* 34: 639–647.

Chapter 20
Enhancement Technologies and Children

J. T. Eberl

Abstract The advent of current and emerging biotechnologies has placed greater levels of control in the hands of parents and prospective parents to shape their children's physical, cognitive, and emotive traits. Ethical questions initially formulated around the selection of embryos or fetuses that have certain desirable versus undesirable traits are now being applied, alongside novel questions, to whether parents have an ethical obligation, or at least a right, to enhance their children to endow them with traits they would not naturally possess. This chapter elucidates how the ethics of enhancement have developed, yet differ in important ways, from the ethics of selection. It then canvasses three primary ethical questions regarding enhancement: (1) whether parents have an ethical *obligation* to enhance their children when it is safe, effective, and feasible for them to do so; (2) if not an obligation, whether parents have a *right* to enhance their children and how their exercise of such a right may alter the nature of the parent/child relationship; and (3) what wider *societal concerns* might mitigate against such a parental right or obligation, at least not without significant socioeconomic restructuring. While this chapter is focused on specific ethical questions raised by the prospect of parents making enhancement choices on behalf of their born or preborn children, the conclusion highlights how such questions arise within a wider debate concerning whether biotechnological forms of human enhancement should be freely allowed, universally restricted, or permitted on a limited basis for specific traits and purposes.

Keywords Enhancement · Disability · Procreative beneficence · Procreative autonomy · Parent/child relationship · Eugenics

J. T. Eberl (✉)
Albert Gnaegi Center for Health Care Ethics, Saint Louis University, St. Louis, MO, USA
e-mail: jason.eberl@slu.edu

N. Nortjé and J. C. Bester (eds.), *Pediatric Ethics: Theory and Practice*,
The International Library of Bioethics 89,
https://doi.org/10.1007/978-3-030-86182-7_20

329

20.1 Introduction

Since the birth of Louise Brown, the first so-called "test tube baby" in 1978 (Dow 2019), medical practitioners, bioethicists, policymakers, and prospective parents have debated what ethical limits there should be, if any, on the use of various forms of biotechnology to *select* for children to have or not have certain traits (Klitzman 2020; Green 2007; Glover 2006; Robertson 1994). For example, fetal screening and diagnostic techniques such as amniocentesis and chorionic villus sampling allow for several congenital conditions—including Trisomies 18 and 21, spina bifida, and cystic fibrosis—to be detected in utero and a choice made whether to continue or terminate the pregnancy (refer to Chap. 19). Selection techniques may also be utilized for various *non-disease traits*, such as biological sex or hair and eye color (Greely 2011). Alongside ethical debates concerning preimplantation—in the case of embryos produced through in vitro fertilization—and prenatal screening/diagnosis, discussed in the previous chapter of this volume, has been ongoing discussion over whether biotechnological means should be used to *enhance* one's children. Enhancement differs from selection insofar as the latter involves choosing whether a particular embryo or fetus will be allowed to develop gestationally due to whether a particular trait is either present or absent, whereas the former involves the use of biotechnology to *endow* one's child with a trait they would not otherwise possess or to *increase* the level of a particular trait they do possess (Juengst 1998). Relevant enhancement technologies include gene editing using CRISPR-Cas9, devices such as cochlear implants, hormonal injections impacting physical growth and development, and pharmaceuticals that affect behavior or cognitive function.[1] It should be noted at the outset that some of the traits for which parents might seek enhancement for their children have multiple bases in the human genome, about which current science is and will remain incomplete for the foreseeable future, not to mention the scientific challenges of understanding the complex pathways from genotype to phenotype. There are also additional concerns regarding clinical research to investigate the safety and efficacy of the above-named interventions. In short, there is no adequate scientific or clinical basis currently and for the foreseeable future for enhancement. Given the *speculative* nature of such interventions, this chapter aims to provide an account of the ethical challenges investigation into enhancement will have to address as the capacity for it is developed in scientific and clinical research.[2]

With these caveats in mind, this chapter will canvas several significant ethical issues concerning whether actual or prospective parents[3] ought to be permitted to

---

[1] These and other enhancement techniques—except CRISPR-Cas9, which had not yet been developed—are canvassed in Savulescu, ter Meulen and Kahane (2011). For explication of CRISPR-Cas9 and its ethical implications, see Doudna and Sternberg (2017).

[2] I am grateful to an anonymous reviewer for stressing the currently speculative nature of enhancement interventions.

[3] From here on, references to "parents" will be understood to include both prospective parents of preconceived or preborn embryos/fetuses and parents of already born children.

enhance their children's[4] physical, cognitive, or behavioral traits. On the one hand, pursuing such enhancements may be viewed as merely an extension of the degree of parental control exercised in preimplantation/prenatal selection; on the other hand, there is arguably a significant moral difference between selecting for or against a child with certain traits and engineering one's child to have particular traits they otherwise would not have, especially if those traits endow them with capacities that transcend typical levels for human beings. The three primary ethical questions that will be addressed in this chapter are: (1) whether parents have an ethical *obligation* to create children with an optimal set of traits; (2) whether parents have a *right* to enhance their children even if they have no obligation to do so; and (3) whether the widespread enhancement of children might lead to a dystopian *eugenic* society characterized by extreme socioeconomic disparities. First, though, certain key concepts which have inherent ambiguities but yet inform relevant arguments must be elucidated.

20.2 Key Concepts

20.2.1 Enhancement

As noted above, enhancement differs from selection insofar as it involves the use of biotechnology to endow someone with traits assessed as contributing toward that person's overall wellbeing. There is tremendous debate over whether enhancements may be distinguished from biotechnological *therapies*, which depends on how concepts such as "disease" or "disability" are defined (Juengst 1998). To avoid getting bogged down in this important but ancillary debate, an enhancement is stipulatively defined in this chapter as either the endowment of a trait that human beings do not naturally possess, the removal of a naturally-possessed trait deemed to be detrimental to overall wellbeing, or the increased level of a naturally possessed trait beyond the level of "normal function" for human beings (Buchanan et al. 2000, 126–130). Determining what constitutes "normal function," however, is highly contestable.

20.2.2 Wellbeing

An essential component of an enhancement is that it contributes to its possessor's overall wellbeing. There are various competing philosophical, psychological, and theological definitions of what constitutes the wellbeing of human persons (Griffin 1986). At one end of the spectrum, wellbeing may be defined purely *subjectively*: each

---

[4] From here on, references to "child/children" will be understood to include both preborn and already born children. This essay takes no particular stance on the vexed question of the ontological or moral status of embryos or fetuses. The ethical views articulated in what follows are equally applicable whether an enhancement intervention is done in vitro, in utero, or postnatally.

individual person defines their own wellbeing. For instance, physical mobility may be viewed as a much more significant contributor to the wellbeing of a marathon runner than of a college professor. Yet, both the athlete and the professor would likely value physical mobility, even if it has different levels of significance to them. Hence, there arguably are at least some *objective* constituents of human wellbeing, including not only various physical capacities that contribute to health, longevity, and the ability to interact with one's environment, but also cognitive and emotive capacities that empower a person to engage in problem-solving, critical thinking, introspective reflection, and artistic endeavors, as well as to relate socially with other persons in satisfying and constructive ways. Furthermore, such capacities may manifest in myriad ways: for instance, if physical mobility is a capacity that objectively contributes to human wellbeing, it does not matter whether it is realized by the use of two biological legs, one or more artificial legs, a wheelchair, etc.

20.2.3   Harm

Just as the concept of human wellbeing is ambiguous and foundationally contested as to whether it can only be defined subjectively or has some objective constituents, the concept of harm is similarly vague and much debated insofar as it refers to the converse of wellbeing (Feinberg 1987–1990). For instance, while the infliction of physical pain on another person may initially appear to be a clearly evident example of harm, there are equally evident counterexamples, such as giving a child a vaccine injection who cries when the needle pierces their skin or individuals who are sexually aroused by painful stimulation. The latter example implies that perhaps what constitutes a harm is acting upon another person without their *consent*; however, the first example involves a person being inflicted by pain who cannot give consent. Furthermore, there are other examples involving persons who are caused pain or other forms of discomfort against their will, such as the punishment of prisoners: justified punishment, especially if it is oriented toward eventual rehabilitation and social reintegration, is not typically considered a harm. For the purposes of this chapter, a harm will be understood as anything that befalls a person, whether accidentally or by the intentional action of another person, that diminishes their overall wellbeing—howsoever defined.

20.2.4   Impairment and Disability

Disability scholarship (Davis 2017) has called into question the long-held *medical model* of disability, in which a person's disability—whether blindness, deafness, being short-statured, etc.—was taken to be the source of any diminishment that person experiences in their overall wellbeing. Disability, on this model, is understood to be essentially a harm. Conversely, the *social model* of disability has gained prominence, in which traits such as being blind, deaf, or short-statured are labeled as "impairments"; but whether such impairments are "disabling" for a person depends on their social context (also refer to Chap. 17). An individual with blindness who has access to audio books, signage in Braille, and other forms of social accommodation will be able to interact effectively with their environment and productively in the economic sphere. Furthermore, certain forms of disability may be a definitive feature of a person's self- or group-*identity*. For instance, many hearing-impaired persons identify themselves as members of the Deaf community, particularly if they communicate using sign language; being deaf, on this view, is as constitutive of one's identity as being Black, Hispanic, or LGBT.

20.2.5 Eugenics

A final term that needs to be disambiguated is "eugenics," which carries a negative connotation such that any form of selection or enhancement to which this label is applied is thereby claimed to be *verboten*. The term itself is innocuous insofar as it is derived from a Greek word meaning "good birth"; in that banal sense, every prospective parent is arguably a eugenicist insofar as they want a healthy and happy child, along with the health care professionals who provide prenatal care (Veit et al. 2021). The negative association goes back to the Eugenics movement of the late nineteenth and early twentieth centuries throughout the U.S. and Western Europe (Kevles 1985). In 1907, the state of Indiana passed what has been historically recognized as the world's first eugenics law, which authorized state-run institutions, such as mental hospitals, to sterilize without consent individuals deemed to be unfit to reproduce due to mental or physical "defects" (Lombardo 2011). This and similar laws in other states were ruled to be constitutional by the U.S. Supreme Court in the 1927 *Buck v. Bell* decision, in which Justice Oliver Wendell Holmes infamously declared that "three generations of imbeciles are enough" (Lombardo 2008). Eugenic practices reached their morally abhorrent peak under the Nazi regime in Germany, including not only non-consensual sterilization, but also involuntary euthanasia and eventually mass exterminations of Jews, Gypsies, homosexuals, and persons with disabilities among others deemed *untermenschen* ("sub-human"). Such practices may be labeled "negative" eugenics insofar as they sought to eliminate the inheritance of traits deemed socially undesirable; however, there were also "positive" eugenic practices aimed toward producing a new generation of persons with socially desirable traits, including "Better Baby" and "Fitter Family" contests held during county and state fairs in the U.S., as well as the Nazi *Lebensborn* experiment (Clay and Leapman 1995). A key question is thus whether the enhancement of children should be considered a form of positive eugenics, with the negative connotation such a label implies.

20.3   Principle of Procreative Beneficence

A purported ethical axiom that informs, implicitly or explicitly, a number of views concerning the permissibility, if not the *mandate*, of parents to engage in selection and, where possible, enhancement of their children is the *principle of procreative beneficence* [PPB]:

> couples (or single reproducers) should select the child, of the possible children they could have, who is expected to have the best life, or at least as good a life as the others, based on the relevant, available information (Savulescu 2001, 415).

While this principle explicitly refers to selection, its underlying logic is also applicable to the use of enhancement technologies provided they are safe and effective with no adverse social consequences beyond the wellbeing of one's individual child. Defenders affirm that PPB covers both disease-bearing and non-disease traits (Savulescu 2001, 414; Savulescu and Kahane 2009, 276) and others have argued that PPB entails a requirement to enhance (Veit 2018).

A more general ethical principle underlying PPB is the putative right of children to an "open future" (Feinberg 1980) (refer to Chap. 3), which stipulates that parents are ethically obligated to ensure, to the extent they are able, that their children develop their inherent capacities to pursue a reasonable range of satisfying lifestyle options. In practice, this means that, while parents are not obligated, because they cannot, ensure that their children become NBA star players or Nobel laureate physicists, they should provide their children with opportunities for physical and intellectual development such that, should their children wish, they could pursue excellence in physical or intellectual endeavors. If this right is taken merely as a *negative* right, it requires parents only to ensure that their children are not born with or later experience anything that would detract from their pursuit of excellence; however, this right is typically construed as *positive*, meaning that parents must seek out opportunities to maximize their children's capacity to pursue excellence. Hence, insofar as certain forms of enhancement may offer a net-positive contribution toward children's pursuit of physical or intellectual excellence, parents would be morally obligated under the PPB to avail themselves of such means of enhancement.

PPB, however, has been subject to criticism on various fronts. A foundational issue is whether PPB entails a moral *obligation* on the part of parents or merely justifies the moral *permissibility* for parents to pursue enhancements of their children should they so choose. Hotke (2014) argues, with respect to PPB's application to selection, if one's potential child is expected have greater wellbeing than another, this would give one *a* reason, but not a morally obligatory reason, to select the first potential child; there is no binding duty to maximize wellbeing, even if maximizing wellbeing gives one a reason to make a particular choice. Saunders (2015) further argues that the more general moral presumption underlying PPB—namely, that one is obligated to do what they have the *most moral reason* to do—would effectively eradicate the category of *supererogatory* acts that go above and beyond one's moral obligations. Parents may have good moral reasons to enhance their children but doing so extends beyond the limits of the more foundational moral obligation to promote their children's right to an open future. Harris (2009) counters that enhancements are indeed a moral obligation so long as they genuinely confer more benefit than risk of harm—both to the enhanced individual and humanity collectively—insofar as to withhold something that constitutes a net-benefit from someone is to harm them.

Harris's contention leads directly to another concern with PPB: How to define what constitutes "benefit" versus "harm" to one's children in order to delineate the limits of what PPB allows or requires parents to do in selecting for or enhancing their children (Holland 2016). As noted above, the concept of human "wellbeing" is notoriously vague. Yet, Buchanan et al. (2000, 167–170) have argued that there are "general-purpose means" which contribute toward one's ability to pursue a vast array of lifestyle plans, thereby maximizing the openness of one's future, as well as certain impediments that would detract from one's ability to pursue nearly any lifestyle plan. A putative example of the former is one's *memory capacity*, enhancement of which would arguably be beneficial for football players, television screenwriters, and college professors alike; conversely, diminishment of one's memory capacity would harm one's ability to succeed in nearly any vocational or career plan. Other putative general-purpose benefits include intelligence, self-discipline, impulse control, foresight, patience, sense of humor, sunny temperament, empathy, and the capacity to live peaceably (Savulescu 2007, 284). This list is intended to help define a set of beneficial traits for each person as an individual, as well as contributing to humanity's collective wellbeing by making our children morally better than ourselves (Persson and Savulescu 2012).

Nevertheless, while it may be difficult to dispute that qualities such as these are good for human beings to have, it does not follow that having *more* of any of these traits would make a person, or humanity in general, better off (Holland 2016, 495). For instance, while patience is often construed as a moral virtue, classical wisdom tells us that virtues lie in the *mean* between extremes of excess and deficiency (Aristotle 1999, 25); hence, an excess of patience transmutes into a vice—consider, for example, a parent who never asserts their authority over their child and "patiently" allows their child to do whatever they want. Enhanced memory could also become problematic if it does not allow someone to forget the details of a traumatic experience they keep reliving in their mind. This speaks to a more general concern of our ability to predict the likelihood of possible effects of enhancement (Tonkens 2011, 279–281); the *speculative* nature of enhancement interventions arguably makes them distinct from therapeutic interventions such that the epistemic gap concerning potential benefits and harms must be narrowed before parents are allowed, let alone *obliged*, to enhance their children.

A final problematic implication of PPB is with respect to persons born with various forms of disability (Glover 2006, 4–36). Whether in reference to selection or enhancement, the valuing of certain traits implies a disvaluing of persons who lack such traits or express them to a lesser extent than others (Eberl, forthcoming). Savulescu (2001, 423) explicitly denies that PPB disvalues the lives of persons with disabilities, only the disabilities themselves. Yet, as noted above, some conditions which Savulescu and others consider disabilities are not viewed as such by those who possess them—e.g., deafness, blindness, or short-stature. Rather, it is due to society's lack of accommodation that having such traits may lead to diminishment of one's quality of life.

20.4 Procreative Autonomy and the Parent/Child Relationship

PPB purports to define an ethical obligation on the part of parents to produce children with the "best chance of the best life" (Savulescu and Kahane 2009); as an *ethical* mandate, however, PPB should not be used to justify *coercing* parents into following it, but rather *persuading* them to make pertinent reproductive choices (Savulescu 2001, 425–426). An equally significant ethical question is whether parent have a *right* to enhance their children, which raises more general questions about the nature of the relationship between parents and children. As Tonkens (2011, 279) notes,

> No parent has ever had this degree of reproductive control. The scope of reproductive choice has hitherto been restricted to things such as the choice of reproductive partner, maternal diet during pregnancy, and choices concerning postnatal development (e.g. pedagogical choices). Although parents can already control which genetic constitutions will combine to produce an embryo and how a particular existing child will be socialized, parents have never been able to choose how specific genes are to be combined to create a specific child.

Enhancement goes beyond selection in not merely determining whether a particular child will be born or not—due to the presence or absence of certain traits—but deciding what traits a particular child may have. The ethical concern here is that children will be *commodified* as objects "designed" to satisfy parental expectations.

Do parents, however, *want* to exercise such control over their children's traits? Writing from her own experience as an ob/gyn and fertility specialist, Klipstein (2017, S31) claims,

> In my experience, we humans have an innate understanding that reproduction leads to an outcome that is inherently uncertain. We are aware that we can predict neither how our children will be at birth nor how they will develop over time. We embrace this uncertainty, and it allows us to strive to raise our children to be the best possible versions of themselves. We rear our children with hope and anticipation, but with the knowledge that their development and strengths and weaknesses are not fully knowable. Such a process would be dampened if much of the future was predetermined, and this would take away much of the joy of raising children. There is a beauty in not knowing and in the randomness of reproduction. Many couples contemplating children wonder if the child will draw from traits possessed by one or the other and look to see how the mixing of their two genomes results in a unique yet recognizable variant of themselves. This desire will not so easily become obsolete.

Nevertheless, just because many, if not most, parents would not seek to exercise a putative right to enhance their children, it does not follow that such a right ought not to be recognized as constitutive of a broader, generally accepted right to *procreative autonomy*.[5]

---

[5] For elucidation and critique of a right to procreative autonomy with respect to enhancement of children, see de Melo-Martín (2017, 62–96).

Consideration of parental procreative autonomy, however, must be balanced with the actual or prospective autonomy of children. While it would clearly be wrong for parents to engineer their children such that their autonomy is essentially stripped from them, Habermas (2003, 53–66) raises the concern that an enhanced child may *perceive* themselves as having been "something made" as an "instrument" of their parents' expectations. Tonkens (2011, 278), while noting that Habermas may ultimately be wrong in his prediction, contends that the *speculative* nature of enhancement interventions requires us to take account of the possibility of enhanced children perceiving themselves in this way. Furthermore, this worry is reinforced by the idea that many parents will very likely not limit themselves to enhancing only one of their child's traits but, if they are allowed to, will aim to enhance *multiple traits at once*, opting for what we might call an "enhancement package." As the imposition of parental intentions to have a *specific child* increases, the child's ability to understand herself as an autonomous agent may be increasingly threatened as well.

Since widescale enhancement has not yet been realized, it cannot be predicted how enhanced children will perceive themselves. Yet, whether one agrees with Habermas's prediction or not, one must consider the potential perspective of the enhanced child in any prognostic ethical assessment of whether parents have a right to exercise their reproductive autonomy in this fashion (Tonkens 2011, 281–283; Habermas 2003, 52–53).

Aside from the concern of whether children's autonomy may be attenuated by virtue of having been enhanced, there is the question of whether parental attitudes and preferences, implemented through particular enhancement interventions, may have an adverse impact on children's overall wellbeing, particularly given what may be idiosyncratic attitudes on some parents' part of what constitutes "wellbeing." Gheaus (2017, 270) cites a "Stepford Children" dystopian vision in which parents have designed their children to be "polite, obedient, hard-working, truth-telling, intelligent, good at sports, and energetic, with no time for television or junk food [but also] deeply dull." As noted above, certain traits that may be considered virtues in one context may become vices in another context or if possessed to an excess degree. For example, while *obedience* to one's parents would certainly be a trait most parents would want to inculcate in their children, a generally obedient attitude toward authority figures may not be a good trait for children to maintain when they become adults and are confronted with authority figures who are corrupt or otherwise abuse their power and thus ought not to be obeyed.

Furthermore, even if parents enhance their children with traits that are objectively defensible and subjectively perceived by the child as contributing to their overall wellbeing, it is nevertheless the case that a certain symmetry in the *power relationship* between parents and children has been disrupted. Currently, while parents can exert a degree of "environmental shaping" of how their children will grow and develop, the fundamental genetic potential of their children is not voluntarily chosen by parents; thus they must learn to adjust to the "spontaneity" of their children's inherent potentialities as each expresses itself. In this way, children exercise a degree of power of their parents, who must adjust their expectations as they discover more and more about their developing child. Gheaus (2017, 279) concludes,

> Genetic shaping allows parents to influence children even before children have any power
> to shape parents – either intentionally or unintentionally – and to react to their attempts
> at shaping. It introduces, in the history of the relationship, a phase when the child had no
> possibility to shape the parent either intentionally or unintentionally. Whereas the parent
> had such a chance and used it.

The picture of the parent/child relationship drawn here does not require parents to have a *conscious desire* to exercise power over their children in a *hubristic* manner, nor does it entail that parents will not *unconditionally* accept and love their children if the chosen enhancements fail to produce the expected outcomes—whether through some malfunction in the enhancement intervention itself or due to the child's choice not to develop the potentialities given to them by the intervention. Hence, it remains defensible that enhancement of children, at least for certain "general-purpose" traits, is *in principle* ethically justifiable; yet, *practical* parental wisdom may dictate that enhancements not be pursued given the present epistemic gap between novel enhancement techniques at the genetic level aimed toward increasing the potential wellbeing of children and tried-and-true environmental methods of maximizing children's wellbeing given their natural inherent potentialities (Tonkens 2011, 288).

20.5   Societal Relationships

A final concern takes us beyond the parent/child relationship to wider *societal* relationships between the enhanced and the unenhanced (refer to Chap. 28). Could enhancement lead to a bifurcated society in which those who have superior genetic endowments are able to compete more effectively in the socioeconomic sphere and thereby attain positions of social, economic, and political influence such that democratic relations among all citizens of a given polity are undermined?

Savulescu (2001, 424; cf. Buchanan et al. 2000, 27–60) rightly notes that the ethical mandate called for by PPB differs from the historical eugenics program insofar as the latter involved state-level interference with individuals' procreative autonomy in order to breed a healthier and happier *population*, whereas PPB aims at bettering the lives of *individual* children by means of ethically-informed, but not state-mandated, exercises of parental autonomy. Holland (2016, 498–499) counters that "social regulation" may replace government regulation in *strongly influencing* parental decision-making about enhancing children by setting social expectations of what traits are valued and will provide *competitive advantages* in the economic marketplace. Increased intelligence, for example, may be both inherently good for the enhanced child, but is also a *positional* good that will likely help them attain higher socioeconomic status than children who have not been so enhanced. This would exacerbate current *inequalities* of opportunity among members of different socioeconomic classes.

This phenomenon is already seen in parents competing to get their children enrolled in top-notch preschools, so they can get into the best (usually private and selective) elementary and secondary schools, so they can then get into the best universities, and so on. While this fact may imply that any social pressure to enhance one's children will be just "more of the same," there is arguably a categorical leap in the degree of pressure to enhance in order to determine socioeconomic outcomes for the children of parents who have the means and the will to enhance them versus those whose parents lack either the means or the will to do so. Sparrow (2019) goes so far as to argue that the generation of largely unenhanced persons, or individuals within a generation of largely enhanced persons, will be rendered "obsolete."

20.6 Conclusion

There is a wider set of ethical concerns regarding human enhancement generally, including whether large-scale enhancement may lead to a dangerous lack of genetic diversity, betray a Promethean drive toward "mastery" over nature, result in irreparable social disruptions, neutralize competition due to the diminishing returns of increasing levels of enhancement, or create a posthuman species who will view and ethically treat us as we treat nonhuman animal species.[6] Debate about how well-founded these and other relevant concerns are has led to two polarized camps. *Transhumanists* seek to maximize "morphological freedom" to reshape ourselves in ways that transcend current biological limitations, even to the point of potentially uploading our minds into a virtual environment freed from bodily constraints (More and Vita-More 2013); whereas *bioconservatives* consider such unbridled freedom to constitute "playing God" by pursuing an illusory ideal of "perfection," and thereby seek regulatory limits on any non-therapeutic biotechnological interventions (President's Council on Bioethics 2003).

Between these camps lie those who favor "truly human" forms of enhancement that do not lead to the creation of posthumans who do not share general human needs and interests (Agar 2014). The vexed question of whether parents have either a right or an obligation to enhance their children in various ways lies within this wider debate, the outcome of which may support either maximal freedom to enhance, a total ban on enhancement, or a limited permissibility to enhance specific traits for certain purposes. Most defensible would be enhancement for "general-purpose" traits that contribute to a defensible objective account of human wellbeing in which access to the means of enhancement are available to any parent who wishes to avail themselves; doubt is cast on such a future system, however, when even basic health care needs are not being met within both economically developed and developing nations.[7]

---

[6] These and other salient issues are comprehensively addressed in the recommended Further Readings below.

[7] I am most grateful to Jacquelyn Cutts for helpful background research that informed this chapter.

20.7   Guiding Principles for the Enhancement of Children

Parental rights and obligations
- Recognize that parents do not have an obligation to enhance their children even if available interventions are demonstrated to be safe and effective
- Acknowledge that parents do not have an unrestricted right to enhance their children
- Safeguard parents from being coerced into making specific enhancement decisions for their children

Perspective of the potentially enhanced child
- Consider the child's perspective when authorizing any potential enhancement interventions
- Focus potential enhancements on general-purpose goods as opposed to idiosyncratic parental choices
- Seek amelioration of societal conditions that may coerce parents into pursuing specific enhancement interventions

References

Agar, N. 2014. *Truly human enhancement: A philosophical defense of limits*. MIT Press.

Aristotle. 1999. *Nicomachean ethics*, 2nd edn. (T. Irwin, Trans.). Hackett.

Buchanan, A., D.W. Brock, N. Daniels, and D. Wikler. 2000. *From chance to choice: Genetics and justice*. Cambridge University Press.

Clay, C., and M. Leapman. 1995. *Master race: The Lebensborn experiment in Nazi Germany*. BCA.

Davis, L. J. (ed.). 2017. *The disability studies readers*, 5th edn. Routledge.

de Melo-Martín, Inmaculada. 2017. *Rethinking reprogenetics: Enhancing ethical analyses of reprogenetic technologies*. Oxford University Press.

Doudna, J.A., and S.H. Sternberg. 2017. *A crack in creation: Gene editing and the unthinkable power to control evolution*. Houghton Mifflin Harcourt.

Dow, K. 2019. Looking into the test tube: The birth of IVF on British television. *Medical History* 63 (2): 189–208.

Eberl, J.T. (Forthcoming). Disability, enhancement, and flourishing. *Journal of Medicine and Philosophy*.

Feinberg, J. 1987–1990. *The moral limits of criminal law*. Oxford University Press.

Feinberg, J. 1980. The child's right to an open future. In *Whose child? Children's rights, parental authority, and state power*, ed. W. Aiken and H. LaFollette, 124–153. Rowman and Littlefield.

Gheaus, A. 2017. Parental genetic shaping and parental environmental shaping. *The Philosophical Quarterly* 67: 263–281.

Glover, J. 2006. *Choosing children: Genes, disability, and design*. Oxford University Press.

Greely, H.T. 2011. Get ready for the flood of fetal gene screening. *Nature* 469: 289–291.

Green, R.M. 2007. *Babies by design: The ethics of genetic choice*. Caravan.

Griffin, J. 1986. *Well-being: Its meaning, measurement, and moral importance*. Clarendon Press.

Habermas, Jürgen. 2003. *The future of human nature*. Polity Press.

Harris, J. 2009. Enhancements are a moral obligation. In *Human enhancement*, ed. J. Savulescu and N. Bostrom, 131–154. Oxford University Press.

Holland, A. 2016. The case against the case for procreative beneficence. *Bioethics* 30 (7): 490–499.

Hotke, A. 2014. The principle of procreative beneficence: Old arguments and a new challenge. *Bioethics* 28 (5): 255–262.

Juengst, E.T. 1998. What does *enhancement* mean? In *Enhancing human traits: Ethical and social implications*, ed. E. Parens, 29–47. Georgetown University Press.

Kevles, D. J. 1985. *In the name of eugenics: Genetics and the uses of human heredity*. Alfred A: Knopf

Klipstein, S. 2017. Parenting in the age of preimplantation gene editing. *Hastings Center Report* 47 (6): S28–S33.

Klitzman, R.L. 2020. *Designing babies: How technology is changing the way we create children*. Oxford University Press.

Lombardo, P.A. 2008. *Three generations, no imbeciles: Eugenics, the Supreme Court, and Buck v. Bell*. Johns Hopkins University Press.

Lombardo, P. A. (ed.). 2011. *A century of eugenics in America: From the Indiana experiment to the human genome era*. Indiana University Press.

More, M., and N. Vita-More. (eds.). 2013. *The transhumanist reader: Classical and contemporary essays on the science, technology, and philosophy of the human future*. Wiley-Blackwell.

Persson, I., and J. Savulescu. 2012. *Unfit for the future: The need for moral enhancement*. Oxford University Press.

President's Council on Bioethics. 2003. *Beyond therapy: Biotechnology and the pursuit of happiness*. HarperCollins.

Robertson, J.A. 1994. *Children of choice: Freedom and the new reproductive technologies*. Princeton University Press.

Saunders, B. 2015. Is procreative beneficence obligatory? *Journal of Medical Ethics* 41: 175–178.

Savulescu, J. 2001. Procreative beneficence: Why we should select the best children. *Bioethics* 15 (5/6): 413–426.

Savulescu, J. 2007. In defence of procreative beneficence. *Journal of Medical Ethics* 33: 284–288.

Savulescu, J., and G. Kahane. 2009. The moral obligation to create children with the best chance of the best life. *Bioethics* 23 (5): 274–290.

Savulescu, J., R. ter Meulen, and G. Kahane. (eds.). 2011. *Enhancing human capacities*. Wiley-Blackwell.

Sparrow, R. 2019. Yesterday's child: How gene editing for enhancement will produce obsolescence—And why it matters. *American Journal of Bioethics* 19 (7): 6–15.

Tonkens, R. 2011. Parental wisdom, empirical blindness, and normative evaluation of prenatal genetic enhancement. *Journal of Medicine and Philosophy* 36: 274–295.

Veit, W. 2018. Procreative beneficence and genetic enhancement. *Kriterion* 32 (1): 75–92.

Veit, W., J. Anomaly, N. Agar, P. Singer, D.S. Fleischman, and F. Minerva. 2021. Can 'eugenics be defended? *Monash Bioethics Review*. https://doi.org/10.1007/s40592-021-00129-1.

Further Readings

Buchanan, A. 2011. *Better than human: The promise and perils of biomedical enhancement*. Oxford University Press.

Parens, E. (ed.). 1998. *Enhancing human traits: Ethical and social implications*. Georgetown University Press.

Savulescu, J., and N. Bostrom. (eds.). 2009. *Human enhancement*. Oxford University Press.

Soniewicka, M., and W. Lewandowski. 2019. *Human genetic selection and enhancement: Parental perspectives and law*. Peter Lang.

Chapter 21
Predicting Childhood Neurologic Impairments: Preparing for or Prejudicing the Future?

P. C. Mann

Abstract Medical providers have attempted to predict neurologic outcomes for parents of infants and children with neurologic injuries or anomalies for many decades. A desire to avoid the clinical outcome of a life lived with substantial neurologic disability has been one of the principle aims of these prognoses. Studies in recent years have increased concerns that these predictions are of marginal benefit and can be harmful when providers overestimate neurologic impairments. Novel technologies continue to emerge and find application for the prediction of an ever-expanding array of neurologic impairments, including mental health disorders. Appropriate ethical frameworks and communication guidelines are needed to protect children and parents from undue disability bias and to assist medical providers in having meaningful, effective and compassionate conversations with families about possible neurologic impairments and outcomes.

Keywords Neuroethics · Disability · Bias · Neurologic outcomes · Prediction

21.1 Introduction

Prevention of neurologic impairments in infants and children has been a longstanding aim of pediatric medicine (Mann 2017). At its best, that goal has led to numerous substantive neuroprotective clinical interventions (e.g. therapeutic hypothermia for encephalopathic infants) and preventive health breakthroughs (e.g. newborn screening for phenylketonuria), which have improved the health and neurologic outcomes of countless children. At its worse, however, the desire to eliminate neurologic impairment in children has been overtly and covertly tantamount to eugenics, pitting providers against parents in attempt to prevent disability and define an acceptable quality of life through a lens of considerable bias (Garland-Thomson 2012) (refer to Chaps. 19 and 20).

P. C. Mann (✉)
Department of Pediatrics, Medical College of Georgia at Augusta University & Center for Bioethics and Health Policy, Augusta University, Augusta, GA, USA
e-mail: pamann@augusta.edu

© Springer Nature Switzerland AG 2022   343
N. Nortjé and J. C. Bester (eds.), *Pediatric Ethics: Theory and Practice*,
The International Library of Bioethics 89,
https://doi.org/10.1007/978-3-030-86182-7_21

P. C. Mann

While the ability to detect neurologic injury and/or difference has continued to advance and find new applications, a definitive understanding of neurodevelopment outcomes for many types of neurologic injuries remains elusive (Kirschen and Walter 2015). Given the considerable neuroplasticity of children, impacts of socioeconomic disparities on neurodevelopmental outcomes, and limitations of neurologic outcome studies, predictions of neurodevelopmental impairments remain challenging and frequently inaccurate (Dennis et al. 2018; Roscigno et al. 2013). Additionally, many predicted neurodevelopmental impairments lack effective treatments, leaving patents burdened with the knowledge of a potential future neurologic disability that has no definitive therapeutic intervention. Children may be harmed by neurologic predictions when inappropriate decisions are made in their medical care based on inaccurate information, or if negatively stigmatized by this prognostication with attitudinal limitations placed on their future potential by others (Roscigno et al. 2011).

This chapter explores the history and current practice of neurologic predictions in pediatric medicine, highlighting present and emerging technologies that are utilized to predict an increasing variety of neurologic impairments. Ethical implications for decision-making in these contexts are explored and best practices for communicating about neurologic prognosis suggested.

21.2 Historical Approaches to Childhood Neurological Impairment Detection, Prediction and Treatment

The early twentieth century gave rise to a political and ideological movement that had devastating consequences for millions of lives. Beginning in the United Kingdom, and spreading throughout Europe and North America, adherents of eugenics espoused the belief that a wide range of disabilities, both physical and mental, were principally hereditary in cause and that the "inferior" individuals who suffered such maladies were reproducing at higher rates than "superior" individuals in the population. This chapter in the history of eugenics would eventually result in widespread sterilization of impaired children and adults in the United States, and state sponsored euthanasia of "lives not worth living" in Nazi Germany (Baker & Lang 2017).

For cognitively impaired children, the eugenics movement resulted in a dramatic rise in residential institutionalization, many who were placed in state sponsored facilities with minimal formal evaluation (Baker & Lang 2017). In 1916, American psychologist and eugenics supporter Lewis Terman published the Stanford Revision of the Binet-Simon Scale to better classify children and adults according to their intellectual abilities. Those with an intelligence quotient (IQ) from 0–25, were classified "idiots"; 25–50, "imbeciles"; and 50–70, "morons". "Mentally defective" children who fell into those categories were candidates for institutionalization, if and when space was available (Cohen 1952). It would be many decades before public opinion of such institutions would change in the wake of ethically troublesome research studies and abuse scandals taking place at organizations like the

Willowbrook State School[1] in New York State, eventually leading to a movement for widespread deinstitutionalization.

By the 1950s, new approaches to diagnosing neurologic and mental impairments in children had become common, including radiographic imaging and electroencephalographic (EEG) studies (Cohen 1952). Experimental treatments had emerged to treat disability, including one surgical intervention aimed at improving cognition and reducing seizure activity in impaired children through brain "revascularization" (Cohen 1952; Adam & Goetz 1970). This pioneering work by prominent American cardiac surgeon and Nobel Prize in Medicine Nominee, Claude Schaeffer Beck, involved creating an arteriovenous fistula[2] between the carotid artery and internal jugular vein to improve cerebral blood flow. While initial studies reported clinical improvement in the mental acuity of 35% of patients operated upon, others were not able to reproduce those results (Adam & Goetz 1970). Many years later, it was reported that up to 70% of patients receiving the operation developed potentially debilitating complications including increased intracranial pressure, pulsating exophthalmos,[3] papilledema,[4] tinnitus and headaches. For others, the procedure decreased their cerebral blood flow, presumably worsening their mental acuity and increasing neurologic impairments.

Neonatology as a specialty was also emerging in the 1950s and a great deal of attention was being paid to the neurologic outcomes of infants born prematurely and offered intensive care (Mann 2017). William A. Silverman, a prolific writer and influential neonatal scientist for decades, was one of the earliest to speak to the significant rates of brain injury in infants surviving premature birth and to recommend a switch in focus from survival alone to an ideal he coined "intact survival". While ambiguously defined, a goal of "intact survival" became firmly entrenched in medical literature beginning in the 1960s, encompassing a substantial range of subjective goals such as the avoidance of specific neurologic sequelae of brain injury (e.g. cerebral palsy), as well more broadly explored utilitarian aspirations, included a goal of ensuring that the children who were born prematurely would ultimately have "usefulness" to society.

Neonatologists formalized attempts to predict and prevent neurologic impairment in premature infants in the 1970s, with widespread adoption of routine neuroimaging studies in order to detect intraventricular hemorrhages (IVH) and other brain injuries (Mann et al. 2013). It was hoped that neuroimaging would clarify neurodevelopmental outcomes for infants, helping to identify those whose impairments were likely to be so severe that withholding or withdrawal life supportive therapies would

---

[1] The Willowbrook State School is an infamous case study in institutionalization of children with developmental disabilities (Weiser 2020). Conditions at Willowbrook were deplorable, eventually leading to class-action lawsuits and public outrage. Some parents, desperate for rapid admission to Willowbrook, felt coerced into consenting to research allowing their child to be intentionally infected with hepatitis as a condition of more expedient acceptance into the institution (Rosenbaum 2020).

[2] An artificial blood vessel connection, created between an artery and vein, used to alter blood flow.

[3] Bulging of one or both eyes.

[4] Swelling of the optic disc at the back of the eye from increased pressure in the brain.

be ethically permissible and could be suggested to parents. As long-term neurode-velopmental follow-up studies became available for infants with IVH, however, a plurality of neurodevelopmental outcomes was demonstrated, even for infants with substantial intracranial bleeding whose outcomes were once thought to be "univer-sally hopeless". Neuroplasticity was postulated as the likely reason for such outcome diversity and cautioned raised for utilizing ultrasonography principally as a tool for selective withdrawal of mechanical ventilator support from premature infants in the first weeks of life.

The 1970s also gave rise to the first civil rights laws preventing discrimination again individuals with disability. The Baby Doe Amendment to the Child Abuse Prevention and Treatment Act of 1974 in mid-1980s would expand those efforts. Baby Doe was an infant born in 1982 with Trisomy 21 and a tracheoesophageal fistula. At the recommendation of their obstetrician, the parents elected to withhold the surgical intervention required to repair the defect, a decision which was upheld by the Indiana Supreme Court. The death of the infant at six days of life outraged many, and the United States government quickly enacted measures that included financial penalties for hospitals that withheld beneficial life-saving treatments from infants because of concerns for future disabilities with the exception of infants thought to be irreversibly comatose. At the peak of enforcement of these laws, posters were placed in neonatal intensive care units across America with a phone number to call to report violators. This created considerable paranoia for providers who believed that their medical decision-making was now subject to scrutinization by the federal government (Mercurio 2009). While this remains law in the United States, these regulations have been rarely enforced since the mid-1980s. Instead, ethical oversight of decisions regarding withholding and withdrawing life-sustaining interventions in infants and children with potential neurologic impairments are currently principally under the local guidance of hospital ethics committees.

21.3   Current and Emerging Approaches to Childhood Neurological Impairment Detection, Prediction and Treatment

Neuroimaging technologies such as computerized tomography (CT), magnetic reso-nance imaging (MRI) and head ultrasonography remain widely utilized in neonatal and pediatric critical care units in order to evaluate brain injury and predict neuro-logic outcomes (Mann et al. 2015). Structural magnetic resonance imaging (MRI) has been joined by advanced imaging techniques including diffusion MRI, functional MRI and magnetic resonance spectroscopy (MRS) to image the injured brain and/or predict future neurologic impairments. Myriad studies have been published linking different patterns of injury, brain metabolites[5] and morphometric biomarkers[6] to neurodevelopmental outcomes for infants and children (Parikh 2016). The results of these studies, however, have been difficult to consistently reproduce in other populations of infants/children and limiting the usefulness of neuroimaging to definitively predict many forms of neurologic impairments including cognitive, executive and behavioral functioning (Mann et al. 2013; Parikh 2016).

Neurocritical care is an emerging multidisciplinary specialty that has coalesced in recent years to standardize the care of infants and children with brain injury and/or at risk for neurodevelopmental impairments (Horvat et al. 2016; Mann et al. 2015). Neurocritical units employ specialized nurses, critical care physicians, neurologists and neurosurgeons to treat children with diseases such as traumatic brain injury, meningitis, status epilepticus, stroke, hypoxic ischemic encephalopathy and spinal cord injury. Neurocritical care is distinct in its approach to clinical management, employing a wide array of neurodiagnostic tools including neuroimaging, continuous and video electroencephalography (EEG), cerebral hemodynamic and oxygenation measurements, intracranial pressure monitoring and sequential neurobehavioral examination with a goal to prospectively optimize neurologic outcomes. It is believed that active cerebral monitoring allows for the detection of clinical changes that may not readily apparent to bedside providers (e.g. seizures), which if not recognized expediently may contribute to secondary brain injury and worsen neurologic outcomes following the initial brain insult. Long-term outcomes for this growing specialty, however, are understudied, and while some studies have indicated improvement in clinical outcomes, few interventions have shown clear benefit (Horvat et al. 2016; Williams et al. 2019). What has become clearer, is that survivors of pediatric neurocritical care units have substantial rates of new disability upon discharge, highlighting the need for appropriate multidisciplinary follow-up to optimize outcomes for the infants and children, and support for their families (Williams et al. 2019).

An emerging frontier for neurologic outcome prognostication in children is focused on predicting mental illness and behavioral disorders (Lane et al. 2020; Lawrie et al. 2019). Research has been conducted into identifying numerous radiographic and genetic biomarkers that can accurately determine the onset of mental and/or behavioral health impairments in infants, children and young adults (refer to Chap. 19). Psycho-radiology has developed as a distinct clinical field combining psychiatry with advanced radiographic imaging techniques to predict the onset mental illnesses such as obsessive–compulsive disorder, depression and schizophrenia with a goal to optimize pharmacological interventions and cognitive treatments. Expansive genome wide association consortiums have enhanced our understanding of psychiatric genomics, identifying numerous genes that enhance the risk of developing mental and behavioral illnesses such as attention deficit and hyperactivity disorder (ADHD), autism and schizophrenia; raising concerns for a new era of eugenics, if applied prenatally (Baker & Lane 2017).

---

[5] MRI can be used to study how the healthy versus injured brain utilizes different substances such as glucose, oxygen, and amino acids.

[6] Measurement of volume and curvature of different areas of the developing brain.

Advances in artificial intelligence have also allowed the development of machine learning algorithms that combine clinical risk factors with neuroimaging and genetic testing results to more accurately quantify psychosis susceptibility (refer to Chap. 25). For those identified as at risk to develop psychosis, research has focused on how to avert or delay its development through preventative pharmacologic therapy. To date, however, no intervention has been definitively identified that can prevent the onset of psychosis. Given the current uncertainties for which patients will definitely develop psychosis, any benefit of preventive pharmacologic therapy will have to overcome the considerable potential risks of harm to patients (i.e. side effects from unwarranted pharmacologic interventions) (Lane et al. 2020).

21.4 Impacts of Neurologic Impairment Prediction on Parents and Children

The twenty-first century has seen many kinds of diversity better recognized and celebrated, however, neurologic differences and disability remain chiefly uncelebrated and undesired. In his book, The End of Normal, Davis (2013) describes the principle design of modern medicine as a pursuit to return aberrant bodies to "normal" and to avoid physical and neurological differences:

> We may want diversity in all things, but not insofar as medicalized bodies are concerned. It is in this realm that "normal" still applies with force. Most people still want normal cholesterol, blood pressure, and bodily functions. The word most people want to hear from an obstetrician after a birth is that the baby is "normal". (p. 7)

Progressive scholars and disability rights activists have challenged this culturally predominate paradigm of ableism (i.e. social prejudice and discrimination against individuals with disability) with a call for society to embrace greater neurodiversity (Baker & Lang 2017). Autism Spectrum Disorder presents a unique case study in the movement to support neurodiversity. From its earliest recognition as a clinical syndrome, children and adults with Asperger syndrome were known to possess superior intellectual abilities in distinct areas of scholarship (e.g. mathematics) despite notable differences in social interactions. Asperger himself mused, "Who among us does not recognize the autistic scientist whose clumsiness and lack of instincts have made him a familiar caricature, but who is capable of extraordinary accomplishments in a highly specialized field?".

Medical providers, however, have historically reflected the culturally predominant viewpoint of all types of neurologic impairment as undesirable, hence the countless technologies that have been developed and used/misused to detect neurologic abnormalities. Despite continuously expanding abilities to detect neurological injury and differences, medical providers remain limited in their abilities to definitively predict holistic neurologic sequela for many diseases, including neurodevelopmental impairments for premature infants with intraventricular hemorrhages and neurologic outcomes for children with traumatic brain injury.

Much of the difficulty in translating meaningful neurologic outcomes from research studies into clinical practice stems from problems with how clinical studies report research endpoints (Janvier et al. 2016) and differences in how parents and providers weigh clinical outcomes (Lemmon et al. 2019). For clinical trials to be adequately powered to achieve statistically significant results, research studies frequently report composite outcomes; for example, survival without neurodevelopmental impairment. In other words, a child dying or a child living a life with significant neurodevelopmental impairments is viewed as the same outcome in many research studies. Even if this is statistically appropriate, it is conceptually problematic. Parents do not view the death of their child in any way equal to life lived with significant future neurodevelopmental impairment (Janvier et al. 2016).

Lemmon and colleagues (2019) demonstrated a clear chasm between parents' and providers' views on death and disability in a study investigating how prognostic information was incorporated into clinical decision-making for infants born extremely premature. In that study, parents focused principally on survival as the goal of care for their infants, infrequently discussing any future neurodevelopmental concerns, and expressing hope that the clinical outcome would be better than predicted. Providers, by contrast, endorsed neurodevelopmental outcomes as the focal point of their prognostic discussions for infants, framing conversations about likely clinical outcomes with parents almost exclusively through a lens of future neurologic impairments and potential suffering. They voiced significant distress that parents seemed to ignore the "reality" of potential disability for their babies. So fixated were medical providers upon neurodevelopmental outcomes, in fact, that they brought them up six times more often than parents in study interviews (Lemmon et al. 2019). Parents, however, are rarely so unattuned to the possibility of neurologic impairments for their child (or "in denial" as they are often pejoratively labeled by medical staff); providers usually make it abundantly clear when they believe that a "bad" outcome is expected. Providers, therefore, need to exercise caution in these contexts, as a focus exclusively on worrisome neurologic outcomes may alienate parents wishing to remain hopeful, making it difficult to maintain a therapeutic alliance and risking avoidant behaviors from families towards "negative" medical providers (Mann 2017).

Even in neurodevelopmental areas where medical providers can frequently accurately predict a particular neurologic impairment due to advancements in technology, (e.g. cerebral palsy), parents do not overwhelmingly endorse a benefit to learning about neurologic deficits for their child. A recent study examining parental perspectives on learning that their child would develop cerebral palsy, reported that 41% of parents did not perceive that it was helpful to be given the information (Guttmann et al. 2018). Additionally, only 24% of parents in that study reported that their child functioned as predicted, with 46% reporting that their child had exceeded expectations. This highlights numerous current difficulties for medical providers attempting to predict neurologic outcomes for parents. Either providers are frequently conflating outcomes to families as worse than they will actually be, or the providers do not factually understand the true incidence of the neurologic outcomes they are predicting. To that point, medical providers in another study overestimated the population prevalence of cerebral palsy by a factor of 10 to 100-fold (Janvier et al. 2016).

Parents of children with traumatic brain injury report similar difficulties in receiving accurate neurologic prognosis from medical providers. In a study by Roscigno et al. (2013), only two of sixteen parents whose children had been diagnosed with traumatic brain injury reported a long-term outcome that was even remotely consistent with the early neurologic prognosis they had been given. For one child whom parents were told that they she "would need a nursing home for the rest of her life", she was reported to be graduating high school, starting community college, engaged to be married and socially active. For another child whose parents were told "on [a] scale of one to ten- ten being brain dead- she was a nine" and "maybe just turn off the machine", she was living independently and attending a four-year university with a "B" average. While there are notable limitations to such studies, such as the possibility that parent's memories of early prognostic interactions might be biased towards negative encounters or that parents might be motivated to continue to partic-ipate in the study because their child did much better than expected, it is extremely worrisome how wrong the predictions turned out to be for these particular families. It is equally worrisome that other families might have withheld or withdrawn beneficial medical therapies based on misleading or inaccurate prognostic information.

Neurodevelopmental impairments, as relayed to parents by providers, include a range of outcomes that a parent might consider highly worrisome (e.g. a child performing three standard deviations below the mean on a neurodevelopmental assessment, severe autism) and those that they might consider quite trivial (e.g. mild cerebral palsy, hearing impairment). To broadly judge or describe a neuro-logic outcome for an infant or a child as likely to be "good" or "bad" is a fictitious paradigm. The ultimate neurodevelopmental outcome for a neurologically injured or at-risk child is comprised of numerous important but distinct neurologic endpoints; motor function, balance and coordination, vision, hearing, cognitive and behavioral outcomes. With the notable exception of brain death and other severe forms of brain injury, providers can infrequently predict neurologic function in all neurodevelop-mental areas with any reasonable degree of accuracy. Additionally, neuroplasticity and numerous socioeconomic factors, such as abilities to participate in rehabilitative therapies and maternal education level, substantially impact neurologic outcomes and can't reasonably be accounted for in initial prognostic discussions with parents (Mann 2017).[7]

Impacts of predictions of neurologic impairments on children are woefully under-studied and underreported. It has been reported, however, that children with traumatic brain injury are very aware that medical providers place limits on their future poten-tial and that this has significant impacts on their health and wellbeing (Roscigno et al. 2011). It has also been reported that children who were born very premature, and living with neurologic injuries such as cerebral palsy, report their quality of life as similar to their peers, and much higher than medical providers and their parents (Mann et al. 2013). It is highly worrisome, therefore, that neurologic prediction has

---

[7] For a case study in how initial prognostic discussions impact families of infants born at extreme premature gestational age, see Aleccia (2015) and Ruthford, et al. (2017).

now moved into the realm of mental and behavioral health. The potential for stigmatization of children who are labeled as high risk to develop a psychiatric disorder as a result of radiographic imaging and/or genetic testing is substantial (Lawrie et al. 2019; Manzini & Vears 2018). Self-image, peer relationships, and future employment could all be negatively impacted by such predictions.

Guiding Principles for Predicting Childhood Neurologic Impairments

---

Compassionate Communication
– Learning that a child may have significant neurologic impairments is a life-altering event for parents. Providers need to thoughtfully plan these conversations and create an appropriate private environment for dialogue. Allow sufficient time to thoroughly answer questions and address concerns. Reassure families that the need for additional discussions is expected

---

Prognostic Honesty
– Providers should not avoid difficult conversations with families when neurologic impairments are suspected. Conversations should be framed from a balanced perspective, describing a range of specific things the child likely will be able do and likely will not be able to do. Avoid the use of negative prognostic extremes (i.e. "your child will never…")

21.5 Meaningful Communication Regarding Neurologic Prognosis

Despite the numerous limitations to the accurate prediction of neurologic prognosis, and possible negative impacts on parents and children, it is essential for medical providers to have these conversations. Parents need to understand that their child is at risk for impairments in order to closely monitor their development and to ensure that they get the services needed to optimize their neurodevelopmental potential. It is critical that providers disclose their neurologic concerns in an effective, sympathetic manner. Mishandling disclosure has been known for decades to result in resentful attitudes towards providers (Cohen 1951) and to impact family's abilities to cope with their child's disability (Graungaard & Skov 2006; Novak et al. 2019; Pearson et al. 2020). Many providers, however, still lack the basic skills necessary to effectively communicate difficult news (Roscigno et al. 2013).

Keys to Effectively Communicating a Worrisome Neurologic Prognosis

(1) *Create an appropriate setting and allot adequate time for disclosure-* Parents clearly remember the moment they first learned that their child may have a disability (Hedderly et al. 2003). Discussions of potential neurologic impairments should be in a quiet, private setting with all parties seated and the infant or child present if desired by the family (Novak et al. 2019). Both parents should ideally be physically present for this conversation to allow them to process the information concurrently and ask questions (Pearson et al. 2020).

It is imperative that parents have adequate time to express grief, explore prognostic implications for their child, and begin to formulate next steps. Schedule a time in the near future to reconvene and address additional concerns.

(2) *Invite parents to help guide the conversation-* Ask open-ended questions to ascertain what parents may already understand or fear regarding their child's future neurologic potential (Novak et al. 2019). Attempt to elicit and address particularly meaningful or worrisome neurodevelopmental outcomes. Depending on their coping style, parents may either actively seek out or try to avoid prognostic information (Harvey et al. 2013). It is important for the provider to understand what the parent's communication needs may be moving forward, and how they best learn and remember information (Kirschen & Walter 2015). In order to avoid confusion and obviate the need for unwanted repetitive information, clearly and explicitly document these conversations in the medical record so that future providers understand comprehensively what prognostic information has already been discussed.

(3) *Provide timely and balanced information-* Parents have described dismissive behaviors from medical providers unwilling to address neurologic concerns and significant delays in receiving a diagnosis of neurologic impairment for their child, even when they suspected that something was wrong (Graungaard & Skov 2006). Avoiding conversations with parents about potential neurologic disability is an abdication of professional responsibility, which can lead to distrust with families and jeopardize a therapeutic alliance moving forward. At the other extreme, pessimistically describing neurologic outcomes as inevitable, using only negative linguistic extremes (e.g. "your child is never going to walk, talk, laugh…"), is equally harmful (Roscigno et al. 2013). Families report benefits to providers leaving room for optimism when delivering bad news and avoiding extinguishing their hope (Roscigno et al. 2012; Lemmon et al. 2019). Possible neurodevelopment impairment should be broadly discussed with parents, giving a range of potential outcomes, but highlighting the most likely outcome (Guttmann et al. 2018). Frame this conversation in "real-world" examples of in both what the child will be able to do and what they likely won't be able to do (Janvier et al. 2016).

(4) *Acknowledge uncertainty-* Uncertainty in neurologic prognosis is unavoidable in many clinical situations and providers frequently disagree about likely neurologic outcomes, especially in the initial days following a brain injury (Kirschen & Walter 2015). Parents can be burdened by these divergent prognostic opinions, and left to wonder why perspectives can by so dissimilar between different providers, who is right, and who can they trust? (Roscigno et al. 2013). If handled appropriately, however, families may benefit from prognostic disagreements by understanding the potential for outcome ambiguity and utilizing that information in their decision-making (Mann et al. 2015).

(5) *Consider relational potential-* Parents frequently choose to embrace neurodevelopmental outcomes in their children that providers would consider a very poor quality of life. Children with profound neurologic disabilities, however,

are deeply loved and supported by their families. Relational potential recognizes "the capacity for a caring relationship to flourish even if it may appear one-sided to an outside observer" (Wightman et al. 2019). Medical providers need to embrace a more inclusive understanding of the value of a child with substantial cognitive impairments in the context of a loving family, supporting parents who chose life-sustaining therapies for those children.

21.6 Conclusions

The troublesome history of neurologic outcome prognostication in pediatric medicine warrants reconsideration of the relative weight of that these predictions have on clinical outcomes for infants and children. Greater humility and compassion from providers are needed when discussing potential disabilities with parents and is better knowledge of actual outcomes for their patients. Medical providers need training in how to communicate neurologic concerns to parents in these contexts and additional research is needed to help elicit a broader understanding of meaningful neurologic outcome. Emerging technologies that could be utilized to predict a broadening array of neurologic outcomes, including mental health, require thoughtful implementation to ensure that their application avoids reinforcing societal norms of ableism and considers potential benefits of neurodiversity.

References

Adam, Y.G., and R.H. Goetz. 1970. Surgically produced carotid-jugular fistula: 18-year follow-up. *Annals of Surgery* 171 (1): 93–97. https://doi.org/10.1097/00000658-197001000-00014.

Aleccia, J. 2015. Evidence complicates decision on when to save preemies. *The Seattle Times.* https://www.seattletimes.com/seattle-news/health/new-evidence-complicates-decisions-on-when-to-save-extremely-premature-babies

Baker, J.P., and B. Lang. 2017. Eugenics and the origins of autism. *Pediatrics* 140(2): e20171419. https://doi.org/10.1542/peds.2017-1419.

Cohen, P. 1952. The problem of the mentally defective child. *California Medicine* 76 (1): 34–37.

Davis, L. J. 2013. The End of Normal: Identity in a Biocultural Era. University of Michigan Press. https://doi.org/10.3998/mpub.5608008.

Dennis, E.L., T. Babikian, C.G. Giza, P.M. Thompson, and R.F. Asarnow. 2018. Neuroimaging of the injured pediatric brain: Methods and new lessons. *The Neuroscientist* 24 (6): 652–670. https://doi.org/10.1177/1073858418759489.

Garland-Thomson, R. 2012. The case for conserving disability. *Bioethical Inquiry* 9 (3): 339–355. https://doi.org/10.1007/s11673-012-9380-0.

Guttmann, K., J. Flibotte, and S.B. DeMauro. 2018. Parental perspectives on diagnosis and prognosis of neonatal intensive care unit graduates with cerebral palsy. *The Journal of Pediatrics* 202: 156–162. https://doi.org/10.1016/j.jpeds.2018.07.089.

Graungaard, A.H. and L. Skov. 2006. Why do we need a diagnosis? A qualitative study of parents' experiences, coping and needs, when the newborn child is severely disabled. *Child: care, health and development* 33(3): 296–307. https://doi.org/10.1111/j.1365-2214.2006.00666.x.

Harvey, M.E., P. Nongena, N. Gonzalez-Cinca, A.D. Edwards, and M.E. Redshaw. 2013. Parent's experiences of information and communication in the neonatal unit about brain imaging and neurological prognosis: A qualitative study. *Acta Pædiatrica* 102 (4): 360–365. https://doi.org/10.1111/apa.12154.

Hedderly, T., G. Baird, and H. McConachie. 2003. Parent reaction to disability. *Current Pediatrics* 13 (1): 30–35. https://doi.org/10.1054/cupe.2003.0406.

Horvat, C.M., H. Mtawh, and M.J. Bell. 2016. Management of the pediatric neurocritical care patient. *Seminars in Neurology* 36 (6): 492–501. https://doi.org/10.1055/s-0036-1592107.

Janvier, A., B. Farlow, J. Baardsnes, R. Pearce, and K.J. Barrington. 2016. Measuring and communicating meaningful outcomes in neonatology: A family perspective. *Seminars in Perinatology* 40 (8): 571–577. https://doi.org/10.1053/j.semperi.2016.09.009.

Kirschen, M.P., and J.K. Walter. 2015. Ethical issues in neuroprognostication after severe pediatric brain injury. *Seminars in Pediatric Neurology* 22 (3): 187–195. https://doi.org/10.1016/j.spen.2015.05.004.

Lane, N. M., S.A. Hunter, and S.M. Lawrie. 2020. The benefit of foresight? An ethical evaluation of predictive testing for psychosis in clinical practice. *NeuroImage: Clinical* 26: 102228. https://doi.org/10.1016/j.nicl.2020.102228.

Lawrie, S.M., S. Fletcher-Watson, H.C. Whalley, and A.M. McIntosh. 2019. Predicting major mental illness: Ethical and practical considerations. BJPsych Open, 5(2): e30. https://doi.org/10.1192/bjo.2019.11.

Lemmon, M.E., H. Huffstetler, M.C. Barks, C. Kirby, M. Katz, P.A. Ubel, S.L. Docherty, and D. Brandon. 2019. Neurologic outcome after prematurity: Perspectives of parents and clinicians. *Pediatrics* 144(1): e20183819. https://doi.org/10.1542/peds.2018-3819.

Mann, P.C. 2017. Partial survivors: The brokenness of intact survival. *Journal of Pediatric Ethics* 1 (1): 29–33.

Mann, P.C., S.M. Gospe, K.J. Steinman, and B.S. Wilfond. 2015. Integrating neurocritical care approaches into neonatology: Should all infants be treated equitably? *Journal of Perinatology*, 35(12): 977–981. https://doi.org/10.1038/jp.2015.95.

Mann, P.C., D.E. Woodrum, and B.S. Wilfond. 2013. Fuzzy images: Ethical implications of using routine neuroimaging in premature neonates to predict neurologic outcomes. *The Journal of Pediatrics* 163 (2): 587–592. https://doi.org/10.1016/j.jpeds.2013.03.055.

Manzini, A., and D.F. Vears. 2018. Predictive psychiatric genetic testing in minors: An exploration of non-medical benefits. *Bioethical Inquiry* 15 (1): 111–120. https://doi.org/10.1007/s11673-017-9828-3.

Mercurio, M.R. 2009. The aftermath of Baby Doe and the evolution of newborn intensive care. *Georgia State University Law Review* 25 (4): 835–863.

Novak, I., C. Morgan, L. McNamara, and A. te Velde. 2019. Best practice guidelines for communicating to parents the diagnosis of disability. *Early Human Development* 139: 104841. https://doi.org/10.1016/j.earlhumdev.2019.104841.

Parikh, N.A. 2016. Advanced neuroimaging and its role in predicting neurodevelopmental outcomes in very preterm infants. *Seminars in Perinatology* 40 (8): 530–541. https://doi.org/10.1053/j.semperi.2016.09.005.

Pearson, T., S. Wagner, and G. Schmidt. 2020. Parental perspective: factors that played a role in facilitating or impeding the parents' understanding of their child's developmental diagnostic assessment. *Child: Care, Health and Development* 46(3): 320–326. https://doi.org/10.1111/cch.12751.

Roscigno, C.I., G. Grant, T.A. Savage, and G. Philipsen. 2013. Parent perceptions of early prognostic encounters following children's severe traumatic brain injury: "Locked up in this cage of absolute horror." *Brain Injury* 27 (13–14): 1536–1548. https://doi.org/10.3109/02699052.2013.831122.

Roscigno, C.I., T.A. Savage, K. Kavanaugh, T.T. Moro, S.J. Kilpatrick, H.T. Strassner, W.A. Grobman, and R.E. Kimura. 2012. Divergent views of hope influencing communications between parents and hospital providers. *Qualitative Health Research* 22 (9): 1232–1246. https://doi.org/10.1177/1049732312449210.

Roscigno, C.I., K.M. Swanson, M.S. Vavilala, and J. Solchany. 2011. Children's longing for every-dayness: Life following traumatic brain injury in the USA. *Brain Injury* 25 (9): 882–994. https://doi.org/10.3109/02699052.2011.581638.

Rosenbaum, L. 2020. The hideous truths of testing vaccines on human. *Forbes*. https://www.forbes.com/sites/leahrosenbaum/2020/06/12/willowbrook-scandal-hepatitis-experiments-hideous-truths-of-testing-vaccines-on-humans.

Ruthford, E., M. Ruthford, and M.L. Hudak. 2017. Parent-physician partnership at the edge of viability. *Pediatrics* 139(4): e20163899. https://doi.org/10.1542/peds.2016-3899.

Weiser, B. 2020. Beatings, burns and betrayal: The Willowbrook scandal's legacy. *The New York Times*. https://www.nytimes.com/2020/02/21/nyregion/willowbrook-state-school-staten-island.html

Williams, C.N., C.O. Eriksson, A. Kirby, J.A. Piantino, T.A. Hall, M. Luther, and C.T. McEvoy. 2019. Hospital mortality and functional outcomes in pediatric neurocritical care. *Hospital Pediatrics* 9 (12): 958–966. https://doi.org/10.1542/hpeds.2019-0173.

Wightman, A., J. Kett, G. Campelia, and B.S. Wilfond. 2019. *Hastings Center Report* 49 (3): 18–25. https://doi.org/10.1002/hast.1003.

Further Readings

Novak, I. M. Thornton, C. Morgan, P. Karlsson, H. Smithers-Sheedy, and N. Badawi. 2016. Truth with hope: Ethical challenges in disclosing 'bad' diagnostic, prognostic and intervention information. In P. L. Racine, E., Bell, and Shevell, M. (2013). Ethics in neurodevelopmental disability. *Handbook of Clinical Neurology* 118: 243–263. https://doi.org/10.1016/B978-0-444-53501-6.00021-4.

Rosenbaum, G. M., Ronen, E. Racine, J. Johannesen, and B. Dan (Eds.), *Ethics in Child Health: Principles and Cases in Neurodisability*, pp. 97–109. Mac Keith Press.

Chapter 22
Ethics of Pediatric Gender Management

K. Moryan-Blanchard, L. Karaviti, and L. Hyle

Abstract The management of gender in pediatrics remains a controversial and challenging topic because the desire to protect children against potential harm may directly conflict with the ability to preserve self-determination and autonomy. Gender development and management will be viewed in this chapter through two lenses. The first is that of individuals born with differences of sexual development, a heterogeneous group of conditions arising from disruption of the typical development of internal and/or external sex organs. The second lens is that of individuals with pediatric gender dysphoria. In both cases, this chapter will examine the ethical issues and potential distress arising from a discordance between the biological sex and personal gender identification of an individual under the age of eighteen years.

Keywords Pediatric gender dysphoria · Differences of sexual development · Pediatric ethics · Gender identity · Paternalism in gender management

22.1 Introduction

Gender is an intimate part of identity that is difficult to define. While sex is defined by the biological anatomy and physiology of the body, gender is an abstract internal identification with a social role or group. Gender is perhaps most easily understood in the context of its difference from biological sex. "Gender dysphoria" is the phrase used to refer to the distress produced by a discordance between biological sex and the personal gender identification of the individual, as might be the case, for example, for

K. Moryan-Blanchard (✉) · L. Karaviti · L. Hyle
Texas Children's Hospital, Houston, USA
e-mail: kristen.moryan-blanchard@utsouthwestern.edu

L. Karaviti
e-mail: lpkaravi@texaschildrens.org

L. Hyle
e-mail: laurelhyle@gmail.com

© Springer Nature Switzerland AG 2022
N. Nortjé and J. C. Bester (eds.), *Pediatric Ethics: Theory and Practice*,
The International Library of Bioethics 89,
https://doi.org/10.1007/978-3-030-86182-7_22

an individual born with male genitalia but identifying as female (American Psychiatric Association 2013). When gender dysphoria occurs in someone under the age of eighteen years, the vulnerability of non-conforming gender identity compounds with the vulnerability of youth and evolving formation of identity, overall development (including cognitive and maturational abilities), and decision-making capacity (Katz and Webb 2016). Given the intensely intimate nature of gender identity and vulnerability of the gender non-conforming and pediatric populations, gender intervention and manipulation by pediatric psychologists, physicians, and surgeons must be undertaken only after significant ethical consideration is given to the ongoing development and wellbeing of the child involved. Parental, family, and community-of-origin issues should be identified and addressed, but above all, the central ethical obligation is to the child, who is both the patient and the most vulnerable.

Gender management in pediatrics has a controversial history marred by paternalism and a wide variety of discrimination grounded in larger societal dynamics (e.g., regarding race, ethnicity, and socioeconomic status) that continues to influence the field. Constructing an ethical framework for current pediatric gender management requires an understanding of the current knowledge of gender development and the complex history of prior medical management.

22.2 Gender Development

The development of gender identity is imperfectly understood. Insights from many case studies and retrospective analyses of individuals with differences of sexual development (DSDs) indicate that gender development of the brain occurs from the prenatal period likely through pubertal development.

Evidence points to critical time periods in brain development during which key exposures—typically hormonal—guide and determine gender identity formation. Many view the prenatal period as a potential "organizing" stage of brain development during which hormonal exposures to varying levels of androgenic or masculinizing hormones influence neural pathway development. Indeed, elevated prenatal androgen exposure is associated with male-type behaviors, but the leap to male gender identity is not as clear. The study of individuals with congenital adrenal hyperplasia[1] (CAH) provides a natural experiment in this theory. Individuals of XX genotype (i.e., those with female sex chromosomes) with CAH are exposed to higher than normal levels of male hormones in utero. This hormone exposure leads to masculinization of the genitalia early in prenatal development and with continued exposure could also influence brain development. Individuals with CAH tend to have greater "male-typed" play behaviors as children, such as a preference for toys more typically associated with boys (cars and planes), and go on to have higher rates of homosexuality and

---

[1] Congenital Adrenal Hyperplasia (CAH) is a condition in which there is decreased enzymatic function in a critical step of adrenal steroid production. This results in both a deficiency of some adrenal hormones as well as excess production of male-type hormones, such as testosterone.

gender dysphoria than the general population. About 5% of XX-CAH individuals later identify as male, versus a rate of approximately 0.5% of gender dysphoria in the general population (Berenbaum and Beltz 2011).

Pubertal-level hormone surges during the first year of life, known as mini-puberty of infancy, define another potential "organizing" period. Data on correlations with long-term gender outcomes are not available, but elevated urine testosterone and penile growth have been correlated with later male-typed behaviors. Such findings indicate that male hormone surges may further influence brain development with regard to gender (Lamminmäki et al. 2012; Pasterski et al. 2015).

Social constructs begin to influence children during the toddler and preschool years with the development of gender labeling ability. Gender labels are concrete at this age and based on early, developing stereotyping. During this time, superficial presentations, such as choice of clothing alone, may be conflated with gender. Gender stereotyping peaks at age five to seven years and is followed by greater flexibility in middle childhood (Drescher and Byne 2012).

Adolescence and the onset of puberty appear to be a critical period in solidifying gender identity. Those who describe the prenatal and mini-puberty periods as "organizational" view the pubertal hormonal surge as the potential "activating" period, putting into action the neural pathways that were developed earlier in development. Socially, gender stereotyping and identification intensify as gender roles become more divergent. This intensification of social roles often also intensifies and worsens gender dysphoria in individuals who do not identify with their assigned sex (Steensma et al. 2013).

All of these findings highlight the ethical tension between the needs and desires of society and those of the child when the child has not reached full maturity and is vulnerable in a specific way in addition to being vulnerable as a member of a minority, underrepresented, and often misunderstood group. This situation also highlights the need for ethicists and the health care community to ensure that structures are in place to protect these children.

22.3 Controversial History of Paternalism in Gender Management

Further evidence of brain gendering can be gathered from studies of controversial cases, particularly that of David Reimer, famously referred to as the "John/Joan" case (Colapinto 1997). David Reimer was born Bruce Reimer, an anatomically and physiologically typical male infant, in 1965. A botched circumcision left him with a severely disfigured penis at the age of eight months, while his twin brother remained unscathed.

Seeking guidance from leaders in the field, his parents traveled to Johns Hopkins Medical Center, where Dr. John Money, a prominent sex psychologist, worked alongside an established group of pediatric endocrinologists. At the strong advice of

Dr. Money and the group of physicians, the family raised Reimer as a girl named "Brenda," alongside his twin brother, which created an experiment in gender development. Dr. Money felt that gender was malleable from a young age and that with steadfast reassurance and social reinforcement, "Brenda" would grow up with typical female gender identity. Along with strong social reinforcement of gender constructs, the family was instructed not to disclose "Brenda's" history to the child, so as not to add confusion during development. Without having reached the maturational and cognitive ability to participate in health care decision-making, the child was subjected to surgery to construct a vagina and remove the testicles and was given estrogen hormone therapy at pubertal age under the guidance of the medical team. "Brenda" had regular visits to Johns Hopkins that involved enduring invasive and often sexually explicit physical and psychological examinations. Although psychological reports and updates from Money and his group regarding the Reimer case were consistent with his theory, Reimer, unaware of this history, struggled with significant gender identity, psychological, and behavioral issues throughout childhood and adolescence.

In 1980, at the age of fifteen, "Brenda" learned of this history from the father and assumed a male gender identity under the name of David. As an adult, David chose to receive testosterone therapy and had multiple surgeries to remove breast tissue and reconstruct a penis. In 1997, with the help of psychologist Dr. Milton Diamond, David Reimer revealed the truth about his case in academic journals. Later that same year, he additionally came forward publicly with his story in a *Rolling Stone* exposé revealing the severe physical, sexual, and psychological trauma he endured at the hands of the prominent medical group and their methods (Colapinto 1997). Unfortunately, after years of psychological harm, David Reimer died by suicide in 2004 (Associated Press 2004).

The David Reimer case not only highlights the early influences of gender on the brain and its establishment, but also the harm that can be inflicted on individuals by medical professionals in the field of gender management. Complex ethical issues arise when psychologists and physicians interfere with gender development, particularly without continued communication and discussion with the individual involved and verification of their continued understanding and informed consent. Clearly such understanding and consent are not available to a child in infancy because this capacity does not mature until significantly later in development. This creates significant ethical concern with regard to surgical intervention in infants and children in the absence of solid evidence that the intervention has clear benefit for the child. Without this evidence, it is difficult—if not impossible—to see how such interventions could achieve an appropriate ethical balance of patient-centered benefit and burden (beneficence and nonmaleficence).

22.4 Introduction to Differences of Sex Development

DSDs are a heterogeneous group of disorders that disrupt the normal development of internal and/or external sex organs. Individuals affected by DSD are often exposed to variable prenatal hormones or may have atypical responses to hormone exposures. As a result, they may be born with ambiguous genitalia and have differences in gender development of the brain. DSD not only offers insights into the influence of variable hormone exposures on resulting sex and gender identification but also raises multiple ethical concerns in terms of autonomous, patient-centered, medical decision-making and gender management. After the birth of a child with a DSD, the family and physicians must make multiple decisions regarding the care of the child in the context of the child's ongoing development without yet knowing how the child may understand their own gender identity.

22.5 Case: Sex Assignment in an Infant with Gonadal Dysgenesis[2]

A newborn presents to a DSD clinic in a major children's hospital for evaluation of ambiguous genitalia. When the doctors evaluate the infant, they discover an enlarged clitoris (or small phallus) with a small, single opening at the base from which urine exits. Medical evaluation suggests a diagnosis of XY gonadal dysgenesis, meaning that the infant has XY chromosomes but because of a genetic abnormality, the child's testes did not develop typically. For this reason, these organs do not produce typical levels of testosterone and other hormones for full masculinization of the external genitalia or fully suppress the development of female internal organs.

On further investigation, this child's identified gonads do not have an ovarian appearance but appear to be dysgenetic (abnormally developed) testes. In addition, the child has some partially formed, rudimentary female internal organs that are not developed enough to suggest a future ability to carry a pregnancy. After the diagnosis is established, the parents, family, and medical team grapple with questions of significant weight for the child's life: how do we decide on a (biological) sex assignment for this child, and how do we address the (social) gender of rearing?

The parents desire a female sex assignment because this was their original expectation from the pregnancy. The medical team agrees that without significant masculinization of the external and internal genitalia, a female sex assignment is a reasonable decision. However, it is unclear currently that there is a consensus medical standard of care or paradigm for whole person ethical decision-making that supports differentiating what is and is not reasonable. The psychologist stresses the gender flexibility and fluidity that the parents must accept as they raise their child as a girl. It is unclear how much a patient-centric approach is discussed or presented to the parents.

---

[2] The cases included in this chapter contain features from multiple real-life patients but are not specific to any particular case. Details have been changed as not to be identifiable to any specific patient.

22.6   Commentary

Although U.S. culture is beginning to be more widely accepting of models of gender fluidity, the population at large still has an expectation of binary sex and gender identification, particularly with the assignment and rearing of infants. "Boy or girl?" is the first question asked of nearly all parents of newborns, and most infant clothing and nursery decorations are highly gendered. However, assigning a sex-ambiguous infant a binary label has several implications.

Binary sex assignment inevitably leads to the rearing and socialization of a child who is likely to be gendered in almost all scenarios. Although certain anatomic features and hormonal and genetic testing may suggest the degree to which masculinizing hormones influenced the development of an infant (female hormones, such as estrogen, appear to be less influential in this regard), no full correlation has been established between testosterone level and future gender identity. The family and medical team must therefore balance beneficence, nonmaleficence, and the degree to which the child's future autonomy can be respected and preserved by delaying unnecessary medical interventions until the child can meaningfully participate.

Many parents worry about their child's ability to "fit in" with societal expectations and view the ability to conform to "normal" as offering the highest beneficial value for their child's future. Historically, within the prior paternalistic model, "normal" was held in the highest regard: assign a sex, surgically match the genitalia to that assignment, raise the individual steadfastly in the prescribed gender from as early as possible, and let no one know their "abnormal" history. What we have learned from the David Reimer case and from many DSD patients and advocates is that our ideas of "normal" as the highest ideal are not necessarily concordant with the concepts of beneficence and nonmaleficence. Indeed, a sex assignment that a child later questions or rejects, coupled with highly gendered inflexible rearing, can lead to significant psychological harm, as the David Reimer case shows. DSD patient groups have instead advocated for more consideration of a child's future autonomy and self-determination (refer to Chap. 3 for a discussion on ROF).

Advocating for autonomy and self-determination in an infant who lacks decision-making capacity can seem difficult at first, but there are a number of ways to navigate this. For instance, where intervention is not medically necessary in infancy, decisions about such interventions can be delayed until the child develops the cognitive and maturational ability to meaningfully participate in decisions about their own body. Before that time, the appropriate decision would be to do nothing and defer the determination of sex and gender until the child can determine for themselves. In the interim, a child could be raised as non-binary, gender neutral, or as a "third gender."

Although many cultures do recognize a non-male or non-female third gender—a category that would include many or most individuals with a DSD—in many of these cultures, the third gender is a marginalized group that is not afforded many basic rights (Hossain 2017). Within the United States, only a handful of states will legally allow a non-binary designation on a birth certificate (Savage 2019). Internationally, Germany, Australia, and parts of Canada allow for an "indeterminate" gender designation on birth certificates, while a handful of other countries offer a third gender option ("Legal Recognition of Intersex People," 2020). So although a gender-neutral approach may be taken and strongly upheld in a child's home, consideration must still be given to the child's place in the wider culture and the psychological impact society can have on the individual. This again focuses our surrogate medical decision-making on the concepts of beneficence and nonmaleficence. How do we optimize benefit and mitigate harm for the child?

22.7   Case: Surgical Intervention in an Infant with Gonadal Dysgenesis

The same infant returns to the DSD clinic at 5 months of age. The parents have been raising the child as a girl and are quite satisfied with this decision. They return to discuss potential surgical interventions for their child. The medical team presents multiple interventions that may be initiated as early as the next month. Removal of the gonads is the first topic introduced, as the risk of malignancy in abnormally formed testes increases over time (Abacl et al. 2015). The doctors additionally introduce options to reduce the size of the clitoris, separate the urethra from the vaginal opening, and construct a vagina.

The parents are shocked by the risk for malignancy and ask that the gonads be removed as early as possible. The mother quickly remarks that she also wants her daughter's genitalia to look "normal" as soon as possible because her masculine appearance has been very distressing to the mother with every diaper change. Furthermore, she thinks that all of the child's surgeries should be done at the same time. "How could we possibly send her to school like this?" The father feels less comfortable with the idea of making permanent changes to his daughter's genitalia, "What if we made a mistake with the sex assignment? We can't commit her to a lifetime in a body she doesn't want.".

After significant counseling of the parents by the medical and psychology team regarding the implications of early versus delayed genital surgery, the parents opt to remove the gonads related to concern of malignancy risk but decide to defer the remaining cosmetic procedures.

22.8   Commentary

Early surgical intervention for individuals with DSD has become one of the most controversial subjects within gender management care. The early paternalistic model, which dominated for multiple decades and has left its imprint on modern management, strongly upheld the concept of matching genitalia to sex assignment early in life (prior to age three) for the purpose of "appropriate" social development (Money et al. 1955). Almost all existing surgical outcome data come solely from this early surgical intervention group. Although surgical techniques have evolved over time, concerns remain regarding the long-term side effects of these interventions, most prominently with regard to decreased or lost clitoral sensation following clitoroplasty and vaginal strictures and poor sexual function following vaginoplasty (Almasri et al. 2018). With uncertainty regarding comparative surgical outcomes between early and deferred surgery, treatment is often pursued primarily to quell the discomfort felt by parents regarding their child's appearance and their fears about future social isolation for their child. Under the direct or implicit recommendation of the medical team, parents often agree to pursue all surgical interventions, some medically necessary[3] and some cosmetic, for their child during infancy, in hopes of closing a difficult chapter in the atypical birth and early life of their child (Timmermans et al. 2018).

For many individuals with DSD, the loss of autonomy and self-determination created by early surgical intervention is just as important as concerns about social integration or potential surgical side effects of delayed surgery. Intersex and DSD advocacy groups have raised significant concerns about the appropriateness of any surgical intervention involving the genitalia prior to the patient's developing full decision-making capacity. Many of these groups have called for all early genital surgeries to be banned and labeled such surgeries "genital mutilation". Several international governing bodies have undertaken ethical inquiries regarding a potential ban on early genital surgeries, equating genital surgery without the individual's direct consent with forced genital mutilation (Carpenter 2016). Medical institutions as a whole still lack a consensus paradigm for addressing ethical concerns regarding early surgical interventions, frequently noting the nuanced variability of individual situations, including patients' anatomic presentations, with some urogenital malformations requiring more immediate medical intervention for protection of the urinary tract. Multiple types of surgeries may be undertaken during early life, and grouping them all together would be overly simplistic because each presents its own ethical concerns. The situation also highlights the historic tension in pediatrics between the desires and choices of the parent(s) on behalf of their minor child and the eventual choices the child (patient) may want or retrospectively have wanted for themselves. Some argue that pediatrics is unduly deferential to parents in such situations and advocate for a more patient-centered model, shifting focus to the best interest of the child rather than the relief of parental discomfort.

As mentioned in the above case, many DSDs are associated with an increased malignancy risk in the abnormally formed testes or ovaries. Standard medical practice has been to perform early gonadectomy (removal of the testes or ovaries) prior to any significant rise in malignancy risk. Depending on the underlying abnormality, malignancy risk can increase by up to 50–60% by the second to third decades of life (Abacl et al. 2015). As surrogate medical decision makers, many parents opt

[3] Some medical considerations include the ability to urinate without obstruction and increased risk for recurrent urinary tract infections. Another medical consideration that is necessary but not time-sensitive during infancy is the ability for menstrual blood to flow without obstruction. The risk/benefit analysis may shift in favor of surgical intervention if faced with potential urinary tract and kidney complications.

for removal to mitigate this risk as early as possible. However, not all disorders are created equal, and data suggest that some disorders may have delayed and overall lower lifetime risk of malignancy development (Abacl et al. 2015). Complicating risk assessment is the poor overall ability to monitor for cancer development in this population, with malignancy only diagnosable on surgical biopsy.

The risk of malignancy development must be analyzed against the loss of potential hormone synthesis and fertility from the abnormally formed testes and ovaries. Despite low functional potential of the abnormal gonads for future hormone production or fertility, the loss of endogenous hormone production and fertility is often viewed as an incredibly devastating personal and intimate loss. The terms "castration" and "sterilization" have such strong connotations that it is difficult to hear the words without associating their meaning with personal harm. And even if the reproductive potential of abnormal testes and ovaries is felt to be extremely low, medical advances may make the possibility of fertility derived from these partially and abnormally formed gonads a reality in the future. When presented with this risk/benefit analysis, surrogate medical decision makers (typically the parents with guidance from medical providers) often view the risk of malignancy as greater than the loss of abnormally formed ovaries and testes. However, we know that in the adult population, a significant number of patients with cancer choose, for example, to forgo radiation therapy in order to maximize their future reproductive capacity. They make this choice even though doing so may result in a lower cure rate for their current condition. Recognizing the importance of reproductive potential, the decision to remove testes and ovaries, regardless of functional potential, should be considered in the context of the limitations this removal places on the autonomy and self-determination of a person with regard to their bodily integrity.

Clitoroplasty, or surgical alteration of the clitoris, presents a much different ethical problem. Infants born with DSD may present on a spectrum between a normal or mildly enlarged clitoris to a small phallus. The clitoris and phallus are developed from the same tissue, and the size and degree of masculinity are influenced by the degree of male hormone exposure during early fetal development (Witchell and Lee 2014). Depending on the history, genetics, and hormonal profile of the child, many children with masculinized-appearing genitalia are assigned female sex, including females born with CAH. As mentioned, later in life, 90–95% of individuals with CAH report female gender identity and 5–10% report male gender identity. Overall, this distribution reflects a majority female-identifying population, but the 5–10% male gender identification in this group represents a much higher rate of gender dysphoria than in the general population. The intent of early clitoroplasty has been to "normalize" the appearance of the genitalia toward a more feminine standard. Preventing psychological harm and promoting female gender identity under the rubric of the 'best interest' of the child have been asserted as the primary motivations for pursuing early clitoroplasty. However, historically, focus has been lacking on the actual preferences that the child may have once they can express them, alongside an almost total focus on the interests and preferences of the parent(s). It is important to note that although parent(s) are generally seen as advocates for their children, the parents' interests are not de facto the same as their child's interests. This is true even if the environment that the parents create and how welcoming and accepting that environment is affect the wellbeing of the child. The ethical paradigm applied should always include a direct analysis of the patient's (here, the child's) interests.

Because early surgical intervention has been performed almost universally for multiple decades, few data exist regarding the psychological and developmental trajectory of children who grow up with ambiguous-appearing genitalia. Although some stakeholders have concerns regarding the understanding and development of gender identity in the setting of genital ambiguity, others express social concern, particularly regarding the early adolescent phase in which a child may be ostracized in the locker room or pool for an atypical appearance (Speiser et al. 2018). The fact remains that we have little data to back up any of these claims, and there is no specific medical necessity for performing clitoroplasty during infancy. Additionally, intersex and DSD advocacy groups have shared concerns regarding the degree of sensitivity remaining in the clitoris following surgery with the more drastic early surgical techniques, which have been compared to female circumcision. Modern surgical techniques aim to preserve the maximal amount of sensitivity in the clitoris; however, data to date still indicate that decreased clitoral sensitivity remains a known side effect of surgery. This outcome can have serious, negative impact on the patient's future sexual life (Almasri et al. 2018). Therefore, the risk–benefit analysis of performing an early clitoroplasty must weigh potential but unknown psychological harm of growing up with ambiguous genitalia against the known risks related to surgery and loss of autonomy and self-determination regarding bodily integrity, specifically for a cosmetic procedure with no clear evidence-based benefit. Respect for future gender identity must also be considered because the individuals who later identify as male gender will have lost critical tissue for a phallus if clitoroplasty is performed during early life.

Vaginoplasty, or creation of a vagina, presents similar cosmetic concerns for masculinized females; however, it also delivers different functional elements that may provide a medical reason for earlier surgery. For many masculinized infants who have been assigned female sex for rearing, the urethra meets with a rudimentary vagina and forms a single confluent opening. Depending on the ability of urine to flow freely, more immediate surgical intervention may be recommended to relieve urinary obstruction or prevent urinary tract infections. Additionally, depending on the individual's anatomy and hormonal potential, menstruation could be obstructed at pubertal age, which would prompt surgical intervention, although that possibility is not of immediate medical concern for an infant. Many argue that a one-step procedure for clitoroplasty and vaginoplasty is preferable, because tissue removed from the clitoroplasty may aid in the creation of a more functional vagina, despite the differing ethical concerns pertaining to each of the two separate procedures. Additionally, normal estrogen exposure during infancy is thought to promote improved healing and produce better surgical outcomes for vaginoplasty (Balgobin et al. 2013). Notably, the same degree of endogenous hormonal exposure will not be available again until pubertal age. Despite the positive effects of estrogen exposure, post-vaginoplasty care requires penetrating vaginal dilation by an individual to produce a sexually functional vagina, an undertaking that is not appropriate until adolescence when the child can self-administer the dilations by their own choice. Although vaginoplasty presents a more complicated clinical picture, which may include differing degrees of medical necessity, it involves a problem similar to that with clitoroplasty regarding future sexual function. If we respect the autonomy, self-determination, and sexual integrity of the child to the degree of delaying vaginal dilation, should we not additionally respect them with regard to the decision about if and when vaginoplasty occurs, assuming there is no urgent medical need?

22.9 Case: Gender Development in a Child with Gonadal Dysgenesis

After gonadectomy, the family does not return to the clinic until the child is nine years old. In the interim, the child has been raised consistently as a girl. During an interview with the child psychologist, the child declares their gender identity as a "boy" because boys are "strong" and don't have to wear makeup. The child denies any desire for breast development and explains that a cousin said that breasts "make it hard to run fast."

On follow-up evaluation two years later at the age of eleven, the child now voices that she is a girl and expresses confidence in this decision. The child and her parents both desire estrogen therapy for induction of puberty. After discussions regarding the permanent effects of therapy and additional meetings with the psychologist, the child starts hormone therapy within six months of returning to medical care. Three months after initiation of therapy, she remains firm in her gender identity and expresses a desire to grow her hair long. She independently states that she does not desire genital surgery.

22.10 Commentary

As discussed previously, gender development is dynamic through childhood (refer to Chap. 2), with stereotyping and concrete ideas regarding gender eventually being replaced by social intensification of gendering and the potential hormonal "activation" of puberty. In the above scenario, the child wavers in their self-identification of gender by the conflation of concrete ideas of gender to gender identity itself. Although this child is not driven by endogenous hormonal activation (the child had their gonads removed during infancy), social intensification has likely influenced their gender development at an early pubertal age.

The concept of gender fluidity is important for individuals with DSD because their fetal and early life hormonal exposures as well as their internal and external genitalia are themselves on a spectrum. Sex assignment may provide a framework for societal integration, but narrow and definitive gendering of a child with DSD may be psychologically harmful.

Choosing when to initiate pubertal hormone therapy in a child who has previously undergone gonadectomy presents another ethical question in terms of the interference with gender development. To mirror natural pubertal development, most hormone therapy is initiated at around age eleven to twelve years. Gender development at this age centers around social intensification, but in the absence of endogenous hormones, any additional hormonal influence on gender identification would have to be from exogenous exposure prescribed by the physician. In a child for whom gender identity has wavered or been ambiguous and whose family and medical team have tried to uphold respect for gender fluidity, the initiation of pubertal hormones represents a hard stop for neutrality. In addition to the potential influence of brain gendering, hormone therapy causes permanent sexually dimorphic changes to the body: estrogen therapy brings breast development and widening of the hips with female fat distribution, whereas testosterone therapy brings development of male hair distribution on the face, chest, and abdomen, change of voice, and male-type musculoskeletal changes.

A child may be able to autonomously voice an opinion about their gender identity at age eleven or twelve years, but it is less clear if their decision-making capacity is mature enough to consent to permanent changes to their body. Prefrontal cortex development, which is necessary for complex critical thinking, and neurological reward centers do not fully mature until young adulthood (Blakemore and Robbins 2012), but fully matured brain development may not be necessary for self-identification of gender. The decision to start pubertal hormones is time sensitive because peers will be going through puberty during the typical window. Thus, this decision is difficult to defer indefinitely because the absence of hormones will be easily noticeable in its effects and additionally has implications for growth and bone health.

22.11   Introduction to Pediatric Gender Dysphoria

Individuals with differences of sex development inform scientists and psychologists regarding the gender fluidity that accompanies the hormonal and anatomic spectrum and reveal a likely gendering of the brain that remains poorly understood. Gendering of the brain discordant from biological sex could produce a transgender identity in what appears to be an otherwise typically developed individual. However, with the lack of clear biological underpinning, the full recognition of transgender identity and accompanying gender dysphoria has not universally been granted to these subjective experiences as warranting hormonal and surgical intervention, particularly within the pediatric population. Gender dysphoria—the distress created by the discordance of biological sex from gender identity—is associated with higher rates of depression, anxiety, and suicide during childhood and adolescence (Connolly et al. 2016). Given the high mental health and safety risks, physicians are now treating gender dysphoria in the pediatric population with hormonal interventions to suppress pubertal development, followed by cross-gender hormonal treatment to align the physical body with gender identity. In an otherwise "healthy" body, the decision to interfere with a minor's natural development in favor of aligning the body with an intangible gender identity raises questions about the strength of the child or adolescent's ability to define their long-term identity, their capacity for mature decision-making in relation to permanent bodily changes, and the effectiveness of medical and surgical intervention.

22.12   Case: "Simon" and Gender Dysphoria in Early Childhood

The parents of a healthy four-year-old child, Simon, notice that the child prefers to play with more typical "girl" toys like baby dolls and princess dress-up accessories, and role plays as the mother during play time. In kindergarten class, the child shows a preference for playing with female rather than male classmates. The parents do not pay any specific attention or concern to their child's behavior and allow Simon to choose activities and playmates freely. At home, the father tries to model for Simon how to urinate standing up because that's "how boys and men pee" and Simon very adamantly tells the father that "I am *not* a boy" and "I am a girl, just like mommy and my friend Sophie."

22.13   Commentary

The approach to expressions of gender fluidity and cross-gender identity in early childhood is both straightforward and complicated. As mentioned earlier, the concept of gender in a preschool-age child is quite concrete and based on stereotype. A young child may conflate the idea of wearing pink clothing, having long hair, and playing with baby dolls as a "girl's" gender identity and may not recognize the variability of expression that can be displayed within each gender. When a young child announces to his or her parents that they feel like the opposite gender, the parents must reconcile different aspects of their child's long-term wellbeing when approaching their response and course of action.

   Without a clear understanding of a child's long-term gender identification, early childhood gender dysphoria is a challenge to approach. Several outcomes are possible, depending on two factors. The first is whether and to what extent the child's social structure is transitioned to validate a cross-gender identity, including a change of name, grooming, dress, and identification at school and beyond. The second is whether the child will persist or desist from the cross-gender behavior. The child might persist in the cross-gender identity and feel physically and psychologically validated and supported by their family and social circle. This outcome offers an obvious benefit with an accompanying reduction in harm. Alternatively, the child may persist in the cross-gender identity but not feel validated or supported. This outcome creates risk for significant social and psychological harm. Finally, the child could desist in the cross-gender identity and transition back to their assigned gender, but in one of two possible subjective states: feeling concordant with their inner identity or discordant with it. If the child transitions back while feeling discordant with their assigned gender, the reasons underlying this dissonance pose a risk for social and psychological harm. Predicting whether a child will persist or desist is difficult. Also difficult is defining the extent to which changes should be made to validate the cross-gender identity, given the unknowns about persistence.

Current data show variability in the long-term persistence of gender dysphoria in prepubescent children who are evaluated for it. Across multiple studies, the rate of persistence of gender dysphoria from prepubescence to adulthood ranges from 2 to 27%. Early childhood gender dysphoria appears to be more highly associated with future homosexual or bisexual orientation, with 18–80% reporting this orientation following puberty. Many report that the first physical signs of puberty along with first sexual feelings and attractions resolved their gender dysphoria. In a retrospective analysis of "persisters," these individuals tended to show higher intensity and conviction. They use stronger verbiage such as "I am" rather than "I wish," and they may show increased intensity of symptoms with the onset of puberty (Steensma et al. 2011).

As a whole, the data may be interpreted to indicate that the majority of prepubescent children will not persist in their gender dysphoria. However, for individuals for whom the gender dysphoria is sustained, a rejection or lack of support of their gender identity can lead to long-term mental health consequences, including increased rates of depression, anxiety, and suicidality (Connolly et al. 2016). At what age or at what degree of intensity and conviction do the self-determination of a young child and risk of harm from inaction override the potential harm of an immature decision and regretted social transition? These are the details that comprise the ethical discussion of beneficence and nonmaleficence regarding the care of children in this population.

Acceptance of a certain degree of gender fluidity and the support of ethicists and child psychologists can help clarify the course of action; however, this does not fully address the complexity of individual cases.

22.14 Case: "Simon/Bella" and Gender Dysphoria in Early Adolescence

As Simon approaches early adolescence, she has remained steadfast in her identification as a girl. Although she has kept the name Simon and more boyish clothing for school, she prefers to be called Bella at home and expresses her femininity at home through her clothes and make-up, which her parents have respected and supported. She frequently remarks about the difficulty of pretending to be a boy at school and pleads with her parents to let her more fully express herself in public. Her parents realize that puberty will soon be approaching and are unsure of how to handle this life- and body-altering transition.

22.15 Commentary

As a prepubescent child with gender dysphoria approaches puberty, the medical option of hormonal pubertal suppression is available and touted as providing several benefits. First, pubertal suppression acts as a means of delaying decision-making regarding final gender identity, offering the child an open future (Feinberg 1980) with greater self-determination and autonomy while respecting the continued development of their decision-making capacity. Pubertal suppression additionally blocks the development of an adult sexually dimorphic appearance.[4] For those who persist in their gender dysphoria, there is clear benefit in avoiding the development of sexual features related to their originally assigned gender. The development of the sexual features of their originally assigned gender itself is typically psychologically distressing to gender-dysphoric individuals and makes them less likely to "pass" as the opposite gender after completion of cross-gender hormone treatment and surgery. When pubertal suppression is used early and immediately followed by cross-gender hormones, the patient undergoes puberty that better reflects their experienced gender and comes closer to achieving a final height consistent with their experienced gender.

However, pubertal hormone changes likely affect brain development and, at least at the time of initiation, may be critical in separating persistent from desistant gender dysphoria. Suppressing puberty from fairly early in development may alter the usual development of the brain, particularly with regard to gender development and critical thinking, which calls into question the usefulness of delaying decision-making if treatment affects the typical course of development. Current medical protocols allow for pubertal suppression at the earliest stages of puberty; however, most of the existing data are from children who initiated therapy later in puberty and had already progressed through several stages of it. Data are limited regarding analysis of the development of the frontal cortex—the location of critical thinking—under pubertal suppression, and such studies are likewise limited by patient groups who were further along in pubertal development at the onset of treatment (Staphorsius et al. 2015).

Pubertal suppression itself is an overall safe medical treatment and has been used for early puberty for years. Bone mineral density may decrease during the treatment

---

[4] Refering to the the secondary sexual characteristics developed in males and females during puberty—male pattern hair growth, broad shoulders and narrow hip vs breasts, widening of hips and fat distribution in females.

period, but subsequent bone mass accrual is preserved, and treatment does not seem to negatively affect peak bone mass (Carel et al. 2009).

Pubertal suppression, however, may not be a practical long-term treatment because peers will in the meantime be progressing through their typical pubertal growth and development. Full suppression of puberty halts individuals in a young, underdeveloped appearance for their age at a time when socialization, particularly gendered socialization, is crucial to psychological wellbeing.

22.16   Case: "Bella" and Gender Dysphoria in Late Adolescence

Bella is approaching her fifteenth birthday and has been on pubertal suppression therapy for over two years. While she is happy that she has not developed an adult male appearance, she is very frustrated because all of her friends have developed breasts and feminine figures and she remains in a childlike body frame. She desperately wants to start estrogen therapy to develop a more feminine body and has even researched online how to obtain hormones without a doctor's permission. She wants to follow the rules but feels like she is being restricted unfairly by the medical system.

22.17   Commentary

The first step in addressing medical intervention in the adolescent population with gender dysphoria is to determine if we have identified the right target for therapy and if we have the right therapeutic agents. The change from "gender identity disorder" in the DSM IV to "gender dysphoria" in the DSM V signified a shift in thought regarding the defined pathology (American Psychiatric Association 2013). Discordant gender identity itself was no longer pathological; rather, it was the resultant dysphoria caused by the discordant identity that was considered pathological. However, if a transgender identity is a variant, it may not warrant treatment in and of itself. Physicians may be over-medicalizing a variant with hormonal therapy and surgical interventions as an indirect way of treating a cultural pathology of unacceptance. If culture is not ready to accept those who do not conform to an imposed gender binary, the resulting discomfort at being outside this binary may lead to over-medicalization[5] that could be omitted in a culture that accepted variant expressions of gender. If dysphoria alone is the pathology, are hormonal therapy and surgical intervention ethical and effective treatment?

---

[5] Over-medicatlization in this context would refer to medical interventions that would otherwise be unnecessary, placing a patient at risk for undesirable side effects of therapy, or potentially creating stigmatization of a condition (Kaczmarek 2018).

The ethical principle of beneficence dictates that a physician should act to optimize the benefit to a patient, which the complementary ethical principle of nonmaleficence compels the avoidance or at least mitigation of patient harm. For the population with gender dysphoria, application of these principles includes mitigating mental health risks. Multiple studies have identified higher rates of depression, self-harm, and suicide attempts in adolescents with gender dysphoria. In a national survey, 41% of transgender adults reported prior suicide attempts, compared with 1.6% in the general population (Grant et al. 2010). Additionally, as we learn more about ongoing stress to the brain and delayed health consequences, health risks may go beyond the mental health effects.

In a 1997 follow-up of young Dutch transgender adults who had undergone cross-gender hormonal treatment and surgery, a Netherlands gender research group found significantly increased social activity, dominance, and self-esteem from baseline. When compared with their adult cohort, the adolescent cohort had superior psychological functioning, suggesting that earlier intervention yielded greater benefit to long-term functioning (Cohen-Kettenis and Van Goozen 1997). In 2014, the Netherlands group did another follow-up of transgender young adults in a cohort that had additionally undergone pubertal suppression prior to cross-gender hormones and surgery. They found that gender dysphoria and body image difficulties persisted through pubertal suppression and became significantly reduced after cross-gender hormones and confirmatory surgery. They additionally showed improvement over time in overall global functioning (de Vries et al. 2014).

Long-term psychological follow-up in U.S. cohorts of young transgender individuals is lacking, however, and the culture surrounding the individual may influence psychological and functional outcomes. It remains unproven that cross-gender hormonal therapy alleviates dysphoria and psychopathology across a variety of cultures that may be less accepting of individuals receiving therapy for a non-traditional gender identity.

Given the introduction of hormones into what we deem to be otherwise healthy bodies, we are ethically obligated to ask if cross-gender hormone treatment itself causes harm. In female-to-male individuals taking testosterone, researchers have noted significant increases in hemoglobin and hematocrit levels and BMI, as well as a decrease in high-density lipoprotein cholesterol levels, metabolic features that align them with other members of their experienced gender. Of the male-to-female individuals taking estrogen, no significant changes were noted in the measured metabolic parameters (Jarin et al. 2017).

Outside limited metabolic risks, one of the more complex side effects of cross-gender hormonal therapy is the potential loss of fertility. For an adolescent consenting to cross-gender hormonal therapy or potential gonadectomy (removal of ovaries or testes), there are ethical concerns about decision-making capacity related to loss of fertility potential. Adolescents may not yet fully appreciate the long-term ramifications of such a decision, may not yet have an established position with regard to their future reproductive capability, and may have little sense of urgency with regard to making such a decision. Decisional regret therefore becomes a concern. If early pubertal suppression proceeds to early cross-gender hormonal therapy, the potential for fertility preservation remains low. To allow for fertility potential, the adolescent must allow natal sex hormones to circulate so that they can develop semen from the testes or eggs from the ovaries. Additionally, semen preservation requires participation of the individual in providing a sample through ejaculation, another sensitive process that an adolescent may wish to avoid if not strongly motivated to preserve fertility. Issues of reproductive capacity not infrequently also involve religious or cultural factors that are important to understand and address.

When assessing the significant decision of allowing an adolescent to consent to permanent body-altering medications and surgeries, the decision-making capacity of the patient must be taken into account and the mature minor doctrine may be considered. Legal precedent from around the world provides various mechanisms through which minors with sufficient maturational and cognitive abilities may consent to medical treatment. This includes treatment that is thought to provide therapeutic benefit accompanied by risk and permanent and irreversible changes and medical treatment that the minor's parents may not agree with (Coleman and Rosoff 2013). However, the personal nature of the particular decisions discussed here hinge on a somewhat intangible concept of identity. Given the fact that adolescence is a time known for identity formation and experimentation with self-expression, there is concern about future decisional regret should an adolescent make a choice with irreversible consequences and later change their mind. Of course, this is a concern in many situations with adults, as well. Parents and physicians often share the concern that a minor allowed to make permanent changes to their body will come to regret their decision as their identity continues to develop into adulthood. However, data from the 1997 and 2014 follow-up studies of young Dutch transgender adults who received cross-gender hormonal therapy indicate that none of them expressed feelings of regret. Although these are limited data, they do provide some information to help inform the decision-making.

Adolescents are additionally prone to impulsive decision-making, as can be seen in increased criminal behavior, drug use, and suicidality during this period. However, many feel that the process of undergoing treatment is not conducive to impulsive decision-making, especially when considering the medical, social, and emotional obstacles adolescents with gender dysphoria face in obtaining cross-gender hormonal treatment (Salter 2017). The impulsivity of adolescents conversely places the transgender adolescent at risk for other rash behaviors if denied treatment, including suicidality, drug use, or illegal hormone use.

Finally, the ethical principle of justice, including equal access to treatment, must be addressed because the transgender community is a marginalized group and minors remain a vulnerable population. A child suffering from a pathology, whether gender dysphoria or a urinary tract infection, deserves access to treatment and should not be presumed incompetent in contributing to decision-making based on age alone. Additionally, there is an ethical obligation to substantively include all patients in medical decisions and the informed consent process to the extent that they are cognitively and maturationally able to participate. The transgender community, an already marginalized group, may be marginalized further if their access to treatment is delayed, dampening their potential to flourish in society.

22.18 Conclusion

The management of gender in pediatrics remains a controversial and challenging topic laden with ethical and medical concerns. The desire to protect against harm may directly conflict with the ability to preserve, until a developmentally appropriate age, an open future that respects self-determination and autonomy. Surrogate medical decision-makers may attempt to maximize psychological benefit and reduce harm during infancy and early childhood; however, given the very personal nature of gender identity, as well as issues of culture, society, religion, and bias, they often encounter difficult and dynamic scenarios. This complex situation is further complicated by the absence of a current consensus standard of care, leaving health care providers untrained and without consensus guidance in helping parents and patients navigate these circumstances. As the child develops, their capacity and ability for full autonomy and self-determination remains difficult to ascertain as decision-making skills and gender identification may remain dynamic across childhood and into adolescence. Therefore, we have an obligation to proceed thoughtfully and carefully, using an evidence-based, patient-centric approach that incorporates ethical principles and principled clinical decision-making.

Guiding Principles for the Ethical Care of Adolescents

Preservation of autonomy and self-determination
- Provide an open future to children to the extent possible taking into account medical and wholistic considerations
- Avoid permanent and irreversible interventions, if not medically necessary for the safety of the child, until: (a) the child can provide consent, or (b) if it is not possible to safely wait until the child can consent, waiting until the child can assent

---

Emerging capacity and consent
- Support children and adolescents to participate in health care decision-making to the full extent possible during the formative period of identity and maturational development, recognizing the concurrent development of decision-making capacity
- Evaluate decision-making capacity and ability to participate maximally in health care decision-making in the context of developmental level, cognitive understanding of risks and benefits, and long-term implications of interventions

---

References

Abacl, A., G. Çatll, and M. Berberoılu. 2015. Gonadal malignancy risk and prophylactic gonadectomy in disorders of sexual development. *Journal of Pediatric Endocrinology and Metabolism* 28 (9–10): 1019–1027. https://doi.org/10.1515/jpem-2014-0522.

Almasri, J., F. Zaiem, R. Rodriguez-Gutierrez, S.U. Tamhane, A.M. Iqbal, L.J. Prokop, … M.H. Murad. 2018. Genital reconstructive surgery in females with congenital adrenal hyperplasia: A systematic review and meta-analysis. *The Journal of Clinical Endocrinology & Metabolism* 103 (11): 4089–4096. https://doi.org/10.1210/jc.2018-01863.

American Psychiatric Association. 2013. *Diagnostic and statistical manual of mental disorders*, 5th edn. Washington, DC.

Associated Press. 2004. David Reimer, 38, Subject of the John/Joan Case. *The New York Times*, Section A, p. 21.

Balgobin, S., T.I. Montoya, H. Shi, J.F. Acevedo, P.W. Keller, M. Riegel, … R.A. Word. 2013. Estrogen alters remodeling of the vaginal wall after surgical injury in guinea pigs. *Biology of Reproduction.* https://doi.org/10.1095/biolreprod.113.112367.

Berenbaum, S.A., and A.M. Beltz. 2011. Sexual differentiation of human behavior: Effects of prenatal and pubertal organizational hormones. *Frontiers in Neuroendocrinology* 32 (2): 183–200. https://doi.org/10.1016/j.yfrne.2011.03.001.

Blakemore, S.J., and T.W. Robbins. 2012. Decision-making in the adolescent brain. *Nature Neuroscience.* https://doi.org/10.1038/nn.3177.

Carel, J.-C., E.A. Eugster, A. Rogol, L. Ghizzoni, and M.R. Palmert. 2009. Consensus statement on the use of gonadotropin-releasing hormone analogs in children. *Pediatrics.* https://doi.org/10.1542/peds.2008-1783.

Carpenter, M. 2016. The human rights of intersex people: Addressing harmful practices and rhetoric of change. *Reproductive Health Matters.* https://doi.org/10.1016/j.rhm.2016.06.003.

Cohen-Kettenis, P.T., and S.H.M. Van Goozen. 1997. Sex reassignment of adolescent transsexuals: A follow-up study. *Journal of the American Academy of Child and Adolescent Psychiatry* 36 (2): 263–271. https://doi.org/10.1097/00004583-199702000-00017.

Colapinto, J. 1997. The case of John/Joan. *Rolling Stone* 54–97.

Coleman, D.L., and P.M. Rosoff. 2013. The legal authority of mature minors to consent to general medical treatment. *Pediatrics* 131 (4): 786–793. https://doi.org/10.1542/peds.2012-2470.

Connolly, M.D., M.J. Zervos, C.J. Barone, C.C. Johnson, and C.L.M. Joseph. 2016. The mental health of transgender youth: Advances in understanding. *Journal of Adolescent Health* 59 (5): 489–495. https://doi.org/10.1016/j.jadohealth.2016.06.012.

de Vries, A.L.C., J.K. McGuire, T.D. Steensma, E.C.F. Wagenaar, T.A.H. Doreleijers, and P.T. Cohen-Kettenis. 2014. Young adult psychological outcome after puberty suppression and gender reassignment. *Pediatrics* 134 (4): 696–704. https://doi.org/10.1542/peds.2013-2958.

Drescher, J., and W. Byne. 2012. Gender dysphoric/gender variant (gd/gv) children and adolescents: Summarizing what we know and what we have yet to learn. *Journal of Homosexuality* 59 (3): 501–510. https://doi.org/10.1080/00918369.2012.653317.

Feinberg, J. 1980. The child's right to an open future. In *Whose child?*, ed. W. Aiken and H. LaFollette, 124–153. Totowa, NJ: Rowman & Littlefield.

Grant, J.M., L.A. Mottet, J. Tanis, J.L. Herman, J. Harrison, and M. Keisling. 2010. National transgender discrimination survey report on health and health care. *National Center for Transgender Equality.* https://doi.org/10.1016/S0016-7878(90)80026-2

Hossain, A. 2017. The paradox of recognition: Hijra, third gender and sexual rights in Bangladesh. *Culture, Health and Sexuality.* https://doi.org/10.1080/13691058.2017.1317831.

Jarin, J., E. Pine-Twaddell, G. Trotman, J. Stevens, L.A. Conard, E. Tefera, and V. Gomez-Lobo. 2017. Cross-sex hormones and metabolic parameters in adolescents with gender dysphoria. *Pediatrics* 139 (5): e20163173. https://doi.org/10.1542/peds.2016-3173.

Kaczmarek, E. 2018. How to distinguish medicalization from over-medicalization? *Medicine, Health Care and Philosophy* 22 (1): 119–128. https://doi.org/10.1007/s11019-018-9850-1.

Katz, A.L., and S.A. Webb. 2016. Informed consent in decision-making in pediatric practice. *Pediatrics* 138 (2): e20161485–e20161485. https://doi.org/10.1542/peds.2016-1485.

Lamminmäki, A., M. Hines, T. Kuiri-Hänninen, L. Kilpeläinen, L. Dunkel, and U. Sankilampi. 2012. Testosterone measured in infancy predicts subsequent sex-typed behavior in boys and in girls. *Hormones and Behavior.* https://doi.org/10.1016/j.yhbeh.2012.02.013.

Legal Recognition of Intersex People. 2020. In Wikipedia. https://en.wikipedia.org/wiki/Legal_rec ognition_of_intersex_people.

Money, J., J.G. Hampson, and J.L. Hampson. 1955. Hermaphroditism: Recommendations concerning assignment of sex, change of sex and psychologic management. *Bulletin of the Johns Hopkins Hospital.*

Pasterski, V., C.L. Acerini, D.B. Dunger, K.K. Ong, I.A. Hughes, A. Thankamony, and M. Hines. 2015. Postnatal penile growth concurrent with mini-puberty predicts later sex-typed play behavior: Evidence for neurobehavioral effects of the postnatal androgen surge in typically developing boys. *Hormones and Behavior* 69: 98–105. https://doi.org/10.1016/j.yhbeh.2015.01.002.

Salter, E.K. 2017. Conflating capacity & authority: Why we're asking the wrong question in the adolescent decision-making debate. *Hastings Center Report.* https://doi.org/10.1002/hast.666.

Savage, R. 2019. Nonbinary? Intersex? 11 U.S. states issuing third gender IDs. Reuters. https://www.reuters.com/article/us-us-lgbt-lawmaking/nonbinary-intersex-11-us-states-issuing-third-gender-ids-idUSKCN1PP2N7.

Speiser, P.W., W. Arlt, R.J. Auchus, L.S. Baskin, G.S. Conway, D.P. Merke, … P.C. White. 2018. Congenital adrenal hyperplasia due to steroid 21-hydroxylase deficiency: An endocrine society* clinical practice guideline. *Journal of Clinical Endocrinology and Metabolism.* https://doi.org/10.1210/jc.2018-01865.

Staphorsius, A.S., B.P.C. Kreukels, P.T. Cohen-Kettenis, D.J. Veltman, S.M. Burke, S.E.E. Schagen, … J. Bakker. 2015. Puberty suppression and executive functioning: An fMRI-study in adolescents with gender dysphoria. *Psychoneuroendocrinology* 56 (February 2014): 190–199. https://doi.org/10.1016/j.psyneuen.2015.03.007.

Steensma, T.D., R. Biemond, F. De Boer, and P.T. Cohen-Kettenis. 2011. Desisting and persisting gender dysphoria after childhood: A qualitative follow-up study. *Clinical Child Psychology and Psychiatry* 16 (4): 499–516. https://doi.org/10.1177/1359104510378303.

Steensma, T.D., B.P.C. Kreukels, A.L.C. de Vries, and P.T. Cohen-Kettenis. 2013. Gender identity development in adolescence. *Hormones and Behavior* 64 (2): 288–297. https://doi.org/10.1016/j.yhbeh.2013.02.020.

Timmermans, S., A. Yang, M. Gardner, C.E. Keegan, B.M. Yashar, P.Y. Fechner, … D.E. Sandberg. 2018. Does patient-centered care change genital surgery decisions? The strategic use of clinical uncertainty in disorders of sex development clinics. *Journal of Health and Social Behavior* 59 (4): 520–535. https://doi.org/10.1177/0022146518802460.

Witchell, S.F., and P.A. Lee. 2014. Chapter 5: Ambiguous Genitalia. In 1147485908 863153204. *Pediatric endocrinology*, ed. M.A. Sperling, 4th ed, 108–145. Philadelphia, PA: Elsevier.

Further Reading

Colapinto, J. 2000. *As nature made him: The boy who was raised as a girl.* New York, NY: HarperCollins.

Grady, D. 2018. Anatomy does not determine gender, experts say. *The New York Times.* Section A, 10.

Mouriquand, P.D.E. 2004. Possible determinants of sexual identity: How to make the least bad choice in children with ambiguous genitalia. *BJU International, Supplement* 93 (3): 1–2. https://doi.org/10.1111/j.1464-410x.2004.04701.x.

Reiner, W.G., and J.P. Gearhart. 2004. Discordant sexual identity in some genetic males with cloacal exstrophy assigned to female sex at birth. *New England Journal of Medicine* 350 (4): 333–341. https://doi.org/10.1056/NEJMoa022236.

Chapter 23
The Child with Cancer: Blurring the Lines Between Research and Treatment

M. Kruger and N. Nortjé

Abstract This chapter focuses on the dual role of the physician-researcher in clinical trials involving childhood cancer. As the cure rate of childhood cancer has improved from less than 10% in the 1950's to nearly 80% currently, it is important to note that this is due to dedicated empirical collaborative clinical trials, without which these achievements would not have been possible. However, there are several ethical issues involved in childhood cancer research as the disease is life-threatening. These ethical issues include the tension between the dual role of being clinician-researcher, recruitment issues, and the translation of clinical research findings into clinical care. Other issues such as proxy consent and therapeutic misconception have been discussed in other chapters.

Keywords Cancer · Research · Recruitment · Clinician-researcher

23.1 Introduction

The success of the current cure rate of nearly 80% of all pediatric cancers are due to therapeutic research through major large national and international collaborative clinical trials over the last 70 years (Reaman 2004). This nearly universal fatal group of rare diseases in the 1950's has changed into curable diseases for almost 80% of children with cancer. This is especially notable for childhood acute lymphoblastic leukemia, the most common childhood cancer, with a history of more than 50 years of randomized clinical trials. These clinical trials have been well controlled, and the goal is cure. There are benefits to participation in these pediatric cancer clinical trials

M. Kruger (✉)
Department of Paediatrics and Child Health, Faculty of Medicine and Health Sciences, Stellenbosch University, Stellenbosch, South Africa
e-mail: marianakruger@sun.ac.za

N. Nortjé
Department of Critical Care Medicine, University of Texas MD Anderson Cancer Center, Houston, USA
e-mail: NNortje@mdanderson.org

© Springer Nature Switzerland AG 2022
N. Nortjé and J. C. Bester (eds.), *Pediatric Ethics: Theory and Practice*,
The International Library of Bioethics 89,
https://doi.org/10.1007/978-3-030-86182-7_23

379

as these trials provide access to collective experience in childhood cancer management, access to novel state-of the art therapies, improved technologies, translation of research discoveries into clinical care, improved stratification according to prognostic features with risk-adjusted therapies and management of long-term effects of childhood cancer chemotherapy.

To address the different ethical issues in pediatric oncology clinical trials for the clinician-researcher, it is necessary to briefly summarize the various phases of clinical trials, as they are traditionally designed as phase 1, phase 2 and phase 3 clinical trials. Phase 1 clinical trials aim to determine the safety, the maximum tolerated dose and pharmacokinetics of a novel medicine. Phase 2 clinical trials investigate the efficacy and adverse events of the novel medicine, while phase 3 studies are conducted when the medicine has been proven to be effective and is the final step in proving efficacy and safety in a large number of research participants before being registered for use. Phase 3 clinical trials also involve studies where an additional single novel medicine is tested against standard of care. Every phase of clinical trials has its unique set of ethical issues in childhood cancer care management. Childhood cancer patients and their parents usually agree to participate in clinical trials as they expect benefit from participation, and it is therefore important to address the ethical issues that may arise.

The empirical clinical trials in pediatric oncology started with phase 1 trials in the 1950's with single medicine studies and evolved to combination treatment protocols, involving various treatment modalities (Reaman 2004). The current practice is to enroll all pediatric patients with cancer in standard treatment protocols, designed as clinical trials with a predefined question. There are also modified clinical trials being conducted in low- and middle-income countries (LMICs), based on results of clinical trials in high income countries (HICs)(Arora et al. 2016) (Israëls et al. 2013). As LMICs often suffer from a lack of health care resources, treatment protocols are modified according to what medicines or supportive care resources, or human resources are available in the local context. This translates into modified treatment protocols, where the efficacy should then be tested through a clinical trial in the local population to improve the survival of children with common childhood cancer in these LMICs and to date these trials have improved the survival of children with cancer.

There important ethical issues in pediatric oncology related to clinical trial research for discussion in this chapter will be the dual role of the treating physician-researcher, recruitment strategies, and translation of clinical research findings into clinical practice.

23.2   Dual Role

The obligations between physicians and researchers are different each from other and therefore leads to a dual obligation in a unique professionalism. Of importance is the fact that, in traditional medical care, the doctor seeks to promote the health and wellbeing of the individual patient. In contrast, the researcher's interest may diverge from the patient's individual health care needs as the research is aimed at "creating generalizable knowledge" through research and in the clinical arena especially through clinical trials (Czoli et al. 2011). The role of the physician has always been to practice medicine for the good of the patient, offering the best standard of care available. According to Aristotle medicine is a techne[1] and its end or telos[2] is health (Oxford Dictionary of Philosohpy 3rd Edition). In clinical care, the doctor uses his techne to cure the ill patient and restore health (Pellegrino 2005). Health is not always attainable, but there is an attempt to restore the patient's disrupted physiology or psychology due to illness. The good clinician therefore aims to care, to alleviate pain or discomfort and to cure if possible. Even palliative care is a form of healing as it provides comfort and alleviates pain, although cure is not possible.

Medical progress, on the other hand, necessitates research to investigate improved medical interventions or medicines for the benefit of patients. This creates a tension between the dual role of the health care worker as clinician and researcher (Czoli et al. 2011) (Boydell et al. 2012). It is important to note that these two roles do not necessarily align with each other and often have divergent goals. As mentioned above the clinician seeks a cure or comfort for the individual patient, while the clinical researcher investigates a scientific question with the hope of creating generalizable knowledge to benefit future patients. Research clinicians may rationalize their research activities as in the best interest of the patient, although this may compromise the research participant's wellbeing, if the research goal overrides the best interest of the patient. In other words, the motivation for the clinician-researcher may compete with the patient's best interest as there are financial incentives or potential career advancement, that are attractive to the clinician-researcher (Bernstein 2003).

It is important for clinicians to reflect on their dual role as clinician-researcher as they need to determine how to optimize trial designs, understand family value systems and needs, and the effects of trial participation on family dynamics (Dupont et al. 2016). An important difference between clinical care and research is that clinical care focus on the individual patient versus clinical trial research which focusses on a group of patients with a certain disease or condition. According to Kant the human species are rational with the ability for autonomous decision-making (Macklin and Sherwin 1975). Kant therefore proposes as a categorical imperative that a person should always be an end goal and never a means to a goal (Kant 1993). The physician's role, according to Kantian ethics will be to restore the good health for the individual patient. For the researcher the research question to be answered is the main objective, which may actually detract from the focus on the individual patient's best interest, in this context health, creating a divergent role. Due to Kant's viewpoint that humans are autonomous beings, the research participant should be seen as a partner in the research process with full disclosure of this dual role in a transparent manner to the child patient and her family. The physician-researcher should never compromise

---

[1] Techne is a term used in philosophy to refer to the making or doing; therefore, the knowledge of how to make or do things (Oxford Dictionary of Philosophy 3Ed).

[2] Telos is the Greek word for purpose or end (Oxford Dictionary of Philosophy 3Ed).

the clinical care of the child patient, even though the child is enrolled in a clinical trial. The physician-researcher should undertake to treat the patient according to valid scientific standard of care with no compromise and with the patient's right to effective medical care to keep the trust of patients, involving no coercion (Bernstein 2003). This is often difficult in pediatric oncology research, as the lines between clinical care and research are often blurred in the search for better medicines for rare childhood cancers, emphasizing the need for transparency and respect to the parents and child patient as partners in the clinical trial.

During clinical trials there are the scientific obligations that should be adhered to. It is important for the physician-researcher to delineate between the aims of clinical management and research. The research protocol should ensure scientific validity with social value and fair selection of research participants. There should be a favorable risk/benefit ratio as harm should be minimized and potential benefit maximized. All research procedures should be rigorously justified, especially if randomization is involved. There is a need for randomization in pediatric cancer clinical trials as knowledge generated, especially with genetic analysis, may assist with delineation between prognostic indicators. Authors have indicated that randomization may lead to a decrease in the personal care and therefore may be ethically difficult (Gordon et al. 2006). The principle of "clinical equipoise" is used to justify a clinical trial with randomization when there is a need to compare different treatment arms and the medical expertise is uncertain about the effect of the experimental therapy under investigation in comparison to the standard of care treatment, with the assumption that the novel medicine or intensified treatment arm will not be inferior to standard of care treatment arm (De Vries et al. 2011; Gordon et al. 2006).

According to Dickert (2019) "role synergy" or the dual role as clinician-researcher may be important since the clinician knows her patient's medical history and situation the best and is therefore in the best position to understand if her patient may be best suited for the research (Dickert 2019). Furthermore, the clinicians is also in the best position to help the patient make a decision as there is already fundamental trust (Dickert and Sugarman 2005). Should the physician take on the dual role she must be cognizant of any underlying personal bias, as well as be able to identify any potential conflict-of-interest prior and during the research process.

23.3 Recruitment

Recruiting participants to participate in a clinical trial is a challenge as the clinician-researcher need to acknowledge that there is a fine balance between harm and benefit; and need to carefully do a physical, emotional, psychological, spiritual, and economical appraisal for the patients, family and society at large (Afshar et al. 2005). The clinician-researcher should be confident that the child's participation in the clinical trial will be in the best interest of the child. She must also take on the extra responsibility to safeguard the vulnerable state of the patient, which arise from their incompetence and fear of the disease, while also promoting the potential benefit of involvement. Research ethics committees (RECs) ought to make sure the trials are well designed and comply with principles of good scientific and clinical practice, based on the guiding principles of the Nuremberg Code; the Declaration of Helsinki, and the Declaration of Geneva. RECs should also review the recruitment strategies to be followed (Pritchard-Jones, 2008).

Conducting pediatric cancer clinical trials is always a challenge due to the small number of patients who are eligible for these trials. This challenge lead to issues recruiting new participants onto pediatric trials. As Pritchard-Jones et al. (2008) point out that this challenge is intensified by an ongoing dilemma, which is the desire for patients to benefit from progress in medical care, while avoiding the risk of harm from such research. The fundamental principle which should be applied when deciding to include a child in a clinical trial, is that she should not be included unless necessary to achieve an important clinical outcome, which is otherwise not likely (Auby and Ivkovic 2020).

The predominant ethical principle that needs to apply in recruiting new participants is that of clinical equipoise, which argues that if there is considerable uncertainty over which treatment (trial or standard of care) is likely in the best interest of the child, then it would be admissible to enroll the child in a clinical trial. However, these enrollments do not come without a challenge from data which questions decision-making pertaining to what may benefit a child, as is shown in work from Snowden et al. (1997) where 25% of parents felt obligated to participate in a study that their primary physician recommended (Snowdon et al. 1997). The reverse is also disconcerting where 32% of parents wanted their primary physician to decide whether their child should be enrolled in the study or not (Zupancic et al. 1997).

It is the opinion of the authors that any effort to recruit participants must pay special attention to the deficit model which argues that those in need of a cure for their child will often misunderstand the principle of randomization which may lead to further ethical issues. All clinical trial recruitment goes hand-in-hand with information brochures, but the efficacy of these have been questioned by some (Pritchard-Jones et al. 2008). Suffice to say 87% of those who were referred to clinical trials were done so by a primary physician underlining the fact that trust is a huge component in recruiting participants into trials and making sure the child as well as the parents are fully informed (Knox and Burkhart 2007).

23.4 Translation of Clinical Research Findings

Although clinical research and medical care are closely related, there are clear boundaries as mentioned above, where clinical care deals with the individual patient while clinical research aims to generate novel knowledge to be used for future patients (Sacristán 2015). It is crucial that these novel scientific and/or clinical research results are implemented into health care practice to ensure that the larger patient pool benefit from these discoveries (Woolf 2008). In this process note the two major definitions for translational research. The first definition refers to "T1 roadblock" which is the translation of new knowledge regarding disease mechanisms, gained in the laboratory setting, into new diagnostic methods, novel therapies and prevention strategies. The second definition refer to as "T2 roadblock" is the translation of clinical study results in day-to-day clinical practice (Sung et al. 2003). Successful interventions should be implemented for the greater number of patients after proven efficacious and safe and include translation of basic science research, epidemiology, behavioral science, pharmacology and psychology. Through translation research we ensure that a larger number of patients receive the health care they need, and this will improve the essential trust relationship between the clinician-researchers and the public at large.

23.5 Conclusion

In conclusion pediatric clinical trials have ensured that childhood cancer has become a curable disease. At all times the child's best interest should guide the scientific rationale for study participation, with expected potential benefit for the individual child, respect for the family's culture and preferences; and joint decision-making between the family and the clinician-researcher. As childhood cancer is usually managed by a multidisciplinary team, it is important to involve the team to assist the family in the decision-making for clinical trial participation.

Guiding Principles for Childhood Cancer Clinical Trials

Dual role
- Recognize the dual role as physician-researcher when conducting clinical trials involving childhood cancer patients
- The good clinician aims to cure or alleviate pain or discomfort and or provide support with palliative care
- The researcher aims to answer a question that can lead to generalizable knowledge and in the context of childhood cancer care will lead to cure
- The physician-researcher should identify and respect the boundaries between these two roles

Important ethical issues
- Recruitment strategies as the child participant is suffering from a life-threatening disease and therefore more vulnerable
- Consent and assent processes are more complex in the face of a life-threatening disease such as childhood cancer
- Translation of clinical research findings into routine clinical care is essential

References

Afshar, K., A. Lodha, A. Costei, and N. Vaneyke. 2005. Recruitment in pediatric clinical trials: An ethical perspective. *Journal of Urology*. https://doi.org/10.1097/01.ju.0000169135.17634.bc.

Arora, R.S., J.M. Challinor, S.C. Howard, and T. Israels. 2016. Improving care for children with cancer in low- and middle-income countries-A SIOP PODC initiative. *Pediatric Blood and Cancer*. https://doi.org/10.1002/pbc.25810.

Auby, P., and J. Ivkovic. 2020. Ethical aspects of research in paediatric psychopharmacology. *Clinical Research in Paediatric Psychopharmacology*. https://doi.org/10.1016/b978-0-08-100616-0.00006-x.

Bernstein, M. 2003. Conflict of interest: It is ethical for an investigator to also be the primary care-giver in a clinical trial. *Journal of Neuro-Oncology*. https://doi.org/10.1023/A:1023959021758.

Boydell, K., R.Z. Shaul, L. D'Agincourt–Canning, M.D. Silva, C. Simpson, C.D. Czoli, …, and R. Schneider. 2012. Paediatric physician–researchers: Coping with tensions in dual accountability. *Narrative Inquiry in Bioethics*. https://www.muse.jhu.edu/article/494852.

Czoli, C., Da Silva, M., Zlotnik Shaul, R., d'Agincourt-Canning, L., Simpson, C., Boydell, K., …, and Vanin, S. 2011. Accountability and pediatric physician-researchers: Are theoretical models compatible with Canadian lived experience? *Philosophy, Ethics, and Humanities in Medicine*. https://doi.org/10.1186/1747-5341-6-15.

De Vries, M.C., M. Houtlosser, J.M. Wit, D.P. Engberts, D. Bresters, G.J. Kaspers, and E. Van Leeuwen. 2011. Ethical issues at the interface of clinical care and research practice in pediatric oncology: A narrative review of parents' and physicians' experiences. *BMC Medical Ethics*. https://doi.org/10.1186/1472-6939-12-18.

Dickert, N., and J. Sugarman. 2005. Ethical goals of community consultation in research. *American Journal of Public Health*. https://doi.org/10.2105/AJPH.2004.058933.

Dickert, N.W. 2019. The importance of listening to patients and to evidence regarding consent for research. *American Journal of Bioethics*. https://doi.org/10.1080/15265161.2019.1572834.

Dupont, J.C.K., K. Pritchard-Jones, and F. Doz. 2016. Ethical issues of clinical trials in paediatric oncology from 2003 to 2013: A systematic review. *The Lancet Oncology*. https://doi.org/10.1016/S1470-2045(16)00142-X.

Gordon, E.J., A.H. Yamokoski, and E. Kodish. 2006. Children, research, and guinea pigs: Reflections on a metaphor. *IRB Ethics and Human Research*.

Israëls, T., J. Kambugu, F. Kouya, N.K. El-Mallawany, P.B. Hesseling, G.J.L. Kaspers, and E.M. Molyneux. 2013. Clinical trials to improve childhood cancer care and survival in sub-Saharan Africa. *Nature Reviews Clinical Oncology*. https://doi.org/10.1038/nrclinonc.2013.137.

Kant, Immanuel. 1993. [1785]. Grounding for the Metaphysics of Morals.*Translated by Ellington, James W*, 3rd ed. Hackett, p. 30. ISBN 0-87220-166-X.

Knox, C.A., and P.V. Burkhart. 2007. Issues related to children participating in clinical research. *Journal of Pediatric Nursing*. https://doi.org/10.1016/j.pedn.2007.02.004.

Pritchard-Jones, K., M. Dixon-Woods, M. Naafs-Wilstra, and M.G. Valsecchi. 2008. Improving recruitment to clinical trials for cancer in childhood. *The Lancet Oncology*. https://doi.org/10.1016/S1470-2045(08)70101-3.

Macklin, R., and S. Sherwin. 1975. Experimenting on human subjects: Philosophical perspectives. *Case Western Reserve Law Review*. https://scholarlycommons.law.case.edu/cgi/viewcontent.cgi?article=3011&context=caselrev

Oxford Dictionary of Philosophy. 3rd edn., p 157, https://www.oxfordreference.com/view/https://www.oxfordreference.com/view/10.1093/acref/9780198735304.001.0001/acref-9780198735304-e-3051?rskey=NYnuAs&result=3126. Accessed 17 Feb. 2021.

Pellegrino, E. 2005. The "Telos" of medicine and the good of the patient. *Clinical Bioethics*. https://doi.org/10.1007/1-4020-3593-4_2.

Reaman, G. 2004. Pediatric cancer research from past successes through collaboration to future transdisciplinary research. *Journal of Pediatric Oncology Nursing : Official Journal of the Association of Pediatric Oncology Nurses*. https://doi.org/10.1177/1043454204264406.

Sacristán, J.A. 2015. Clinical research and medical care: Towards effective and complete integration. *BMC Medical Research Methodology*. https://doi.org/10.1186/1471-2288-15-4.

Snowdon, C., J. Garcia, and D. Elbourne. 1997. Making sense of randomization; responses of parents of critically ill babies to random allocation of treatment in a clinical trial. *Social Science and Medicine*. https://doi.org/10.1016/S0277-9536(97)00063-4.

Sung, N.S., W.F. Crowley, M. Genel, P. Salber, L. Sandy et al. 2003. Central challenges facing the national clinical research enterprise. *Journal of the American Medical Association*. https://doi.org/10.1001/jama.289.10.1278.

Woolf, S.H. 2008. The meaning of translational research and why it matters. *JAMA—Journal of the American Medical Association*. https://doi.org/10.1001/jama.2007.26.

Zupancic, J.A.F., P. Gillie, D.L. Streiner, J.L. Watts, and B. Schmidt. 1997. Determinants of parental authorization for involvement of newborn infants in clinical trials the online version of this article, along with updated information and services, is located on the World Wide Web at : Determinants of parental authorization form. *Pediatrics*. https://doi.org/10.1542/peds.99.1.e6.

Further Readings

Afshar, K., A. Lodha, A. Costei, and N. Vaneyke. 2005. Recruitment in pediatric clinical trialsXE "Clinical trial": An ethical perspective. *Journal of Urology*. https://doi.org/10.1097/01.ju.0000169135.17634.bc.

De Vries, M. C., M. Houtlosser, J.M. Wit, D.P. Engberts, D. Bresters, G.J. Kaspers, and E. Van Leeuwen. 2011. Ethical issues at the interface of clinical care and research practice in pediatric oncology: A narrative review of parents' and physicians' experiences. *BMC Medical Ethics*. https://doi.org/10.1186/1472-6939-12-18.

Dupont, J.C.K., K. Pritchard-Jones, and F. Doz. 2016. Ethical issues of clinical trialsXE "Clinical trial" in paediatric oncology from 2003 to 2013: A systematic review. *The Lancet Oncology*. https://doi.org/10.1016/S1470-2045(16)00142-X.

Chapter 24
Reproductive Controversies: Fertility Preservation

J. Taylor, L. Shepherd, and M. F. Marshall

Abstract Fertility preservation is increasingly available to pediatric and adolescent populations whose future fertility is threatened. These reproductive technologies raise questions about the interests of younger children in future fertility, parental interests and influences on adolescents, and the interests of persons no longer living. Ethical and legal analyses of specific case examples highlight key issues of parental permission and minor assent, emerging adolescent autonomy, and postmortem gamete retrieval.

Keywords Fertility preservation · Adolescent · Child · Reproductive rights · Transgender · Posthumous reproduction

24.1 Introduction

Over the last decade, advances in reproductive technologies have expanded options for preserving fertility in pediatric and adolescent patients. Concurrently, outcomes for childhood illnesses (cancer, sickle cell disease, organ failure resulting in transplantation, etc.) have improved, and attention has turned toward the impact of life-saving therapeutics on survivors of childhood illnesses, including reproductive and fertility outcomes. Partly due to new reproductive technologies, greater attention

J. Taylor (✉)
Department of Pediatrics, School of Medicine Center for Health Humanities and Ethics, University of Virginia, Charlottesville, VA, USA
e-mail: JFT4P@hscmail.mcc.virginia.edu

L. Shepherd
Department of Public Health Sciences, School of Law, School of Medicine Center for Health Humanities and Ethics, University of Virginia, Charlottesville, VA, USA
e-mail: LLS4B@hscmail.mcc.virginia.edu

M. F. Marshall
Department of Public Health Sciences, School of Medicine Center for Health Humanities and Ethics, School of Nursing, School of Law, University of Virginia, Charlottesville, VA, USA
e-mail: MFM@hscmail.mcc.virginia.edu

© Springer Nature Switzerland AG 2022
N. Nortjé and J. C. Bester (eds.), *Pediatric Ethics: Theory and Practice*,
The International Library of Bioethics 89,
https://doi.org/10.1007/978-3-030-86182-7_24

is being paid to the reproductive needs and outcomes of sex-and-gender-diverse persons. Given the importance of reproductive concerns and the burgeoning field of fertility preservation for children and adolescents, professional guidance recommends discussing future fertility and the impact of potentially gonadotoxic[1] therapeutics or surgeries and fertility preservation options with patients and families at the outset of treatment (Campo-Engelstein et al. 2017; Hembree et al. 2017; Klipstein et al. 2020; Oktay et al. 2018; Sutter 2009). As a rapidly expanding field of study and clinical practice, fertility preservation in pediatric and adolescent populations raises important practical, ethical, and legal questions that warrant careful thought and discussion. This chapter describes available fertility preservation options for pediatric and adolescent patients and uses three case examples to highlight key ethical considerations.

Options for fertility preservation are available for persons with testes and persons with ovaries, but differences in pubertal timing, reproductive maturation, and gamete-retrieval processes result in variable acceptability, invasiveness, availability, and costs.

24.2   Evolution of Reproductive Technologies

Human sperm was first cryopreserved in 1953 by Sherman and Bunge, who froze glycerol-treated sperm with dry ice (Ombelet & Van Robays 2015; Robey 2015). Embryo creation and in vitro fertilization became available to adults soon after and resulted in the first "test tube baby" in 1978 (Ombelet & Van Robays 2015). In 1980, Dr. Cappy Rothman performed the first recorded case of sperm retrieval postmortem (Robey 2015). Cryopreservation of oocytes and embryos developed around the same time, with the first births from a cryopreserved embryo and oocyte reported in 1983 and 1986, respectively (Gook 2011). Cryopreservation and assisted reproduction techniques have advanced to allow pre-pubertal children and adolescents to preserve surgically removed testicular and ovarian tissue for future use (Klipstein et al. 2020). There are no reported live births from re-implantation of testicular tissue. In 2004, the first live birth following auto-implantation (re-implantation into the person from whom it was removed) of cryopreserved-then-thawed ovarian tissue from an adult was reported, with over 100 live births since (American Society of Reproductive Medicine [ASRM] 2019). Whether similar results can be achieved with pre-pubertal tissue preservation remains to be seen; research on immature ovarian and testicular tissue function and development has raised questions about the long-term viability and reproductive potential of pre-pubertal samples (Klipstein et al. 2020).

---

[1] Gonadotoxic means toxic or harmful to the primary reproductive glands (ovaries and/or testes).

Table 24.1 Fertility-preservation options for persons with testes

Preservation technique	Description	Experimental	Other considerations
Pre-pubertal			
Gonadal shielding	Testicles are protected from radiation with a lead shield or radiation rays are focused on a small area	No	Routinely performed if testes are proximal to radiation field
Testicular tissue cryopreservation	Testicular tissue containing cells that make sperm is removed, frozen, and stored for later use	Yes	Offered prior to treatment, generally in a research protocol, to individuals at highest risk of infertility
Post-pubertal			
Sperm collection by masturbation	Mature sperm are collected and frozen for later use	No	Available before treatment. Often > one sample needed for sufficient viable sperm
Assisted sperm collection	Non-surgical sperm retrieval methods i.e., electro-ejaculation, vibratory, retrograde ejaculation	Yes	If masturbation is unacceptable or unsuccessful
Surgical sperm extraction	Requires anesthesia (local or general) and either surgical aspiration or an incision to collect sperm	No	Provides options for individuals without sperm present in ejaculate
Gonadotropin-releasing hormone analogs	Medications that suppress gonadal activity	–	No longer recommended in persons with testes due to lack of effectiveness

Costs[2] associated with fertility preservation vary by the method used, location, insurance status, qualification for financial assistance or grant programs, and whether the procedure is offered as part of a clinical trial. Cryopreservation techniques for sperm or testicular tissue demonstrate a wide price range (In 2020, in US dollars: sperm cryopreservation cost $500–1,000 for initial testing and freezing, $1000–1500 for surgical sperm extraction, $2,500–5,000 for testicular tissue retrieval, $150–500 yearly for storage). Preservation of ovarian tissue, oocytes, and embryos requires a more substantial financial commitment (In 2020, in US dollars: $10,000–15,000 for initial procedure and testing, ~$4000 for medications, $300–$500 yearly for storage). Many preservation procedures for children and adolescents will require

[2] Cost data derived from the following sources: *Paying for Treatments.* (n.d.). Alliance for Fertility. Retrieved September 25, 2020, from https://www.allianceforfertilitypreservation.org/paying-for-tre atments/. *Decision-making Aids.* (n.d.) Fertility Preservation in Pittsburgh. Retrieved September 25, 2020, from https://fertilitypreservationpittsburgh.org/fertility-resources/decision-making-aids/#2.

Table 24.2 Fertility-preservation options for persons with ovaries

Preservation technique	Description	Experimental	Other considerations
Pre-pubertal			
Ovarian transposition or Shielding	*Ovaries and/or fallopian tubes surgically moved or protected from radiation by a lead shield over the pelvis*	No	Uterine damage may preclude pregnancy (without use of a surrogate) even if ovarian function is preserved Used post-pubertally if other fertility-sparing procedures not possible
Ovarian tissue cryopreservation	An entire ovary or portions of an ovary are surgically removed, frozen, and stored for later use	No	Re-implantation not recommended in some malignancies In vitro maturation of oocytes not established to date Used post-pubertally if other fertility-sparing procedures not possible
Post-pubertal			
Mature oocyte cryopreservation	One or more unfertilized eggs retrieved under via needle aspiration through vaginal wall, frozen, stored for later use	No	May require hormonal ovarian stimulation; contraindicated in some individuals with certain cancers/comorbidities
Embryo cryopreservation	After in vitro fertilization (retrieved eggs are combined with sperm in a laboratory to form embryos), embryos are frozen and can later be thawed and implanted into a healthy uterus	No	Requires a committed male partner or donor sperm
Gonadotropin-releasing hormone analogs	Medications that suppress gonadal activity	Yes	Should not replace established therapies for which individual is a good candidate May have other benefits such as reducing menstrual bleeding

future costly ($5000–10,000) assisted-reproductive techniques to produce biologi-
cally related children, even without a gestational surrogate ($60,000–125,000). The
tables below outline current fertility-preservation procedures (Tables 24.1, 24.2).[3]

---

[3] For more detailed information on fertility preservation procedures, see (Klipstein et al. 2020;
ARSM 2019; Oktay et al. 2018).

24.3   Case Discussions

The following is a series of fictional cases with discussions of unique ethical challenges arising in pediatric and adolescent fertility preservation. The role of parental decision-making in future fertility, benefit-and-risk calculations for different populations, respect for emerging adolescent autonomy, children's interests, and how death (or poor prognosis) affect decisions about retrieval and disposal of reproductive material are examined.

24.4   Case 1: Young Cancer Patient

Jasmine is a 4-year-old child, recently diagnosed with rhabdomyosarcoma (a tumor arising from primitive muscle cells). She is scheduled to start chemotherapy after surgery to remove as much of the tumor as possible. Reviewing the consent form for chemotherapy, her parents read that she will be at high risk of infertility. After investigating online, they request fertility preservation for Jasmine. The hospital where she receives care does not have a surgeon who performs ovarian tissue cryopreservation, so her parents inquire about a transfer to a hospital that offers the procedure.

24.4.1   Discussion

Fertility preservation is an ethically permissible choice for parents to make even if it poses some risks because they are holding their child's fertility as a "right in trust," taking steps to preserve their child's ability to reproduce (Jadoul et al. 2010), as we argue in our analysis below.

Equity and justice. One ethical consideration here is equity and access to fertility-related care. Jasmine's parents are requesting a fertility-preserving procedure that involves surgical removal of part (or all) of an ovary for sectioning and freezing which requires expertise unavailable in many locations. Issues of justice and access arise based on geographic location and ability to pay since these procedures aren't always covered by insurance. Some families can afford to pay for them absent insurance coverage, but many cannot, making future fertility available only to some. Cost is a significant factor in such decisions (Oktay et al. 2018). There are upfront costs for the initial procedure and annual storage fees requiring payment for decades until a pediatric patient is ready to consider reproduction. It is impossible to know whether Jasmine will want the cryopreserved tissue for reproduction and, if she does, whether she can afford covering storage costs and the higher costs of assisted reproduction when she reaches the age of legal majority.

Interests of the child. A second ethical concern is the potential for Jasmine's current interests to conflict with her future reproductive interests. For a child this age, developmentally unable to participate in complex medical decisions, medical professionals rely on parental understandings of their child's current and long-term interests to guide care. Ovarian-tissue cryopreservation after cancer diagnosis is considered a low-risk procedure when performed simultaneously with other procedures (e.g., central-line placement), reducing risks of surgery and anesthesia. If Jasmine has already undergone these procedures, ovarian tissue preservation will involve additional risks from another surgery and additional anesthesia (which may include risks of organ damage, scar tissue, arrhythmias, allergic reactions). Another important risk to consider is the effect of delayed chemotherapy on Jasmine's prognosis. Hospital transfer for ovarian tissue preservation could delay Jasmine's cancer treatment. Rhabdomyosarcoma[4] is generally curable; 70% of children with localized disease receiving treatment will achieve 5-year survival, but treatment delays may affect Jasmine's treatment response and survival (Rhee et al. 2020).

Assuming Jasmine's tumor has favorable characteristics for her survival, the permissibility of pursuing preservation partially depends on whether her interest in future biologic parenthood and the likelihood of successful reproduction outweigh the burdens of transferring care prior to her cancer treatment. Ovarian-tissue cryopreservation, based primarily on safety and outcome data from adults, is no longer considered experimental (ASRM 2019). However, the success of ovarian-tissue transplantation on fertility outcomes or assisted reproduction using pre-pubertal ovarian tissue in children as young as Jasmine is still under investigation (Poirot et al. 2019). For pre-pubertal children with ovarian failure after chemotherapy, another potential, but not established benefit of storing ovarian tissue is re-implantation to induce puberty (Wallace et al. 2016). With uncertain benefit and the potential for harm, the question remains whether Jasmine's parents should pursue fertility preservation. If her risk for infertility were lower or uncertain, the burdens of undergoing another procedure may be too great.

If Jasmine were older (as in Case 2) and willingly undertook the risks in order to preserve her ability to have biologically related children, that would be permissible, assuming she's provided sufficient information for an informed choice. In this case it's unlikely that Jasmine has strong parenting desires that would be heavily weighted in her parents' decision. Instead, they are weighing her risk of infertility and the importance of Jasmine's ability to make her own fertility choices in the future, her "right to an open future" (Cutas & Hens 2015) against her more immediate short-term need to start chemotherapy and the risks of delaying treatment. While protecting a child's future fertility and future reproductive choice should be taken into account, parents are not obligated protect a child's future fertility at all costs but must instead carefully weigh the benefits versus the burdens given the particulars of the clinical situation.

It is conceivable Jasmine's parents can afford to pay out-of-pocket and place such value on her future fertility and her ability to decide for herself in the future that they seek transfer and, if the process is coordinated such that there is no substantial delay in her care, this is an ethically-permissible option (refer to Chap. 3). It is also possible

---

[4] Rhabdomyosarcoma is a tumor arising from primitive muscle cells.

to imagine a time when fertility preservation is widely available, affordable, and no longer experimental, creating an ethical obligation for parents to pursue preservation options that do not otherwise impact expected survival.

Interests of parents. What are Jasmine's parents' interests and how much weight should they be given? Are they interested in perpetuating a family lineage or in future grandchildren? Are they unable to have other children? Ethicists disagree on the extent to which parents should include interests other than the child's in decision-making (refer to Chap. 9). Jasmine's interests, however, should be primary.

We don't have additional information about factors that could adversely alter Jasmine's prognosis (unfavorable location, residual tumor after resection). A poor prognosis alters the evaluation of her parents' request for fertility preservation (Daar et al. 2019). Low likelihood of survival would shift the ethical analysis from preserving Jasmine's future fertility to questions about posthumous reproduction and gamete disposal. As discussed in Case 3, parental requests for retrieval, storage, and use of their minor child's gametes are generally not honored (Daar et al. 2019; Klipstein et al. 2020).

24.5   Case 2: Transgender Youth

Maria is a 14-year-old transfemale (assigned male at birth) interested in starting gender-affirming estrogen therapy. Her parents support this desire and future interventions, including sex reassignment surgery, to complete her transition. However, they would like Maria to preserve her fertility by storing reproductive gametes for future use. Maria has rejected the idea. Her parents, having struggled with infertility prior to adopting Maria, are fearful that she will change her mind in the future. They insist she meet with a trans-friendly fertility specialist to discuss options.

24.5.1   Discussion

This case examines whether a parent may ethically request that their child undergo additional fertility counseling before permitting her to start gender affirming hormones.

Interests of the adolescent. Several factors bear consideration; how firm are her parents in their insistence—are they nudging Maria or are they conditioning their permission for hormone therapy on further counseling? How acutely is Maria experiencing distressing gender dysphoria (refer to Chap. 22) and will it be exacerbated by fertility preservation? What kind of counseling has already occurred–e.g., does Maria understand the associated benefits and risks of fertility preservation and the methods that would be available? Have those conversations felt supportive and trans-friendly? Has Maria been able to speak privately with her physician to discuss her reservations?

Infertility and reduced fertility are potentially irreversible changes associated with the use of hormones or surgeries that affect reproductive organs (Hembree et al. 2017). Gamete-preservation techniques accessible to a transitioning adolescent—sperm and oocyte preservation—are well-established and non-experimental. Given the potential benefit, health care providers have an ethical duty to counsel about fertility preservation before starting adolescents on pubertal suppression or gender-affirming hormones and to provide referrals to appropriate specialists (Ethics Committee, ASRM 2015). Maria's parents may feel responsible for holding her fertility as a right in trust so that it is not irreversibly lost at an age when Maria may not be ready to consider long-term reproductive interests.

Adolescents' decision-making processes are different from adults and should be considered. Studies show that although adolescents may be able to understand and reason like adults, their susceptibility to peer influence is greater, they tend to weigh immediate considerations over longer-term ones, and they are more inclined to engage in risky behaviors and make impulsive decisions. (Diekema 2020). In our case, it would not be surprising if Maria did not want to think about children now and is more worried about the fact that her peers are all undergoing pubertal changes while she is not. But when Maria is 30, this may all have changed, and she might desire fertility and the option of having children. The longer-term benefits of gamete preservation for an individual patient will depend upon that person's future interest in biologically-related children and the ease with which gamete storage and later use of assisted-reproductive technology can be financially borne and, if needed, a partner or gestational surrogate found to carry a pregnancy. These benefits are difficult to assess at this time.

In this case we do not know why Maria is resisting further fertility counseling; the impasse suggests the need to examine what has gone before—what has Maria's experience been regarding gender-related care and fertility preservation? What are her concerns, and can they be addressed? To pursue fertility preservation, an adolescent must temporarily cease puberty blockers or delay gender-affirming hormone therapy, causing further progression of natal puberty (if brief) and the potential for intensified gender dysphoria (Nahata et al. 2018). For transfemales, sperm collection via masturbation may be distressing, although testicular sperm extraction is an option (Chen et al. 2017). For transmales, oocyte retrieval is costly, physically invasive and uncomfortable, requiring daily hormone injections for 10–14 days, transvaginal ultrasounds and oocyte retrieval (Chen et al. 2017).

Although research is limited, Maria's resistance to the idea of fertility preservation is not uncommon (Chen et al. 2017). In a 2016 retrospective study of transgender youth in a large gender clinic, only 2 of 73 pubertal or post-pubertal patients attempted fertility preservation; all but one had received fertility counseling prior to starting hormone therapy (Nahata et al. 2017). The most common reasons for rejecting fertility preservation were the patient's plan to adopt children (45%) or their disinclination to have children (21%). Expense was a concern for some (8.2%); smaller percentages reported concerns about delaying hormone treatment or discomfort about the process.

The rate of fertility preservation observed in this group of transgender patients is significantly lower than among adolescent and young adult cancer survivors (Nahata et al. 2017). Transgender adults report much greater willingness to have considered fertility preservation if they had been given the opportunity (Nahata et al. 2017). The authors hypothesize that these differences might be explained by the following: the prevalence of mental health co-morbidities in transgender youth may affect their decision-making about fertility preservation; body dysphoria may foster reluctance to consider interventions associated with gender anatomy; they may perceive social pressures about what a family should look like and/or a sense of urgency to begin hormone therapies such that it crowds out other considerations (Nahata et al. 2017).

Parental and adolescent rights and responsibilities. Because Maria is a minor she may not be able to pursue hormone therapy without her parents' permission. The age for consent for medical treatment varies across European Union members from 14 to 18, although some consider the minor's maturity instead (Dubin et al. 2020). In the U.S., minor consent is acceptable for a variety of reproductive health concerns (testing/treatment of sexually transmitted illnesses, family planning and contraceptives), but gender-affirming treatment and fertility preservation have not fallen under these statutory carve-outs to the age of consent for health care. Judicial precedent relating to minors' abortion rights may have similar applicability here, allowing gender-affirming treatment on the basis of the minor's maturity or the minors' best interest.

Although in most places Maria's parents will need to provide permission for her gender-affirming hormone therapy, it is widely recognized that minors should be involved in medical decision-making—and their assent (Leikin 1983) to medical treatment decisions obtained—to the extent they are capable of understanding and evaluating treatment options. Eliciting and honoring Maria's wishes are important for respecting her as a person, developing and maintaining her trust in her health care providers, and advancing her wellbeing. Minors may be best positioned to know what will benefit or burden them; being included and respected in the process is, in itself, a likely benefit.

Equity and justice. Even transyouth with supportive families and friends face significant barriers to routine as well as gender-related care. They may lack adequate insurance coverage for needed care or face challenges finding nearby transgender-medicine specialists or even clinicians who are trans-friendly and trans-informed. These barriers are often greater for fertility preservation.

Cost is also a barrier to access in many places, perhaps more so in the United States than in countries providing more universal health coverage (Tishelman et al. 2019). In some U.S. states (Kyweluk et al. 2018) and Australia's public health care system (Telfer et al. 2018) fertility-preservation coverage mandates may include oncology but not transgender patients. While there is no morally relevant difference between pediatric oncology patients and transitioning adolescents in the cause of impaired fertility—an iatrogenic complication of a medical intervention—transyouth are more likely to face barriers to appropriate fertility-preservation care.

24.6   Case 3: Posthumous Gamete Retrieval
and Posthumous Reproduction

Jason is a 17-year-old male who had an unwitnessed collapse with prolonged resuscitation efforts and now meets death by neurologic criteria. His mother and fiancée, Dana (engaged for one month with plans to marry after high school graduation), are at the bedside. During a discussion of organ donation, they raise the question of sperm retrieval for Dana's future use in assisted reproduction before removing Jason's organ support.

24.6.1   Discussion

Because Jason has not reached the age of majority, in most countries a request by his mother or fiancée would likely not be honored. Guidance from international professional associations and analyses in the academic literature speak primarily to the ethical issues of postmortem retrieval from adults but is also instructive here.

The birth of Liam Blood, the second child conceived via posthumous reproduction, occurred following a protracted legal battle. In 1995, 30-year-old Stephen Blood was left comatose after contracting bacterial meningitis. At his wife's request, sperm samples were taken while he was comatose and just prior to declaration of his death. The Infertility Research Trust, where the samples were stored, refused to release them to his widow. The Civil Division of England's Court of Appeals ultimately allowed Ms. Blood to export the samples to Brussels where she conceived Liam via assisted reproduction (*R v. Human Fertilisation and Embryology Authority, ex parte Blood* 1997; Sabatello 2014). Three years later she conceived a second son, Joel, also with her deceased husband's sperm. The Human Fertilisation and Embryology Act of 1990 precluded Stephen Blood from being named as the boys' father on their birth certificates. It was amended in 2003 to allow children conceived after their father's death to have his fatherhood documented on their birth certificates (Sabatello 2014).

Posthumous gamete retrieval and reproduction have become more commonplace in recent years, although legal questions about parenthood and entitlement to inheritance or survival benefits continue, as do ethical questions regarding the circumstances under which requests for these services should be honored.

Respect for autonomy. The primary ethical concern is respect for Jason's former autonomy. Here, the vexing question arises of what, if any, rights or interests a decedent has. Can a decedent be harmed or wronged? Directives about organ donation, wills articulating the disposition of one's estate, and instructions about funeral arrangements typify the range of legal and other interests that individuals and society at large have in honoring premortem communications.

We do not know Jason's wishes regarding sperm retrieval postmortem or posthumous reproduction. This raises questions about his right to bodily integrity; would he have consented to invasive surgical procedures to retrieve sperm? We also do not know what, if any, wishes Jason may have had regarding posthumous reproduction. Professional organizations have relatively consistent guidance on these questions. The most stringent requirement is for written documentation by the decedent declaring his wishes regarding sperm retrieval postmortem and posthumous reproduction (Pennings et al. 2006; Ethics Committee ASRM 2018). Other guidance allows for proxy consent by the surviving spouse or partner assuming that they also have authority to make decisions about organ donation and disposal of the decedent's body (Ethics Committee ASRM 2018; Robey 2015). Some independent authors argue for presumed consent to allow a surviving partner to fulfill a desire (presumably shared by the decedent) to create a child who shares the genes of both partners. All, however, caution against allowing parents to make these decisions out of concern for potential harms with the creation of a "commemorative child" to somehow replace the decedent or simply to perpetuate a genetic legacy (Klipstein & Fallat 2020; Pennings et al. 2006; Ethics Committee ASRM 2018).

Unless they are in high-risk professions, most people (including Jason) do not have written advance directives regarding posthumous reproduction. Even if Jason had been clear that he wanted a future family, that wish is generally not considered sufficient justification for sperm retrieval postmortem and posthumous reproduction because he may not have wanted a child that he would not live to parent. Also, there is no opportunity for him to consent to an invasive procedure for sperm retrieval postmortem, though this again raises the perplexing question of whether a decedent can be harmed or wronged.

Even if the hospital agreed to retrieve sperm (it isn't required to do so), it is not obligated to participate in future assisted reproduction. Both circumstances rest on the issue of whether the decedent would have given permission for each of the posthumous procedures. Although it seems intuitive that willingness to perform posthumous removal of gametes implies a willingness to perform assisted reproduction, this is not the case. Different clinicians from different departments within the same hospital are generally involved in the two procedures. While a urologist may be willing to retrieve posthumous sperm, a reproductive endocrinologist may not be willing to provide attempts at assisted reproduction. It may, however, also be ethically appropriate for clinicians/hospitals to provide both services. Hospitals should have policies with guidelines that specify their willingness (or lack thereof) to engage in posthumous gamete retrieval and subsequent attempts at assisted reproduction (Ethics Committee ASRM 2018).

Interests of the Surviving Partner. A second ethical issue involves the procreative liberty and interests of the surviving partner. Dana, as with most persons who lose a partner due to sudden death, may be mired in grief or emotional trauma—not an ideal circumstance in which to make life-altering reproductive decisions. If sperm were retrieved, recommendations (and some requirements) are for Dana to wait a period of six months to one year before attempting assisted reproduction. This would allow her time to grieve and thus make a less emotional, more reflective and autonomous decision. Her initial counseling should include pragmatic realities such as the significant costs of sperm retrieval postmortem (the procedure; freezing, storing, then releasing the samples) as well as the costs and probability of successful assisted reproduction. Depending on the country or state, such costs could be covered by insurance, completely out of pocket, or some combination of the two. Only a small fraction of surviving partners ultimately pursue assisted reproduction (Bahadur 2004; Kramer 2009; Robey 2015).[5]

Interests of a potential child. A third ethical issue surrounds the potential child's welfare. Concerns include being raised in a single-parent family (generally culture specific), whether the child will feel stigmatized by the manner in which they were conceived, and legal and social questions of legitimacy or inheritance (Knapp et al. 2011). Will the child be able to secure social security or other survivor benefits, or rights to an estate? A common apprehension is whether the child will be viewed by the parent or other family members as a "commemorative child" or as a "symbolic replacement of the deceased" (Pennings et al. 2006). This rationale undergirds the proscription in several countries against allowing the decedent's parents to give proxy consent for posthumous reproduction. Other countries are more permissive, some placing central religious or cultural value on continuing a genetic legacy. In a precedent-setting case in Israel (and likely the world) in 2011, a magistrate court gave permission for the parents of Chen Aida Ayish (a 17-year-old) to retrieve and freeze her oocytes. The court, however, denied their request for fertilization and embryo storage as the parents were unable to prove that Chen would have wanted children (Simana 2018). In 2018, a U.K. couple successfully advocated for posthumous reproduction after their 26-year-old son died in a motorcycle accident. They retrieved their deceased son's sperm postmortem but were unable to pursue assisted reproduction in the UK. They shipped the sperm samples to a clinic in California and engaged a gestational surrogate for in vitro fertilization resulting in the birth of a grandson, whom they are raising (Simana 2018).

24.7 Conclusion

Guiding principles for fertility preservation decisions for children/adolescents

Equity and justice
– Provide opportunities for health (including reproductive health) to all, regardless of age, gender/gender identity, ethnicity, place of birth or residence, citizenship, socioeconomic or insurance status, political beliefs or religion

(continued)

[5] Data obtained from the California Cryobank show that attempted conception occurred in only 2 of 148 cases of sperm retrieval postmortem and subsequent banking of at least two samples (Robey 2015); Badahur reports similar data; 21 cases of postmortem sperm retrieval and no use of those samples; Kramer reports 13 cases and no subsequent pregnancies in a 4-year period (Bahadur 2004; Kramer 2009).

(continued)

Equity and justice
- Provide opportunities for health (including reproductive health) to all, regardless of age, gender/gender identity, ethnicity, place of birth or residence, citizenship, socioeconomic or insurance status, political beliefs or religion

Respect for emerging autonomy and interests of children and adolescents
- Recognize and respect the wishes of adolescents as they develop and mature
- Identify and mitigate potential constraints on adolescent decision-making: lack of lived experience, coercion, privacy and confidentiality, etc
- Keep future options open to allow children to exercise their autonomy upon reaching adulthood
- Carefully consider present and future interests of children and adolescents, which may sometimes be in conflict; include consideration of their interests in participating in decision-making processes

As each case highlights, the question of fertility preservation for pediatric and adolescent patients requires careful consideration of the appropriate limits of parental decision-making, the weight of future interests and potentially mutable desires and preferences in benefit-versus-burden analyses, and the impact of emerging autonomy on reproductive choices. Equitable access not only to information about fertility outcomes and preservation, but also to the preservation techniques themselves should be prioritized for all children and adolescents whose treatments threaten future fertility.

References

Bahadur, G. 2004. Ethical challenges in reproductive medicine: Posthumous reproduction. *International Congress Series* 1266: 295–302. https://doi.org/10.1016/j.ics.2004.01.105.

Campo-Engelstein, L., D. Chen, A.B. Baratz, E.K. Johnson, and C. Finlayson. 2017. The ethics of fertility preservation for pediatric patients with differences (disorders) of sex development. *Journal of the Endocrine Society* 1 (6): 638–645. https://doi.org/10.1210/js.2017-00110.

Chen, D., L. Simons, E.K. Johnson, B.A. Lockart, and C. Finlayson. 2017. Fertility preservation for transgender adolescents. *Journal of Adolescent Health*.

Cutas, D., and K. Hens. 2015. Preserving children's fertility: Two tales about children's right to an open future and the margins of parental obligations. *Medicine, Health Care and Philosophy* 18 (2): 253–260.

Daar, J., J. Benward, L. Collins, J. Davis, O. Davis, L. Francis, E. Gates, E. Ginsburg, S. Gitlin, W. Hurd, S. Klipstein, L. McCullough, R. Reindollar, G. Ryan, M. Sauer, S. Tipton, L. Westphal, and J. Zweifel. 2019. Fertility treatment when the prognosis is very poor or futile: An Ethics Committee opinion. *Fertility and Sterility* 111 (4): 659–663. https://doi.org/10.1016/j.fertnstert.2019.01.033.

Diekema, D.S. 2020. Adolescent brain development and medical decision-making. *Pediatrics*, 146 (S1), S18–S24. https://pediatrics.aappublications.org/content/146/Supplement_1/S18.

Dubin, S., M. Lane, S. Morrison, A. Radix, U. Belkind, C. Vercler, and D. Inwards-Breland. 2020. Medically assisted gender affirmation: When children and parents disagree. *Journal of Medical Ethics* 46 (5): 295–299. https://doi.org/10.1136/medethics-2019-105567.

Ethics Committee of the American Society for Reproductive Medicine. 2015. Access to fertility services by transgender persons: An ethics committee opinion. *Fertility and Sterility* 104(5): 1111–1115.

Ethics Committee of the American Society for Reproductive Medicine. 2018. Posthumous retrieval and use of gametes or embryos: An ethics committee opinion. *Fertility and Sterility* 110(1): 45–49.

Gook, D.A. 2011. History of oocyte cryopreservation. *Reproductive BioMedicine Online* 23 (3): 281–289. https://doi.org/10.1016/j.rbmo.2010.10.018.

Hembree, W.C., P.T. Cohen-Kettenis, L. Gooren, S.E. Hannema, W.J. Meyer, M.H. Murad, S.M. Rosenthal, J.D. Safer, V. Tangpricha, and G.G. T'Sjoen. 2017. Endocrine treatment of gender-dysphoric/gender-incongruent persons: An endocrine society clinical practice guideline. *The Journal of Clinical Endocrinology & Metabolism.*

Jadoul, P., M.M. Dolmans, and J. Donnez. 2010. Fertility preservation in girls during childhood: Is it feasible, efficient and safe and to whom should it be proposed? *Human Reproduction Update* 16 (6): 617–630. https://doi.org/10.1093/humupd/dmq010.

Klipstein, S., M.E. Fallat, S. Savelli, Committee On Bioethics, Section On Hematology/Oncology, Section On Surgery. 2020. Fertility preservation for pediatric and adolescent patients with cancer: Medical and ethical considerations. *Pediatrics*, 145(3). https://doi.org/10.1542/peds.2019-3994.

Knapp, C., G. Quinn, B. Bower, and L. Zoloth. 2011. Posthumous reproduction and palliative care. *Journal of Palliative Medicine* 14 (8): 895–898. https://doi.org/10.1089/jpm.2011.0102.

Kramer, A.C. 2009. Sperm retrieval from terminally ill or recently deceased patients: A review. *The Canadian Journal of Urology* 16 (3): 4627–4631.

Kyweluk, M.A., A. Sajwani, and D. Chen. 2018. Freezing for the future: Transgender youth respond to medical fertility preservation. *International Journal of Transgenderism* 19 (4): 401–416. https://doi.org/10.1080/15532739.2018.1505575.

Leikin, S.L. 1983. Minors' assent or dissent to medical treatment. *The Journal of Pediatrics* 102 (2): 169–176.

Nahata, L., A.C. Tishelman, N.M. Caltabellotta, and G.P. Quinn. 2017. Low fertility preservation utilization among transgender youth. *Journal of Adolescent Health.*

Oktay, K., B.E. Harvey, A.H. Partridge, G.P. Quinn, J. Reinecke, H.S. Taylor, W.H. Wallace, E.T. Wang, and A.W. Loren. 2018. Fertility preservation in patients with cancer: ASCO clinical practice guideline update. *Journal of Clinical Oncology* 36 (19): 1994–2001.

Ombelet, W., and J. Van Robays. 2015. Artificial insemination history: Hurdles and milestones. *Facts, Views & Vision in ObGyn* 7(2): 137–143.

Pennings, G., J. Cohen, P. Devroey, and B. Tarlatzis. 2006. ESHRE task force on ethics and law 11: Posthumous assisted reproduction. *Human Reproduction* 21 (12): 3050–3053. https://doi.org/10.1093/humrep/del287.

Poirot, C., L. Brugieres, K. Yakouben, M. Prades-Borio, F. Marzouk, G. de Lambert, H. Pacquement, F. Bernaudin, B. Neven, A. Paye-Jaouen, C. Pondarre, N. Dhedin, V. Drouineaud, C. Chalas, H. Martelli, J. Michon, V. Minard, H. Lezeau, F. Doz, …, and J.-H. Dalle. 2019. Ovarian tissue cryopreservation for fertility preservation in 418 girls and adolescents up to 15 years of age facing highly gonadotoxic treatment. Twenty years of experience at a single center. *Acta Obstetricia et Gynecologica Scandinavica* 98(5): 630–637. https://doi.org/10.1111/aogs.13616.

Practice Committee of the American Society for Reproductive Medicine. 2019. Fertility preservation in patients undergoing gonadotoxic therapy or gonadectomy: A committee opinion Fertility and Sterility 112(6): 1022–1033.

R v. Human fertilisation and embryology authority, ex parte Blood, (February 6, 1997).

Rhee, D.S., D.A. Rodeberg, R.M. Baertschiger, J.H. Aldrink, T.B. Lautz, C. Grant, R.L. Meyers, E.T. Tracy, E.R. Christison-Lagay, R.D. Glick, P. Mattei, and R. Dasgupta. 2020. Update on pediatric rhabdomyosarcoma: A report from the APSA cancer committee. *Journal of Pediatric Surgery.* https://doi.org/10.1016/j.jpedsurg.2020.06.015.

Robey, C. 2015. *Posthumous Semen Retrieval and Reproduction: An Ethical, Legal, and Religious Analysis.* Wake Forest University.

Sabatello, M. 2014. Posthumously conceived children: an international and human rights perspective symposium: The legal and ethical implications of posthumous reproduction. *Journal of Law and Health* 27 (1): 29–67.

Simana, S. 2018. Creating life after death: Should posthumous reproduction be legally permissible without the deceased's prior consent? *Journal of Law and the Biosciences* 5 (2): 329–354. https://doi.org/10.1093/jlb/lsy017.

Sutter, P.D. 2009. Reproductive options for transpeople: Recommendations for revision of the WPATH's standards of care. *International Journal of Transgenderism* 11 (3): 183–185.

Telfer, M.M., M.A. Tollit, C.C. Pace, and K.C. Pang. 2018. Australian standards of care and treatment guidelines for transgender and gender diverse children and adolescents. *Medical Journal of Australia* 209 (3): 132–136. https://doi.org/10.5694/mja17.01044.

Tishelman, A.C., M.E. Sutter, D. Chen, A. Sampson, L. Nahata, V.D. Kolbuck, and G.P. Quinn. 2019. Health care provider perceptions of fertility preservation barriers and challenges with transgender patients and families: Qualitative responses to an international survey. *Journal of Assisted Reproduction and Genetics* 36 (3): 579–588. https://doi.org/10.1007/s10815-018-1395-y.

Wallace, W.H.B., T.W. Kelsey, and R.A. Anderson. 2016. Fertility preservation in pre-pubertal girls with cancer: The role of ovarian tissue cryopreservation. *Fertility and Sterility* 105 (1): 6–12. https://doi.org/10.1016/j.fertnstert.2015.11.041.

Further Readings

Ethics Committee of the American Society for Reproductive Medicine. 2015. Access to fertility services by transgenderXE "Transgender" persons: An ethics committee opinion fertility and sterility 104(5): 1111–1115.

Klipstein, S., M.E. Fallat, S. Savelli, Committee On Bioethics, Section On Hematology/Oncology, Section On Surgery. 2020. Fertility preservation for pediatric and adolescent patients with cancer: medical and ethical considerations. *Pediatrics* 145(3).

Pennings, G., J. Cohen, P. Devroey, and B. Tarlatzis. 2006. ESHRE Task Force on Ethics and Law 11: Posthumous assisted reproduction. *Human Reproduction* 21 (12): 3050–3053.

Chapter 25
The Ethical Principles that Guide Artificial Intelligence Utilization in Clinical Health Care

W. A. Hoffmann and N. Nortjé

Abstract Artificial Intelligence is regarded as a disruptive technology that increasingly affects and challenges traditional interpretation and communication practices, knowledge systems, professional relationships and customer/client engagements. It is widely expected that AI will cause significant and lasting social and economic change on a global scale in the coming years and will also influence pediatric medicine significantly. This chapter promotes an ethical framework, based on important ethical principles, to guide decision-making in order to benefit individual and societal wellbeing, as well as contribute to global development and innovation.

Keywords Artificial Intelligence (AI) · Accountability · Explainability · Transparency · Privacy

25.1   Introduction

Artificial intelligence (AI) brought about fundamental changes in the way that humans think and interact with technology (Duan et al. 2019; UNESCO 2020). As such, AI is regarded as a disruptive technology that increasingly affects and challenges traditional interpretation and communication practices, knowledge systems, professional relationships and customer/client engagements. It is widely expected that AI will cause significant and lasting social and economic change on a global scale in the coming years (Geis et al. 2019). These disruptions raise many ethical questions and challenges regarding the moral status of AI and the human values that AI should, or ought to, be aligned with (ASILOMAR 2017). It also raises challenges

W. A. Hoffmann (✉)
Department Business Excellence, Chief Technology Office, Philips Research, Eindhoven, The Netherlands
e-mail: braam.hoffmann@philips.com

N. Nortjé
Department of Critical Care Medicine, University of Texas MD Anderson Cancer Center, Houston, USA
e-mail: NNortje@mdanderson.org

© Springer Nature Switzerland AG 2022   403
N. Nortjé and J. C. Bester (eds.), *Pediatric Ethics: Theory and Practice*,
The International Library of Bioethics 89,
https://doi.org/10.1007/978-3-030-86182-7_25

regarding our individual and collective moral obligations and responsibilities to duly consider the ethics of data use by AI systems, as well as how to conduct ourselves as professionals in an AI-enabled world (Geis et al. 2019; Van Belkom 2020).

The impact of AI is especially relevant in contexts of decision-making and the skills to make good decisions (UNESCO 2020; European Commission 2019; Farisco et al. 2020). *Decision-making* is an inherently human activity (Phillips-Wren 2012) and refers inter alia to the selection of a viewpoint or course of action among multiple available or potential alternatives, which then often leads to, or at least influences, particular actions. In general, human decision-making is informed by the person's knowledge, skills, viewpoints and values, as well as by social and cultural preferences and beliefs (Geis et al. 2019). In contexts where AI systems are increasingly applied to make, support or influence decisions, it is thus imperative to ensure that these decisions are, and remain to be, in line with personal, professional, social, cultural and societal values and expectations (European Commission 2019).

In health care/clinical contexts, AI systems are increasingly implemented and used to make decisions in all stages of the health care continuum, including care routing (i.e. triage) and care services (i.e. data-driven diagnosis, image-based diagnosis, clinical decision support, medication compliance monitoring and AI-facilitated self-care) (USAID 2019). AI systems can potentially enable clinicians to more accurately, in more detail, much faster and even at an earlier stage when utilizing predictive algorithms, analyze a patient's complex symptoms and health data (e.g. radiological images). This in turn, can assist clinicians and health care professionals to earlier detect diseases and abnormalities (e.g. tumors), to make more accurate diagnoses, to provide targeted and personalized preventive interventions and treatments, and to continuously conduct in-patient and out-patient (remote) patient monitoring in order to alert them of potentially problematic values and patterns in patients' conditions (European Commission 2019; USAID 2019). However, AI systems also influence the nature of the clinician-patient interaction and the relationship in clinical practice, particularly around issues involving trust, privacy, transparency and autonomy, to name a few (Geis et al. 2019).

Despite the clear advantages of using AI systems in clinical decision-making contexts, one should not lose sight of the fact that clinician and patient decisions are not solely based on the physiological and anatomical parameters that underpin AI models and algorithms. Patient decisions are often also influenced by emotional considerations (e.g. fear of pain), personality traits, social factors (e.g. family roles) and cultural values (e.g. decision-making hierarchy; religious beliefs). These factors and considerations form part of the multidimensionality of human intelligence. At the moment AI systems cannot adequately deal with these "soft" issues, which in turn highlights an important limitation of AI systems in the health care decision-making context (Farisco et al. 2020).

25.2 Definition of Artificial Intelligence

There is no single definition of AI that is accepted by all. In this chapter we will use the term *AI systems* which is defined as technological systems (software models and algorithms) that have the capacity to process information (i.e. to achieve a given goal) in a way that resembles intelligent behavior. It typically includes aspects of learning, perception of the environment through sensors, reasoning, and making predictions, recommendations and decisions based on data (UNESCO 2020; Farisco et al. 2020). In other words, AI systems do not merely involve a computational process based on specific instructions, but it rather displays intelligent behavior by adaptively and flexibly analyzing the environment and taking actions and decisions to achieve specific goals (Boucher 2020; Farisco et al. 2020). AI systems inter alia include the following approaches and technologies: machine learning, machine reasoning, knowledge representation, and the processing of data collected by sensors (UNESCO 2020; European Commission 2019). Traditional AI has as its focus the improvement of analytical effectiveness and efficiency, while more enhanced augmented intelligence focuses on decision-making. For the purpose of this chapter we will refer to AI as the latter of the aforementioned.

Several role players who are involved in AI systems are generally recognized, namely developers, deployers and end-users. In this chapter, we will mostly focus on the roles and responsibilities linked to the deployers and end-users within a health care context. *Deployers* refer to public or private entities (e.g. organizations and institutions) that either use AI systems within their specific business processes or that offer AI systems and services to others. They are also sometimes referred to as *AI practitioners*, although the latter term generally covers a wider group of role players than only deployers. The responsibilities of the deployers include the following: (i) to ensure that the AI systems they use or offer meet minimum operational and functional requirements; and (ii) to deploy and implement AI systems that adhere to the relevant ethical principles (European Commission 2019). The *end-users* are those individuals and groups that directly or indirectly engage (use) with the AI systems (e.g. health care providers). The responsibilities and expectations of the end-users include the following: (i) to be aware and informed about the relevant AI systems' operational and functional aspects, and (ii) to be aware and informed about the application of the relevant ethical considerations (European Commission 2019).

Given that society is presently immersed in the fifth wave of innovation, namely the rise of knowledge and information systems (Silva and Serio 2012), it is not surprising that health care knowledge is not left out. Densen (2011) indicates that between 1900 and 1950 medical knowledge doubled once, between 1950 and 1980 it doubled approximately every seven years; and by 2010 every 3.5 years. It is estimated that in the 2020s medical knowledge will double approximately every 73 days (Densen 2011). The result is that knowledge is expanding faster than human capacity to assimilate and apply effectively to patient care. AI holds the ability to assimilate the vast base of knowledge and data, and to align it to potential diagnoses. However, it stands to reason that any technology that is developed to help with this important task should be developed with a clear and thoughtful reflection on ethical principles.

25.3 Ethical Frameworks and Principles

On the one hand, AI systems raise similar ethical issues to that of technology in general. On the other hand, it raises unique ethical issues because AI systems perform activities and functions that were previously only, or almost exclusively, possible for humans. As such, AI systems pose significant challenges to the sense and value of dignity and autonomy amongst all role players (moral agents) in the health care context (e.g. health care managers, clinicians, patients and caregivers). The impact of AI systems in the health care context is further complicated by the increased interaction, especially in the area of decision-making, between AI systems, deployers and end-users (UNESCO 2020). In this section, we specifically aim to raise awareness and to stimulate ethical reflections on the most important ethical principles associated with AI systems in the pediatric health care context.

AI ethics is a sub-field of applied ethics that specifically focuses on the practical perspectives and implications of ethical issues raised by the development, deployment and use of AI systems (Van Belkom 2020). AI ethics principles, reflections and perspectives are primarily concerned about how AI can either advance the good life of individuals, groups, communities and societies in terms of quality of life and the respect for fundamental human rights, or how AI poses challenges, or even undermine, these aspects (European Commission 2019). AI systems should ideally align with a respect for human dignity, the recognition of human rights and freedoms, and awareness of cultural diversity and other ethical principles at all stages of its operation, i.e. from development to its application by end-users (ASILOMAR 2017).

Numerous role players and organizations, from scientists and religious groups to governments and global organizations have already developed and formulated AI ethics guidelines. To name a few: Google's Perspectives on Issues in AI Governance, the ASILOMAR AI Principles of the Future of Life Institute, the European Commission's Ethics Guidelines for Trustworthy AI by the High-Level Expert Group on Artificial Intelligence, and UNESCO's draft text on the Ethics of Artificial Intelligence (Van Belkom 2020; Boucher 2020; ASILOMAR 2017; European Commission 2019; UNESCO 2020).

Most AI ethics principles are derived from three prominent ethical approaches, namely deontology, utilitarianism and virtue ethics. *Deontology* (principle ethics) holds that a principle is always used as a starting point, for example respect for human dignity and non-maleficence. When confronted by an ethical challenge, the solution is to be found in upholding all these principles at all times regardless of the consequences (Van Belkom 2020). The following four deontological principles are described in the rest of this chapter section: respect for human dignity; autonomy (including vulnerability); non-maleficence; and non-discrimination.

In contrast, *utilitarianism* (consequential ethics) holds that the overall consequences of a given action, rather than holding onto a universal principle, determine whether or not a specific solution is 'right'. In other words, the moral value ('rightness') of an action is determined by the extent that it advances the common good and human prosperity (Van Belkom 2020; ASILOMAR 2017). The following two utilitarian principles are described in the rest of this chapter section: beneficence; justice (fairness, equality, equity and inclusiveness).

In the case of *virtue ethics*, the focus is not on certain principles or outcomes, but on the inherent character of the person performing the action. In order to engage in morally 'good' actions, one must act in accordance with certain positive character traits (virtues). In other words, the focus of virtue ethics is not on any individual action, but on the character and intentions of the person involved (Van Belkom 2020). The following five virtues are described in the rest of this chapter section: trustworthiness; transparency (explicability); explainability; accountability (responsibility); and ethical stewardship.

It is noteworthy that in a recent study on AI ethics codes, Van Belkom (2020) found that the following principles are indicated in more than half of all the consulted sources: (1) transparency, (2) justice and fairness, (3) reliability (trustworthiness), (4) accountability (responsibility) and (5) privacy. In the written evidence submissions to the Select Committee on Artificial Intelligence in the House of Lords, United Kingdom (2018), the principle of transparency (explicability) was mentioned most frequently (>600 times), followed by privacy (557), non-maleficence (risk, harm) (522), beneficence (benefit) (495), trustworthiness (trust, reliability) (305), justice (fairness, equality, equity, inclusiveness) (250) and accountability (242). The least frequently mentioned principles were vulnerability (5 times), stewardship (11), human dignity (14), explainability (44) and autonomy (91).

25.3.1   Respect for Human Dignity

Respect for human dignity is the most important ethical principle to guide any reflection, analysis and application of the impacts and interactions of AI systems within the health care context (UNESCO 2020). It refers to the idea that every human being has an equal, inalienable intrinsic moral worth, and should never be regarded as mere objects with instrumental value (European Commission 2019). This intrinsic worth should not depend on any person's gender, age, citizenship, religion, language, political affiliation, legal status, socio-economic position or any other categorization (UNESCO 2020). Human dignity should be respected and upheld at all times; it should not be disregarded, compromised or violated by other persons and groups, or by any form of technology (European Commission 2019; UNESCO 2020).

The research, design, development, implementation and use of AI systems should reflect respect for human dignity (i.e. being human-centric) in its focus to serve and protect humans' physical and mental integrity, quality of life, sense of identity and essential needs (European Commission 2019; UNESCO 2020). This is especially important given the increasing global interconnectedness of all humans as facilitated by various modern technological developments in inter alia communication, health care and education (UNESCO 2020). Fundamentally, a disrespect for human dignity may result in the objectification, exploitation, discrimination or dehumanizing of individuals, groups, communities and societies (UNESCO 2020). So, AI in pediatric medicine should always focus on the health care provider, child, and parent(s) as moral agents worthy of dignity.

25.3.2 Autonomy

Autonomy (self-determination) refers to the recognition of an individual's capacity and moral responsibility to hold viewpoints, to make authentic decisions, to take actions based on personal values and beliefs, to take responsibility for one's actions, and to determine one's moral destiny. In the context of AI systems, autonomy specifically refers to the ability and freedom to take decisions and actions regarding one's interaction and use of the technology. It also includes the ability and freedom to control the right to private life and privacy, as well as the freedom to choose how and whether to delegate decisions to AI systems (UNESCO 2020; European Commission 2019; ASILOMAR 2017).

The end-users of AI systems should be able to maintain full and effective autonomy over their lives and decisions, including their cognitive and social decision-making skills. In addition, end-users should be empowered with knowledge and skills to understand, interact and assess AI systems. In order to uphold this principle, AI systems should not unjustifiably subordinate, coerce, deceive, manipulate or condition them. Instead, AI systems should enable meaningful opportunity for human choices and decisions (European Commission 2019). This is even more important in contexts where end-users are affected by or part of decisions based on automated processing and the decisions have legal implications (European Commission 2019). One solution is to introduce oversight mechanisms (e.g. national and international governance agencies) that exclude/prohibit the use of an AI system in a particular situation. These mechanisms should explicitly establish levels of human discretion during the use of AI systems, or ensure that decisions made by AI systems can be challenged and/or revoked (European Commission 2019). However, the principle of personal autonomy should keep individuals aware and empowered to not solely rely on these oversight mechanisms to the extent that they uncritically accept the decisions and solutions of so-called 'cleared/approved' AI systems in the health care sector.

The rapid developments in the capabilities of AI systems pose significant challenges to end-users to either underestimate or overestimate its decision-making accuracy and solutions. These decision-making developments hold clear threats to human autonomy, especially when end-users (e.g. patients and health care professionals) arrive at a realization that certain AI systems are becoming superior to humans at decision-making. One can then start to entirely rely on the AI systems' decisions, solutions and recommendations (Boucher 2020; Van Belkom 2020). In this regard, end-users should also be vigilant of automation bias, which refers to the human predisposition to increasingly and eventually uncritically favor and accept machine-generated decisions, solutions and recommendations, while ignoring data and human decisions that indicate alternative options (Boucher 2020; Geis et al. 2019). This may result in errors of omission and commission. Omission errors occur when humans fail to notice, or disregard, the incomplete or inaccurate decisions and recommendations of an AI system. Commission errors occur when humans accept or implement an AI system's decisions and recommendations despite clear evidence, knowledge or expert opinions to the contrary (Geis et al. 2019).

25.3.3 Privacy

Privacy refers to the personal freedom from intrusion from others, and to have control over sensitive and intimate personal information, opinions and behaviors. The respect for and the protection of privacy is essential to the respect for human dignity. It is also closely linked to the principle of informed consent for data use in research and the development and use of AI systems (UNESCO 2020).

In the context of health care-related AI systems, privacy is relevant with regards to the right to access, manage and control the data that systems use and generate. This is especially the case during the development of AI systems that rely on large clinical datasets for machine learning processes; larger and detailed datasets increase the accuracy and power of AI systems (ASILOMAR 2017). As such, privacy considerations require inter alia adequate data governance mechanisms (e.g. the strict General Data Protection Regulation that was introduced in the European Union in 2018). Such mechanisms are meant to uphold and protect the quality, integrity and relevance of the data used in AI systems, as well as to ensure that private information is processed by AI system developers in a manner that respects and protects privacy. Ultimately, these mechanisms create a context that allows individuals to trust the data gathering and data processing process, as well as to ensure that the health care data collected and processed by AI systems will not be used for unlawful, unethical or unjust purposes (European Commission 2019; USAID 2019).

25.3.4 Non-maleficence

Non-maleficence refers to the moral obligation to not inflict harm. *Harm* in this regard means the potential for and impact of injury to an individual, family, community or society. The scope of harm is much more that only physical harm; it inter alia includes psychological harm, moral harm, social harm and financial harm. The principle of non-maleficence means that AI systems should neither cause nor worsen any type of harm (European Commission 2019). So, all types of harms and risks associated with the development, deployment and use of AI systems should be duly recognized, avoided and actively mitigated when unavoidable (UNESCO 2020; ASILOMAR 2017; European Commission 2019). This includes clear processes and reporting mechanisms to determine to what extent the data, the AI algorithm, the skills of the end-users or the available resources to manage complex AI systems are responsible or contributed to the harm (Geis et al. 2019; ASILOMAR 2017). This is especially true in pediatric medicine as health care systems deploying/implementing AI need to be cognizant of the vulnerable population they serve. Only then will AI practitioners, health care professionals and patients be in a position to trust that AI systems are safe to use in specific contexts (ASILOMAR 2017).

In the context of non-maleficence, health care professionals have a specific moral obligation towards vulnerable patients who, due to power asymmetries, may experience or are prone to potentially adverse impacts as a result of AI systems' decisions, solutions and recommendations (European Commission 2019). In this regard, one can think of the significant negative clinical impact of false positive/negative cancer diagnoses and the subsequent incorrect clinical care pathways, as well as the psychological and social harms inflicted by personal/family traumatic experiences and life course changes for pediatric patients (European Commission 2019). These kinds of risks for catastrophic or existential harms, must be subject to adequate control measures and mitigation efforts (ASILOMAR 2017).

25.3.5   Non-discrimination

Non-discrimination refers to actions and attitudes that avoid and counteract unjust categorization, intolerance of differences, isolation, exclusion and devaluation based on for example gender, ethnicity, health status, literacy level, socio-economic status and social class. The development, implementation and use of AI systems should be sensitive and responsive to the needs of all, as well as the inclusion and empowerment of different groups in order to avoid potential discrimination and bias, whether explicit or implicit (UNESCO 2020). In addition, AI practitioners should actively strive to respect diversity and inclusiveness at all levels in order to render AI systems applicable and relevant to all persons and groups. This will then align with the moral and legal requirements embedded in international human rights law, standards and principles (UNESCO 2020).

25.3.6   Beneficence

Beneficence refers to the moral obligation and desire to advance the interests and wellbeing of patients, families, communities and society; i.e. to advance the common good. In the health care context it refers to the obligation to act in the best interest of patients, for example to provide relevant information, to initiate appropriate treatment plans, to prevent and/or remove avoidable harm, and to weigh and balance an action's possible good against its cost and possible harms. A primary objective of AI developers and deployers should be to offer beneficial intelligence that, in turn, can be implemented in contexts that benefit as many persons and groups as possible (ASILOMAR 2017). From an ethical perspective, the direct and indirect benefits offered by AI systems should not only be accessible to socio-economically privileged persons, groups and communities but should be shared with and empower many people, groups and communities around the world (ASILOMAR 2017). In other words, the developers and deployers of AI systems should strive to maximize the global benefits of AI systems, inter alia by embedding trustworthy, robust and user-centric AI in their systems, products and services (European Commission 2019; ASILOMAR 2017).

25.3.7 Justice (Fairness, Equality, Equity & Inclusiveness)

The justice principle refers to the moral obligation to fairly develop, deploy and use AI systems. It implies a commitment to ensure an equal and just distribution and access to the benefits and opportunities (i.e. distributive justice), as well as the commitment to ensure that individuals, groups and communities are not unfairly treated due to any form of deception, bias, discrimination and stigmatization. Justice further includes the obligation to establish mechanisms and conditions in which AI users can raise concerns, challenge and seek effective redress against decisions made by AI systems and AI practitioners (European Commission 2019).

Modern-day society is characterized by widespread social inequality and exclusion, which in turn, underpins a key challenge for AI system role players. The challenge is to ensure that AI systems are not inherently biased due to the specific training datasets that were used in its development; so-called algorithmic biases. Such datasets may then inevitably reflect the inherent structural inequalities and exclusions in broader society (Boucher 2020). So, in terms of equity and inclusiveness, all AI role players should strive to minimize and avoid practices and applications that reinforce or perpetuate biases based on unfair categorization, for example racial, ethnic, gender, age and cultural factors (UNESCO 2020). Ultimately, equal respect for the moral worth and dignity of all human beings must include commitments and efforts to ensure equity and inclusiveness. In an AI context, equality is evident in the development and deployment of AI systems that don't generate unfairly biased outputs, i.e. by using datasets to train AI systems that are as inclusive as possible within the context it is implemented. This also requires adequate respect for and the inclusion of training data from potentially vulnerable persons and groups, such as women, persons with disabilities, ethnic minorities, children and other groups/communities at risk of exclusion (European Commission 2019). In addition, the developers of AI systems should provide deployers and end-users with an explicit description of the relevant demographic information of the training datasets in order to be aware and informed of its inclusiveness.

25.3.8 Trustworthiness

In general, *trust* refers to the willingness of one moral agent to rely on the actions and intentions of an exchange partner in whom one has confidence (Moorman et al. 1993). However, trust is not a present-absent condition. In the acceptance of AI technology, trust can range from distrust when there is little or no trust in the capabilities and implementation of AI systems, to calibrated trust when the AI system capabilities and uses appropriately match the expected outcomes and experiences. It may even turn into over-trust when trust exceeds an AI system's capabilities which may then result in misuse, overreliance and later rejection of the AI system and/or technology (Lee and See 2004).

The concept *trustworthy AI* is widely used to refer to the actions and intentions of AI developers and deployers to improve individual flourishing and collective well-being through equality in the distribution of socio-economic, education, health care and other advantages. This is achieved through the development and implementation of AI systems that can contribute to achieving a fair, equal and inclusive society, inter alia by increasing persons' health and wellbeing in ways that foster equality in the distribution of socio-economic opportunities (European Commission 2019).

Trust, whether it be on personal, community or societal level, is essential in all contexts impacted by rapid technological change, i.e. disruptive technologies. Confidence and trust in AI systems' development, deployment and application depend on clear and comprehensive efforts and mechanisms to achieve and support its trustworthiness (European Commission 2019). Many role players, inter alia the developers, deployers and end-users (including clinicians and patients in the health care context) contribute to the overall trustworthiness of AI technology and AI systems (European Commission 2019). On the one hand, the responsible actions and outcomes associated with each of these role players can inspire trust in the use of AI systems, while on the other hand it may infringe on trust when characterized by low accountability and non-transparency (UNESCO 2020).

Trustworthiness includes confidence in the benefits that AI technology and AI systems offer, as well as confidence that adequate measures are taken to identify and mitigate risks (UNESCO 2020). The European Commission's Ethics Guidelines for Trustworthy AI (2019), have identified three aspects that are fundamental to the trustworthiness of AI, namely: (1) it should be lawful (i.e. complying with all applicable laws and regulations); (2) it should be ethical (i.e. in alignment with ethical principles and values as discussed in this chapter); and (3) it should be technically robust. All three aspects are essential to ensure trustworthy AI. However, in reality, there may exist tensions between them, for example when national and international law is not aligned with one or more ethical principles (European Commission 2019).

25.3.9 Transparency (Explicability)

Trust in AI systems is underpinned by the level of transparency and explainability appropriate to the use context (UNESCO 2020; Geis et al. 2019; European Commission 2019). This specific principle is regarded as one of the most important AI ethics principles (Van Belkom 2020; House of Lords 2018; Bostrom and Yudkowsky 2014).

Transparency, sometimes also called explicability, refers to the context where the processes regarding the logic, purpose, reliability and capabilities of AI systems are clear and openly communicated as far as possible and within the limitations of reality. It includes that the decisions and solutions offered by AI systems, even when it proves to be sub-optimal or erroneous, can be adequately explained to affected persons, groups and communities in terms that they understand. However, in some cases, for example deep learning models and algorithms capable of unsupervised learning, it is not readily possible to provide an explanation as to why the AI system has generated a particular output or decision. In these cases, other measures are required to maintain acceptable levels of trustworthiness, for example traceability, auditability and transparency regarding its capabilities. Ultimately, the degree to which transparency (explicability) is important depends on the context in which it is used and the impact of potentially inaccurate and/or erroneous decisions, solutions and predictions on affected individuals, groups and communities (European Commission 2019; Geis et al. 2019; Boucher 2020; Van Belkom 2020).

The increased complexity of deep learning models and algorithms pose significant challenges in terms of its reduced, or even lack of, transparency regarding the full extent of its positive and negative impacts, outcomes and safety, both in the long- and short-term. Not surprisingly, this may result in lower levels of trust in the application and use of AI systems (Boucher 2020; UNESCO 2020; Van Belkom 2020).

Transparency is closely linked to autonomy as it enables persons, groups and communities to make informed decisions and to take appropriate actions regarding the development, deployment and use of AI systems (UNESCO 2020). This includes AI users' right to be informed that they are interacting with an AI system or that an AI system is contributing to decisions and solutions that have a direct impact on them.

25.3.10 Explainability

The concept *explainability* is closely related to transparency. Both principles are linked to AI system outcomes that are comprehensible, traceable, accountable and appropriate in the specific context. Explainability specifically refers to the ability and responsibility to provide information and insights into inputs (e.g. clinical data), process and outputs (i.e. decisions, solutions and predictions) of AI systems (UNESCO 2020; ASILOMAR 2017). As in the case of transparency, AI systems underpinned by deep learning algorithms and models pose a significant challenge to meet the principle of explainability. Such systems are often so complex that it is almost impossible for anyone except AI experts to come to a reasonable level of understanding and insight in how decisions, solutions and predictions are made. This might not be a significant ethical concern when the potential risks and impacts on persons, groups and communities are low. However, as the severity, significance and extent of the potential risks (e.g. incorrect or inaccurate decisions) and impacts increase, so does the ethical obligation for increased levels of explainability (Van Belkom 2020). As such, explainability is closely linked to accountability (Doshi-Velez and Kortz 2017).

25.3.11   Accountability (Responsibility)

Accountability is both a legal and an ethical concept. It refers to the attribution of legal and moral responsibility of AI system oversight. In other words, AI practitioners, whether they are developers, deployers or end-users, should accept responsibility and accountability for the development, deployment and use of AI systems. They should recognize that they are moral agents who are morally accountable for the implementation and use of AI systems in decision-making contexts. As such, AI systems should be treated as technological instruments that can assist humans in decision-making and AI-guided actions, but it should never replace any of the AI role players' responsibilities in terms of legal and ethical obligations (UNESCO 2020; ASILOMAR 2017). In other words, the responsibility and accountability for decisions and actions based on AI system decisions, solutions and predictions should remain attributable to the relevant human role players (UNESCO 2020). Furthermore, all AI role players need to recognize that they have the opportunity and responsibility to contribute to current and future moral discussions regarding the role and position of AI systems in local and global contexts.

25.3.12   Ethical Stewardship

In a virtue ethics framework, *ethical stewardship* refers to the virtue of being consistent with ethical principles, to promote the fair application of those principles in all contexts, as well as to be aware of one's own biases and preferences. The global nature of AI ethics requires of all role players to be vigilant to several ethical and practical challenges in the promotion of justice, diversity, inclusiveness, equality and non-discrimination in AI systems (UNESCO 2020). In actual fact, ethical stewardship calls on all AI role players to actively disclose, report and/or address any form of stereotyping, unfair discrimination and bias in AI systems. It also calls for the promotion of equal access for all to the benefits that AI systems offer in health care and educational contexts, including access for various types of vulnerable and marginalized groups (UNESCO 2020).

In the context of clinical decision-making, ethical stewardship is especially important with regards to a phenomenon known as *automation bias*. The regular and frequent use of AI-based decision support tools that yield high levels of accurate decisions and predictions may gradually result in such high levels of user confidence that the AI results are implicitly and uncritically trusted. Such trust in AI systems may even surpass the trust that health care professionals have in their own clinical interpretations and decisions. The implication is that the role of AI systems to support human decision-making is replaced by a situation in which health care professionals gradually and unwittingly surrender clinical advice and decisions to AI systems. In some cases, this may even result in the reversal of clinically correct decisions to align with an AI system that provides a less correct decision or solution. From an ethical perspective, automation bias undermines end-user autonomy and increases the potential for harm, especially for vulnerable persons and groups. Lastly, it may gradually result in the loss of clinical experience and skills, both of which are essential to effective clinical decision-making (Boucher 2020).

25.4 Discussion

In light of the prominence of AI systems in many sectors of society, it is important to keep in mind that AI technology is not and should not be an end in itself. Rather, it should be regarded as a promising technology that should enhance individual and societal wellbeing, as well as contribute to global development and innovation (European Commission 2019).

Respect for human dignity is widely recognized as the cornerstone of most, if not all, ethical frameworks. However, the operationalization of this abstract principle in the context of AI systems has been met with various technical and contextual challenges and complexities. This situation calls for sound ethical reasoning, reflections, deliberations and dialogues between all AI role players, including developers (e.g. government agencies, corporate businesses and start-up ventures), deployers and end-users, including vulnerable groups, representing various contexts and geographical locations in developed and developing countries. Ethical reflections can lead to a better understanding and guidance in order to identify what *should be* done in various contexts rather than what currently *can be* done with AI technology (Boucher 2020; European Commission 2019).

In the complex world of AI systems and its application in the health care context, it is reasonable to expect that the ethical principles described in this chapter will at times come in conflict with each other. Especially in the clinical decision-making context, these ethical challenges should be approached with ethical reasoning skills and evidence-based reflections rather than intuition or personal convictions (European Commission 2019). This requires the wide-spread development of an AI ethics culture and mind-set through public debate, global bioethics education efforts and individual awareness and understanding (European Commission 2019). Thus, specific communication and education strategies are required to ensure global awareness, literacy and knowledge of the potential roles and impact of AI systems. It almost goes without saying that ethicists need to play a central role in all these efforts and activities (European Commission 2019).

The rapidly increasing application of AI systems, especially with regards to deep learning models, in the health care context poses potentially unique challenges for clinician-patient interactions. Specific human qualities have traditionally been accepted as necessary and fundamental in this interactional space. However, the use of AI systems may have a profound effect on this space when patients increasingly perceive and accept that AI algorithms and models are capable of more clinically "intelligent" activities than clinicians. In addition, clinicians might come to question what it means to be a "clinician" (which phenomenologists would argue creates value discrepancy) and what their role and skills in the patient-clinician interaction in the new AI ecosystem should and could be (Farisco et al. 2020; Geis et al. 2019). The only certainty is that AI systems will continue to impact the health care context and patient-clinician interactions in ways we cannot yet imagine or anticipate. This includes the expectation that it will be accompanied by new ethical challenges (i.e. fundamental, normative and practical challenges) and the development of new or revised AI ethics codes and guidelines (European Commission 2019; Geis et al. 2019; Farisco et al. 2020). Within this context of disruptive technologies, Topol (2019) indicates that AI may actually make health care "human" again by virtue of encouraging human contact between health care professionals and patients. This is in contrast to current contexts in which health care professionals spend significant amounts of time behind computer screens to work through laboratory test reports and radiological images and scans, which in turn leave little time for the human component of empathy and true connection with patients. If deep learning models and algorithms can provide trustworthy diagnostic, predictive, intervention and prognostic options to health care professionals, they may have more time to spend with their patients and to address their uncertainties, fears, hopes and informational needs. Thus, the utilization of trustworthy AI applications can increasingly facilitate health care professionals to "make eye contact" again and to restore the human "care" aspect in health care (Topol 2019).

25.5   Summative Table: Guiding Principles for the Ethical Utilization of Artificial Intelligence in Clinical Health Care

Respect for human dignity
- Recognize that AI systems should be human-centric in its focus to serve and protect patients' physical and mental integrity, quality of life, sense of identity and essential needs
- Avoid the utilization of AI systems in clinical practices that result in the objectification, exploitation, discrimination or dehumanizing of patients and caregivers

Autonomy
- Ensure that patients and caregivers maintain full and effective autonomy over their decision-making contexts and skills
- Empower patients and caregivers with knowledge and skills to understand and interact with AI systems
- Be vigilant of automation bias to uncritically favor and accept machine-generated decisions, solutions and recommendations, while ignoring clinical data and clinical decisions that indicate alternative options

Non-maleficence
- Be aware of the care obligation towards vulnerable patients and caregivers who experience adverse impacts because of AI systems' false positive/negative diagnostic decisions, solutions and recommendations

Non-discrimination
- Respect diversity and inclusiveness at all levels by utilizing AI systems that are applicable and relevant to all persons and groups

Beneficence
- Maximize the benefits of AI systems by embedding trustworthy, robust and user-centric AI in clinical systems and practices

Justice & Fairness
- Avoid clinical practices and AI applications that reinforce or perpetuate biases based on unfair/unjust categorization

Transparency (Explicability) and Explainability
- Be informed about the logic, purpose, reliability and capabilities of AI systems as far as possible and within the limitations of reality
- Inform patients and caregivers when they are interacting with an AI system that contributes to their health care decisions and solutions

Accountability
- Clinicians, patients and caregivers should recognize their moral accountability for the implementation, use and acceptance of AI systems in decision-making contexts

Ethical stewardship
- Be aware that the use of highly accurate AI-based decision support tools can result in automation bias to implicitly and uncritically trust AI systems
- Be aware to not unwittingly surrender clinical advice and decisions to AI systems; it undermines end-user autonomy and result in the loss of clinical decision-making skills
- Develop skills to assess and integrate the ethical application of algorithm-driven decision-making solutions in clinical practice and clinical decisions

25.6   Conclusion

Pediatric medicine is filled with unpredictability and a large component of emotive behavior which makes patient-clinician encounters unique. While AI and deep learning models have brought significant advantages to the clinical context, clinicians have been challenged to develop skills in assessing and integrating the ethical application of these algorithm-driven decision-making solutions in their clinical practice and clinical decisions. In such clinical encounters and contexts there remains a moral obligation on clinicians and other health care professionals to ensure due protection and care of vulnerable patients (children) and the children's parents/caregivers (European Commission 2019). This protection and care should extend to clear knowledge, strategies and actions to empower and enable the patients and their parents/caregivers with skills and knowledge regarding AI technology in general and AI systems in particular.

Ultimately, AI technology in the pediatric context should promote human wellbeing (beneficence), respect human dignity, minimize harm (non-maleficence), respect autonomy and privacy, be appropriately transparent, explainable and trustworthy, and avoid bias and unfair discrimination. Accountability should remain with the developers, deployers and users of AI systems in order to promote justice in the clinical decision-making context (Geis et al. 2019).

References

ASILOMAR. 2017. *AI Principles*. https://futureoflife.org/ai-principles/. Accessed 10 Aug 2020.

Bostrom, N., and E. Yudkowsky. 2014. The ethics of artificial intelligence. In *Cambridge handbook of artifical intelligence*, ed. K. Frankish, W. Ramsey. New York: Cambridge University Press.

Boucher, P. 2020. *Artificial intelligence: How does it work, why does it matter, and what can we do about it?* Brussels: European Parliamentary Research Service. https://www.europarl.europa.eu/thinktank/en/document.html?reference=EPRS_STU(2020)641547. Accessed 30 June 2020.

Densen, P. 2011. Challenges and opportunities facing medical education. *Transactions of the American Clinical and Climatological Association* 122: 48–58.

Doshi-Velez, F., and M. Kortz. 2017. Accountability of AI under the law: The role of explanation. Berkman Klein Center Working Group on Explanation and the Law, Berkman Klein Center for Internet & Society working paper. http://nrs.harvard.edu/urn-3:HUL.InstRepos:34372584. Accessed 18 Oct 2020.

Duan, Y., J.S. Edwards, and Y.K. Dwivedi. 2019. Artificial intelligence for decision-making in the era of big data—Evolution, challenges and research agenda. *International Journal of Information Management* 48: 63–71.

European Commission. 2019. *Ethics guidelines for trustworthy AI*. Brussels: European Commission. https://ec.europa.eu/digital-single-market/en/news/ethics-guidelines-trustworthy-ai. Accessed 8 Apr 2019.

Farisco, M., K. Evers, and A. Salles. 2020. Towards establishing criteria for the ethical analysis of artificial intelligence. *Science and Engineering Ethics*. https://doi.org/10.1007/s11948-020-00238-w. Accessed 25 Sept 2020.

Geis, J.R., A.P. Brady, C.C. Wu, J. Spencer, E. Ranshaert, J.L. Jaremko, S.G. Langer, A.B. Kitts, J. Birch, W.F. Shields, R. van den Hoven, E. van Genderen, J.W. Kotter, T.S. Gichoya, M.B.

Cook, A. Morgan, N.M. Safdar. Tang, and M. Kohli. 2019. Ethics of artificial intelligence in radiology: Summary of the Joint European and North American multisociety statement. *Canadian Association of Radiologists Journal* 70 (4): 329–334.
House of Lords, United Kingdom. 2018. *Select committee on artificial intelligence—Collated written evidence volume.* https://old.parliament.uk/business/committees/committees-a-z/lords-select/ai-committee/publications/. Accessed 27 Sept 2020.
Lee, J.D., and K.A. See. 2004. Trust in automation: Designing for appropriate reliance. *Human Factors* 46 (1): 50–80.
Moorman, C., R. Deshpande, and G. Zaltman. 1993. Factors affecting trust in market research relationships. *Journal of Marketing* 57 (1): 81–101.
Phillips-Wren, G. 2012. AI tools in decision-making support systems: A review. *International Journal of Artificial Intelligence Tools* 21 (2): 1–13.
Silva, G., and L.C. Serio. 2012. The sixth wave of innovation: Are we ready? *RAI Revista De Administracao e Inovacao* 13 (2): 128–134.
Topol, E. 2019. *Deep medicine—How artificial intelligence can make health care human again.* New York: Basic Books.
UNESCO. 2020. *Outcome document: First version of a draft text of a recommendation on the Ethics of Artificial Intelligence (SHS/BIO/AHEG-AI/2020/4 REV).* https://unesdoc.unesco.org/ark:/48223/pf0000373434. Accessed 6 Sept 2020.
USAID. 2019. *Artificial intelligence in global health: Defining a collective path forward.* https://www.usaid.gov/cii/ai-in-global-health. Accessed 22 Sept 2020.

Further Reading

Van Belkom, R. 2020. *AI no longer has a plug.* Rotterdam: The Netherlands Study Centre for Technology Trends. https://detoekomstvanai.nl/wp-content/uploads/2020/06/AInolongerhasaplug_vanBelkom.pdf. Accessed 30 June 2020.
European Commission. 2019. *Ethics guidelines for trustworthy AI.* Brussels: European Commission. https://ec.europa.eu/digital-single-market/en/news/ethics-guidelines-trustworthy-ai. Accessed 8 Apr 2019.
House of Lords, United Kingdom. 2018. *Select committee on artificial intelligence—Collated written evidence volume.* https://old.parliament.uk/business/committees/committees-a-z/lords-select/ai-committee/publications/. Accessed 27 Sept 2020.
Lee, J.D., and K.A. See. 2004. Trust in automation: Designing for appropriate reliance. *Human Factors* 46 (1): 50–80.
Phillips-Wren, G. 2012. AI tools in decision-making support systems: A review. *International Journal of Artificial Intelligence Tools* 21 (2): 1–13.
Topol, E. 2019. *Deep medicine—How artificial intelligence can make health care human again.* New York: Basic Books.

Chapter 26
When Should Society Override Parental Decisions? A Proposed Test to Mediate Refusals of Beneficial Treatments and of Life-Saving Treatments for Children

Allan J. Jacobs

Abstract Health care workers or others may wish to override parental decisions because of their impact on the health or safety of a child or others. Justification of such an action requires two types of principle: an authority principle that designates the process for reversal, and an intervention principle that specifies the grounds for reversal. It is generally accepted that states may overrule parents' decisions for good cause. I argue that the role of the state is to provide sufficient protection against parental malfeasance. Parental malfeasance can be construed as either exposing a child to harm or as insufficient defense of the child's interests. I propose a test to determine what sorts of parental decisions might trigger intervention. I also propose constraints on government action to minimize government unfairness in applying the test. I show how this plays out in application.

Keywords Intervention principle · Best interest · Harm principle · Sufficientarianism · Modus vivendi · Political realism

26.1 Introduction

Consider the paradigmatic situation in which Galen, a physician[1] and Eve, a parent disagree on treatment[2] of Seth, Eve's child. They disagree on whether Seth should

---

[1] For reasons of clarity and style I will use the words 'physician' and 'doctor' to encompass all health care professionals. Similarly, a 'parent,' in this paper, is a legal guardian serving as default surrogate decision-maker for a child.

[2] What I say here about medical treatment may apply to non-medical situations, such as maintaining unsanitary home conditions.

A. J. Jacobs (✉)
Coney Island Hospital, Brooklyn, NY, USA
e-mail: Allan.jacobs@nychhc.org

SUNY Downstate School of Medicine, Brooklyn, NY, USA

© Springer Nature Switzerland AG 2022   421
N. Nortjé and J. C. Bester (eds.), *Pediatric Ethics: Theory and Practice*,
The International Library of Bioethics 89,
https://doi.org/10.1007/978-3-030-86182-7_26

visit Galen every 3 or every 6 months to monitor a stable condition. This might be resolved by (1) Galen's accommodation to Eve; (2) parental education for Eve; (3) providing more convenient services such as a clinic closer to home; (4) Eve finding another physician who will agree to see Seth every 6 months; or (5) ideally, discussion leading to mutual understanding. Galen is unlikely to seek state intervention to force the 3-month regimen. But what if Seth had newly diagnosed acute lymphoblastic leukemia, which chemotherapy cures in about 90% of children. Eve, however, wants to use only dietary modification and prayer. Galen wants to sue to compel chemotherapy and may well prevail (refer to Chap. 11). Most situations are neither as trivial as the first case or as momentous and clear-cut as the second.

This chapter asks when, and how, governments[3] should resolve impasses between physicians and parents over health-related treatments.

26.2   Conceptual Clarification/Definitions

Parents are the default decision-makers for their minor children. Their decisions ordinarily should enhance children's interests. The ethics of parental decisions has been much debated and is discussed throughout this book. However, once the power of the state is invoked, political as well as interpersonal issues are in play, invoking considerations of political theory. Also, legal realities constrain solutions to real-life questions, whether or not they affect theoretical analysis. Thus, law restricts advice given by clinical ethicists in an institutional setting. State policy is determined by three goals. First is promotion of the health, safety, and welfare of the population, termed *police power* in American jurisprudence (*Home Building & Loan Association v. Blaisdell*, 290 U.S. 398, 437 (United States Supreme Court, 1934)). Second, is protection of vulnerable persons against injury from other parties, including their parents. Anglo-American law calls this the *parens patriae* doctrine (*Prince v. Massachusetts*. 1944. 321 U.S. 158, United States Supreme Court (1944)). Finally, there is a state interest in general stability and prosperity. This interest requires healthy citizens who are productive, and not economically dependent on the state (Szreter and Woolcock 2004; Black 2008). Unlike the police power and *parens patriae* doctrines, this state interest is not explicitly constitutionalized.

26.3   The Role of the State

Ethical analysis of what the state should allow and prohibit cannot result in a satisfactory principled solution. Value pluralism demonstrates that many desirable values accepted in liberal societies are in conflict. There are tradeoffs between justice

---

[3] States in this essay are entities with sovereign authority over a geographically defined area; governments are organizations (or their agents) that legitimately exercise state power.

and mercy, safety and freedom, liberty and equality, etc. Theoretical approaches to this problem cannot yield a principled consensus on how to prioritize conflicting values on which people agree, let alone providing a means to validate controversial values approaches (Bellamy 1999). Furthermore, acceptable resolution of these value conflicts may vary with circumstance. People may assign safety priority over freedom in times of increased ambient danger, for example. People will always disagree regarding political principles, and "we do not have an uncontroversial method to assess which [moral] claims are 'true'" (Waldron 1999, 164). Second, liberal states confront the dilemma of whether to tolerate illiberal practices by domestic groups. If these are tolerated, then some citizens are being denied the benefits of liberalism. But a state that suppresses such groups is itself acting illiberally. Governments must navigate this paradox. Rawls's (1971) view of justice as consisting of political liberty with economic equality or prioritarianism, all enforced by a powerful state has not resolved these dilemmas to general satisfaction.

In any event, John Rawls was depicting an ideal political entity, but the world is not ideal, and humans may be incapable of being good Rawlsian leaders and citizens. Governments and those who administer them are imperfectly ethical (Shklar 1984). Governments do have the capacity, though, to enforce their will through coercive means that extend to violence. Government work may attract people with narcissistic and antisocial tendencies. Those who desire power are attracted to careers that provide power, just as those who desire wealth are attracted to commerce, or those concerned with health to medicine. Even if government officials are, on the whole, no worse than the rest of us, they still possess the same human flaws—hypocrisy, snobbery. betrayal—and especially cruelty—as we do. They inevitably have a biased perspective; furthermore, their actions may be arbitrary or capricious. Judith Shklar's insights sculpt government as inevitably "cruel", which grounds her "liberalism of fear" (Id., 238). One cure for government cruelty is to restrain the scope of state interference in private lives. The idea of limiting government to protect its citizens has more traction in the Anglophone world than in European nations influenced by the traditional French view of the state as the guarantor of individual prerogatives. Observers of modern history and of current events can judge for themselves whether powerful governments have safeguarded individuals and promoted flourishing better than limited governments.

Voters tend to be instinctive consequentialists, most concerned about whether their government satisfies Thucydides' classic triad of fear, honor, and interest. Citizens readily turn to other leaders if they are not satisfied. Politicians therefore are likely to accommodate popular opinion. A desire to achieve successful policies and to remain in office is not corrupt, nor necessarily distasteful. However, actual politics is closer to a marketplace of interests than to a dialog of ideas. Issues are negotiated to reach a modus vivendi[4] based on the relative strength of factions within the state

---

[4] A *modus vivendi* has the following elements. It (1) is an expedient resolution of differences, (2) negotiated under the aegis of a legitimate governing body, which resolution (3) reflects the importance of parties' interests to the parties themselves, (4) as well as the relative strength of the various parties. A *modus vivendi* agreement is prudential and is not based on moral concerns.

and government, and on importance of various goals to these factions. These ideas been termed *political realism* (Philp 2010). Modus vivendi need not be the sort of naked coercion exemplified by two wolves and a lamb voting on the dinner menu. Berlin, Shklar, Montesquieu, Madison, Galston and other liberal political realists have advocated institutional limitation on the scope and power of government. This includes limiting the ability of any political modus vivendi process to undermine the dignity of individuals.

Western liberals disagree both with Aristotle that children are parental property and with the Platonic notion that children should be raised by the state. It is generally accepted that most children will live in a family with one or more parents, who ultimately are responsible for family decisions. Arrangements to the contrary such as Israeli kibbutzim and British boarding schools for young children are unusual, and generally impermanent. Lainie Friedman Ross (1998) and Ana Iltis (2010) have written sophisticated expositions of interactions among parents and children within family units. They emphasize the child situated as a member of a family all of whose members, including other children, have needs. Ross relies on Allen Buchanan and Dan Brock's rationale to justify primary parental authority over minor children. First, parents best understand their children's interests, and are best positioned to make decisions that will serve their children's welfare. Second, since parents bear many of the consequences of choices made for their children, they should have some control over these choices. Third, there is a prima facie parental right to raise children according to the parents' own values. Fourth, the family itself is a valuable social institution. Mark Navin and Adam Wasserman (2017) points out other advantages in the health care context. First, parents can advocate for their children when their care is suboptimal. Second, parents are more likely to comply with treatment plans if they are involved with and agree with the plan; much long-term treatment requires parental action. Second, parental assertiveness sets an example for children to develop autonomous behavior as they mature. I add two additional justifications for parental decisional authority. First, the chance to perpetuate one's physical existence and interests, including one's values, is a prime motivation for having children A system that allows wide discretion to parents will make parenthood more rewarding than will a system that regards parents as mere caretakers of their children acting on behalf of the state, under the state's close supervision. Engaged parents are likely to be better parents. Reproduction is essential to society. Second, the vast number of quotidian parental decisions parents make cannot be supervised without undue intrusion. Iltis hypothesizes (as a *reductio ad absurdum*) home visits by the food police and the book police to ensure that children receive nutritious diets and exposure to educational books.

The parent-child relationship is in a class by itself; there are no analogous relationships. Although McCullogh (2010), among others, characterizes parents as fiduciaries, and as co-fiduciaries with physicians regarding the health of their children, I believe that this is inaccurate. Certainly, parents' obligations, created through procreation, are far greater than those found in an ordinary arm's-length interpersonal relationship. Parents are not, however, true fiduciaries. A fiduciary relationship is a limited one, within whose bounds the fiduciary must prioritize the interests of the beneficiary. The relationship between parents and children, however, spans the whole of the child's experience. Fiduciaries are compensated through a fee schedule; parents are largely responsible financially as well as physically for their children and expend a great deal of money on raising their children. Furthermore, a fiduciary, unlike a parent, may not be an active participant in matters involving conflicts of interests between the beneficiary and any third parties (See, *Cinerama v. Technicolor, Inc.*, 663 A. 2d 1156 (Del. 1995)). Such conflicts are always present in family situations. Parents not only must balance the interests of all their children but may prioritize their own interests at times (Ross 1998). Other unique characteristics of the parenting relationship include the love between parent and child; permanency of the relationship; children's long-term dependency; resources required to raise a child; closeness induced by frequent physical contact, etc. Parents make multiple quotidian and momentous decisions involving their children. Parents' beliefs and habits inevitably will suffuse these decisions. Parents interact with and understand the totality of their children's environment and their needs, as well as reconciling each child's interests with those of other family members—including other children in the family. Physicians, on the other hand tend to focus on maintaining the health of the child who is their patient, the greatest concern usually being physical health. The fact that some parents do an inadequate job of parenting is not sufficient justification for giving states broad license to regulate parenting to its own advantage.

This view is not ethical relativism. Theorists such as Shklar and Sen correctly point out that it is easier to recognize gross injustice than to specifically characterize justice. Thus, pluralism is not relativism. "[P]luralism lies in the inability to *defend* any position categorically; whereas relativism [is] the inability to *condemn* any position categorically." (Torcello 2011, 89; emphasis in original). Piercing an infant's ears seems trivial to most people but binding her feet or selling her into slavery is unconscionable.

One need not be a libertarian or a conservative (and most political realists are not) to embrace the idea that government should be circumspect in interfering with the family. (Ross 1998). Primacy of the family has scholarly opposition, though. Critical studies scholars have viewed families as incorporating unjust relationships and perpetuating unjust views (Fineman 2019). Some such scholars call for sweeping state-driven revision of family structure or demand intense state scrutiny of parental behavior. They regard parents as merely stewards of the child, accountable to the state for their performance. Dwyer (1994) stated that parents have no rights other than those necessary to their performance of parental *duties*. Dwyer (2009) explicitly regards parentage as a relationship created by the state. This is an extreme position, a polar opposite from the notion that children belong to parents as quasi-property. However, it has been propounded by many scholars, if not as starkly as Dwyer states it. These include Samantha Brennan, Colin MacLeod, Harry Brighouse, Adam Swift, and Martha Fineman.

These views fail to recognize that even though families may be a source of injustice, it does not follow that plenary government regulation of familial relationships will grant children either greater justice or more personal flourishing. Some governments have used children as informers against parents and have removed children from homes for political reasons. Even democratic Anglophone nations systematically removed aboriginal children from their families well into the 20th Century, ostensibly for the children's' benefit. Remaking families according to state policy is hardly liberal; admittedly, critical studies scholars generally eschew liberalism. In any event, I see no reason to believe that governments can devise comprehensive mechanisms to that will successfully coerce families to behave in a manner that is just by their lights.

Legal systems must tailor legal rights to balance conflicting interests of parents and children. European rights law is largely based on the European Convention on Human Rights (ECHR), which is enforced by the European Court of European Rights (ECoHR) and supersedes their laws of 47 member nations of the Council of Europe. (COE). The ECHR guarantees more comprehensive rights than analogous American Constitutional law as stated in Amendments 1–10 and 14. The United States Supreme Court has constitutionalized extensive parental rights to raise their children as they see fit, even where parents' values conflict with social norms and values (*Reno v. Flores*, 507 U.S. 292, pp. 303–304). European law is generally less protective of parents' prerogatives rights than American law. It is likely, for example, that the parents in the Charlie Gard case would have prevailed in an American court (Paris et al. 2017). Europe is less deferential to parents, generally following the United Nations Convention on the Rights of the Child. The United Kingdom's children's welfare, or paramountcy, doctrine, is a virtually unlimited view of the child's claims against parents and others: "When a court determines any question with respect to the upbringing of a child…the child's welfare shall be the court's paramount consideration" (UK Children's Act, 1989, Sect. 1). This is modified by the ECHR guarantees of everyone's "right to respect for his private and family life" (Art. 8. §1).

Besides the parent and the child, parental decisions may affect third parties, including other children in the family, unrelated citizens (e.g., when an unvaccinated child transmits a serious communicable disease) or the state itself (e.g., by failing to raise children as productive citizens). State police power and the ethical principle of justice allow the state some power to regulate citizens' behavior for the benefit of the general welfare and the welfare of its citizens. In the United States this power is based on common law, as protected by the 10th Amendment. Europe incorporates this power into rights documents as explicit limits on liberty. For example, the ECHR right to domestic privacy (Art. 8, §1) is constrained in §2:

> There shall be no interference by a public authority with the exercise of this right *except such as is in accordance with the law and is necessary in a democratic society in the interests of national security, public safety or the economic wellbeing of the country, for the prevention of disorder or crime, for the protection of health or morals, or for the protection of the rights and freedoms of others* (Emphasis supplied).

Such language is a basis for undermining parental prerogative. This has occurred in cases such as *Wunderlich v. Germany* (2019; ECoHr no. 18925/15) removing children from a home in which a family insisted on home schooling them.

It is desirable to maximize transparency and consistency when imposing limitations on individual prerogatives. It is desirable to minimize arbitrariness and inconsistency when adopting rules and making decisions. This ensures predictability of legal processes, maximizes fairness, and furthers state legitimacy. Rules should permit minority cultural practices which compare in harm and benefit to permissible majority practices. A legal regime that allows children's competitive lacrosse while banning ice hockey should be suspect.

With this background, I will propose a test that could serve as a modus vivendi for resolving state-parent conflict in the health area and in other arenas. I expect that it would be widely acceptable, as it (1) is compatible with much current law; (2) minimizes externalities that harm parties outside the affected families; and (3) reduces inconsistencies.

26.3.1 Ethical Considerations

Buchanan and Brock (1989) laid out several categories of principle that have to be addressed in determining who is a surrogate for individuals lacking decisional competence, and how those decisions are to be made. Their *authority principle*— who makes decisions—defaults to parents, with the possibility of state intervention if parents do not perform this task adequately. *Guidance principles* and *intervention principles* are detailed in Chaps. 4 and 7. I have suggested that violation of a guidance principle does not automatically justify government intervention. I will suggest when government may supersede the *authority* of the family and inquire when such *intervention* is acceptable.

Intervention standards should, as much as possible, incorporate three features: determinativeness, ethical coherency, and compatibility with constitutional and social norms. *Determinative* rules provide clear guidance and limits for state decision-makers. They allow for a high degree of consistency and predictability in decision-making. Analogous cases should lead to similar outcomes. Admittedly, borderline cases, individual differences in judging, and new problems will prevent complete elimination of indeterminacy. The opposite of a determinative test is a *conclusory* test—where judicial or administrative outcome define the supposedly determinative terms, rather than the reverse. An example of a conclusory rule is a Connecticut statute supposedly defining the "best interest" of a child as the standard for assigning custody. The statute directs the judge to balance 16 different factors, and instructs the court that it "is not required to assign any weight to any of the factors that it considers" (*Connecticut General Statutes* § 46b—56(c). Obviously, 'best interest' here is what the judge says it is; the judge's conclusion, rather than the statute, actually defines "best interest," subject only to the forgiving standard of judicial discretion.

Second, the interventional rule must be *ethically coherent*, meaning that a test should arrive at results compatible with a local ethical consensus, if one exists. In a diverse community such consensus requires an approach that is compatible with a broad range of metaethical systems. Even a rule derived through modus vivendi cannot seem grossly immoral. Many consequentialists, secular deontologists, care-based ethicists, and practitioners of most religious systems common in the West should feel comfortable using the test that I will propose. In the Western world, a modus vivendi will prioritize children's needs while also addressing the needs and interests of all relevant parties (United Nations 2013; Jacobs and Arora 2015). Ethical coherence will not allow child welfare to be used as a pretext for discrimination against minorities (Jacobs and Arora 2015). The costs of government coercion should be proportionate to the costs of targeted parental behavior.

Finally, the approach must be *compatible with the constitutional and social norms* of the state and nation. Even decisions that are reasonable and ethical are toothless without the force of law. On the other hand, strongly held mores and values (such as disapproval of marketing horse meat or of uncovered genitalia in public places) may be codified even with little justification in ethics or expediency.

The two principal approaches to formulating interventional principles regarding children, the *best interest standard* (BIS) and the *harm principle* (HP), have been discussed at length earlier in Chap. 4. Both approaches minimize the interests of the broader community. Neither approach takes explicit account of the needs and interests of other members of the family. Erica Salter (2012) emphasizes that a child's best interest cannot be determined objectively. Though physicians tend to focus on a child's immediate health interest, children also have economic, psychological, and spiritual interests. Their interests may be present- or future-directed, so that a current harm can foreseeably bring about future gain. Interests may be self- or other-directed. A child may wish to sacrifice direct self-interest to benefit someone she loves, as by being an organ donor for a sibling.

Two of the constraints on state action in Douglas Diekema's elaboration of HP are need for immediacy of harm, and absence of less intrusive option. These, especially the first, seem unreasonably strict. If this iteration of HP demands too little of parents, BIS (construed literally) demands too much. Both HP and BIS incorporate sufficient indeterminacy to allow different outcomes in identical cases (Birchley 2016). As long as humans with different experiences and biases are interpreting laws and standards, some indeterminacy is inevitable, but I believe it can be reduced below what BIS and HP provide.[5]

No legal standard is absolute. BIS is the stated American standard for assigning parental care, but terminating parental custody generally requires parental neglect or abuse. Furthermore, a series of Supreme Court cases guarantee broad parental prerogatives based on Constitutional interpretation. The Charter of Fundamental Rights of the European Union (2009) supersedes the 27 EU nations' laws. Article 24 states the importance of children's interests without assigning it trump status:

> *Children shall have the right to such protection and care as is necessary for their wellbeing.*
> *… age and maturity…In all actions relating to children, whether taken by public authorities*

---

[5] Ross (1998) and Iltis (2010), among others, have proposed standards intermediate between BIS and HP. They would require that parents provide an upbringing that allows their children to become autonomous adults, and their trigger for intervention is set accordingly. Space does not allow a full discussion of their work. Some feminist legal scholars have expressed parallel ideas in legal terms. Ross' term, *constrained parental autonomy*, has received a lot of traction as allowing parents to balance interests of all family members provided that decisions do not harm a child.

or private institutions, the child's best interest must be a primary consideration. (Emphasis supplied).[6]

Policies and guidelines proffered by medical organizations tend to reflect the viewpoint of local physicians as passed through an ethical filter. The most recent AAP statement on consent bases the limits of parental authority chiefly on HP and constrained parental authority, though it acknowledges BIS as a guidance principle (Katz et al. 2016). The statement explicitly subordinates children's religious and cultural interests to their physical wellbeing. A policy recommendation adopted by several organizations of critical care physicians recommends that doctors refuse to provide treatments that "have no chance of achieving the intended physiologic goal" (Hayes et al. 2015, p. 1696). This statement proposes that requests for treatment in which the burden on the patient is greater than anticipated benefit should be settled, if possible, through dialog. If that is unsuccessful, the discussion can be escalated to hospital committees and, ultimately, referred to the courts.

My proposal is somewhat more deferential to children than American standards and less so than European. It also abandons both BIS and HP.

26.3.2 A Sufficientarian Approach

Feinberg (1984, 33) defines harm as "thwarting, setting back, or defeating an interest." This definition characterizes interests and harms as two sides of the same coin. It is harmful to limit realization of an interest, while any harm attacks some interest. The doctrine of *sufficientarianism* provides a way to finesse the difference between HP and BIS, and also to place the issue of intervention principles in the arena of political modus vivendi. *Sufficientarianism* holds that a just distribution requires *only* that everyone possesses the sufficient amount or degree of any relevant resource, termed a *currency* (Frankfurt 1987).[7] Currencies are not fungible, so that if there is more than one currency, there must be sufficient distribution of each currency. A surplus of education does not overcome insufficiency of health care resources, and surplus of cancer therapy does not overcome insufficiency of immunization services. Currencies need not be restricted to material or tangible goods. The amount or quantity of a currency adequate to define sufficiency is determined politically; there is no objective way to do this.

Parental malfeasance need not be intentional; it can be negligent, or due to erroneous parental perceptions. For reasons apparent from the previous paragraph, it is

---

[6] Both the Treaty of Lisbon and the ECHC are binding on the 27 EU nations, all of which also belong to the Council of Europe and are subject to the ECoHR. It is theoretically possible that the ECoHR and the EU's European Court of Justice could issue conflicting opinions. This does not seem to have happened (ECoHR 2011).

[7] *Pluralistic sufficientarianism*, allows other considerations to apply above the sufficiency level, but is not germane to the present argument.

not necessary to label parental malfeasance either as a harm or as a failure to serve interests. Therefore, we do not need to decide between BIS and HP.

I propose that *protection against parental malfeasance* be regarded as such a currency. We now can define as a political modus vivendi a sufficient degree of protection against parental malfeasance.

I now have proposed sufficient theoretical framework to be able to analyze whether government action against parents is appropriate. I will state this in the form of a test that I believe is reasonably determinative, ethically coherent, and compatible with the mores of a liberal democratic state. The test can inform both ethical and legal decision-making. It grounds decisions in a political context, allows the state to meet society's needs, and limits unfair discrimination.

26.4   The Test

This two-pronged test to serve as an intervention principle to determine when states may override parental decisions. It modifies previous versions of this test (Jacobs 2013; Jacobs and Arora 2015; Levin et al. 2016; Jacobs and Arora 2018):

> *A state may limit a parental decision that jeopardizes a child's health or safety if either of the following two bases in the first prong exists, but only if none of the constraints in the second prong applies.*
>
> **First prong: bases for limiting parental decisions.** *It is warranted for a state to overturn a parental decision if the decision's consequences are likely to create unreasonable burdens either for:*
>
> **Basis 1: Direct Effects Basis.** *Children to whom the decision applies, by: creating a substantial chance of death or of major tangible harm, such as malnutrition or major psychological morbidity. (In situations where both a disease and a proposed treatment carry such risk of harmful effects, intervention is permissible if seek care whose risk/benefit ratio greatly exceeds the risk/benefit ratio expected from the management choice proposed by* health care *professionals.)*
>
> **Basis 2: Indirect Effects Basis.** *Society as a whole or members of society beside the involved child.*
>
> **Second prong: constraints on government action.** *Notwithstanding the bases in the First Prong, a state should restrict parental action only if none of the following constraints apply:*
>
> **First constraint: Likelihood of Effect Constraint.** *The putative harm underlying the restriction must be actual, rather than hypothetical; and likely, rather than rare;*
>
> **Second constraint: Comparability of Effect Constraint.** *The harm that the restriction seeks to avoid must be of a magnitude greater than harms typically tolerated for comparable practices;*
>
> **Third constraint: Benefit/Harm Constraint.** *Benefits of the restriction to all parties concerned should foreseeably exceed overall harms.*

26.5   The Test as Applied

Government intervention should be a last resort and should be sought only when attempts at counseling and mediation have failed. Physicians and clinical ethicists must try to understand why parents refuse medical treatment. Refusal may involve expense or needed parental commitment of the treatment itself or parental inability or unwillingness to participate in expensive, difficult, or time-consuming care of a chronic condition. Various authors have identified other factors behind parental refusal. These included (1) the impact of the treatment on the child's quality of life during and after treatment; (2) parental powerlessness combined with suspicion of the health care system and motives of those treating the children; (3) idiosyncratic or religiously-motivated belief in non-standard treatment; (4) poor insight into the severity of the disease and (5) immersion in a culture in which traditional medicine is normative, and Western medicine alternative. Clearly, only some of these factors are amenable to resolution through dialog. However, litigation does not necessarily secure the care that physicians seek. The family may defy the court, hide, or leave the state.

Sometimes, though, physician may desire to seek judicial intervention to resolve an impasse. This test can be used as a guideline for physicians seeking intervention, and by courts for deciding whether to intervene.

What are unreasonable burdens for society and its members? An obvious application involves communicable diseases. It is unreasonable to bring a child with measles to a theme park, or to send a child with head lice to school. Another paradigm involves parents that demand that society commit extensive resources to their children with no chance of therapeutic benefit. Still a third is parental tolerance of aggressive behavior that may cross the line into bullying; or risky behavior that may pressure other children to do the same. Fourth, the boundaries of acceptable behavior vary from place to place. Recently, in Akko, Israel, I asked a resident about a group of boys who were jumping off a stone seawall about 10 m into the water. He replied, "You're not an Akko man if you don't jump off the wall when you are a boy." In contrast, most Western parents and governments would not allow their children to jump off high seawalls. Fourth, state needs will guide states in requiring parents to cooperate in health and fitness programs. States with a requirement for a comprehensive selective service may emphasize fitness, for example.

The Direct Effects Basis will be more uniform across localities, as there is likely to be relative agreement regarding the criteria for "major disruption of a physiological function, or ... other objectively severe harmful effects." Consider clashes over life-saving treatment. This can take the form of parents refusing treatment proposed by physicians, as when Jehovah's Witness parents refuse to authorize a transfusion for their children. It can also take the form of parents demanding treatment that physicians deem inappropriate, as in the Charley Gard dispute. For the mature child, the British approach seems reasonable. A child demonstrating sufficient maturity and intelligence may consent to a procedure (*Gillick v West Norfolk and Wisbech* AHA. AC 112 ((HL)) 1986). If such a child refuses medical advice, though, the parents may overrule the child if the consequences of refusal are sufficiently grave (Griffith 2016). I would use the language of the Direct Effects Basis to define the intervention threshold. If refusal or demands by a parent or mature child meets the criteria of the Direct Effects basis, physicians can go to court on behalf of the child, seeking protection against parental malfeasance or reevaluation of the child's competence. Also, besides being parental malfeasance, idiosyncratic parental treatment refusal that transgresses Direct Effect Basis standards may be evidence of parental incapacity.

If the child is not mature, the parent's wishes do not prevail if the Direct Effects Basis is breached. If parents refuse blood transfusions for hemorrhage, appendectomy for possible ruptured appendix, or antibiotics for tibial osteomyelitis, then physicians should appeal to courts as a last resort,[8] and courts ordinarily should order treatment. If potential harm to the child does not rise to the level stipulated by the Direct Effects Basis, then parents should have the authority to refuse a medical suggestion. A situation sufficient to invoke intervention need not be immediate or urgent. If parents do not take steps to control a child's type 2 diabetes mellitus through appropriate dietary, exercise, and pharmaceutical measures, then physicians may ethically seek intervention by state agencies, whether social service or, ultimately, the courts. That is not to say that the courts will always agree with physicians seeking intervention.

Parental demands for inappropriate action follow the same principles. A dramatic form of this is caregiver-fabricated illness in the child (Flaherty et al. 2013). Physicians should not accede to providing risky treatments lacking benefit. Furthermore, if repeated parental calls for care are likely to cause severe harmful effects—even psychological—it may be warranted for physicians to seek intervention from public agencies. This might be the case if a delusional or obsessive parent is persistently convinced or concerned in the face of evidence to the contrary that her child is being subjected to physical or sexual abuse.

Now consider a different sort of decision. Consider a situation in which physicians propose a treatment that is difficult to tolerate and that has a high rate of serious complications for a child with an illness that is likely to be fatal or disabling. Different specialties currently employ different survival thresholds for requesting court intervention (Gerdes and Lantos 2020). Cardiac surgeons will accept refusal if the expected survival is <90%, while neonatologists use 25–50% as a survival floor. Neither the rigor of the treatment as experienced by the child nor long-term quality of life issues account for these differences. Criteria for seeking and granting intervention thus are not determinative. This would be alleviated by uniform application of the test as an intervention rule, with greater overall consensus on the benefit/risk calculus that warrants intervention. Perhaps neonatologists would have to defer to parents more than at present, while the reverse might be true of cardiac surgeons. Clinical ethicists are in an excellent position to elucidate the facts and interpret them using this or any other test that provides reasonable specificity.

---

[8] The consensus of the literature on the subject recommend ethics consultation with full exploration of reasons for parental refusal of physician recommendation. A shared decision is preferable to a coerced solution.

Parents and physicians may attempt to game disputes to avoid an unfavorable outcome. Parents may resort to doctor-shopping if they do not like the opinion they receive. Physicians may try to prevent this, by using claims of medical child abuse to invoke court action (Eichner 2016).

Competent persons are free to ignore the wishes of professionals or to discharge them. This is true if professionals claim fiduciary status, including physicians. The doctor may be wrong. The patient may lose confidence because she is not getting better, or because there was a complication. There may be a personality incompatibility. The patient may also choose inappropriate, but easy treatment, such as enema therapy for a cancer that is curable with surgery. I propose that parents should be free to change children's doctors under circumstances in which a reasonable adult might change doctors. This is a stricter standard for parents than a competence standard; it would allow transfer of care when a child is not getting better, on the basis of a reasonable second opinion that appeals to the parent, or even if the parent dislikes the doctor. It does not permit inappropriate care. Furthermore, if the parent resorts to visits to many doctors as an excuse to avoid treatment, that might be tantamount to refusal of reasonable care.

I now turn to the three constraints. The Likelihood constraint precludes states from interfering with a procedure because of hypothetical or rare considerations. It is a check on application of an extreme form of the precautionary principle. This constraint assigns the state the burden to prove that the practice is dangerous, rather than on the parent to prove that it is safe. The relative safety of common vaccinations or surgical correction of cleft lip prevent the fact of unusual complications from being used to argue against employing these procedures. This constraint applies to disclosure as well as to treatment. The rebuttable presumption should be that information is disclosed to children. The notion that disclosure of negative information causes undue distress is often hypothetical. This may even apply to disclosure of genetic information to sufficiently mature children. (McGowan et al. 2018). "Rarity" in this test should be titrated to the severity of the consequences. Thus, a 0.1% chance of hospitalization may be considered rare, while a 0.1% chance of death is not.

The Comparability Constraint prohibits pretextual use of health decisions to act against unpopular practices. It is based on a United States Supreme Court decision (*Church of the Lukumi Babalu Aye, Inc. v. City of Hialeah,* 508 U.S. 520 (1993)) overturning laws that prohibited animal sacrifice, while allowing other killing of animals. Majorities should not impose restrictions on those with different cultural practices and beliefs that they would not impose on themselves. This constraint will be a factor in practices associated with minority cultures, such as circumcision of minors (Jacobs and Arora 2015).

Finally, the Harm/Benefit constraint requires that action against a practice acknowledged to be bad does not foreseeably make matters worse. Punitive measures require resources to institute and enforce. They harm not only to those punished, but to their families, including to the children affected by the act for which punishment is inflicted. If a government punishes unpopular parental practices resulting in minor harms, this is likely to create greater harm than the parents inflict. These concerns are insufficient to override or bypass parents who would preclude their minor children from obtaining contraception, pregnancy termination, and treatment of sexually transmitted infections. However, parental authority could not be overridden when the issues at stake are minor.

26.6 Conclusion

This chapter is not about how parents should treat their children; it is about when states should restrict parents who are performing this role inadequately. The intervention standard is not parental perfection. Rather, there are limits to the permissible limits of state action, since there are possible harms from state intrusion. On the other hand, parties other than the child have interests which states should consider. These interests may warrant state interference with parental choice. The principles expounded in this chapter are summarized in Table 26.1.

I propose a test based on democratic political process informed by liberal pluralism and sufficientarianism, and constrained by constitutional guarantees. Constraints on state action are needed to protect parents and minority groups against bias, malice, and overreach, as well as to protect children. The test's threshold for state action is broader than the harm principle would allow, but not as intrusive as a literal construction of the best interest standard, taken as a guidance principle. The test is susceptible application both my clinical ethicists and by courts. It embodies safeguards against

Table 26.1 Guiding principles for resolving conflicts between parents and society

Status of parents vis-à-vis the state
Families and parent-child relationship are more fundamental than the state
The state should not reverse parental decisions or punish parents, except in cases of gross malfeasance, such as abuse or neglect
Coercive measures should be considered a last resort, to be employed only if lesser measures, such as persuasion, education, or nudging, fail to avoid malfeasance
Coercive measures are justified if a parental decision or action creates unreasonable burdens for either
A child to whom the decision applies, by creating a substantial chance of death or of major tangible harm
Society as a whole, or members of society other than the affected child
Notwithstanding the burdens of parental actions, states may take coercive action against parents only if none of these constraints applies
The harm must be actual, rather than hypothetical; and likely, rather than rare
The magnitude of the harm must be greater than harms typically tolerated for comparable mainstream practices
Benefits of the restriction to all parties concerned should foreseeably exceed overall harms

using child welfare as a pretext for unfair discrimination. Without completely eliminating indeterminacy, the test would improve consistency in decision-making. It offers a fair and consistent basis on which to balance parental and minority rights with governmental obligations to protect the health of children and the welfare of society.

Acknowledgements I am grateful to Drs. Nico Nortjé and Johan Bester for inviting this contribution, and for their valuable comments. Much of the work in this chapter was performed in collaboration with Dr. Kavita Shah Arora and Hillel Y. Levin, as cited. Finally, I would like to express my gratitude to Dr. Pamela Ravin Jacobs, my intellectual partner, life partner, sounding board, and inspiration.

The opinions herein are my own, and do not necessarily reflect the opinions of the New York City Health and Hospitals Corporation, Physician's Affiliates Group of New York, State University of New York, or any division of these organizations.

References

Bellamy, Richard. 1999. *Liberalism and pluralism: Towards a politics of compromise*. London: Routledge.

Birchley, Giles. 2016. Harm is all you need? Best interest and disputes about parental decision-making. *Journal of Medical Ethics* 42: 111–115.

Black, Carol. (2008). *Working for a healthier tomorrow. Dame Carol Black's Review of the health of Britain's working age population*. London: TSO.

Buchanan, Allen E., and Dan W. Brock. 1989. *Deciding for others: The ethics of surrogate decision-making*. Cambridge: Cambridge University Press.

Dwyer, James G. 1994. Parents' religion and children's welfare: Debunking the doctrine of parents' rights. *California Law Review* 82 (6): 1371–1447.

Dwyer, James G. 2009. Constitutional birthright: The state, parentage, and the rights of newborn persons. *UCLA Law Review* 56 (4): 755–835.

ECoHR. 2011. European Court of Human Rights. *Research Report: Child Sexual Abuse and Child Pornography in the Court's Case-Law*. https://www.echr.coe.int/Documents/Research_report_child_abuse_ENG.pdf.

Eichner, Maxine. 2016. Bad medicine: Parents, the state, and the charge of "medical child abuse." *University of California Davis Law Review* 50 (1): 205–320.

Feinberg, J. (1984). *Harm to others, Vol 1 of the moral limits of the criminal law*. New York: Oxford University Press.

Fineman, M. 2019. Vulnerability and social justice. *Valparaiso University Law Review* 53 (2): 341–370.

Flaherty, Emalee G., Harriet L. MacMillan, and the American Academy of Pediatrics Committee on Child Abuse and Neglect. 2013. Caregiver-fabricated illness in a child: A manifestation of child maltreatment. *Pediatrics* 132 (3): 590–597. https://doi.org/10.1542/peds.2013-2045

Frankfurt, Harry. 1987. Equality as a moral ideal. *Ethics* 98 (1): 21–43.

Gerdes, Hannah, and John Lantos. 2020. Differing thresholds for overriding parental refusals of life-sustaining treatment. *HEC Forum* 32: 13–20. https://doi.org/10.1007/s10730-019-09384-6

Griffith, R. 2016. What is gillick competence?" *Human Vaccines and Immunotherapeutics* 12(1):244–7. https://doi.org/10.1080/21645515.2015.1091548

Hayes, Margaret M., Alison E. Turnbull, Sandra Zaeh, Douglas B. White, Gabriel T. Bosslet, Kevin C. Wilson, and Carey C. Thomson. 2015. Responding to requests for potentially inappropriate treatments in intensive care units. *Annals of the American Thoracic Society* 12 (11): 1697–1699.

Iltis, Ana S. 2010. Toward a coherent account of pediatric decision-making. *Journal of Medicine and Philosophy* 35 (5): 526–552.

Jacobs, Allan J. 2013. The ethics of circumcision of male infants. *Israel Medical Association. Journal: IMAJ* 15 (1): 60–64.

Jacobs, A.J., and K.S. Arora. 2015. Ritual male infant circumcision and human rights. *American Journal of Bioethics* 15 (2): 30–39.

Jacobs, Allan J., and Kavita Shah Arora. (2018). When may government interfere with religious practices to protect the health and safety of children? *Ethics in Medicine and Public Health* 5 (1): 86–93.

Katz, Aviva L., Sally A. Webb, and the Committee on Bioethics, American Academy of Pediatrics. (2016). Informed consent in decision-making in pediatric practice. *Pediatrics* 138 (2): e20161485.

Levin, Hillel Y., Allan J. Jacobs, and Kavita Shah Arora. 2016. To accommodate or not to accommodate: (When) Should the state regulate religion to protect the rights of children and third parties? *Washington and Lee Law Review* 73 (2): 915–1017.

McGowan, Michelle L., Cynthia A. Prows, Melissa DeJonckheere, William B. Brinkman, Lisa Vaughn, and Melanie F. Myers. (2018). Adolescent and parental attitudes about return of genomic research results: Focus group findings regarding decisional preferences. *Journal of Empirical Research on Human Research Ethics* 13 (4): 371–382.

Navin, Mark C., and Joel Wasserman. 2017. Reasons to amplify the role of parental permission in pediatric treatment. *American Journal of Bioethics* 17 (11): 6–14.

Paris, J.J., J. Ahluwalia, B.M. Cummings, and D.J. Wilkinson. 2017. The Charlie Gard case: British and American approaches to court resolution of disputes over medical decisions. *Journal of Perinatology* 37 (12): 1268–1271. https://doi.org/10.1038/jp.2017.138.

Philp, M. 2010. What is to be done? Political theory and political realism. *European Journal of Political Theory* 9 (4): 466–484.

Rawls, John. 1971. *A theory of justice*. Cambridge, MA, USA: Belknap Press.

Ross, L.F. 1998. *Children, families, and health care decision-making*. Oxford: Clarendon Press.

Salter, E.K. 2012. Deciding for a child: A comprehensive analysis of the best interest standard. *Theoretical Medicine and Bioethics* 33 (3): 179–198.

Shklar, J.N. 1984. *Ordinary vices*. Cambridge, MA: Belknap Press.

Szreter, S. M., and Woolcock. 2004. Health by association? Social capital, social theory, and the political economy of public health. *International Journal of Epidemiology* 33 (4): 650–667. https://doi.org/10.1093/ije/dyh013

Torcello, L. 2011. Sophism and moral agnosticism, or, how to tell a relativist from a pluralist. *The Pluralist* 6 (1): 87–108.

United Nations. (2013). United Nations Committee on the Rights of the Children. *General Comment No. 14 on the right of the child to have his or her best interest taken as a primary consideration (art. 3, para. 1)*. http://www.crin.org/en/docs/GC.14.pdf

Waldron, J. 1999. *Law and disagreement*. Oxford; New York: Oxford University Press.

Further Readings

Buchanan, A., and D.W. Brock. 1989. *Deciding for others: The ethics of surrogate decision-making*. Cambridge: Cambridge University Press.

Dwyer, J.G. 2014. Who decides? In *The nature of children's wellbeing: Theory and practice*, ed. Alexander Bagattini and Colin MacLeod. Chapter 10, 157–175. Dordrecht, Netherlands: Springer.

Katz, Aviva L., Sally A. Webb, and the Committee on Bioethics, American Academy of Pediatrics. 2016. Informed consent in decision-making in pediatric practice. *Pediatrics* 138 (2): e20161485.

Ross, L.F. 1998. *Children, families, and health care decision-making*. Oxford: Clarendon Press.

Chapter 27
Vaccine Ethics: Ethical Considerations in Childhood Vaccination

J. C. Bester

Abstract The vaccination of children raises a host of unique ethical issues. This chapter considers the moral status of vaccination drawing on various arguments and ethical perspectives and explores policy implications of the moral implications of vaccines. Vaccination is at once an individual medical decision for a child and a public health intervention with broader implications for society. Arguments are presented that draw on the best interest standard, the unfairness of free riding a public good, utilitarianism, and theories of justice. All these arguments ground vaccination as a moral imperative, something that ought to be guaranteed for children. Because vaccination is morally obligatory, clear policy implications can be defined. Society must guarantee sufficient access to vaccines and must adopt policies to ensure adequate uptake of vaccination. Clinicians have an important role to play in addressing parental concerns, building trust relationships with parents, and encouraging the uptake of vaccination.

Keywords Vaccination ethics · Vaccination policy · Public health ethics · Immunization · Vaccines

27.1 Introduction: Vaccine Ethics—Ethical Considerations in Childhood Vaccination

Vaccinations form a central part of the preventive medical care of children, priming the child's immune system to develop protection against a host of serious viral infections. Vaccinations are widely recommended and used in the care of children, and has been credited with saving countless lives, preventing disease, and providing savings in health care expenditure.

Despite this, there is a host of reasons that children may go unvaccinated. One important reason is the continued presence of vaccine hesitancy and vaccine refusal

J. C. Bester (✉)
Kirk Kerkorian School of Medicine at UNLV, University of Nevada, Las Vegas, USA
e-mail: johan.bester@unlv.edu

© Springer Nature Switzerland AG 2022   437
N. Nortjé and J. C. Bester (eds.), *Pediatric Ethics: Theory and Practice*,
The International Library of Bioethics 89,
https://doi.org/10.1007/978-3-030-86182-7_27

in many countries and many communities. While vaccine resistant persons are generally in the minority, they significantly affect the uptake of vaccines for children. At the very least, vaccine hesitant or vaccine resistant parents may refuse vaccination for their own children. Another issue is access to vaccination. Another important consideration that should not be forgotten, is societal distributions of health care and wealth that may place barriers to vaccine access before some children. Some children may lack access to medical care or to a vaccination program or live in such poverty that their parents cannot afford to pay for vaccinations. Parents may sometimes have to make trade-offs about where the money goes or whether to keep a job or take a child for preventive care, and the result may be that children miss the opportunity to be vaccinated.

Consequently, the vaccination of children raises a number of significant ethical questions. In this chapter, I explore some of these ethical questions and issues. I do so by reference to two main questions. First, what is the moral status of childhood vaccines? Second, in the light of the moral status of childhood vaccines, which vaccination policies should be instituted?

27.2   The Moral Status of Childhood Vaccines

The first issue is to consider whether childhood vaccination has specific moral implications for society, clinicians, and parents. That is, whether vaccination establishes moral claims and considerations that should guide our vaccination practice and policy. To answer this question, we need to think about the empirical attributes of vaccination (the facts about vaccination) and then apply a moral framework to these facts. This will allow us to draw conclusions about the moral implications of childhood vaccines, who have which moral responsibilities, and what our general approach to vaccination practice and policy should be.

27.2.1   Relevant Facts About Vaccines

The facts relevant to ethical discussion of recommended childhood vaccines can be summarized in a number of statements. These are:

(1)   Vaccines prevent infectious diseases that have serious complications.
(2)   Vaccines are in general effective, safe, and cost-effective.
(3)   Vaccines often create herd immunity, an important concept in protecting an entire population.
(4)   There is always a vulnerable group, a subsection of people within a population who do not receive direct protection from vaccination.

Childhood vaccines recommended by the CDC and WHO provide protection against serious infectious diseases such as polio, measles, diphtheria, and pertussis

(CDC 2015). Prior to the widespread use of vaccines, these diseases were responsible for thousands of deaths, hospitalizations, and permanent loss of function among children each year. Vaccines have turned the tide, and removed many of these illnesses to the point that it is now uncommon for a general pediatrician to see a case of measles, polio, or diphtheria.

One analysis examined the impact of 20 years of use of the recommended vaccination schedule in the United States between 1994 and 2013 (Whitney et al. 2014). During this time, 78.6 million children were born in the United States. Use of the recommended vaccine schedule in this birth cohort resulted in the prevention of over 322 million infectious disease cases, 21 million hospitalizations, and 731,700 deaths. The vaccine schedule was remarkably cost effective in this birth cohort, and was estimated to result in net savings of $295 billion in direct health care spending and $1.38 trillion in social costs. Some of the most significant estimates of prevention of death and disease related to specific illnesses were:

- measles (over 70 million cases, 8 million hospitalizations, and 57,000 deaths prevented);
- diphtheria (over 5 million cases, 5 million hospitalizations, and 507,000 deaths prevented)
- pertussis (over 54 million cases, 2 million hospitalizations, and 20,000 deaths prevented)
- hepatitis B (over 4 million cases, 600,000 hospitalizations, and 59,000 deaths prevented).

Childhood vaccines are highly effective (CDC 2015). For example, diphtheria vaccine is given as 4 doses, and its clinical efficacy is 97%. Measles vaccine is given in two doses, and has 95–98% effectiveness. Pertussis vaccine effectives is between 70 and 90%.

Vaccines are also known to be safe, with a favorable adverse event profile. Relatively common adverse events from vaccines are temporary and not serious, including wrist pain, rash, or fever. Serious adverse events are rare, and are significantly outweighed by benefit from vaccination. For example, serious adverse events from measles vaccine include serious allergic reaction (less than 1 in a million doses), and this can be mitigated by readiness to respond to allergic reactions and avoiding vaccination in children known to have allergy to vaccine components. Numerous studies and systematic reviews of studies have confirmed the safety and effectiveness of recommended childhood vaccines (CDC 2015; Di Pietrantonj et al. 2020; IOM 2011; Maglione et al. 2014; SBU 2009).

Another important concept is herd immunity. If sufficient numbers within a population are vaccinated and become immune, the spread of the disease within the community can be decreased or halted. This provides some protection for those within the population who have not developed immunity against infection. Herd immunity refers to protection against disease for unimmunized persons derived from high immunity rates within a population (John and Samuel 2000). Herd immunity is an important concept; there is always a group of people in every society who cannot be vaccinated because of medical reasons and need herd immunity to be established.

Neonates and infants who have not yet been vaccinated, those who have a medical contra-indication against vaccination, and those who do not become or remain immune from the vaccine depend on herd immunity for protection against disease. Take measles vaccine as an example (Bester 2017, 2018). Measles is highly infectious, but the spread of measles within a population can be interrupted. This requires vaccination rates of 93–95% with two doses of vaccine. There are various persons who are not protected directly by vaccination: those who are immune compromised, those who are under 1 year of age (first vaccine given at 12 months), those who lose immunity over time, and those with vaccine allergy. This means there is always a vulnerable group within the population, who depend on high vaccine uptake for protection against measles.

27.2.2   Moral Considerations in Childhood Vaccination

I will here consider four different ways in which the moral implications of vaccines can be analyzed.

27.2.2.1   The Best Interest Standard

One way to think of vaccines is as a medical decision for an individual child. Parents and clinicians have to make decisions for children under their care, and government policies affect such decision-making for individual children. If we approach vaccination as a medical decision about an individual child, we may use the best interest standard as ethical guideline. (Bester 2017; Dawson 2005).

The best interest standard (BIS) (refer to Chap. 4) is a broad ethical guideline in medicine with particular application in pediatrics. The BIS grounds various policies, procedures, and ethical obligations in pediatrics. It sets standards and grounds more specific rules that clinicians follow in treating their patients. But it can also be used, when applied correctly, to analyze specific individual decisions. When used in this way, the BIS states that in decisions made about a child the option should be chosen that best protects or promotes the interests of the child. Interests refer to things that are important for the child's wellbeing; to promote interests is to promote the wellbeing of the child, to hinder interests is to set back the wellbeing of the child. This means that for any specific individual medical decision, available options should be weighed in light of how they affect the interests of the child.

What is in the best interest of young children regards vaccination can be determined by weighing benefits and harms of vaccination against foregoing vaccination (Dawson 2005). This requires us to consider the empirical attributes of vaccination and vaccine-preventable disease, and consider the potential outcomes for the individual child. The option that best promotes and protects the interest of the child is morally preferred by reference to the best interest standard.

For the vaccines that are on recommended childhood vaccination schedules, such as vaccines against measles, mumps, diphtheria, and pertussis, the facts are clear. Vaccination protects a child against the devastating consequences of a serious infectious disease. These vaccines are safe and effective, with a very favorable adverse effect profile. Risk of serious harm is very low, and pales in comparison to the risk of harm from infectious disease should the child be left unvaccinated. The benefit from vaccination is substantial, and the risk of harm minimal. When looked at this way, it is clear that vaccination is in the best interest of the individual child.

One caveat to this is that there should be no medical contra-indication to vaccination present. If a child is allergic to vaccine components, or if a child has a medical condition that would precipitate a serious adverse reaction to vaccination, harms from vaccination outweigh benefits and vaccination is not in the best interest of the child. For example, a child with immune deficiency can suffer serious harm from a measles vaccine and should not receive measles vaccine. In children who do not have medical contra-indication for vaccination, it can be said that vaccination is in the best interest of the child.

A counter-argument may be made in highly vaccinated societies that non-vaccination is in the best interest of the child. Some may say that a child is at low risk of getting the infectious disease in a society where herd immunity is established, so that risk of harm from vaccine-preventable disease is low. In such a situation, a child may forego vaccination and be protected from disease through herd immunity, while not being exposed to risk of vaccine adverse effects. So, if this holds, the individual child's wellbeing is better protected through non-vaccination than through vaccination.

Some respond to this argument by reference to the unfairness of free riding, which I will explore further down. But just from the perspective of the best interest of the child, this argument has serious flaws (Bester 2017).

- First, herd immunity is not as reliable a protection against vaccine-preventable disease as is direct vaccination. Even highly vaccinated communities see outbreaks of vaccine-preventable disease through imported cases that spread through susceptible clusters in the population.
- Second, herd immunity is dependent on the actions of others, and there is no telling when population behavior may change so that herd immunity is lost.
- Third, some of these vaccine-preventable diseases are so contagious that very high levels of vaccine uptake are needed to maintain herd immunity. This means that very few can be left unvaccinated, given that there are some who already cannot get the vaccine to begin with.
- Fourth, those who get these infections later in life tend to suffer a higher rate of complications; this has been demonstrated with measles, with a complication rate 4.5 times higher in highly vaccinated societies.
- Fifth, this argument does not work for diseases where no herd immunity is established, such as tetanus.
- Lastly, and most importantly, the argument is self-defeating. If we leave a child unvaccinated because it protects this child best, we should realize that all other

decision-makers will do the same for their children. This will lead to the loss of the very herd immunity that the argument depends on. This way of protecting the interests of the child leaves the child vulnerable to the inevitable outbreaks that will happen.

In light of these considerations, the only way to reliably protect and promote the interests of the child against vaccine-preventable disease is through vaccination. The best interest standard would stipulate that it is a moral requirement to vaccinate the individual child unless there are medical contra-indications present that would require foregoing vaccination.

27.2.2.2 The unfairness of free-riding a public good

Another way to think about the moral implications of childhood vaccinations is to view childhood vaccination as a public good, and to view vaccine refusal as unfair free-riding of a public good (Dawson 2011; van den Hoven 2012). A public good is something that benefits members of society and is created by the actions of many people. A public good cannot be divided up, and is available for use by everyone in the society. Free-riding refers to making use of a public good without contributing to its creation or preservation. This is generally seen as unfair and immoral, exploiting the efforts of others without contributing one's fair share.

Vaccination creates a public good: herd immunity. This herd immunity protects those who are not themselves immune to disease, and is available to everyone in society who is non-immune. To establish herd immunity depends on high vaccination rates, often above 90%, so that the collective action of many people is required. Herd immunity cannot be broken up or divided, and is freely available to all. Herd immunity therefore meets the criteria for a public good (Dawson 2011; van den Hoven 2012). There are further benefits to society from vaccination programs that establish herd immunity: prevention of hospitalizations, which decreases pressure on infrastructure; and significant cost savings in both direct health care spending and societal spending. Everyone in society benefits from these savings, so that it is evident that high levels of vaccination create intangible public goods.

A parent who refuses vaccination for their child while appropriating these benefits to themselves and their child is thought to free-ride the public good. That is, they unfairly make use of the benefits of the public good without contributing to it. The unfairness of this situation is even more striking when it is considered that the act required to contribution of the public good is consenting to vaccination for a child, which counts as a benefit to the child concerned in the first place. Further, not contributing to the public good may place the existence of the public good in jeopardy, since very high vaccination rates are needed to establish herd immunity for some infections and there are always a group that cannot be directly vaccinated.

One possible objection may be that herd immunity is different than other kinds of public goods, in that once one has contributed to the existence of herd immunity one no longer benefits from it. If a child receives a vaccine, becomes immune, and is protected against disease, the child no longer depends on herd immunity for protection against the disease. There are a number of replies to this. First, that the child may over time lose immunity (as happens sometimes in a percentage of vaccinated people), and will then depend on herd immunity for protection. Second, that there are other intangible benefits such as savings to health care and societal costs that benefit the child (and others) as a public good. Third, that the child may have friends, family, and neighbors who depend on the public good. Lastly, that the vaccine program that exists to maintain herd immunity is central to the public good, and this vaccine program makes vaccination and its benefits reasonably available to the child.

Overall, there is a moral case to be made that children ought to be vaccinated because it contributes to a public good, and relying on the public good without paying one's share is unfair. It is not the child who is the free-rider; rather, this speaks to parents, health care workers, and broader society, indicating that we have good moral reasons that justify provision of vaccination to children who have no medical contra-indication.

27.2.2.3 The greatest good for the greatest number: Vaccines and utility

It is also possible to view vaccination in terms of its overall outcomes for the public using the moral framework of utilitarianism (Dare 1998; Guibilini et al. 2018; Field and Caplan 2012). Utilitarianism is concerned with consequences. The morality of actions or policies are determined in so far as they lead to the greatest good for the greatest number of people. When concerned with a specific action choice or formulating of policy, the moral choice is the one that leads to the greatest overall good for the greatest number of people. Classic utilitarianism was focused on happiness as the greatest good, but in public health and in health care the good aimed at tends to be wellbeing. There is also a distinction between act-utilitarianism and rule-utilitarianism. In act-utilitarianism, we are concerned with evaluating the morality of a specific act or set of actions by examining the consequences of the action. An action that maximizes the good is morally right. Rule-utilitarianism, in contrast, states that those rules must be instituted that over time would lead to the greatest good for the greatest number. Utility then forms the basis of a set of moral rules which, if followed by everyone in society, make people in society the best off they can be.

From the perspective of vaccinations in children, we therefore have to see what the consequences would be from non-vaccination versus vaccination, and which distribution of vaccination leads to the greatest good for the greatest number of people. If we take the empirical facts about vaccination, the utilitarian argument almost writes itself. Vaccine-preventable infectious diseases are serious illnesses with significant effects on health and wellbeing. Vaccines are safe, effective, and cost-effective. If we vaccinate enough children, we achieve herd immunity which protects those who cannot receive the vaccine themselves.

If we consider the individual act of vaccination, we have something similar to the arguments presented under the best interest standard: a child is better off being vaccinated unless there is some medical contra-indication to vaccination. If we consider it from a societal perspective, we achieve the greatest benefits for the greatest number by instituting a vaccination program that vaccinates in a way that the most possible medically eligible children receive vaccination, and herd immunity is established. This way, children who receive direct vaccination are protected from disease through the vaccine, and people who cannot receive vaccination are protected through herd immunity. This strategy leads to protection of health and savings in health care and societal costs in the best way possible when compared to other vaccination distributions. For instance, if we establish herd immunity but leave children who are eligible for vaccination unvaccinated, those children remain susceptible to imported outbreaks of disease, and also will be susceptible if herd immunity is lost. The greatest good for the greatest number is served by vaccinating every child medically eligible. Further, if we vaccinate in a way that does not establish herd immunity, those children who were not vaccinated remain vulnerable to disease and the harm that disease may bring, as do the people who depend on herd immunity for their protection.

The best outcomes in terms of health, wellbeing, and societal spending are reached when vaccination programs or schedules are instituted which leads to vaccinating all children who have no medical contra-indication, and by the establishment of herd immunity. If we follow the direction of act-utilitarianism, it is a moral imperative to establish herd immunity and to choose vaccination when making decisions for the individual child. If we follow rule-utilitarianism, we recognize the need to institute and follow moral rules, policies, and vaccination programs that would lead to vaccination uptake that would establish herd immunity, and that would enable vaccination for every medically eligible child.

27.2.2.4 Vaccination owed to children: Vaccination of children as justice obligation

The appropriate starting point for thinking of the moral implications of vaccination is to ask what we owe children. That is: what are the moral claims that children have on us as society, and on those who stand in significant relationships with them. From this perspective, we may recognize that children are dependent on others for their wellbeing. Children cannot protect or advance their own interests, and require others to do so for them. The question arises – does vaccination fall into the category of things that parents and society owe children? Is there any moral obligation on parents, health care workers, and society in general to guarantee vaccination for children?

I've defended an argument which concludes just that: that vaccination is something morally owed to children (Bester 2018). The argument draws on justice for its moral force, and therefore creates obligations on society, its institutions, and on those who stand in morally significant relationships with children. If the argument holds, children have a moral claim on us to be protected against vaccine-preventable disease, and we should ensure that vaccination happens. The original argument focuses solely on measles vaccination. Here I adapt it to apply to childhood vaccination that incorporates the recommended vaccines as on the CDC or WHO schedules, as follows:

(1) A just society is obligated to institute measures that protect the wellbeing of children against readily preventable insults.
(2) Vaccine-preventable infectious disease is a serious threat to the wellbeing of children.
(3) A society can protect its children against vaccine-preventable infectious disease through use of a childhood vaccination program that is safe, effective, and cost-effective. This requires vaccination in a way that protects those who can receive vaccination through direct vaccination, and those who cannot be vaccinated through herd immunity.
(4) (Conclusion 1): A just society is obligated to protect its children against vaccine-preventable infectious disease through use of a vaccination program that vaccinates those who are eligible for vaccination, and establishes herd immunity to protect those who are not eligible.
(5) The obligations of a just society to its children are discharged by citizens and institutions situated in relevant positions to do so.
(6) Vaccination requires cooperation of parents, health care workers, vaccine manufacturers, and government.
(7) (Conclusion 2) Parents, health care workers, vaccine manufacturers, and government have obligations to ensure adequate use of vaccination to protect children against vaccine-preventable diseases.

If the premises are more likely true than their denials, and the conclusion follows from the premises, we have a sound argument and can have confidence in the truth of the conclusion. Most of the premises follow from empirical observations and known facts about vaccination and the process of manufacturing, delivering, and scheduling of vaccinations. There are possibly two premises that stand in need of more thorough justification. Premise (1) and premise (5).

Premise (1) can straightforwardly be grounded by reference to a theory of justice such as the wellbeing-justice theory of Powers and Faden, or the capabilities-justice theory of Nussbaum and Sen. These theories stipulate justice obligations on society to promote and protect the wellbeing of its members (or the protection of capabilities to attain wellbeing), with particular focus on protecting the health of children. This is enough to ground this premise, but consider also what the denial of this premise looks like: a just society does not institute measures to protect the wellbeing of children against preventable insults. This seems implausible on any theory of justice, so that we can say premise (1) is more likely true than its denial.

Premise (5) can be thought possibly controversial, as there is some disagreement about which justice obligations fall on which people within society. But it seems clear that society consists of individuals and institutions. And if society has obligations to be a certain way and to do certain things in order to be a just society, it is up to the individuals and institutions of society to act on these obligations. It will certainly fall to those institutions and people who are placed in positions to act on these obligations to do the work of justice that society requires. For example, if we say that a just society makes sure children have access to food, it is not controversial to think that there is an obligation on parents to provide food. If parents cannot, or need assistance, others may step to assist, inclusive of society's institutions such as government, churches, clubs, and schools. There is simply no other way for us to discharge society's justice obligations than through the actions of individuals and institutions that are appropriately placed to do so.

The implication of this argument is that children are owed vaccination, and that society ought to ensure that they get it. Parents have to take their children to be vaccinated, health care providers have to provide it, manufacturers and government have to ensure a safe and adequate supply of vaccination with easy access, and government and its delegates have to set up reasonable vaccination programs. This means that vaccination is morally speaking not a matter of parental discretion, but of obligation.

27.3   Which Vaccination Policies Should Be in Place?

We have good reasons to think that vaccination of children is morally obligatory. This conclusion is reached by a variety of different approaches to ethical analysis. Society in general and those who stand in relationships with children like parents and health care provides therefore have obligations to ensure that children are vaccinated. The goal is to protect as many children as we can through direct vaccination, and to protect those who cannot get vaccination indirectly through herd immunity.

A next question is one of policy. Government has special obligations when it comes to public health interventions like vaccination, because of how it is situated within society (Childress et al. 2002). Government has unique powers to write and enforce policy, to direct public spending, and is uniquely situated to coordinate the responses of various institutions and individuals. If society has an obligation to vaccinate its children, it follows that government has a special and urgent set of obligations that are among other actions discharged by setting and enforcement of public policy.

There are really two things that policy should accomplish. First, it should guarantee a safe and sufficient supply and distribution of vaccinations so that children have ready access. Second, it should respond to vaccine hesitancy and vaccine refusal in ways that encourage vaccination uptake and ensure ongoing vaccination of children.

27.3.1 Sufficient Access to Safe Vaccines

I will spend little time on the first and least controversial policy area. Government should ensure that a safe, sufficient supply of vaccines are available to society's children. This includes focus on areas such as manufacturing of vaccines, distribution of vaccines, and vaccine administration. Quite simply stated, government should set standards for vaccination safety, and has the obligation to create vaccination programs that ensure safe and fair access to vaccines. There are many ways in which these can be accomplished.

By way of example, consider how these goals have been pursued in the United States (Orenstein et al. 2005). Standards for vaccine safety are regulated by a governmental organization, the Food and Drug Administration (FDA). These set high standards for testing, manufacture, distribution, and administration of vaccines. Sufficient vaccine production is ensured by a variety of public–private partnerships. The government ensures public funding mechanisms that can be used by states to purchase vaccine, and in so doing guarantees a market and encourages the production and distribution of vaccines at sufficient levels. In the US, vaccines are usually delivered at point of care by a doctor or nurse and paid for by insurance of privately, but a vaccination program called Vaccines for Children ensures access to vaccines for those with financial barriers. Vaccine use, effectiveness, and safety are monitored by governmental agencies such as the CDC and FDA.

There are other policies and means to reach these goals, but these are examples of the kinds of things governments can do to respond to the obligation to ensure access to sufficient and safe vaccination for society's children.

27.3.2 Response to Vaccine Hesitancy and Vaccine Refusal

Vaccine hesitancy and subsequent parental vaccine refusal decrease vaccination uptake and creates risks for disease outbreaks that endanger the wellbeing of children (Siddiqui et al. 2013). Since there is a moral imperative to vaccinate children, vaccine hesitancy and vaccine refusal represent major challenges that should be addressed through vaccine policy. The World Health Organization (WHO) identified vaccine hesitancy as one of the ten biggest threats to global health in 2019 (WHO 2019).

Vaccine hesitancy has been well described in the academic literature (Edwards et al. 2016; Siddiqui et al. 2013). Vaccine hesitancy is a lack of confidence in vaccinations or doubts about vaccinations that may lead to delay or refusal of vaccines. Vaccine-hesitant parents are not all the same, and may hold diverging views and uncertainties about vaccinations. Factors that may induce or support vaccine hesitancy include false information about vaccine safety for example on the internet, lack of trust in experts and in government, and given the absence of circulating vaccine-preventable disease the over-estimation of the risks and under-estimating of the need for vaccines. With support from a trusted clinician, many refusing parents will eventually accept vaccination for their children; studies estimate rates of between 30–47% of parents changing their minds after attempts at persuasion by clinicians (Edwards et al. 2016). There is a small group of parents who will maintain their refusal of vaccinations, and it is with this in mind that policy options should be crafted.

Elsewhere, I've suggested that there are three components to a policy response to vaccine refusal that should be considered (Bester 2015). The three components work together to maximize vaccination uptake, and should be seen as three legs of a stool working together to keep vaccination rates up in the face of vaccine hesitancy. These are mandatory policies, education, and building of trust.

27.3.2.1   Mandatory Policies

Mandatory policies enforce vaccination by imposing a penalty on parents for refusing vaccination. For example, vaccinations may be mandated as a condition for school entry for children. No vaccination—no school. Such policies generally encourage and maintain high vaccination levels, and may be particularly strong if it assumes a default position of vaccination with the need for parents to opt-out using some sort of onerous process. But there are some concerns that need to be considered. First, the penalty for non-vaccination needs to be carefully thought through and weighed in the light of various alternate ethical considerations. For example, putting parents in jail for non-vaccination would place significant burdens on otherwise good parents and on children meant to be protected by the policy to enforce vaccination. Huge financial fines would impact the financial security of the parent and eventually have the potential to negatively impact the wellbeing of the child. It becomes hard to think of a penalty that does not have the ultimate draw-back of imposing burdens on the child that is meant to be protected by the policy. Second, if mandatory policies are too stringent and the penalty too severe, it has the potential to foment backlash and resistance, ultimately hindering the goals the policy was set in place for. Mandatory policies can be justified given the moral imperative to vaccinate, but careful thought should be given to the penalty imposed and the effect on competing values. In the light of respecting competing values, something like a mandatory policy as requirement for school attendance with narrowly defined opt-out clauses can be justified.

27.3.2.2   Education

Since a main contributing factor to vaccine hesitancy is the copious amount of vaccine misinformation that is readily available, it seems logical to think that correcting misinformation through public education can fix the problem. Certainly, there is some merit to the idea, and using educational public health interventions in a variety of forms such as mass media or social media is sure to contribute to combating misinformation. But, educational interventions can backfire if not correctly employed. For example, those who hold very stringent anti-vaccine views may be strengthened in their views if challenged on their beliefs by an educational intervention. It is therefore important that public education campaigns be carefully studied and designed for the correct audience before they are employed in public.

27.3.2.3 Trust

An important factor in vaccine acceptance is trust. Clinicians who provide care for children play an important role in vaccine acceptance. Parents who are in a trusting clinical relationship with a clinician can resist anti-vaccine messaging, and ultimately may decide to vaccinate their child despite initial hesitancy or refusal. In this regard, clinicians have important obligations to foster trusting relationships, to take parental concerns seriously, and to communicate in a skilled and compassionate way with parents. Because trust and a trust-relationship with a clinician play such an important role in ensuring vaccination, it is important that the system which delivers medical care to children be designed to foster trusting parent-clinician relationships, and that this forms a building block of society's efforts to ensure sufficient vaccine uptake. This may seem like a small or straight-forward point, but further reflection will show how radical a suggestion it is. The whole health care system should be reformed in such a way that each child is guaranteed a primary care clinician that can build a trust relationship with their parent.

27.4 Conclusion

There are sound arguments from different ethical approaches that justify a moral imperative to vaccinate children. Vaccination should be employed in such a way that the most children possible are protected through direct vaccination, and those who are unable to be vaccinated are protected through herd immunity. This creates justification for policy responses, and a three-legged policy response to ensure high levels of vaccinations should be in place: mandatory policies, educational interventions, and measures that build trust such as trusting clinician-parent relationships. Vaccination is of immense moral importance to those who are entrusted with protecting and promoting the wellbeing of children.

Guiding principles in the ethics and policy of childhood vaccination

---

Important ethically relevant facts about recommended childhood vaccines
- Vaccines prevent childhood infectious diseases that have serious complications
- Vaccines recommended for children are in general effective, safe, and cost-effective
- Vaccines can create herd immunity, an important concept in protecting an entire population against disease
- There is always a vulnerable group, a subsection of a population who do not receive direct protection from vaccination

---

(continued)

(continued)

Moral considerations in childhood vaccination

- Best interest standard: Vaccination better protects the interests of children than non-vaccination, unless a child has a medical contra-indication to vaccination
- Herd immunity as a public good: Children who are vaccinated contribute to the public good of herd immunity. This public good protects those who cannot be vaccinated themselves
- Utility: The best outcomes for the greatest number of people in terms of health, wellbeing, and social spending are reached when uptake of vaccinations is maximized and herd immunity is established
- Vaccines owed to children as a societal justice obligation: Those who stand in morally significant relationships with children have obligations to ensure adequate use of vaccination to protect children against vaccine-preventable disease

Guiding principles for vaccination policy

- Policies should exist that ensure that a safe, sufficient supply of vaccines are available to society's children
- Policies should exist that maximize vaccination uptake. This includes a societal policy response to vaccine hesitancy and vaccine refusal

References

Bester, J.C. 2015. Vaccine refusal and trust: The trouble with coercion and education and suggestions for a cure. *The Journal of Bioethical Inquiry* 12 (4): 555–559.

Bester, J.C. 2017. Measles vaccination is best for children: The argument for relying on herd immunity fails. *Journal of Bioethical Inquiry* 14 (3): 375–384.

Bester, J.C. 2018. Not a matter of parental choice but of social justice obligation: Children are owed measles vaccination. *Bioethics* 32 (9): 611–619.

Centers for Disease Control and Prevention (CDC). 2015. *Epidemiology and prevention of vaccine-preventable diseases*, 13th ed., ed. J. Hamborsky, A. Kroger, and S. Wolfe. Washington, DC: Public Health Foundation.

Childress, J.F., R.R. Fade, R.D. Gaare, L.O. Gostin, et al. 2002. Public health ethics: Mapping the terrain. *Journal of Law, Medicine, and Ethics* 30: 170–178.

Dare, T. 1998. Mass immunization programmes: Some philosophical issues. *Bioethics* 12 (2): 125–149.

Dawson, A. 2005. The determination of 'best interest' in relation to childhood vaccinations. *Bioethics* 19 (2): 72–89.

Dawson, A. 2011. The moral case for the routine vaccination of children in developed and developing countries. *Health Affairs* 30 (6): 1029–1033.

Di Pietrantonj, C., A. Rivetti, P. Marchione, M.G. Debalinin, and V. Demicheli. 2020. Vaccines for measles, mumps, rubella, and varicella in children. *Cochrane Database of Systematic Reviews* (4), Art. No: CD004407. https://doi.org/10.1002/14651858.CD004407.pub4

Edwards, K.M., et al. (Edwards, KM, Hackell, JM, and The Committee on Infectious Diseases, The Committee on Practice and Ambulatory Medicine, American Academy of Pediatrics). 2016. Countering vaccine hesitancy. *Pediatrics* 138 (3): e20162146.

Field, R.I., and A.L. Caplan. 2012. Evidence-based decision-making for vaccines: The need for an ethical foundation. *Vaccine* 30: 1009–1013.

Guibilini, A., T. Douglas, and J. Savulescu. 2018. The moral obligation to be vaccinated: Utilitarianism, contractualism, and collective easy rescue. *Medicine, Health Care, and Philosophy* 21: 547–560.

Institute of Medicine (IOM). 2011. *Adverse effects of vaccines: Evidence and causality*. Washington, DC: The National Academy Press.

John, T.J., and R. Samuel. 2000. Herd immunity and herd effect: New insights and definitions. *European Journal of Epidemiology*. 16 (7): 601–606.

Maglione, M.A., L. Das, L. Raaen, A. Smith, et al. 2014. Safety of vaccines used for routine immunization of US children: A systematic review. *Pediatrics* 134 (2): 325–337.

Orenstein, W.A., R.G. Douglas, L.E. Rodewald, and A.R. Hinman. 2005. Immunizations in the United States: Success, structure, and stress. *Health Affairs* 24 (3): 599–610.

Siddiqui, M., D.A. Salmon, and S.B. Omer. 2013. Epidemiology of vaccine hesitancy in the United States. *Human Vaccines and Immunotherapeutics* 9 (12): 2643–2648.

Swedish Council on Health Technology Assessment (SBU). 2009. Vaccines to children: Protective effect and adverse events. A systematic review. SBU Yellow Report No. 191. https://www.ncbi.nlm.nih.gov/books/NBK447995/pdf/Bookshelf_NBK447995.pdf.

Van den Hoven, M. 2012. Why one should do one's bit: Thinking about free riding in the context of public health ethics. *Public Health Ethics* 5 (2): 154–160.

Whitney, C.G., F. Zhou, J. Singleton, and A. Schuchat. 2014. Benefits from immunization during the vaccines for children program era—United States, 1994–2013. *Morbidity and Mortality Weekly Report* 63 (16): 352–355.

World Health Organization (WHO). 2019. Ten threats to global health in 2019. https://www.who.int/news-room/feature-stories/ten-threats-to-global-health-in-2019. Accessed 9 July 2020.

Further Reading

Bester, J.C. 2018. Not a matter of parental choice but of social justice obligation: Children are owed measles vaccination. *Bioethics* 32 (9): 611–619.

Dawson, A. 2011. The moral case for the routine vaccination of children in developed and developing countries. *Health Affairs* 30 (6): 1029–1033.

Edwards, K.M., et al. (Edwards, KM, Hackell, JM, and The Committee on Infectious Diseases, The Committee on Practice and Ambulatory Medicine, American Academy of Pediatrics). 2016. Countering vaccine hesitancy. *Pediatrics* 138 (3): e20162146.

Guibilini, A., T. Douglas, and J. Savulescu. 2018. The moral obligation to be vaccinated: Utilitarianism, contractualism, and collective easy rescue. *Medicine, Health Care, and Philosophy* 21: 547–560.

Chapter 28
Society's Obligations to Children

F. X. Placencia

Abstract Children are dependents and therefore a uniquely vulnerable group. Due to this vulnerability and their age, they are exquisitely sensitive to the social determinants of health. Various models of social justice have been used to tackle the special status of children with varying degrees of success. Powers and Faden's moderate essentialist theory is ideally suited to address the needs of children. According to their theory, these needs include threshold levels of health and reasoning. To provide these threshold levels, children need access to a decent minimum of health care. This access can be justified by appealing to the principles of public health ethics. The principle of interdependence seeks to balance individual rights and the acknowledgment that the health of some is intertwined with the health of others. However, because of the dependent nature of children, we must recognize their right to an open future and therefore the need to provide them with health care to protect that right. The principles of fundamentality and justice also oblige us to provide all children with access to health care in order to address the primary causes of disease. The best interest standard, the guiding principle of pediatric medical ethics, is also used as an ideal to guide social policy. The United Nations and other organizations have stated that the protection of children's interests is a positive right, and with it include the rights to life, nutrition, education, and health care.

Keywords Social justice · Social determinants of health · Best interest · Children's rights · Access to health care

28.1 Introduction

> "I believe the children are our are future
>> Teach them well and let them lead the way".
>> —Greatest Love of All—Whitney Houston

F. X. Placencia (✉)
Baylor College of Medicine—Texas Children's Hospital, Houston, USA
e-mail: fxplacen@bcm.edu

© Springer Nature Switzerland AG 2022
N. Nortjé and J. C. Bester (eds.), *Pediatric Ethics: Theory and Practice*,
The International Library of Bioethics 89,
https://doi.org/10.1007/978-3-030-86182-7_28

Children are a uniquely vulnerable group. Despite making up nearly a quarter of the global population, children hold no political power and thus possess no control over their current condition or their future. The very term *dependent* highlights their fragile political, economic, moral, and ethical status. Nevertheless, any society that hopes to endure must promote the welfare of its children.

This chapter will explore what society owes to its children. In the first section I will explore the social determinants of health and discuss how their distribution and the policies that govern their distribution reflect the priorities of a society with respect to its children and then explore how different theories of justice would address disparate outcomes. The second section will delve further by analyzing the arguments surrounding the access to health care for children from the perspective of public health. This chapter concludes with an examination of society's obligations to its children by focusing on the best interest standard as an ideal by which to guide social policy.

28.2 Social Determinants of Health and Social Justice

Ask a child why a tiger has stripes, and they will tell you that the stripes help the tiger to hide in the jungle from its prey. Likewise, they will tell you that polar bears are white to blend into the snow. No less than animals, humans are shaped by their environment. The conditions in which they are born, live, work, and play have a strong impact on their health and are referred to as the social determinants of health. Examples of social determinants are so broad that a comprehensive list is impossible to create, however the US Office of Disease Prevention and Health Promotion's Healthy People 2020 initiative (2020) organizes social determinants of health into five key areas:

– Economic Stability
– Education
– Social and Community Context
– Health and Health care
– Neighborhood and Built Environment.

These social determinants of health are shaped by how money, power, and other resources are distributed within a society, and can be altered by policy choices. This puts children in a particularly precarious position. As dependents they are neither able to directly influence policy, nor do they have control over the resources that shape the social determinants of health. Disparities in health outcomes can reveal uneven distribution in a society's resources, as well as the priorities that a particular society has in the policies that allocate those resources. This is true within a country; recall the frequently cited fact that an individual's zip code is strongly correlated with their life expectancy and other health indicators (Thomas et al. 2006). It can also reveal different priorities between countries, such as the how the lack of a robust social safety net results in poorer perinatal mortality rates in the US compared to

Europe (Chen et al. 2016). How these resources should be distributed, and what policies should be enacted is a question of social justice.

Whether in architectural design, law, or biomedical engineering, children are almost always an afterthought. Unfortunately, this is no less true for theories of justice. Most theories assume rational agents, social contracts, or atomistic individuals acting without regard for the relationships which both define society and influence moral behavior. Nevertheless, some approaches to social justice are better equipped to handle the special status of children. I will begin by exploring libertarianism, a theory of justice with considerable appeal in Western political philosophy, especially in the United States. I will then juxtapose three of the more compelling theories of social justice which support providing children with a decent minimum of health care.

Libertarianism has long influenced Western political philosophy, having arisen as a reaction to monarchical systems of government. The underlying conceit of most forms of libertarianism is that each individual is self-owning. Corollaries that follow this postulate include that each individual has an extensive set of negative liberty rights, including the right to own property, and that the power of government to use coercion is limited to protecting these liberty rights. A libertarian society should be thought of as dynamic web of voluntary associations between individuals. Libertarianism thus is closely related to Adam Smith's (1937) laissez-faire capitalism and Kantian deontology (Beck 1960).

It would be incorrect to say that libertarianism prohibits social programs such as universal access to health care. Indeed, individuals can voluntarily agree to participate in such mutual aid societies, but they cannot be mandated to do so. Nor can individuals be compelled to fund such a system, the belief behind such libertarian slogans as "taxation is theft." In general, health care in a libertarian society would be obtained through a free market by those willing and able to pay for it. Thus, children would be reliant on their parents' ability and willingness to purchase access to care for them, until they are capable of doing so for themselves. Charitable organizations may be able to assist those in need, but cannot be relied upon to do so, nor can they be mandated to choose their beneficiaries impartially.

Children's access to health care in a system based on libertarian principles would therefore be tenuous, reliant upon the social standing of their parents and the resources they could marshal. In such a system, the social determinants of health can be reliably predicted to exacerbate health inequalities, especially amongst vulnerable populations. To mitigate their unique vulnerability as dependents, the health care system must be founded upon a different theory of justice.

Utilitarianism, the theory first popularized by Jeremy Bentham and subsequently expounded upon by John Stuart Mill (Mill and Sher 2009), is founded on the precept that a just distribution of resources rests on the principle of utility. This principle holds that utility should be distributed to maximize the good to the maximum number of individuals. It requires that these terms be specified. For example, what is the utility being distributed? Is it health care dollars, physicians, or some other resource? And what is the good in question? Life expectancy, infant mortality rate, and quality adjusted life years are all valid indicators of health but measure vastly different things. Nonetheless, due to its consideration of all affected individuals, utilitarianism is ideally suited as a moral framework for public health.

Moral theorists have proposed variations on utilitarianism, a few of which bear discussion here (Ozoliņš 2015). Instead of judging individual acts, rule utilitarianism weighs whether a rule or policy maximizes the good. Consider the laws permitting minors to consent for medical care related to their sexual health, such as contraception or sexually transmitted illnesses. A frequent justification for these laws is that minors may not seek medical care if they have to involve their parents, whether out of fear of reprisals or because they were suffering abuse at hands of their parents. Thus, the policy is justified because in general it maximizes the good. In contrast, rather than maximize the good, negative utilitarianism seeks to minimize harm. Thus, an act or rule that seeks to minimize deaths within a population would be judged better than one which resulted in more deaths overall but benefited the survivors to such an extent that the overall number of years survived is higher. Therefore, a negative rule utilitarianism would strongly support access to basic health care for children due to the likelihood that such a policy would in general minimize harm to this population.

As a moral framework for public health, utilitarianism is not without its drawbacks. It permits extreme imbalances as long as those imbalances maximize the good. Such an appeal was used to justify the wealth imbalances caused by supply-side economics. Utilitarianism can be used justify an environment in which marginalized groups repeatedly bear the negative outcomes of a particular policy or social structure if it sufficiently benefits the mainstream population. Furthermore, without limitations, strict utilitarianism may override individual rights or liberties. In the midst of a public health emergency, this may be appropriate, however infringing on the liberty rights of individuals should never be taken lightly.

John Rawls proposed a theory on social justice that he described as "justice as fairness" (Almegren 2013). It is an ideal theory, in that it does not reflect reality, but instead postulates what society should be like in an ideal setting. His theory is a form of social contract model, in which all members of society start out in an "original position" from which the members agree to the rules governing society from a "veil of ignorance" as to what their social status in their society will be. Therefore, it can be assumed that the interests of children would be protected, as everyone making the rules for society is veiled from knowing if they will end up being a child once the veil is lifted.

There are two principles of social justice in Rawls' theory. The first is the equal basic liberties principle, which is that are certain basic liberties that are universal and indefeasible. Everyone possesses these liberties equally. The difference principle is second, and allows for social and economic inequalities, but only under two circumstances. The first is that these inequalities are attached to positions that are open to everyone under fair conditions of equality of opportunity and the second circumstance is that these inequalities benefit the most vulnerable members of society.

Bearing these two principles in mind, a society based on Rawlsian principles will ensure that its members have the requisite benefits necessary for a just society. The most important is an equal distribution of primary goods. Rawls describes these as "rights and liberties, opportunities and powers, and income and wealth." Though he didn't address health care in his initial works, Rawls eventually conceded that health care had to be one of these primary goods. Any inequalities in the distribution of these primary goods, such as the prestige or higher salary often afforded to physicians, must serve to benefit their most vulnerable patients. That these positions must be open to everyone under equality of opportunity supports a social environment with considerable social mobility and attention to social welfare including a right to health care.

Nevertheless, Rawls is not without his critics. In developing his capabilities approach, Amartya Sen (1992) argued that the Rawlsian fixation on the equal distribution of goods ignores the variation in need between different individuals. Similarly, according to Madison Powers and Ruth Faden, the chief concern of social justice should be the wellbeing of individuals, not whether they are afforded equal opportunities or that the primary goods are distributed equally. It is to their theory that I now turn.

In Powers and Faden's (2006) theory of justice, the wellbeing of individuals involves multiple dimensions. The authors identify six dimensions of wellbeing, any of which they believe a significant deficiency of threatens the wellbeing of the individual. For this reason, they describe their theory as a moderate essentialist theory, for although only a threshold amount of each dimension is required, nevertheless these dimensions are essential for wellbeing. The six dimensions identified are:

- Health
- Personal Security
- Reasoning
- Respect
- Attachment
- Self Determination.

The justice of social structures can be evaluated by how they promote or diminish the dimensions of wellbeing. For the purposes of this chapter, I will naturally be focusing on the dimension of health. However, it is important to note how the different dimensions may interact with each other. For example, a lack of personal security may negatively impact a person's health. Likewise, poor health may endanger the development of attachments or the capability to reason. Therefore, deficiency in one dimension may impact another. Powers and Faden argue that due to their importance, the six dimensions could serve as the origin for basic human rights. I will explore this idea further, especially as it pertains to children and their access to health care.

28.3 Access to a Decent Minimum of Health Care

There is some aspect of our common morality, that awareness of certain moral norms that is innate to most people (Veatch 2003), which intuits that children should have access to basic health care. Even in the United States, with its historically dysfunctional approach to health care, there has been bipartisan support to ensure that all (citizen) children have access to health care through the approval of the SCHIP program and its various reauthorization bills.[1] Certainly, the last three theories of justice explored in the previous section strongly support policies guaranteeing access to a decent minimum of health care for all children. In the following section, I will first establish why libertarianism cannot adequately justify or even provide access to health care for children. Since providing access to health care is a matter of public health, I will then explore how the principles of public health justify providing children access to a decent minimum of health care.

The first step in justifying children's access to a decent minimum of health care is to explore the unique position that children are in (also refer to Chap. 1 for an international perspective). I laid the groundwork for this argument in the introduction, but I will revisit it here. Children should be considered to have full, independent moral status, meaning that when decisions are made that affect the child, we are obligated to consider their interests (Jaworska and Tannenbaum 2018). However, children often lack the cognitive capacity to effectively advocate for their own interests. Even when they do possess this capacity, as legal dependents children do not possess the legal competence to do so. Furthermore, they generally lack the resources, financial and otherwise, to act on their interests. It is for these reasons that a libertarian approach to health care, which relies on the voluntary actions of individuals to obtain care for themselves, cannot adequately handle the position of children. A society in which children must rely on their guardians to obtain the care they need and thus safeguard their medical interests cannot value children as having full, independent moral status. In such a state, children become, in fact, closer to the position of chattel, their interests subject to the whims and fortunes of their parents. Thus, if their guardians cannot provide access to at least a decent minimum of health care, it falls upon the state to do so.

Having established that children possess full, independent moral status, I now turn to the justifications for providing them access to a decent minimum of health care. I have established that ensuring that children have access to at least a decent minimum health care is the responsibility of the state, it may be helpful to analyze the justifications for its provision through a public health lens. In order to provide straightforward ethical guidance, many authors have put forth principles of public health ethics. Taking their cue from Tom Beauchamp and James Childress' Georgetown mantra, Geoffrey Swain and colleagues (2008) have distilled these principles down to four. For the purposes of this chapter, we will utilize these principles as they represent normative arguments that are widely accepted in the literature without being overly derivative of Beauchamp and Childress' (2013) Four Principles. These principles are: interdependence, community trust, fundamentality, and justice.

Interdependence is the principle that seeks to balance respect for individual rights with the understanding that the health of some is often intertwined with the health of

---

[1] Although not immune to partisan wrangling, support for SCHIP and its reauthorization bills has been largely bipartisan and signed into law by both Democratic and Republican presidents. For more information, go to the Kaiser Family Foundation Report at https://www.kff.org/wp-content/uploads/2013/01/7743-02.pdf.

others. This principle neatly reflects the unique position of children. Dependent on their families and communities to provide for their basic needs, children are enmeshed in their environment to such a degree that their health is exquisitely sensitive to the health of those around them and the social determinants of health.

Balanced against this concern are the traditional liberty rights. However, these rights go beyond freedom from coercion. Otherwise, society would be forced to balance the health of children against a libertarian worldview which I have already shown cannot adequately address the need to provide children with access to a decent minimum of health care. To more fully consider the moral interests of children, we must also consider the child's "right to an open future." Joel Feinberg argued that children have their own anticipatory liberty rights which are "violated when children's future options are prematurely closed, and respected when children's future options are kept open" (Mills 2003). In short, parents should not make decisions that foreclose future options for their children (refer to Chap. 3). Mianna Lotz (2006) developed this concept further, arguing that the right to an open future is more than a negative liberty right; the right creates positive obligations on the part of the parents to "provide adequate conditions for a child's emerging autonomy." This emerging autonomy cannot be protected without access to at least a decent minimum of health care, parents are obligated to provide this access, and when they cannot, this obligation falls to the state. Thus, interdependence obliges us to balance negative liberty rights with the right to an open future and the fact that the health of children is intertwined with the health of their family and community.

The development and maintenance of **community trust** is the second principle. It reflects the need to engage the community and obtain its consent for participation in public health measures, to ensure transparency and confidentiality when appropriate, and to do so in a culturally appropriate way. While this principle is useful in guiding society in *how* to provide access to health care for children, it is less useful in justifying *why* such care should be provided. It is included here not merely for the sake of completeness however, but also because it underscores that without community trust, effective, and therefore just, health care cannot be provided to children.

The third principle, **fundamentality**, demands that we focus on the fundamental causes of disease. To do so requires attention to the core requirements for health. Although there are many such requirements including the physical and social environment in which children are raised, for the purposes of this chapter I will focus on the provision of health care. There is ample data to show that increasing children's access to care improves their health outcomes (Leininger and Levy 2015). Furthermore, Paul Farmer (2004) argues that the lack of access to medical care is a form of institutional violence against the poor. When coupled with existing structural inequalities, such as those resulting from inequalities in the physical and social environment, i.e. the social determinants of health, this leads to excess mortality amongst the most vulnerable, what he calls "stupid deaths." Since it is the children of the poor who are most likely to lack access to a decent minimum of care, fundamentality obliges society to provide it to prevent needless morbidity and mortality.

The final principle is **justice**. Although related to Beauchamp & Childress' version, Swain and colleagues' (2008) public health principle of justice unsurprisingly focuses on social justice. They state that justice obliges us to ensure that the conditions necessary for health are accessible to all, including the vulnerable and the disenfranchised. In describing the principle of fundamentality, I made the case that the provision of access to a decent minimum of health care is necessary to address the underlying causes of disease. Likewise, in order to ensure that the children of the poor are able to enjoy the conditions necessary for health, we must ensure that all children are provided such access.

Due to the initial assumptions in libertarian theory, it is unable to justify providing all children access to at least a decent minimum of health care. Yet our common morality tells us this is an unjust scenario. However, by appealing to public health ethics, particularly the principles of interdependence, fundamentality, and justice, strong justification for providing such access can be found. Having also found support for providing children access to a decent minimum of care within theories of social justice, I will now turn to describing society's obligations to protect their interests.

28.4   What Society Owes Children: Society's Obligations to Protect Childhood Interests

"What do I have to do for my child? What kind of world will I leave for them?" These are questions that keep many a new parent up at night. But what do we owe children on a societal level? In this final section, I will explore society's obligations to children and their interests.

I have written elsewhere that the best interest standard (refer to Chap. 4) has long been the guiding principle of pediatric medical ethics (Placencia and McCullough 2011). However, Loretta Kopelman (1997), after examining how this principle has been used beyond medical ethics, argues that the best interest standard is also used as an ideal to guide social policy. Following this ideal creates a social obligation to protect children. For example, children must be protected from neglect and abuse and other harms which may impair their development. This is the inspiration behind child-labor laws or environmental laws such as those banning the use of lead paint. This ideal goes farther however, by creating an obligation to provide children with certain goods in order to promote their wellbeing and enable them to realize their potential. I would argue that chief amongst these goods are food, education, and health care. By providing these goods, we enable children to transition from dependents, reliant on others to safeguard their wellbeing and promote their interests, to fully individuals prepared to exercise their autonomy.

Nowhere is this commitment to children's interests better exemplified than in the resolutions passed by international world bodies. These include the Geneva Declaration of the Rights of the Child (1924), the UN Declaration of the Rights of the Child (1959), and the UN Convention on the Rights of the Child (1989)(refer to Chap. 5).

The League of Nations passed the Geneva Declaration in 1924 in response to the tragedies suffered by children during World War I. It outlines five rights which must be provided to children. These include (1) the means necessary for normal development, (2) food, shelter, special education for those in need, and health care, (3) the priority for relief in times of distress, (4) the opportunity to earn a livelihood and be protected from exploitation, and (5) a commitment to use their talents to serve everyone. These rights were affirmed by the United Nations in 1959 and expanded upon. The rights added in 1959 include the right to racial and religious equality, a name and nationality, love and understanding by parents and society, and the right to play.

Finally, in 1989, UN signed the Convention on the Rights of the Child. The Convention is a far more comprehensive document than its antecedents, with 54 articles identifying different types of rights. These rights can be grouped into four categories (Kosher et al. 2016). The first category includes the right to life, and to rights essential to survival, and development, including health care and education. Non-discrimination rights make up the second category. The third category enshrines the child's best interest as the guiding principle of all actions concerning them. Lastly, are self-determination and participation rights aimed at enabling children to become active members of society.

The obvious conclusion to be reached from these landmark documents is that there is widespread acceptance that all societies owe a great deal to children. By acknowledging their rights to health care, education, food and shelter, we acknowledge that we owe them the opportunity to grow into functional adults, to explore and develop their capacities and interests, and to become functional members of their communities. But perhaps more importantly, by acknowledging their rights to be free from abuse, to play, and to their parents' love, we acknowledge that we owe them the right to be children.

28.5 Conclusion

As dependents, children are by their nature a particularly vulnerable group. This vulnerability is only heightened by their susceptibility to the environment in which they are raised. In this chapter, I have examined how the social determinants of health impact children. By exploring different models of social justice, I have highlighted the shortcomings in libertarianism when it comes to children's health, and have proposed that utilitarianism, Rawlsian liberalism, and the moderate essentialist theory of Powers and Faden are better equipped to address the needs of children. Of these, I believe Powers and Faden most compellingly justify the need to protect the interests of children by identifying the six domains they claim are required for human flourishing and tying these domains to basic human rights.

The need to provide children with access to a decent minimum of health care can also be justified by appealing to the principles of public health. Of these principles, interdependence seeks to balance respect for individual rights with the acknowledgement that the health of some are tied to the health of others. When applied to children, this obliges us to consider a child's right to an open future when ascertaining that balance. Protecting that right creates a positive obligation to protect the health of children. The principles of fundamentality and justice also oblige us to provide all children, especially those most at risk to suffer from disparities in the social determinants of health, with access to health care in order to address the primary causes of disease.

The United Nations and other organizations have long relied upon the best interest standard as an ideal by which to guide their policies on children. Inspired by this standard, these organizations have identified rights, both negative and positive, essential to the safeguarding of the interests of children. Foremost amongst these rights are the right to life, food, education, and health care. No less important however, is the right to just be children.

Guiding principles for society's obligations to children

Social determinants of health
– The conditions in which people are born, live, work, and play in, they are shaped by how resources are distributed within society. These conditions have a strong impact on their health and can be altered by policy choices
Theories of Social Justice
– How resources are distributed is a question of social justice. Libertarianism allows for disparities in health outcomes, which is exacerbated by the dependent nature of children. Utilitarianism strives to either maximize health care outcomes or minimize harm, but also allows for health care disparities. Rawlsian justice promotes equality of access to health care. Powers and Faden's moderate essentialist theory goes a step further arguing that social justice requires the provision of a minimum threshold of six dimensions of wellbeing, one of which is health
The moral status of children
– Children have full, independent moral status, and thus they are entitled to access to a decent minimum of health care. This protects their right to an open future
Principles of public health
– Swain and colleagues have identified interdependence, community trust, fundamentality, and justice as the four principles of public health. These principles justify the provision of a decent minimum of care to all children
Best interest standard
– As an ideal to guide social policy, the best interest standard creates a social obligation to protect children and their interests. This protection includes the provision of goods that promote their wellbeing, including access to health care

References

Almegren, G. 2013. A primer on theories of social justice and defining the problem of health care. In *Health care politics, policy, and services. A social justice analysis*, 2nd ed., ed. G. Almegren, 1–41. New York, NY: Springer USA.

Beauchamp, T., and J. Childress. 2013. *Principles of biomedical ethics*. New York, NY: Oxford University Press.

Beck, L.W. 1960. *A commentary on Kant's critique of practical reason*. Chicago, IL: University of Chicago Press.

Chen, A., E. Oster, and H. Williams. 2016. Why is infant mortality higher in the United States than in Europe? *American Economic Journal. Economic Policy* 8 (2): 89–124. https://doi.org/10. 1257/pol.20140224.

Farmer, P. 2004. *Pathologies of power*. Oakland, CA: University of California Press.

Jaworska, A., & Tannenbaum, J. 2018. *The grounds of moral status*. https://plato.stanford.edu/ent ries/grounds-moral-status/. Accessed 9 Oct 2020.

Kopelman, L. 1997. The Best interests standard as threshold, ideal, and standard of reasonableness. *The Journal of Medicine and Philosophy* 22: 271–289.

Kosher, H., A. Ben-Arieh, and Y. Hendelsman. 2016. *Children's rights and social work*. Cham: Springer International Publishing AG.

Leininger, L., and H. Levy. 2015. Child health and access to medical care. *The Future of Children* 25 (1): 65–90.

Lotz, M. 2006. Feingber, Mills, and the child's right to an open future. *Journal of Social Philosophy* 37 (4): 537–551.

Mill, J., and G. Sher. 2009. *Utilitarianism*. Indianapolis, IN: Hackett Publishing Company Inc.

Mills, C. 2003. The child's right to an open future? *Journal of Social Philosophy* 34 (4): 499–509.

Office of Disease Prevention and Health Promotion. 2020. *Social determinants of health*. HealthyPeople.gov.   https://www.healthypeople.gov/2020/topics-objectives/topic/social-determ inants-of-health.

Ozoliņš, J.T. 2015. Rationality in utilitarian thought. In *Foundations of health care ethics: Theory to practice*, ed. J. Grainger and J.T. Ozoliņš, 102–119. Australia: Cambridge University Press.

Placencia, F.X., and L.B. McCullough. 2011. The history of ethical decision-making in neonatal intensive care. *Journal of Intensive Care Medicine* 26 (6): 368–384.

Powers, M., and R. Faden. 2006. *Social justice: The moral foundations of public health and health policy*. New York, NY: Oxford University Press.

Sen, A. 1992. *Inequality re-examined*. Oxford: Clarendon Press.

Smith, A. (1937). *An inquiry into the nature and causes of the wealth of nations*. Reprinted, New York, NY: Random House (Original work published 1776)

Swain, G.R., K.A. Burns, and P. Etkind. 2008. Preparedness: Medical ethics versus public health ethics. *Journal of Public Health Management and Practice* 14 (4): 354–357.

Thomas, A.J., L.E. Eberly, G.D. Smith, and J.D. Neaton. 2006. ZIP-code-based versus tract-based income measures as long-term risk-adjusted mortality predictors. *American Journal of Epidemiology* 164 (6): 586–590. https://doi.org/10.1093/aje/kwj234.

Veatch, R.M. 2003. Is there a common morality? *Kennedy Institute of Ethics Journal* 13 (3): 189–192. https://doi.org/10.1353/ken.2003.0024.

Further Readings

Patel, K., & Rushefsky, M.E. 2014. *Health care politics and policy in America*. Armouk, NY: M.E. Sharpe.

Sandel, M.J. 2007. *Justice: A reader*. New York, NY: Oxford University Press.

Young, I.M. 2011. *Justice and the politics of difference*. Princeton, NJ: Princeton University Press.

Chapter 29
Pediatric Resource Allocation, Triage, and Rationing Decisions in Public Health Emergencies and Disasters: How Do We Fairly Meet Health Needs?

D. J. Hurst and L. A. Padilla

Abstract Issues of resource allocation, triage, and rationing decisions are common in the context of disasters and public health emergencies, such as pandemics. However, to date, the majority of the literature focuses on an adult population with very little attention given to a pediatric population or to a population that may be mixed: adults and children. Furthermore, decisions of rationing scarce resources do not only occur during disasters and other wide-scale emergencies. Such decisions are commonplace in pediatric organ transplantation and can creep into areas of experimental medicine, such as xenotransplantation. This chapter explicates differences between pediatric and adult triage decisions by first looking at the trifold issues of resource allocation, triage, and rationing decisions mainly in the context of a disaster situation or public health emergency. However, the issues of pediatric organ transplantation, health disparities, and experimental medicine will also be covered.

Keywords Allocation · Disaster medicine · Pediatric · Public health emergency · Rationing · Triage

29.1 Introduction

In many Western societies, health care resources are relatively abundant. Yet, even in contexts where resources are plentiful, to some extent, resources must still be rationed or allocated limitedly due to the finite budgets of governments and other public goods and goals of society that health care must compete against (e.g. social welfare resources, education). Yet, in emergency or disaster situations, resource allocation

D. J. Hurst (✉)
Department of Family Medicine, Rowan University School of Osteopathic Medicine, Stratford, NJ, USA
e-mail: hurst@rowan.edu

L. A. Padilla
Department of Epidemiology and Surgery, University of Alabama at Birmingham/Children's of Alabama, Birmingham, AL, USA
e-mail: lpadilla@uabmc.edu

© Springer Nature Switzerland AG 2022   465
N. Nortjé and J. C. Bester (eds.), *Pediatric Ethics: Theory and Practice*,
The International Library of Bioethics 89,
https://doi.org/10.1007/978-3-030-86182-7_29

becomes especially acute. The field of disaster medicine is accustomed to issues of resource allocation, triage, and rationing decisions, but children are rarely treated as the subject of these discussions or given any specific attention (Antommaria et al. 2010). Rather, adults are the focus, either explicitly or indirectly. A mass critical care task force noted the paucity of guidelines that explicitly refer to children, stating "the area most desperately in need of future study is pediatric triage" (Devereaux et al. 2008).

Rationing of scarce medical resources also routinely occurs outside disasters and emergency situations. For instance, resource allocation decisions are commonplace in pediatric organ transplantation and can creep into areas of experimental medicine, such as xenotransplantation. This chapter will examine differences in pediatric and adult triage decisions by first looking at the issues of resource allocation, triage, and rationing in the context of a disaster situation or public health emergency. However, pediatric triage and resource allocation will be broadened to examine the issues of pediatric organ transplantation, health disparities, and experimental medicine.

29.2   Case Study

In a severe pandemic there is the worry that demand for certain health care resources, such as intensive care, will exceed what the health care system in a particular context can handle, which is termed a surge. In the wake of the COVID-19 pandemic of 2020, many ethicists, hospitals, and professional organizations developed policies regarding the allocation of ventilators and other resources in the event that the demand for these resources surged, creating a situation in which demand outpaced supply. A cross-sectional study of children admitted to pediatric intensive care units at one point during the height of the pandemic in early 2020 found that 38% of children required ventilation during their admission (Shekerdemian et al. 2020). In a high-density context with a large case load, a situation in which the demand for pediatric ventilators, as well as persons trained to manage their settings, could plausibly have outpaced supply. In such a case, how would triage and allocation of these scarce resources have occurred?

Older adults and children are commonly labeled as vulnerable populations, yet the majority of ventilator and resource shortage discussions have only been held nationally in adult clinical units. Disproportionately listening to adult over children's needs during the pandemic may lead to an availability crisis and thus have an effect on triage. Because of such cognitive bias, one may believe that the priority for care is higher for a COVID positive 70-year-old individual with respiratory distress and who requires a ventilator than that of a five-year-old patient. A simple utilitarian ethic does not help us out of this dilemma. One must consider and weigh issues such as that the five-year-old has much life left to live. The various items to consider when approaching this sort of ethical question will be kept in focus throughout this chapter.

29.3 Disasters and Public Health Emergencies

Disasters and public health emergencies are dynamic events. There is often a high degree of uncertainty with these events, and their unpredictable nature limit response and the options that are available, practically and ethically (Afolabi 2018). This section begins by defining the terms that will be used in the chapter and presenting the ethical theory that underpins the content herein. From there we move on to the differences in resource allocation, triage, and rationing decisions in a pediatric population as opposed to an adult population.

29.3.1 Defining Our Terms

Decisions of resource allocation, triage, and medical rationing appear most acutely in times of disaster and public health emergencies. Triage decisions in the midst of a disaster or public health emergency can be overwhelming and leave health care workers with moral distress and other feelings. Yet, this also points to the need for established rules and guidance prior to the emergency situation and for health care workers who may respond to such situations to be well trained in these areas. Though these plans are certainly within the realm of medicine, we must not miss that they also reflect moral values, which must be adequately articulated and justified. While the terms resource allocation, triage, and medical rationing are similar and may even be used synonymously at times, there are nuances that need to be noted.

Triage is one way to ration medical resources (e.g., supplies, personnel) when not everyone's needs can be met at the same time and to the same degree (Childress 1997). The term *triage* was originally used in agriculture and then spilled over into military usage and lexicon. The French army under Napoleon sorted casualties in war according to a triage system, yet they did not term it this. During World War I, Allied armies used the terms *triage* and *sorting* to refer to the systematic classification of casualties according to the needs of the patient (Childress 1997). Because of this history, there has been a lot written on triage in mass casualty scenarios within a military context (Gross 2006).

We can define triage as a system of allocation or rationing under crisis or emergency circumstances in which a decision must be made immediately about the treatment or non-treatment of patients (Childress 1997). Categories within a system of disaster triage may differ but generally follow a schema such as: immediate (patient is critical; death or serious impairment if not treated immediately), urgent (likely to survive if care is provided, but not immediate treatment), nonurgent (minor injury; care can be delayed while treating others) (Childress 1997; Kipnis 2002). Triage systems may also include a category of those who are dead or not expected to survive even with treatment in the field.

Within a system of triage, decisions of medical rationing and resource allocation must be made. Medical rationing is the idea that plentiful medical resources to treat all needs likely do not exist and, thus, resources must be allotted in a specific way. As noted by Dan Brock resource scarcity leads to resource allocation, when resources are scarce then what naturally follows is some degree of health care rationing—denying a potentially beneficial health care resource to some who may benefit from it (Brock 2004). This is where a system of triage would be useful in deciding how and on whom medical resources are utilized.

29.3.2 Ethical Theory

Much of the literature on resource allocation, triage, and rationing decisions within bioethics takes place within the two subfields of public health ethics and disaster ethics (O'Mathuna 2014). Either explicitly or implicitly, all systems of triage have a utilitarian foundation (Childress 1997). While that much can be agreed upon, how to implement triage systems varies considerably, and there is a lack of universal agreement. This has been seen most prominently in the myriad triage guidelines, each with nuances, that have been offered during the COVID-19 pandemic. Nonetheless, this section offers some of the ethical theory that underpins these systems.

The first systematic accounts of utilitarianism were developed independently by Jeremy Bentham (1748–1832) and John Stuart Mill (1806–1873) (Pence 2017). While there were distinctions to each version, the essence of utilitarianism is that the greatest amount of utility (literally: happiness, wellbeing) for the greatest number of people should be sought. Utilitarianism forms the basis for triage and the belief that if not everyone can be saved then the maximization of wellbeing should be sought.

Realizing that in a public health emergency or disaster event that not everyone may be able to receive care, let alone receive what would typically be considered a normal standard of care, may strain our moral sensibilities. To be blunter, when faced with the decision to set aside the most seriously injured to not receive care so that resources can be focused on those who may still be able to be saved, the moral distress created may be severe. This points to the need that in order for triage to be seen as a necessity then it must be explicitly grounded and properly justified in ethics. Philosopher Ken Kipnis has noted that there are powerful ethical arguments in support of triage. Kipnis argues that (1) triage produces the best outcomes as it saves lives and prevents deaths; (2) in a hypothetical situation, rational persons, would choose triage as it gives them a greater chance of survival; and (3) in a situation where resources are limited, triage prevents waste and is a good steward of what is available (Kipnis 2002).

Kipnis' rationale rests upon a utilitarian ethic that seeks good outcomes for the greatest number of individuals, as well as the preservation or stewardship of critical resources. These points are widely agreed upon in the literature in regard to triage.

Yet, a utilitarian ethic alone may also be insufficient for us without other considerations. Jeffrey Burns and Christine Mitchell (2011) explain the difficulty of solely using a utilitarian ethic as it does not help us determine if we should save two elderly adults who might live five more years each versus one child who might live an additional 70 years. Should health care professionals seek to preserve the most lives in such a scenario or the patient with the most potential life-years remaining (Burns and Mitchell 2011)? Empirical data on the viewpoints of society may play a role in our decision-making, but it is also just one aspect to consider. For instance, a 2010 poll by the American Academy of Pediatrics found that 76% of Americans believe that, during a disaster situation, if resources are limited, children should be given priority versus adults for life-saving treatments, and 75% believe children should be treated first for the same condition (American Academy of Pediatrics 2010). However, can we directly translate these viewpoints into an ethical justification and into actual policy?

Burns and Mitchell (2011) have provided one of the only comprehensive accounts on resource allocation and triage in pediatric populations, offering five different criteria that have been proposed for how to deal with triage. We provide each of these criteria in Table 29.1 along with a short explanation of each.

Each criterion in Table 29.1 is not without its critics. Social value criteria may be critiqued for its impreciseness and seeming discrimination. In critiquing a set of triage guidelines published by the Spanish Society of Critical Intensive Care Medicine and Coronary Units in the wake of COVID-19, Herreros and colleagues

Table 29.1 Criteria to determine triage

Social value criteria	Some have proposed that the allocation of scarce health care resources should be based upon social value criteria. That is, the contributions a person offers to society—including past contributions and potential for future contributions—should determine their priority for receiving scarce health care resources
Instrumental value criteria	The instrumental value criteria is a more narrow construction of the social value criteria and refers to a person's ability to perform a specific (usually important or specialized) function within a given society (e.g. health care workers; emergency personnel)
Life-cycle criteria	Also known as "fair innings" or "years life saved," there are varying versions of this criteria. A strict view would grant priority primarily based on age, where a 6-month-old child would take precedence over a 12-month-old. However, a more nuanced version of these criteria seeks to allocate resources so that each person has equal opportunity to live through various stages of a normal lifespan. Younger persons would generally be favored, as they have not had the opportunity to progress through older stages
Conservation criteria	The focus in conservation criteria is on the quantity of resources that are required for an individual person. The argument here is that resources should be maximized by focusing on persons who will likely benefit with the use of minimal resources
Egalitarianism	Egalitarianism argues for equal regard for individuals. Rather than seeking to maximize benefit for the greatest number of persons (utilitarianism), egalitarianism argues that each person should be given equal chance to survive

noted that social value criteria are abstract, subjective, and do not lend themselves to easily being agreed upon, which may increase discrimination against certain persons. Furthermore, Herreros et al. argue that if a society could agree on social value criteria then health care professionals should not be the ones to carry out the assessment due to lack of expertise and the fear that such determinations by physicians could undermine society's trust in the profession (Herreros et al. 2020).

Instrumental value criteria can be critiqued for its ambiguity and the social hierarchy it creates. Certain persons are deemed of greater instrumental value to society and should receive care first. Often these personnel are labeled "essential" and may include frontline health care workers, police officers, and emergency medical service personnel. However, in a public health emergency, such as COVID-19, the essentiality of other persons to support those deemed "essential" and the working of society has been highlighted, such as grocery store workers, gas station attendants, primary school educators, and other persons we rely upon for a normally functioning societal system. Society is joined by interacting entities that work as a unified whole. Changes that affect any of its pieces regardless of its role within the system can have a domino effect on others. Hence, one would argue that all pieces within a system are valuable and the notion of "instrumental value" is not without its challenges.

The life-cycle criteria can also be critiqued for its ambiguity. If one does not take a strict view of whether to save the 6-month-old versus the 12-month-old who are in relatively similar physical conditions and at relatively similar life stages, how does one decide? The answer is not altogether clear.

Conservation criteria focuses on resource utilization, noting that resources should be maximized by focusing on persons who will likely benefit with the use of minimal resources. This may have the effect of neglecting persons who may a high likelihood of survival, yet their treatment may exhaust a significant volume of supplies.

Egalitarianism argues that all persons should be given equal chance to survive. However, this may result in an inordinate amount of critical resources (medical supplies, health care personnel, time) being allocated to persons who are unlikely to survive. A potential result is that persons who have a higher likelihood of survival may not receive treatment as immediately. Egalitarianism would also favor a random lottery system for treatment of persons with similar conditions, which may seem arbitrary (Bannon and Farmer 2015).

29.3.3   Guidance from Professional Codes and Organizations

Professional organizations have issued guidance in regard to ethics in the midst of disaster or emergency situations. The World Medical Association (WMA) has issued the guidance on medical ethics in the event of disasters, noting that triage systems must be solely based on the patient's medical status and predicted response to the treatment, which may lead to those persons who are the most seriously injured receiving only symptom control such as analgesia, when available. While demanding, per WMA, such systems "are ethical provided they adhere to normative standards"

(World Medical Association 2017). Furthermore, WMA makes such advice based upon utilitarian notions, which is clear here:

> 8.2.1 It is ethical for a physician not to persist, at all costs, in treating individuals "beyond emergency care", thereby wasting scarce resources needed else-where. The decision not to treat an injured person on account of priorities dictated by the disaster situation cannot be considered an ethical or medical failure to come to the assistance of a person in mortal danger. It is justified when it is intended to save the maximum number of individuals. However, the physician must show such patients compassion and respect for their dignity, for example by separating them from others and administering appropriate pain relief and sedatives, and if possible, ask somebody to stay with the patient and not to leave him/her alone.

> 8.2.2 The physician must act according to the needs of patients and the resources available. He/she should attempt to set an order of priorities for treatment that will save the greatest number of lives and restrict morbidity to a minimum. (World Medical Association 2017)

The American Academy of Pediatrics (AAP) has published a set of training materials—"Planning and Triage in the Disaster Scenario"—to aid emergency preparedness planning (Ugarte et al. 2011). AAP highlights the "unique physiological, psychological, and developmental needs" of the pediatric population and offers guidance on how to design emergency preparedness plans to respond to a disaster situation (Ugarte et al. 2011).

29.3.4 Pediatric Differences

The uniqueness of the resource allocation, triage, and rationing decisions discussed in this chapter lie not so much with the topic as with the nature of the subject who is the focus here: the pediatric population. The Pediatric Emergency Mass Critical Care Task Force has noted that there exists particular needs and vulnerabilities—physical and psychological—in the pediatric population that differ from an adult population (Antommaria et al. 2011). Bioethical discourse, especially in areas of research ethics and global bioethics, expresses concerns about the susceptibility of a population to being injured, and terms this *vulnerability*. While in one sense all human beings are vulnerable to one injury or another, there is a special sense in which certain populations of people, or certain situations, can make amplify vulnerability and signal their need for further protections. Vulnerability here is simply defined as the susceptibility of being harmed or wounded, which may include physical, emotional, and/or psychological harm (ten Have 2016). Compared to adults, children may be disproportionately vulnerable to harm during a crisis (Burns and Mitchell 2011). For the pediatric population, in a public health emergency or disaster event, physical vulnerabilities may include being more susceptible to hypothermia or in an environment with low herd immunity they are at higher risk of an infection (Antommaria et al. 2011, S164).

Depending on their age and level of development, in a disaster situation a child may not be able to physically leave the affected area. Furthermore, the lack of published literature that has gone into the subject of pediatric triage may be reflective of a lack of overall planning for pediatric public health emergencies and disasters, which is a further vulnerability for children. Lack of planning may increase the risk of pediatric morbidity and mortality which could have otherwise been prevented had adequate planning been conducted (Antommaria et al. 2011). Moreover, children may not easily communicate their needs and rely on their primary caregiver to voice or attend them. The unique vulnerabilities of children create a moral duty for certain sectors of society (e.g. governments, emergency medical personnel, hospital systems) to create triage and resource allocation plans that incorporate pediatric needs and adequately train those who are most likely to care for them.

In addition to physical vulnerabilities, children also face unique psychological vulnerabilities. Antommaria and colleagues (2011) have noted that children, in a disaster situation, may be unable to communicate clearly about their symptoms and needs. Furthermore, they are vulnerable due to limited ability to recognize danger and react when danger is present. These vulnerabilities may be increased in children with disabilities (Antommaria et al. 2011).

These particular risks should be accounted for in planning for the response to disasters and public health emergencies that will likely include a pediatric population. Children operate most often within a familial network in which their parents or guardians make many of their decisions for them. Respect for the role of the family unit during a crisis may be critical to the success of disaster management (Maves et al. 2020). In an infectious disease public health emergency, such as COVID-19, institutions have to strike a balance between the needs of family-centered care vs the risk of spreading an infectious disease (Maves et al. 2020). As at least one parent or guardian typically travels with a pediatric patient and stays in close contact during hospitalization, in the case of an infectious disease then such caregiver(s) should be treated as presumably exposed and should be further monitored for symptoms and tested if available (Maves et al. 2020).

As this section has highlighted, there are several factors to consider in disaster and public health emergencies. Table 29.2 provides some guiding principles for pediatric resource allocation, triage, and rationing in these situations.

Table 29.2 Guiding principles for the ethics of pediatric resource allocation, triage, and rationing resources

Triage is a necessary process for resource allocation and rationing resources. However, one must consider the application of egalitarianism and the social, instrumental, life cycle and conservation criteria to appropriately determine triage

Aspects to consider when triaging, allocating, and rationing resources
- patient's medical status
- years of potential life loss
- patient's ability to respond to treatment
- assess if the decision will have risks to others
- resource must be rationed in a manner that seeks to maximize total benefit
- be explicitly grounded and properly justified in ethics

It should be noted that these aspects to consider may change in the setting of novel or experimental medical treatment or during public health crises

29.4   Allocating Scarce Resources Outside of a Disaster Event

Decisions regarding resource allocation, triage, and rationing do not only occur in states of disaster or public health emergencies. Scarce resources, such as organs, must regularly be rationed in a manner that seeks to maximize total benefit, amongst other considerations (American Medical Association 2016; Persad et al. 2009). However, disparities exist in how resources are allocated, which leads to the question of whether scarce resources, such as organs, are distributed fairly. This section looks at these myriad considerations using the allocation of organs in a pediatric population as the main focus.

29.4.1   Pediatric Health Disparities in Resource Allocation, Triage, and Rationing Decisions

In the US, around 2,000 children under the age of 18 are on the national transplant waiting list. Overall mortality rates and outcomes depend on a number of variables, such as patient age and comorbidities. Organs are scarce resources, yet some are scarcer than others, such as pediatric hearts. As we have noted previously, few options exist for infants with complex cardiac malformations. An example of this is hypoplastic left heart syndrome (HLHS). Children with HLHS who receive an allo-transplantation fare moderately well, with 60% 25-year graft and patient survival. However, the problem becomes supply; oftentimes a compatible deceased human donor organ is not available when the child needs it. Because of this scarcity, the mortality for an infant on the cardiac waitlist is approximately 35–40% (Hurst et al. 2020).

   In the event that two pediatric patients are equally suitable candidates for one pediatric heart, how should the resource be allocated fairly? Different systems have been proposed for allocating scarce resources outside a disaster event, such as a first-come, first served-system, lottery system, or the urgency of need.

   The health care disparities gap constantly changes as advances in medical care are made (Mechanic 2005). Children with congenital heart disease bear one of the highest organ waitlist mortalities when compared to other age groups. This dejected fact is only worse for racial/ethnic minorities, whose wait-list mortality is even higher than White children, even after taking into account clinical variables like condition at listing, underlying diagnosis and blood type (Singh et al. 2009). The current organ allocation system intends to be clinically egalitarian through the use of a computer that constantly examines and evaluates the list for a potential match. The factors considered vary by organ but remain mostly based on the current medical status and need of the patient, size of the organ, blood type compatibility and distance from donor hospital (the heart has the shortest organ preservation time at 4–6 h) (U.S. Department of Health & Human Services). Despite this attempt of being impartial,

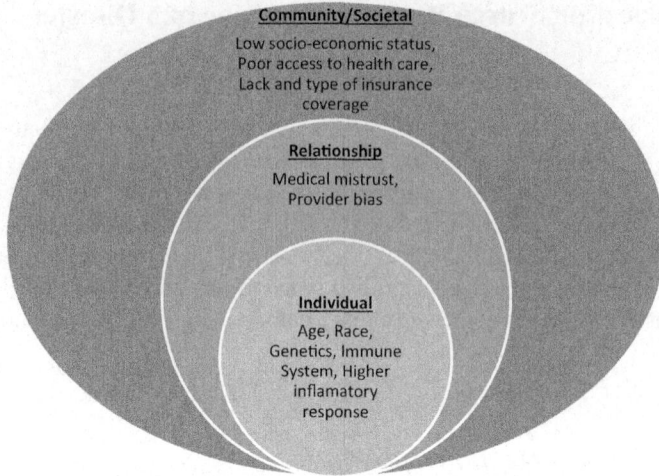

Fig. 29.1 Social-ecological model (adapted from the U.S. Centers for Disease Control and Prevention)

Black children face higher mortality and a higher risk of graft loss when compared to other races (Mahle et al. 2005). Non-individual factors of the social-ecological model exist that may be more controllable and that contribute to pediatric heart transplant disparities that affect resource allocation, triage, and rationing decisions within this population. Some examples using the social-ecological model as described by the US Centers for Disease Control and Prevention are noted in Fig. 29.1 (U.S. Centers for Disease Control and Prevention). However, these socio-economic factors seem to only play a small role in the racial disparities seen. Factors that contribute to this racial disparity for Black children seem to stem from uncontrollable inherent characteristics such as genetic or immunological differences (Mahle et al. 2005). Therefore, should the current organ allocation system have a way to be racially equitable and not equal? Should race be a factor that influences your place on a waitlist especially if the resource has life-saving potential?

29.4.2 Resource Allocation, Triage, and Ratioing Decisions in Experimental Medicine

This final section examines resource allocation, triage, and rationing decisions that may be made in areas of experimental medicine. Experimental medicine is the category of medicine that encompasses a broad array of medical research, from benchwork to clinical trials. Our main focus here will be xenotransplantation, though some principles may be applicable elsewhere.

Xenotransplantation (specifically, pig-to-human transplant) is, at the time of this writing, in pre-clinical trial stages with orthotopic pig-to-baboon kidney and heart transplants having been conducted with some success (Reichart et al. 2020). It is expected that kidney xenotransplantation clinical trials with human participants could begin as early as 2022, though many factors could affect this date. Nonetheless, the hope is that by genetically modifying pig organs that not only do they function as effectively as an allotransplant, but the risk of rejection could even be less than with a human donation. If xenografts were to become a clinical option, they would likely not replace allografts entirely; thus, both organ pools would exist. This raises the question of if/when medicine advances to the point that xenografts are relatively equal to allografts in terms of safety and efficacy, how will the allocation of who receives a xenograft versus an allograft proceed? Some persons will undoubtedly have objections, particularly early on, regarding receiving a xenograft versus an allograft. It is assumed that allografts will continue to be a scarce resource, so how will this be allocated?

The organ allocation preference cannot be decided by the recipient child or, in this case, their primary caregiver, as the assumption would be that if both sources (human and pig) have equal outcomes one would preferentially prefer the allograft. But if the time-to-receive an organ is shorter for those who would opt for a xenograft would that be an incentive for adoption and in itself balance allocation? Organ allocation can also be left to a governing body like the US Food and Drug Administration (FDA), Organ Procurement and Transplantation Network (OPTN), or the United Network for Organ Sharing (UNOS). However, ultimately it would still be up to the parents/guardians to provide consent to receive the genetically engineered pig organ or continue to take their chances on the allograft wait-list. Should they be allowed to hold their spot on the list after declining the pig organ alternative in order to wait for a human organ? Allocation may become even more complex if the hypothetical outcome of xenotransplantation is not comparable to that of a human organ due to a diminished life expectancy of the organ and if its use will be limited only as a bridge to allotransplant. This hypothetical scenario only opens further discussions on health care resource utilization and clinical decision-making. Unlike in adults with acquired heart disease, the current "bridging" options for children like ventricular assist devices have suboptimal outcomes for children. Would we submit our children with heart failure and that of our health care system to undergo two organ transplant surgeries? A cost–benefit analysis of the effects of using a xenograft as a bridge on the child, the family and on the health care system would certainly be warranted.

29.5 Conclusion

Pediatric triage has received little attention in the ethics literature to date. Nonetheless, there are distinct ethical issues when encountering triage in the pediatric population when compared to an adult population. While triage is oftentimes viewed as a system to be implemented during times of disaster or public health emergencies, its use in allocating scarce resources has also been examined with particular focus on the pediatric population with organs needed for transplant.

References

Afolabi, M. O. 2018. *Public health disasters: A global ethical framework*. Springer.

American Academy of Pediatrics. 2010. Poll: Children's needs should be prioritized in disaster planning, response and recovery efforts. http://web.archive.org/web/201205101 90746/. https://www.aap.org/en-us/about-the-aap/aap-press-room/Pages/Poll-Children's-Needs-Should-be-Prioritized-in-Disaster-Planning,-Response-and-Recovery-Efforts.aspx

American Medical Association. 2016. AMA code of medical ethics: Allocating limited health care resources. https://www.ama-assn.org/delivering-care/ethics/allocating-limited-healthcare-resources. Accessed 6 Aug

Antommaria, A.H., J. Sweney, and W.B. Poss. 2010. Critical appraisal of: Triaging pediatric critical care resources during a pandemic: Ethical and medical considerations. *Pediatric Critical Care Medicine* 11 (3): 396–400. https://doi.org/10.1097/PCC.0b013e3181dac698.

Antommaria, A.H., T. Powell, J.E. Miller, M.D. Christian, and C. Task Force for Pediatric Emergency Mass Critical. 2011. Ethical issues in pediatric emergency mass critical care. Pediatric Critical Care Medicine 12 (6 Suppl): S163–168. https://doi.org/10.1097/PCC.0b013e318234a88b

Bannon, M. P., and J.C. Farmer. 2015. Triage: Ethics in the field. In *Encyclopedia of trauma care*, ed. P. J. Papadakos and M.L. Gestring. Springer. https://doi-org.ezproxy.rowan.edu/10.1007/978-3-642-29613-0_227

Brock, D. 2004. Ethical issues in the use of cost effectiveness analysis for the prioritization of health resources. In *Handbook of bioethics: Taking stock of the field from a philosophical perspective*, ed. G. Khushf, p. 353. Kluwer. https://doi.org/10.1007/1-4020-2127-5_2

Burns, J.P., and C. Mitchell. 2011. Resource allocation and triage in disasters and pandemics In *Clinical ethics in pediatrics: A case-based textbook*, ed. D.S. Diekema, M.R. Mercurio, and M.B. Adam, 199–204. Cambridge University Press.

Childress, J.F. 1997. *Practical reasoning in bioethics*. Indiana University Press.

Devereaux, A.V., J.R. Dichter, M.D. Christian, N.N. Dubler, C.E. Sandrock, J.L. Hick, T. Powell, J.A. Geiling, D.E. Amundson, T.E. Baudendistel, D.A. Braner, M.A. Klein, K.A. Berkowitz, J.R. Curtis, L. Rubinson, and C. Task Force for Mass Critical. 2008. Definitive care for the critically ill during a disaster: A framework for allocation of scarce resources in mass critical care: From a Task Force for Mass Critical Care summit meeting, January 26–27, 2007, Chicago, IL. *Chest* 133 (5 Suppl): 51S-66S. https://doi.org/10.1378/chest.07-2693.

Gross, M.L. 2006. Wartime triage. In *Bioethics and armed conflict: Moral dilemmas of medicine and war*, 137–173. MIT Press.

Herreros, B., P. Gella, and D. Real de Asua. 2020. Triage during the COVID-19 epidemic in Spain: Better and worse ethical arguments. *Journal of Medical Ethics* 46 (7): 455–458. https://doi.org/10.1136/medethics-2020-106352.

Hurst, D.J., L.A. Padilla, D.K.C. Cooper, D.C. Cleveland, and W. Paris. 2020. Clinical trials of pediatric cardiac xenotransplantation. *American Journal of Transplantation*. https://doi.org/10.1111/ajt.16151.

Kipnis, K. 2002. Triage and ethics. *Virtual Mentor* 4 (1). https://doi.org/10.1001/virtualmentor.2002.4.1.puhl1-0201

Mahle, W.T., K.R. Kanter, and R.N. Vincent. 2005. Disparities in outcome for black patients after pediatric heart transplantation. *Journal of Pediatrics* 147 (6): 739–743. https://doi.org/10.1016/j.jpeds.2005.07.018.

Maves, R.C., J. Downar, J.R. Dichter, J.L. Hick, A. Devereaux, J.A. Geiling, N. Kissoon, N. Hupert, A.S. Niven, M.A. King, L.L. Rubinson, D. Hanfling, J.G. Hodge Jr., M.F. Marshall, K. Fischkoff,

L.E. Evans, M.R. Tonelli, R.S. Wax, G. Seda, J.S. Parrish, R.D. Truog, C.L. Sprung, M.D. Christian, and A.T.F.f.M.C. Care. 2020. Triage of scarce critical care resources in COVID-19 an implementation guide for regional allocation: An expert panel report of the task force for mass critical care and the American College of Chest Physicians. *Chest*. https://doi.org/10.1016/j.chest.2020.03.063.

Mechanic, D. 2005. Policy challenges in addressing racial disparities and improving population health. *Health Affairs (Millwood)* 24 (2): 335–338. https://doi.org/10.1377/hlthaff.24.2.335.

O'Mathuna, D. 2014. Disasters. In *Handbook of global bioethics*, ed. H. ten Have and B. Gordijn, 619–639. Springer.

Pence, G. 2017. *Medical ethics: Accounts of ground-breaking cases*, 8th ed., 9–10. McGraw Hill.

Persad, G., A. Wertheimer, and E.J. Emanuel. 2009. Principles for allocation of scarce medical interventions. *The Lancet* 373 (9661): 423–431. https://doi.org/10.1016/s0140-6736(09)60137-9.

Reichart, B., M. Langin, J. Radan, M. Mokelke, I. Buttgereit, J. Ying, A.K. Fresch, T. Mayr, L. Issl, S. Buchholz, S. Michel, R. Ellgass, M. Mihalj, S. Egerer, A. Baehr, B. Kessler, E. Kemter, M. Kurome, V. Zakhartchenko, S. Steen, T. Sjoberg, A. Paskevicius, L. Kruger, U. Fiebig, J. Denner, A.W. Godehardt, R.R. Tonjes, A. Milusev, R. Rieben, R. Sfriso, C. Walz, T. Kirchner, D. Ayares, K. Lampe, U. Schonmann, C. Hagl, E. Wolf, N. Klymiuk, J.M. Abicht, and P. Brenner. 2020. Pig-to-non-human primate heart transplantation: The final step toward clinical xenotransplantation? *Journal of Heart and Lung Transplantation* 39 (8): 751–757. https://doi.org/10.1016/j.healun.2020.05.004.

Shekerdemian, L. S., N.R. Mahmood, K.K. Wolfe, B.J. Riggs, C.E. Ross, C.A. McKiernan, S.M. Heidemann, L.C. Kleinman, A.I. Sen, M.W. Hall, M.A. Priestley, J.K. McGuire, K. Boukas, M.P. Sharron, J.P. Burns, and C.-P.C. International. 2020. Characteristics and outcomes of children with coronavirus disease 2019 (COVID-19) infection admitted to US and Canadian pediatric intensive care units. *JAMA Pediatrics*. https://doi.org/10.1001/jamapediatrics.2020.1948

Singh, T.P., K. Gauvreau, R. Thiagarajan, E.D. Blume, G. Piercey, and C.S. Almond. 2009. Racial and ethnic differences in mortality in children awaiting heart transplant in the United States. *American Journal of Transplantation* 9 (12): 2808–2815. https://doi.org/10.1111/j.1600-6143.2009.02852.x.

ten Have, H. 2016. *Vulnerability: Challenging bioethics*. Routledge.

U.S. Centers for Disease Control and Prevention. 2020. The social-ecological model: A framework for prevention. https://www.cdc.gov/violenceprevention/publichealthissue/social-ecologicalmodel.html. Accessed 20 Aug.

U.S. Department of Health & Human Services. Organ procurement and transplantation network. How organ allocation works. https://optn.transplant.hrsa.gov/learn/about-transplantation/how-organ-allocation-works/. Accessed 21 Dec 2020.

Ugarte, C., J.A. Tieffenberg, R. Amsalu, L.E. Romig, and T.T. Vu. 2011. Planning and triage in the disaster scenario. American Academy of Pediatrics. https://www.aap.org/en-us/Documents/disasters_dpac_PEDsModule3.pdf. Accessed 30 June.

World Medical Association. 2017. WMA statement on medical ethics in the event of disasters. https://www.wma.net/policies-post/wma-statement-on-medical-ethics-in-the-event-of-disasters/. Accessed 16 June.

Further Reading

Antommaria, A.H., T. Powell, J.E. Miller, M.D. Christian, and C. Task force for pediatric emergency mass critical. 2011. Ethical issues in pediatric emergency mass critical care. Pediatric Critical Care Medicine 12 (6 Suppl): S163–168. https://doi.org/10.1097/PCC.0b013e318234a88b

Burns, J.P., and C. Mitchell. 2011. Resource allocation and triage in disasters and pandemics In *Clinical ethics in pediatrics: A case-based textbook*, ed. D. S. Diekema, M. R. Mercurio, & M. B. Adam, 199–204. Cambridge University Press.

Childress, J.F. 1997. *Practical reasoning in bioethics*. Indiana University Press.

Index

A

Access to healthcare, 4, 5, 14, 16, 18, 230, 453–455, 457–459, 462

Accountability, 305, 407, 412, 414, 417, 418

Adolescents, 10, 17, 25, 29–33, 35, 47, 49, 61, 102, 111–113, 119, 129, 130, 155–167, 171, 192, 199, 208, 209, 215–217, 316, 317, 319, 321, 366, 369, 372–375, 387–389, 391, 393–395, 398, 399

Advanced directives, 84, 282, 397

Africa, 5, 88, 211, 265

Aftercare, 221

Against Medical Advice (AMA), 219, 284, 320

Age appropriate, 61, 207

Age of consent, 10, 18, 115, 159, 317, 395

AI, ethics, 406

Allow natural death, 227, 294

American Academy of Pediatrics, 24, 41, 83, 117, 118, 129, 157, 193, 210, 260, 283, 321, 469, 471

Amniocentesis, 102, 315, 325, 330

Artificial Intelligence (AI), 348, 403–418

Artificial nutrition, 280

Assent, 5, 25, 27, 28, 30, 34, 61, 83, 90, 111–113, 117–119, 122, 123, 129, 130, 134, 135, 155–157, 160–162, 187, 192, 193, 195, 196, 199, 203–205, 217, 218, 222, 313, 314, 318, 320, 326, 375, 384, 387, 395

Authority, 6, 10, 17, 23, 27, 30, 58, 59, 62, 64, 65, 68, 69, 71, 74, 80, 111–119, 122, 123, 144, 148–150, 152, 157, 158, 177, 192, 195, 203, 215, 221, 232, 239, 241–246, 251, 252, 259, 261, 262, 266, 277, 278, 281, 282, 293, 298, 335, 337, 396, 397, 421, 422, 424, 426–428, 432

Authority principle, 421, 427

Automation bias, 409, 415, 417

Autonomy, 7, 8, 10, 16, 17, 24, 29, 32, 35, 39–46, 49, 50, 52–54, 60, 61, 65, 69, 72, 79, 81–84, 88, 95, 102, 105, 111–114, 116, 119, 122, 123, 129–132, 134, 139, 143, 144, 155, 156, 158, 161, 164, 166, 167, 191, 192, 195, 205, 207, 231–233, 236, 239, 240, 242, 243, 254, 258, 261–264, 267, 278, 282, 302, 303, 314, 318, 319, 321, 326, 337, 338, 357, 362, 364–367, 371, 375, 387, 391, 396, 399, 404, 406–408, 413, 415, 417, 418, 459, 460

Autopsy, 221

Autosomal recessive, 319, 323, 324

Availability, 215, 216, 229, 230, 234, 293, 388, 466

B

Baby Doe, 104, 259, 260, 275–278, 346

Baby K, 278

Beauchamp and Childress, 60, 62, 80–82, 87, 89, 176, 177, 206, 227, 240, 458

Belgium, 10, 263, 296, 297, 299

Beliefs, 6, 8, 10, 13, 16, 27, 29, 35, 40, 80, 82, 87, 93, 94, 96, 99, 118, 133, 137, 138, 174, 175, 179, 180, 187–189, 191, 193, 195–199, 212, 231, 243, 263, 279, 280, 282, 286, 301, 306,

© Springer Nature Switzerland AG 2022
N. Nortjé and J. C. Bester (eds.), *Pediatric Ethics: Theory and Practice*,
The International Library of Bioethics 89,
https://doi.org/10.1007/978-3-030-86182-7

318, 344, 398, 399, 404, 408, 425, 431, 433, 449, 455, 468
Belmont Report, 14, 79, 80, 85, 86
Beneficence, 13, 79, 84, 85, 87, 95, 102, 105, 129, 130, 158, 175, 181, 182, 195, 206, 210, 230, 233, 235, 236, 258, 300–303, 314, 334, 360, 362, 363, 370, 373, 407, 410, 417, 418
Beneficial treatment, 61, 65, 69, 72, 143, 144, 146, 148, 149, 151–153
Benefit/harm constraint, 430, 433
Bentham, Jeremy, 455, 468
Best interest, 7, 8, 11, 13, 34, 51, 58, 71, 84, 87, 88, 112, 114, 116, 119, 120, 123, 136–138, 144, 150, 152, 161, 164, 171, 176, 179, 180, 193, 194, 205, 207, 211, 216, 222, 225, 226, 229, 231–237, 247, 249, 275, 276, 278, 281, 282, 285, 298, 300, 302, 303, 316, 320, 364, 365, 381–384, 395, 410, 427–429, 440, 441, 461
Best Interest Standard (BIS), 39, 41, 57–71, 73–75, 84, 87, 101, 105, 149, 164, 187, 193–195, 247, 281, 282, 313–315, 428–430, 434, 437, 440, 442, 444, 450, 453, 454, 460, 462
Bias, 83, 314, 343, 375, 382, 410, 411, 414, 418, 434, 466
Bioconservative, 339

C
Canada, 9–12, 15, 80, 158, 159, 208, 215, 216, 261, 262, 296, 297, 362
Canada Health Act, 9
Cancer, 8, 24, 119, 128, 131, 133, 134, 136, 138, 161, 163, 171–174, 176, 180, 181, 195, 242, 317, 319, 365, 379, 380, 382–384, 387, 390–392, 395, 410, 429, 433
Capacity, 10, 11, 17, 24, 25, 28, 30–34, 41–43, 50, 53, 54, 57, 83, 90, 111–113, 115, 117, 119, 129, 130, 143, 144, 155, 157–161, 164–166, 192, 203, 206, 211, 213, 215, 216, 229, 240–242, 256, 280, 285, 313, 314, 318, 320, 321, 330–332, 334, 335, 353, 360, 365, 369, 374, 375, 405, 408, 423, 458, 461
Carrier screening, 322, 324
Carrier status, 41, 323, 324, 326
Carrier testing, 323, 324
Casuistry, 79, 80, 85, 86, 95
Cause effect, 25, 432

Chaplaincy, 287
Charlie Gard, 8, 69, 262, 266, 281, 426
Child abuse, 85, 104, 116, 162, 213, 281, 282, 346, 433
Child Abuse Protection and Treatment Act, 260
Child protective services, 69, 151, 153, 260, 281, 282
Child's interests, 9, 49, 51, 54, 60, 62, 69, 73–75, 111, 114–116, 119, 122, 123, 130, 134, 147, 150, 152, 213, 239, 243, 245, 246, 281, 365, 421, 425
Child's rights, 134, 145
China, 6, 7, 264, 265, 269
Circumcision, 42, 359, 366, 433
Clinical ethics, 13–15, 17, 50, 101, 119, 153, 164, 167, 171, 172, 174, 175, 182, 183, 314
Clinical trial, 5, 6, 161, 163, 164, 218, 349, 379–384, 389, 474, 475
Clinician-patient relationship, 61, 62, 129
Coercion, 85, 113, 158, 192, 206, 212, 215, 316, 382, 399, 424, 428, 455, 459
Co-fiduciaries, 62, 111, 114, 119, 120, 122, 424
Cognitive, 10, 23–35, 46, 60, 115, 117, 155, 156, 158, 160, 161, 193, 206, 229, 233, 240, 242, 277, 279, 283, 285, 329–332, 347, 350, 353, 358, 360, 362, 374, 375, 408, 458, 466
Cognitive development, 25, 26, 28, 31, 33, 34, 50, 160
Comfort care, 84, 280, 285
Commemorative child, 397, 398
Common ground, 122, 172, 183
Communication, 26, 27, 50, 57, 61, 62, 139, 140, 150, 151, 155, 162, 167, 176, 187, 189, 191, 196, 198, 199, 209, 211, 221, 225, 229, 232, 236, 242, 262, 267–269, 276, 279, 286, 287, 314, 326, 343, 351, 352, 360, 396, 403, 408, 416
Community of concern, 95–98, 100, 101, 103
Comparability of effect constraint, 430
Competent, 13, 24, 32, 40, 87, 112, 113, 121, 160, 166, 187, 204, 206, 211, 218, 222, 231, 235, 262, 292, 294, 326, 433
Confidentiality, 35, 61, 65, 90, 102, 116, 155, 156, 161, 162, 165–167, 199, 205, 216, 217, 221, 233, 236, 318, 319, 322, 399, 459

Conflict, 12, 32, 35, 66, 71, 81, 89, 98, 116, 119, 130, 131, 137, 150, 151, 161, 162, 175, 187, 188, 195, 199, 211, 229, 233, 243, 254, 256, 257, 261, 263, 266–269, 300, 314, 319, 326, 357, 375, 382, 391, 399, 416, 422, 423, 425–427, 434

Consent, 10, 11, 13, 14, 16, 24, 32, 34, 35, 73, 83, 90, 102, 112, 113, 116–118, 129, 155–159, 165–167, 192, 193, 203–207, 210, 211, 213–219, 221, 222, 231, 241, 243, 277, 281, 297, 300, 314, 320, 324, 326, 332, 333, 360, 364, 368, 374, 375, 379, 384, 391, 395, 397, 398, 429, 431, 456, 459, 475

Consequentialism, 234

Consilience, 177

Constrained Parental Autonomy (CPA), 72–74, 194, 428

Convention on the Rights of the Child, 6, 18, 79, 119, 192, 460, 461

Council of Europe, 188, 426, 429

Counseling, 102, 283, 293, 314, 317, 321–323, 363, 393, 394, 397, 431

COVID-19, 7, 8, 14, 16, 208, 209, 220, 466, 468–470, 472

CRISPR, 330

Critical care, 225, 226, 228, 230–232, 234, 235, 238, 257, 258, 260, 346, 347, 429, 466, 471

Cross cultural, 187, 188, 196, 198, 200

Cruzan, 296

Cultural, 5–7, 15, 16, 23, 25, 27, 30, 46, 49, 54, 86, 89, 116, 118, 131, 133, 138, 157, 188–190, 195, 196, 198, 199, 201, 215, 229, 240, 243, 248, 269, 275–278, 372, 374, 398, 404, 406, 411, 427, 429, 433

Cultural competence, 187, 198, 200

Cultural, humility, 187, 198, 199

Cultural, safety, 187, 198

Culture, 8, 13, 15–17, 30, 50, 60, 80, 115, 131–133, 155–157, 159, 179, 187–189, 196–199, 241, 261, 266, 277, 294, 295, 362, 363, 372, 373, 375, 384, 398, 416, 431, 433

D

Deadlock, 175

Death, 5, 7, 8, 12, 69, 71, 89, 104, 135, 164, 174, 180, 182, 183, 195, 213, 227, 233, 235, 240, 241, 243, 247–249,

256, 259, 263–266, 277, 278, 280, 282, 283, 285, 292–296, 298, 302–304, 306, 346, 349, 350, 391, 396, 430, 433, 434, 439, 456, 459, 467, 468

Decision-making, 5–9, 13–17, 23–35, 39, 41, 42, 49–52, 54, 57–59, 61–65, 68–71, 73–75, 82–84, 90, 97, 99, 104, 111–115, 117–123, 127–131, 134, 136, 140, 144, 148–153, 155–158, 160, 161, 164, 165, 184, 189, 191–194, 196, 199, 200, 208, 209, 211, 220, 225–227, 229–236, 240, 242–244, 249–251, 254, 256, 257, 259, 261–264, 267, 268, 275–284, 286, 287, 291, 298, 300, 314, 320, 321, 323, 338, 344, 346, 349, 352, 360, 361, 363, 369, 371, 374, 375, 381, 383, 384, 391, 393–395, 399, 403–406, 408, 414–418, 427, 430, 435, 440, 469, 475

Decision-Making Capacity (DMC), 23–27, 29–31, 33, 34, 60, 82, 83, 90, 102, 144, 155–157, 160, 164, 167, 192, 199, 232, 257, 282, 294, 295, 297, 302, 358, 362, 364, 368, 371, 373–375

Degradation, 179

Deontological, 134, 176, 406

Detention center, 13

Develop, 18, 23–28, 30, 31, 45, 47, 60, 68, 90, 93, 96, 102, 115, 118, 120, 122, 134, 145, 160, 167, 199, 231, 236, 255, 267, 282, 294, 317, 318, 321, 326, 330, 334, 336–338, 348, 349, 351, 361, 362, 372, 374, 375, 399, 411, 417, 418, 424, 437, 461

Development, 6, 7, 12, 16, 23–25, 27–32, 34, 40, 42, 46, 47, 50, 52–54, 58, 62, 71, 88, 96, 116, 130, 145, 147, 156–158, 161, 179, 192, 205, 229, 230, 257, 276, 283, 284, 299, 300, 317, 320, 321, 323, 326, 330, 334, 336, 348, 351, 357–363, 365–369, 371, 372, 375, 388, 403, 406–416, 457, 459–461, 471

Difference of Sexual Development (DSD), 358, 360–368

Direct effect basis, 432

Direct to Consumer Testing (DTC), 322, 326

Disability, 11, 87, 99, 103, 205, 228, 230, 233, 240, 261, 275, 276, 278–283,

287, 303, 304, 331–333, 335, 343–
    349, 351–353, 411, 472
Disclosure, 7, 16, 53, 61, 65, 83, 121, 127–
    131, 133–140, 162, 167, 229, 233,
    313, 314, 320, 321, 326, 351, 381,
    433
Discrimination, 50, 87, 228, 229, 233, 236,
    318, 326, 346, 348, 358, 406, 408,
    410, 411, 414, 417, 418, 428, 430,
    435, 461, 469, 470
Disruptive technology, 403, 412, 416
Dissent, 25, 27, 28, 30, 34, 130, 135, 160,
    161, 199, 204, 205
Distribution of resources, 13, 455
Donation after cardiac death, 12
Do Not Attempt Resuscitation (DNAR),
    241, 250, 294
Do Not Resuscitate (DNR), 228, 239–242,
    244–246, 250, 294
Double effect, 249
Down syndrome, 276, 277
Dual role, 379–382, 384
Dual systems model, 31
Duty to inform, 319
Duty to warn, 319

E
Effectiveness, 369, 389, 405, 439, 447
Elective pediatric surgeries, 39, 41, 42
Emancipated minor, 112, 129, 157–159
Emerging capacity, 35, 90, 102, 155–157,
    160, 167, 199, 375
Emotional, 23–30, 32, 35, 40, 50, 52–54, 96,
    121, 131, 155, 157, 158, 161, 175,
    189, 206, 209, 215, 229, 243, 246,
    248, 251, 374, 382, 397, 404, 471
End-of-life, 7, 8, 69, 71, 106, 121, 164, 226,
    228, 231, 239, 240, 242, 245–247,
    249–251, 258–264, 266, 269, 291,
    292, 298, 300, 304–306
Enhancement, 329–331, 333–340
Equity, 391, 395, 398, 399, 407, 411
Ethical disagreement, 279
Ethical principles, 57, 58, 60, 61, 66, 68,
    70, 75, 79–81, 85, 95, 129, 152, 158,
    162, 222, 226, 230, 235, 249, 306,
    319, 320, 334, 373–375, 383, 403,
    405–407, 412, 414, 416, 426
Ethical relativism, 425
Ethics, AI, 406
Ethics, committee, 15–17, 98, 100, 102–104,
    106, 217, 258, 266, 325, 346, 394,
    397

Ethics, consult, 102, 122, 286
Ethics, consultation, 105, 151, 167, 432
Ethics, pediatric, 3, 4, 6, 8, 39, 40, 48, 60,
    71, 72, 74, 79, 80, 82, 85, 86, 89, 112
Ethics, research, 13, 14, 16, 17, 79, 80, 83,
    217, 218, 383, 471
Ethics, virtue, 79, 406, 407, 414
Eugenics, 297, 331, 333, 338, 343, 344, 347
Europe, 8, 9, 86, 87, 209, 217, 253, 258, 261–
    263, 266, 285, 298, 322, 333, 344,
    426, 455
European Courts of European Rights, 426
Euthanasia, 104, 291–293, 295–297, 299–
    302, 304, 305, 333, 344
Euthanasia, neonatal, 291, 292, 299–306
Expert, 86, 100, 121, 160, 174, 175, 180,
    199, 243–245, 291, 304, 305, 317,
    406, 409, 414, 447
Explainability, 407, 413, 414, 417
Explicability, 407, 413, 417
Exploitation, 5, 50, 60, 228, 235, 408, 417,
    461
Extreme emergency, 204

F
Faden, Ruth, 457
Faith healing, 191
Familial relationships, 115, 425
Family centered, 16, 157, 227
Family-centered decision making, 194, 278
Feinberg, Joel, 41, 43–46, 48, 60, 133, 135,
    178, 332, 334, 371, 429, 459
Fertility preservation, 42, 374, 387–395,
    398, 399
Fetus, 283, 315, 324, 326, 330
First person authority, 113, 114
Framework, 13, 17, 23, 24, 26, 39–41, 45,
    50–54, 58, 59, 62, 65, 71, 74, 148,
    155, 156, 158, 171, 172, 177, 181–
    184, 198, 199, 240, 343, 358, 367,
    403, 406, 414, 415, 430, 438, 443,
    456
Freedom, 41, 61, 72, 115, 123, 143, 145, 178,
    179, 205, 230, 243, 339, 406, 408,
    409, 423, 426, 459
Free-riding, 437, 441, 442
Futility, 13, 172–174, 177, 180, 182, 253–
    259, 261–266, 268, 269
Futility, physiological, 173, 174, 253, 254,
    268
Futility, qualitative, 173, 180, 254, 256–258,
    262, 263

Futility, quantitative, 173, 174, 180, 256, 258, 262
Future autonomy, 45, 47, 50, 51, 318, 320, 325, 362

G
Gamete retrieval, 387, 396, 397
Gender, dysphoria, 357–359, 365, 368–374, 393, 394
Gender, identity, 11, 358–362, 365–373, 375, 398, 399
Gender, management, 358–361, 363
Gender reassignment, 11
Gene editing, 330
Genetic ethics, 41, 42
Genetic testing and screening, 313–315, 322, 327
Genetic variants, 314
Gillick test, 32
Goal dissonance, 253, 267
Goals, 5, 24, 25, 28, 35, 45, 46, 57, 89, 104, 111, 113–115, 117, 119–123, 136, 144, 150, 151, 153, 158, 160, 162, 166, 173, 175, 198, 214, 216, 218, 232, 236, 247, 249, 250, 254, 256, 258, 266–269, 275, 276, 278–280, 282, 286, 287, 292, 296, 302, 305, 315, 316, 320, 326, 343, 345, 347, 349, 379, 381, 405, 422, 424, 429, 446–448, 465
Groningen protocol, 291–293, 299, 300, 302, 306
Guidance principle, 65, 70, 427, 429, 434

H
Harm, 5, 9, 45, 49, 53, 54, 60, 67–71, 73, 74, 84, 85, 89, 90, 100, 101, 105, 106, 111, 114, 115, 123, 129, 131, 132, 134, 136, 137, 140, 145, 149, 157, 161, 166, 172, 173, 175–183, 187, 192–195, 207, 210, 217, 230, 231, 233, 235–237, 239–248, 250, 251, 254, 255, 258, 259, 269, 282, 285, 302–306, 313, 314, 316, 318, 321, 322, 325, 332, 335, 348, 357, 360, 362, 363, 365, 366, 370, 373, 375, 382, 383, 392, 397, 407, 409–411, 415, 418, 421, 427–430, 432–434, 440, 441, 444, 456, 460, 462, 471
Harm Principle (HP), 68–71, 74, 428–430
Harm threshold, 176, 182

Health disparities, 9, 14, 17, 465, 466, 473
Herd immunity, 438–446, 449, 450, 471
High risk pregnancy, 278
Hope, 18, 99, 103, 138, 180, 182, 189, 200, 220, 228, 254, 258, 262, 267, 268, 279, 280, 286, 287, 292, 336, 349, 352, 364, 381, 416, 454, 475
Human dignity, 93–99, 101, 104, 205, 222, 406–409, 415, 417, 418
Hydration, 239, 241, 246–248, 251, 260

I
Impairment, 25, 275, 276, 278, 280–283, 287, 332, 333, 343–353, 467
Incidental findings, 320, 324
Indirect effect basis, 430
Infant, 10, 14, 26, 51, 84, 99–106, 117, 212, 259–261, 276, 277, 283, 291, 292, 294, 297–299, 301–306, 315, 343, 345–351, 353, 359–363, 365, 366, 425, 440, 455, 473
Informed consent, 5, 6, 12, 14, 27, 28, 61, 65, 85, 111–113, 115, 118, 119, 122, 123, 129, 130, 158–161, 187, 192, 204, 232, 293, 313, 314, 320–325, 360, 374, 409
Injured children, 203, 210, 212, 220, 225, 267
Institutional Review Board (IRB), 85, 217
Insurance, 84, 85, 215, 281, 318, 320, 322, 323, 326, 389, 391, 395, 398, 399, 447
Interest, 7–9, 11, 12, 28, 34, 39–41, 49–54, 58–62, 64, 65, 67, 68, 70–75, 84, 86, 112, 114–116, 118, 120, 121, 123, 131, 139, 143, 145–153, 158, 174, 175, 178–180, 183, 205–208, 232, 235, 239, 240, 243–245, 251, 254, 259, 261, 269, 281, 282, 286, 298, 300, 301, 322, 326, 339, 365, 366, 381, 382, 387, 391–394, 396–399, 410, 422–426, 428–430, 434, 440, 442, 444, 450, 453, 456, 458–462
Intervention principle, 65, 193, 421, 427, 429, 430

J
Japan, 7, 8, 264, 266, 283, 296
Jehovah Witness, 13
Justice, 79, 85, 87, 89, 94, 95, 129, 146, 148, 158, 233–236, 239, 240, 262, 314, 333, 374, 391, 395, 398, 399,

407, 411, 414, 417, 418, 422, 423, 425, 426, 429, 437, 444–446, 450, 453–458, 460, 462

K

Kantian, 52, 72, 81, 87, 175, 381, 455
Karyotyping, 315

L

Left-behind children, 7
Legally emancipated, 32
Legal standard, 32, 58, 59, 65, 129, 158, 282, 428
Libertarianism, 455, 458, 461, 462
Life expectancy, 277, 293, 454, 455, 475
Life-prolonging interventions, 256, 259, 260, 284, 294–296, 298
Life-prolonging technologies, 295
Life-sustaining therapy, 227, 228, 231, 233–235, 353
Life-sustaining treatment, 12, 195, 235, 257, 264, 265, 292, 304
Likelihood of effect constraint, 430
Little emperors, 6
Long-term survival, 173

M

Majority, 7, 10, 12, 15, 119, 120, 123, 144, 158, 159, 216, 221, 240, 254, 256, 284, 299, 301, 318, 321, 365, 370, 391, 396, 427, 433, 465, 466
Marginalized, 18, 85, 88, 89, 362, 374, 414, 456
Maturation, 23, 26–31, 35, 115, 116, 388, 390
Mature minor, 10, 11, 32, 35, 83, 129, 144, 158, 159, 164, 167, 216, 217, 220, 374
Mechanical ventilation, 103, 226, 227, 236, 246, 278, 304
Mediating, 68, 71, 111, 122, 175
Medical Assistance in Dying (MAiD), 10, 11
Medically Provided Nutrition and Hydration (MPNH), 295, 296, 298
Medical model of disability, 332
Medical paternalism, 256, 277, 278
Mental health, 82, 90, 102, 112, 116, 158, 161, 166, 167, 199, 203, 205, 214, 216, 217, 343, 353, 368, 370, 373, 395

Metaphysical harms, 176
Mexico, 12, 13
Microarray, 315, 325
Mid-level principles, 79, 81
Milestones, 23, 25, 26, 256
Mill, John Stewart, 44, 68, 71, 85, 176, 455, 468
Minor, 5, 8, 10, 11, 27–30, 32–34, 52, 74, 82–85, 87, 89, 90, 111–113, 116, 119, 120, 122, 127–129, 131, 136–138, 140, 150, 155, 157–160, 192, 193, 195, 196, 199, 203–210, 213–217, 220–222, 280, 281, 286, 294, 318, 320, 321, 324, 326, 364, 369, 374, 387, 393, 395, 422, 424, 433, 434, 456, 467
Minor, emancipated, 129, 157–159
Minor, mature, 10, 11, 32, 35, 83, 129, 144, 158, 159, 164, 167, 216, 217, 220, 374
Model of disability, medical, 332
Model of disability, social, 332
Moderate essentialist theory, 453, 457, 461, 462
Modus Vivendi, 423, 424, 427–430
Moral distress, 323, 467, 468
Moral philosophy, 45, 176
Morphological freedom, 339
Multicultural, 195
Multidisciplinary team, 139, 140, 195, 227, 228, 233, 384

N

Natural death, 233, 279
Neo-casuistry, 93–95, 97–100, 103, 106
Neonatal, euthanasia, 291, 292, 299–306
Netherlands, The, 8, 10, 291, 292, 296, 297, 299, 301, 304, 306, 373
Neurodevelopment, 344, 352
Neurological impairment, 264, 344, 346
Newborn, 88, 98–100, 102–105, 209, 259, 260, 275–284, 286, 287, 293, 300, 301, 304, 305, 313–315, 323, 324, 343, 361, 362
New Zealand, 15, 16, 198
Nondiscrimination, 318
None beneficial, 69, 100, 104, 105, 118, 172, 268
Nonmaleficence, 130, 258, 360, 362, 363, 370, 373
Normative judgments, 113
Normative preferences, 111, 113, 116, 119

Nudges, 212, 213
Nutrition, 12, 14, 60, 86, 105, 246–248, 260, 453

O

Obligation, 8, 33, 47, 57, 58, 60–62, 64, 66, 71–75, 95, 101, 115, 116, 120, 128–130, 137, 139, 140, 143–148, 151, 152, 156, 162, 171, 172, 177, 178, 180–182, 191, 215, 227, 243, 250, 259, 262, 265, 268, 294, 298, 300, 301, 303, 305, 319, 323, 325, 329, 331, 334–336, 339, 340, 358, 374, 375, 380, 382, 393, 404, 409–411, 414, 417, 418, 424, 435, 440, 444–447, 449, 450, 454, 459, 460, 462
Organizational integration, 93, 94, 97–99, 101, 105, 106
Orthothanasia, 265
Other-regarding, 113–116, 122, 123
Overlapping consensus, 183
Overrule, 68, 69, 72, 82, 148, 151, 152, 194, 205, 421, 432

P

Palliative sedation, 239, 241, 249, 250, 295, 304
Paralytics, 291, 292, 304, 305
Parental authority, 58, 69–72, 74, 115, 116, 123, 130–132, 143–146, 148, 152, 178, 194, 243, 259, 266, 303, 424, 429, 434
Parental, choices, 9, 43, 176, 340, 434
Parental discretion, 13, 70, 79, 84, 87, 147, 152, 280, 446
Parental, freedoms, 176
Parental permission, 61, 68, 72, 112–114, 146, 159, 211, 216, 387
Parent–child relationship, 149
Parents, 4–6, 8, 9, 11, 13, 14, 16, 18, 24, 27, 28, 32–34, 40–42, 44–48, 50, 53, 54, 58–64, 66–75, 82–85, 87–90, 98, 100–105, 111, 112, 114–123, 127–134, 136–140, 143–153, 155–159, 161–167, 178, 179, 181, 182, 187–196, 199, 203–205, 207–215, 217–222, 226, 229, 231–233, 236, 239–246, 248, 251, 252, 254, 259, 260, 266–269, 276–287, 292–294, 296, 298, 300, 301, 303–305, 314, 316–318, 320, 321, 323–325, 329–331, 333–340, 343, 345, 346, 348–353, 359, 361–367, 369, 371, 374, 375, 380, 382, 383, 391–395, 397, 398, 408, 418, 421, 422, 424–428, 430–434, 437, 438, 440, 442–449, 455, 456, 458–461, 472, 475
Pater familias, 9
Paternalism, 9, 80, 82, 97, 205, 206, 261, 263, 267, 269, 277, 358, 359
Permission, 72, 111, 117, 122, 130, 158, 165, 193, 204–207, 212, 214, 218, 222, 232, 243, 372, 393, 395, 397, 398
Phronesis, 93, 94, 99, 106
Physician Assisted Dying (PAD), 292, 296–298, 302
Physician-researcher, 379–382, 384
Police power, 422, 426
Policy, 6, 7, 11, 16, 18, 24, 58–61, 64, 75, 85, 88, 89, 144, 148, 149, 159, 165, 167, 182, 189, 192, 198, 225, 226, 228, 234, 246, 257, 261, 264, 266, 268, 281, 318, 323, 397, 422, 423, 426, 429, 437, 438, 440, 443, 444, 446–450, 453–456, 458, 460, 462, 466, 469
Political instability, 5, 18
Political realism, 424
Population health, 15
Postmortem retrieval, 396
Postnatal, 275, 336
Potential harmful treatments, 172
Power relationships, 337
Predictions, 280, 337, 343, 344, 346, 348, 350, 351, 353, 405, 413–415
Predictive genetic testing, 39–41, 53, 54, 318
Preferences, 9, 13, 14, 26–28, 34, 35, 42, 43, 100, 101, 105, 118–121, 131, 140, 188, 191, 210, 231, 232, 239, 240, 242–245, 248, 251, 280, 281, 298, 318, 337, 358, 365, 369, 384, 399, 404, 414, 475
Prefrontal cortex, 26, 29, 31, 161, 368
Prenatal, 32, 275, 313–315, 323–326, 330, 331, 333, 358, 359, 361
Prima facie obligations, 59, 75, 138
Principle, authority, 421, 427
Principle, guidance, 65, 70, 427, 429, 434
Principle, intervention, 65, 193, 421, 427, 429, 430
Principle of Procreative Beneficence (PPB), 334–336, 338
Principles, limiting, 63–65, 68, 70, 71

Privacy, 16, 35, 61, 90, 102, 116, 167,
    178, 179, 199, 205, 209, 216, 217,
    221, 233, 236, 295, 322, 399, 404,
    407–409, 418, 426
Private healthcare, 177, 234
Process vision, 93, 94, 97–100, 103, 106
Procreative autonomy, 336–338
Professional integrity, 137, 139
Prognosis, 98–100, 102, 104, 121, 128, 129,
    132, 135, 136, 138, 279, 281, 283,
    284, 293, 300, 314, 343, 344, 350–
    352, 391–393
Prognostic, 117, 131, 134, 135, 257, 337,
    349–352, 380, 382, 416
Protect, 5, 11, 28, 34, 42, 49, 52, 53, 57–
    60, 62–64, 67, 70–72, 75, 82, 84, 88,
    89, 131, 132, 135, 137, 140, 143–146,
    148–152, 194, 205, 216, 236, 301,
    343, 357, 359, 375, 392, 407, 409,
    417, 423, 434, 435, 440–446, 450,
    453, 460–462
Protector, 181
Providers, 15, 24, 25, 27, 33–35, 51, 61, 111,
    112, 114–123, 134, 138, 139, 155,
    156, 162, 164, 166, 167, 173, 175,
    181, 187–189, 191, 197, 205, 206,
    208, 220, 232, 233, 236, 264, 266,
    277, 279–282, 286, 287, 297, 298,
    303–305, 315, 318–322, 325, 326,
    343, 346–353, 365, 375, 394, 395,
    405, 408, 446
Proxy, 117, 182, 282, 379, 397, 398
Psychosocial risks, 316
Pubertal timing, 388
Public good, 437, 442, 443, 450, 465
Public health, 12, 15–17, 85, 86, 112, 116,
    156, 234, 395, 437, 443, 446, 448,
    453, 454, 456, 458–462, 465–468,
    470–473, 475

Q
Quality, 42–44, 47–49, 51, 86, 96, 99, 134,
    138, 140, 157, 162, 173, 180, 182,
    227–231, 233, 235, 244, 249, 251,
    255, 301, 335, 409, 416, 455
Quality of life, 24, 84, 101, 104, 163, 164,
    225, 230, 231, 236, 240, 243, 247,
    282, 301, 303, 336, 343, 350, 352,
    406, 407, 417, 431, 432
Quinlan, 294, 295

R
Rationing, 465–468, 471–474
Rawls, John, 172, 176, 177, 183, 423, 456,
    457
Reasonable, 9, 25, 42, 51, 53, 66, 69, 75,
    82, 84, 115, 119, 135, 138, 148, 152,
    157, 160, 164, 167, 171–175, 180,
    194, 210, 216, 230, 239, 244, 245,
    248, 251, 258, 259, 267, 300, 302,
    316, 334, 350, 361, 414, 416, 428,
    431–433, 446
Reasonable medicine, 174
Reasonableness, 63, 114, 116, 172, 174, 177,
    182, 193
Reflective equilibrium, 177
Refugees, 86, 203, 214, 215
Refusal of treatment, 115, 146, 147, 149,
    235, 263
Rehabilitation Act, 277, 278
Relationships, power, 337
Relationships, societal, 338
Religious preference, 188, 190
Reproduction, maturation, 388
Reproduction, posthumous, 393, 396–398
Reproductive issues, 203, 216, 217
Research ethics, 13, 14, 16, 17, 79, 80, 83,
    85, 217, 218, 383, 471
Resource allocation, 12, 15, 193, 281, 465–
    469, 471–474
Resource constraints, 15
Respect for Children, 8, 79, 83, 88
Respectful care, 199
Respectful children, 192
Resuscitation, 105, 219, 241, 242, 244–246,
    250, 258, 261, 283, 284, 294, 396
Resuscitation Council, 258, 283
Retrieval, gamete, 387, 396, 397
Retrieval, postmortem, 396
Rhetorical power, 173
Right, 7, 9, 11, 13, 23, 25, 39–45, 47–50,
    52–54, 58, 62, 63, 69, 71, 72, 74, 80–
    84, 86–90, 105, 112, 129, 132, 134,
    137, 143–146, 149, 151, 156, 158–
    160, 162, 163, 166, 171, 172, 177–
    182, 193, 194, 199, 206, 211, 227–
    237, 240, 241, 243, 245, 246, 260,
    277, 278, 285, 295, 302, 314, 316,
    318, 323, 324, 326, 329, 331, 334,
    336, 337, 339, 340, 346, 348, 352,
    362, 372, 382, 391, 394–396, 398,
    406–410, 413, 424–426, 428, 435,
    443, 453, 455–462

Right to an Open Future (ROF), 39–49, 52–54, 133, 135, 137, 162, 318, 335, 362, 392, 453, 459, 462
Right to be heard, 88, 229
Risk perception, 29
Risks and benefits, 26, 29, 90, 113, 119, 148, 161, 164, 220, 222, 278, 281, 286, 314, 375
Rule of sevens, 23, 24, 28, 192

S
Safety, 60, 87, 217, 330, 368, 375, 380, 392, 413, 421–423, 426, 430, 433, 439, 447, 454, 475
Seken, 7, 8
Self-determination, 43, 44, 46, 47, 51, 112, 145, 231, 236, 280, 300, 357, 362, 364–367, 370, 371, 375, 408, 457, 461
Self fulfillment, 43, 44, 46–48, 51, 101
Self-interest, 62, 84, 175, 428
Self-regulatory, 29
Sexual abuse, 88, 213, 432
Sexual development, 42, 357, 358
Sexually transmitted infections, 158, 204, 221, 434
Shared Decision Making (SDM), 111, 117, 119, 120, 122, 155, 161, 166
Social, 6, 7, 11, 12, 17, 23–30, 32, 34, 35, 40, 45, 46, 58, 85, 86, 88, 89, 96, 139, 140, 148, 151, 153, 156, 157, 174, 177, 179, 182, 189, 190, 199, 212, 214, 217, 226–232, 236, 240, 248, 262–264, 275, 277, 280, 295, 316, 318, 332–334, 338, 339, 348, 357, 359–361, 363, 364, 366–370, 373, 374, 382, 395, 398, 403, 404, 408–411, 424, 426–428, 432, 439, 448, 450, 453–457, 459, 460, 462, 465, 469, 470, 472, 474
Social determinants of health, 17, 453–455, 459, 461, 462
Social emotional, 23–30, 32, 35, 157, 374
Socialized, 177, 336
Social justice, 13, 453–457, 460–462
Social model of disability, 332
Societal relationships, 338
Sociocultural, 131–133, 135, 137
Somebodyness, 94, 96, 98, 99, 101, 104, 106
Sophia, 93, 94, 106
South America, 16, 17, 265
Sperm, 388–390, 394, 396–398

Spiritual care, 151, 153
Sterilization of minors, 39, 41, 42, 44, 52
Stewardship, 407, 414, 415, 417, 468
Sudden death, 221, 397
Sufficientarianism, 429, 434
Surrogate, 65, 84, 85, 98, 111, 112, 114, 158, 194, 232, 261, 262, 280, 282, 300, 320, 363–365, 375, 390, 394, 398, 421, 427
Survival, 5, 24, 84, 88, 103, 173, 179, 229, 246, 256, 258, 260–263, 276, 277, 280, 281, 283, 284, 345, 349, 380, 392, 393, 396, 432, 461, 468, 470, 473
Sustainable Developmental Goals (SDG), 5
Symptom control, 239, 247, 249–251, 470
Systems, 7, 9, 12, 13, 15–17, 31, 89, 97, 101, 113, 135, 161–163, 188–193, 195–198, 215, 220, 228, 241, 248, 279, 281, 314–316, 339, 372, 381, 395, 403–418, 424, 426, 427, 431, 437, 449, 455, 466–468, 470, 472–475

T
Teenagers, 83, 111, 119, 203, 209, 215, 217, 221
Termination of pregnancy, 204, 301
Therapeutic misconception, 5, 379
Therapeutic privilege, 129, 132, 133
Time-limited trials, 279, 284
Titrated clinician directives, 267, 268
Toddlerhood, 26
Tracheostomy, 11, 227, 230, 234, 276, 278, 280
Transgender, 42, 368, 372–374, 393–395
Transhumanist, 339
Transient exuberance, 26
Transparency, 96, 183, 305, 322, 382, 404, 407, 412, 413, 417, 426, 459
Transplant, 73, 118, 285, 473–476
Triage, 208–210, 219, 244, 404, 465–475
Trilateral relation, 112, 116
Trisomy 13, 103–105, 280
Trust, 6, 33, 44, 45, 61, 84, 90, 112, 121, 129, 132, 134–138, 145, 151, 152, 162, 165, 166, 171, 172, 181, 189, 208, 211, 213, 253, 269, 286, 287, 314, 322, 326, 352, 382–384, 391, 394–396, 404, 407, 409, 410, 412, 413, 415, 417, 437, 447–449, 458, 459, 462, 470
Trustworthy, 6, 406, 411, 412, 416–418

Truth, 95, 127–129, 133, 134, 136–138, 140,
    196, 233, 236, 245, 337, 360, 445
Truth-telling, 128, 129

U
United Nations (UN), 6, 18, 58, 79, 80, 86,
    87, 89, 145, 146, 149, 214, 216, 228,
    232, 240, 242, 428, 453, 460–462
United Nations Convention on the Rights of
    the Child, 58, 86, 177, 228, 235, 314,
    426
United States, 13, 14, 58, 86, 131, 158, 159,
    177, 294–298, 321, 322, 324, 326,
    344, 346, 362, 395, 422, 426, 433,
    439, 447, 455, 457
Universal access, 5, 133, 455
Unreasonable, 32, 59, 82, 174, 175, 207, 430,
    431, 434
Utilitarian, 52, 81, 87, 176, 180, 234, 236,
    345, 407, 443, 466, 468, 471

V
Vaccine, 13, 61, 149, 332, 437–445, 447–450
Vaccine hesitancy, 437, 446–448, 450
Vaccine refusal, 149, 437, 442, 446–448, 450
Value pluralism, 422
Values, 7–10, 15, 16, 24, 27, 29, 30, 33–35,
    42, 46–49, 51, 53, 61, 63–69, 71, 73,
    74, 80, 82, 85, 89, 97, 98, 100, 101,
    105, 111, 113–123, 127, 128, 131,
    135, 137, 138, 143, 144, 151, 152,
    157, 158, 164, 171, 172, 174, 188–
    193, 195–198, 205–207, 222, 229,
    231–233, 236, 240, 242, 248, 253,

    254, 257, 268, 277–280, 282, 285–
    287, 300, 332, 353, 362, 381, 382,
    392, 398, 403, 404, 406–408, 412,
    416, 422–424, 426, 428, 448, 458,
    467, 469, 470
Vaping addiction, 217
Veracity, 65, 128
Virtue ethics, 79, 80, 406, 407, 414
Voluntary Stopping of Eating and Drinking
    (VSED), 295, 298
Vulnerable, 5, 7, 8, 13, 25, 57, 59, 60, 70,
    85, 89, 146, 157, 166, 218, 228, 235,
    244, 261, 301, 323, 358, 359, 374,
    382, 384, 410, 411, 414, 415, 417,
    418, 422, 438, 440, 442, 444, 449,
    453–457, 459–461, 466, 471, 472

W
Wellbeing, 6, 12, 13, 29, 42, 44–46, 49–51,
    53, 54, 57–64, 66–68, 70–73, 75, 111,
    114, 115, 117, 143–153, 156, 191,
    193, 194, 231, 248, 258, 331, 332,
    334, 335, 337–339, 350, 358, 366,
    369, 372, 381, 395, 403, 410, 412,
    415, 418, 426, 428, 429, 440, 441,
    443–445, 447–450, 457, 460, 462,
    468
Withdrawal treatment, 235, 263
Withholding information, 127–129, 131,
    133–135, 137, 138

Y
Young adult, 123, 242, 347, 373, 395

Printed in Great Britain
by Amazon